Proceedings

of the 5th International Yellow River Forum on Ensuring Water Right of the River's Demand and Healthy River Basin Maintenance

Volume Ⅲ

Yellow River Conservancy Press

图书在版编目(CIP)数据

第五届黄河国际论坛论文集/尚宏琦,骆向新主编.—郑州:黄河水利出版社,2015.9

ISBN 978-7-5509-0399-9

Ⅰ.①第… Ⅱ.①尚… ②骆… Ⅲ.①黄河-河道整治-国际学术会议-文集 Ⅳ.①TV882.1-53

中国版本图书馆CIP数据核字(2012)第314288号

出 版 社:黄河水利出版社

地址:河南省郑州市顺河路黄委会综合楼14层　　邮政编码:450003

发行单位:黄河水利出版社

发行部电话:0371-66026940、66020550、66028024、66022620(传真)

E-mail:hhslcbs@126.com

承印单位:河南省瑞光印务股份有限公司

开本:787 mm×1 092 mm　1/16

印张:149.75

印数:1—1 000

版次:2015年9月第1版　　印次:2015年9月第1次印刷

定价(全五册):960.00元(US$155.00)

Under the Auspices of

Ministry of Water Resources, People's Republic of China

Sponsored & Hosted by

Yellow River Conservancy Commission (YRCC), Ministry of Water Resources, P. R. China

China Yellow River Foundation (CYRF)

Editing Committee of Proceedings of the 5th International Yellow River Forum on Ensuring Water Right of the River's Demand and Healthy River Basin Maintenance

Director: Chen Xiaojiang

Vice Directors: Xu Cheng, Xue Songgui

Members: Shang Hongqi, Luo Xiangxin, Zhao Guoxun, Hou Quanliang, Ma Xiao, Sun Feng, Sun Yangbo

General Editors: Shang Hongqi, Luo Xiangxin

Deputy Editors: Ma Xiao, Sun Feng, Sun Yangbo

Members of Editors (Translators): An Chao, Chang Shan, Chang Xiaohui, Chen Nan, Chen Yuanshen, Dang Suzhen, Dang Yonghong, Dong Wu, Du Yuanzheng, Fan Jie, Ge Zhonghong, Ge Zijia, Gong Zhen, Hu Huiyong, Huang Feng, Huang Yufang, Jiang Xiaohui, Jiang Zhen, Jiao Minhui, Lei Xin, Li Gaolun, Li Peng, Li Yuehui, Lin Binwen, Liu Chenmu, Liu Juan, Liu Yun, Lu Qi, Luo Ruiqi, Lv Xiuhuan, Ma Lina, Pang Hui, Peng Shaoming, Shao Ying, Song Jing, Song Ruipeng, Su Dan, Sun Kai, Sun Penghui, Tian Kai, Tong Yifeng, Wang Meizhai, Wang Zhongmei, Wen Huina, Wu Hongxia, Wu Xiaochen, Xi Huihua, Xiao Peiqing, Xie Hong, Xing Jianying, Yang Jianshun, Yang Juan, Yang Libin, Yang Ming, Yang Xue, Yang Yingjun, Yang Yisong, Yao Zhenguo, Yu Baoliang, Yu Hang, Zhang Dangli, Zhang Huaxing, Zhang Meili, Zhang Xiaoli, Zhang Xulan, Zhao Baoyin, Zhao Xin, Zheng Falu

Executive Editors: Tong Yifeng, Li Yuehui, Zhang Huaxing, Yue Dejun, Wen Yunxia, Zhu Guangya, Yang Wenhui, Ban Peili

Executive Proofreaders: Lan Yunxia, Yang Xiuying, Lan Wenxia, Zhang Caixia

Cover Designer: Xie Ping

Exceutive Printing: Chang Hongxin, Wen Hongjian

Welcome

(preface)

The 5th International Yellow River Forum (IYRF) is sponsored by Yellow River Conservancy Commission (YRCC) and China Yellow River Foundation (CYRF). On behalf of the Organizing Committee of the conference, I warmly welcome you from over the world to Zhengzhou to attend the 5th IYRF. I sincerely appreciate the valuable contributions of all the delegates.

As an international academic conference, IYRF aims to set up a platform of wide exchange and cooperation for global experts, scholars, managers and stakeholders in water and related fields. Since the initiation in 2003, IYRF has been hosted for four times successfully, which shows new concepts and achievements of the Yellow River management and water management in China, demonstrates the new scientific results in nowadays world water and related fields, and promotes water knowledge sharing and cooperation in the world.

The central theme of the 5th IYRF is "Ensuring water right of the river's demand and healthy river basin maintenance". The Organizing Committee of the 5th IYRF has received near one thousand paper abstracts. Reviewed by the Technical Committee, part of the abstracts are finally collected into the Technical Paper Abstracts of the 5th IYRF.

An ambience of collaboration, respect, and innovation will once again define the forum environment, as experts researchers, representatives from national and local governments, international organizations, universities, research institutions and civil communities gather to discuss, express and listen to the opportunities, challenges and solutions to ensure the sustainable water resources management.

We appreciate the generous supports from the co – sponsors, including domestic and abroad governments and organizations. We also would like to thank the members of the Organizing Committee and the Technical Committee for their great supports and the hard work of the secretariat, as well as all the experts and authors for their outstanding contributions to the 5th IYRF.

Finally, I would like to present my best wishes to the success of the 5th IYRF, and hope every participant to have a good memory about the forum!

Chen Xiaojiang
Chairman of the Organizing Committee, IYRF
Commissioner of YRCC, MWR, China
Zhengzhou, September 2012

Contents

F. Structural and Non – structural Measures with New Technology to Ensure Water Right of the River's Demand

G. Water Security, Water Transfer and Advanced Water Saving Technology and Monitoring Equipment

F. Structural and Non - structural Measures with New Technology to Ensure Water Right of the River's Demand

Structural and Non – structural Measures for Increasing Water Use Efficiency: Results of the Yongding River Basin Management Pilot Project

Starkl M. [1], *Bisschops I.* [2], *Chen J.* [3], *Li P.* [4], *Wang Y.* [4], *Yang Y.* [5], *Yu F.* [3], *Yu X.* [6], *Zhang X.* [5], *Zeng S.* [3] and *Zhong Y.* [4]

1. Competence Centre for Decision Aid in Environmental Management, University of Natural Resources and Life Sciences, Vienna, Austria
2. Lettinga Associates Foundation, Wageningen, the Netherlands
3. Tsinghua University, Beijing, 100053, China
4. Development Research Centre, Ministry of Water Resources, Beijing, 100083, China
5. Ministry of Environment, Foreign Economic Cooperation Office, Beijing, 100035,China
6. MEP – FECO & Chinese Academy for Environmental Planning, Beijing, China

Abstract: The project "Managing water scarcity: intelligent tools and cooperative strategies", funded by the European Commission, has aimed at pointing out the potential for increasing water use efficiency in the Yongding River Basin, a sub basin of the Hai River Basin in North – East China. It has been implemented over the last 4 years in collaboration with the following partners from China: Foreign Economic Cooperation Office of the Ministry of Environment, Development Research Centre of the Ministry of Water Resources, Tsinghua University and the Tianjin Environmental Protection and Technology Development Centre. this paper presents the main results of this project: the project has studied structural and non – structural measures for increasing water use efficiency in 3 sectors (Agriculture, industry and domestic use); Agricultural water use: the project studied the potential and limitations of micro – irrigation technologies in the Yongding River Basin, based on case study evaluations. Industrial water use: the project studied the potential for increasing water use efficiency in the most relevant industries in the Yongding River Basin: textile, chemical and power industry. For each industry 1 ~ 2 factories were selected and strategies to improve water use efficiency elaborated. Domestic water use: domestic water use is comprised of showers, toilets, washing machines and other uses. In Beijing, the largest water users are showers (34%), followed by toilets (26%) and washing machines (22%). Water saving types are available for each of those and the potential and constraints for mainstreaming the most water saving types have been investigated based on user surveys. Rainwater harvesting (RWH): RWH is considered to have a large potential for the Yongding River Basin, but it has only been applied to a very limited extent.

Those studies have shown that water use efficiency can be considerably increased in the Yongding River Basin through implementing water saving technologies in each of the studied sectors. To facilitate their implementation non – structural measures such as policy instruments can substantially help. Experiences with such policy instruments have been surveyed and improved policy instruments suggested. Then, those policy instruments have been discussed with various stakeholders in the Yongding River Basin and based on the feedback policy recommendations were elaborated.

Key words: domestic water use, irrigation, policies, recycling, water saving

1 Introduction

the Yongding River Basin is an important sub – basin of the larger Hai River Basin. The Hai River Basin is in the economic centre of northern China, occupying 3.3% of the national land surface, containing 10% of the population and producing 12% of the country's GDP (Wei et al,

2008). The basin suffers from low rainfall and high potential evaporation, a very high anthropogenic water demand, and severe water pollution. It depends on water transfers from other river basins, and the water withdrawal amounts to over 250% of the local resources (Wang and Jin, 2006).

Overexploitation and intensive damming of rivers for electricity production and irrigation have greatly reduced the water discharge to the sea, from 24×10^9 to 1×10^9 m^3 between the 1950 s and 2001 (Wei et al, 2008). The alarming water quantity and quality situation causes ecological problems such as wetland deterioration and rapidly declining groundwater levels, and hinders the social and economic development in the river basin (Wei et al, 2008; Wang et al, 2010a). It is paramount that water consumption and pollution are reduced, calling on all involved sectors to make a large effort.

This paper presents a study on the potential of structural and non - structural measures for water saving in agriculture, industry and the domestic sector in the Yongding River Basin.

2 Structural measures

Structural measures are a technical response to the decrease water scarcity. The potential of technical improvements in the domestic sector, industry and agriculture was investigated. At household level, the installation of water saving devices, in industry, the use of clean technologies, and in agriculture, the implementation of water saving irrigation technologies can reduce water consumption.

2.1 Domestic sector

Studies have shown that showers, toilets and washing machine cause the largest amount of residential water consumption, namely 34%, 26% and 22%, respectively (Zhang et al, 2007). Therefore, these three devices were selected for detailed assessment. The survey included a literature review and interviews with key technology providers in the Yongding River Basin to understand to what extent water saving appliances are already used in the case study areas and hence what is the potential for future applications. With this information the distribution of different types of water saving devices across the Yongding River Basin was estimated. Based on this information the potential maximum (theoretic) water saving amount for the case that all existing non - water saving devices are replaced with the currently available most - water saving devices.

2.1.1 Results

With the maximized promotion of water saving appliances, based on the statistics in 2008 (Ministry of Housing and Urban - Rural Development of P. R. China, 2009), residential water consumption in the Yongding River Basin will decrease from 95.4×10^6 m^3 to 72.7×10^6 m^3, equal to 23.8% water conserved. Improving the efficiency of showerhead, toilet and washing machine will cut down water use by 30.0%, 27.3% and 30.3%, respectively. Especially, the promotion of water - saving showerheads can achieve an obvious potential of 9.7×10^6 m^3. The vast amount is a result from the large market share of low efficiency showerheads. If 20% of households would use dry toilets, which means no water would be consumed for toilet flushing, the water saving potential would improve from 27.3% to 41.8%.

2.1.2 Discussion

When people have little awareness on why they should save water, the installation of water saving devices can end up in offsetting behaviour such as taking longer showers can result in lower reductions or even an increased water use (Cobacho et al, 2004). The effort that people are willing to make in saving water is mainly linked to their awareness, understanding and appreciation concerning water and the environment. In most cases water has a very low cost compared to other household spending, which is why there is hardly an economic incentive to save water (Willis et al, 2011).

Even though the implementation of water - saving appliances would save around 30% of the water use for those devices, popularity of these appliances is not high. Users were asked for measures that would help them to change to and adopt water saving devices. Among the most important were a lower price for such devices, good quality of the products and governmental support. This points out that some sort of financial support at least for poorer families to implement water saving devices, in particular for showers, could help to increase the distribution of water saving devices and hence to better utilize the huge water saving potential form such devices. Other measures such as market control as already practices for toilets could also contribute to utilize better this potential, and producers need to try to develop products that better meet the demands of the consumers.

2.2 Agriculture

The case study areas for the agriculture investigation were Huailai County and Ying County. They are situated in Hebei province and Shanxi province, respectively.

Based on the collected information, the water saving potential of applying water saving irrigation technologies in Yongding River Basin was calculated. The water saving irrigation technologies include canal seepage control, irrigation by low pressure pipes, sprinkler irrigation and micro – irrigation.

The total planting area in Huailai is 38,353 hm^2, of which 10,433 hm^2 are irrigated by using water saving irrigation. The share of sprinkler irrigation and micro – irrigation low: only 2.6% of the water saving irrigation is occupied by these two technologies. In Ying County, of the 36,000 hm^2, 10,000 hm^2 are cultivated using water – saving irrigation, of which sprinkler irrigation and micro – irrigation area occupy 10.6 %.

2.2.1 Results

In accordance with traditional wild flooding irrigation mode, 900 m^3 water is required for each hectare every time annually, irrigation is performed 4 times, and the irrigation water is 3,600 m^3. After micro – irrigation is adopted, irrigation will be performed 4 times, 2,850 m^3 water can be saved per hectare every year, and 1,275 watt – hour can be saved; the benefit from saving water and power is about 1,545 yuan/hm^2.

Based on the collected information, the water saving potential of applying water saving irrigation technologies in Yongding River Basin was calculated. As there is no integrated river basin management institution, there are no official statistics on river basin level. The water saving area of different water technology in Yongding River Basin was calculated based on the water saving irrigation area of the Haihe River Basin from the study of agriculture water use and water saving in the Haihe River Basin.

According to the agriculture water saving investigation in Huailai and Yingxian, the water use of different water saving technology was obtained. With reference to the water use of different water saving technology in the Haihe River Basin, the water use of different water saving technology in the Yongding River Basin was determined, see Tab. 1.

Tab. 1 Water use of different irrigation technologies in Yongding River Basin and saving compared to flood irrigation

	m^3/mu	m^3/hm^2	Water saving compared to flood irrigation (m^3/hm^2)
Flood irrigation	240	3,600	0
Canal seepage control	140	2,100	1,500
Pipeline	150	2,250	1,350
Sprinkler	95	1,425	2,175
Micro	50	750	2,850

If all irrigated areas are changed to micro irrigation than around 40% ~60% of the current agricultural water use could be saved.

2.2.2 Discussion

For the case study counties different constraints for the introduction of modern water saving irrigation technology were identified. It is expected that the situation in Huailai and Yingxian is mostly representative for other agricultural areas in the Yongding River Basin.

Constraints can be roughly divided in four groups: natural, financial, cultural and organisational factors were identified. Natural factors such as windy conditions, hot groundwater and high concentrations of minerals can indeed limit the applicability of certain technologies, but if the incidences are localised and by choosing the right technology it may be possible to mitigate the limitations. Financial constraints were found to exist in both insufficient funds for investment in water saving irrigation, and a lack of financial stimulus to save water. The agricultural water price is too low, and in some villages water is charged based on land area, not on water use. In the case of financial constraints, changes in funding schemes or water charges could make a difference.

The current village culture includes trust in the authority and management capabilities of village committees. Farmers seem to have no enthusiasm and capability for self – management, and only a weak awareness of it. In addition, they mostly lack the sense of protecting the irrigation facilities, especially for those which are installed with investments by the state. Next to this low interest shown by farmers in the systems, there is the problem of "concept constraint". They favour technologies with a low technical level, even though the systems with a higher technical level such as drip – irrigation have an obviously better water – saving effect. These constraints involve a change of mind of the farmers. It will be difficult and take a long time to overcome these issues. Education into the benefits of self – management and more knowledge on irrigation technologies would be needed.

2.3 Industry

Data available to the Chinese Ministry of Environmental Protection was used to identify the industries with the greatest water use and pollution discharge in Yongding River Basin. The data comprised 33 industry types, with a total of 309 companies (SEPA – FECO 2007). A limited number of industries was chosen to serve as case studies for the identification of improvement options. For these industries one factory each was selected for the evaluation. As the situation among all factories within one particular industry type is usually similar, it is assumed that the results of the studied factory can be extrapolated to other factories in the case study area.

2.3.1 Results

The textile industry, chemical industry (nitrogen fertiliser) and heat electric power industry were chosen as case studies; one factory was selected per category. For each case studies measures to reduce water consumption and pollution load were elaborated (Yang et al, 2011 for details).

By multiplying the obtained reductions with the number of similar factories in the river basin, an estimation of the reduction in water use and pollution that could be achieved were made Tab. 2 shows the extrapolated values for the entire river basin.

Tab. 2 Extrapolation of case study results to entire the Yongding River Basin

Improvement potential		Textile industry	Chemical industry	Power plant
Visited factory	Water (1,000 m^3/a)	1,105	5,032	3,612
	COD (t/a)	2,592	466	
River basin	Number of factories	13	47	25
	Water (1,000 m^3/a)	14,369	236,513	90,300
	COD (t/a)	33,696	21,902	

The total water – saving and pollution reduction potential for the three studied industries in the Yongding River Basin is 3.41×10^8 m^3/a and 55,598 t COD/a, respectively. Compared to the total industrial water demand in the Yongding River Basin this could save around 95% of the current industrial water used. Considering the severe water scarcity of the Yongding River Basin this is a huge amount, which should not be left unused.

2.3.2 Discussion

The case studies have shown that certain measures can save substantial amounts of water and also reduce the pollution potential but also create economic benefits with a short amortisation period. For such measures monetary policy incentives such as low interest loans or non – monetary incentives such as awareness raising and propaganda may support the implementation of such measures. Also regulations that would make the use of water and pollution reducing state of the art technologies mandatory should be envisaged. For more expensive infrastructure such as RO plants subsidies or special funds to help factories to cope with the financial burden may have a positive effect.

3 Non – structural measures

Chapter 2 has identified how technological interventions can help to make water management in each sector more efficient, but has also shown that structural measures need a enabling environment to be implemented on larger scale. Non – technical measures such as policy instruments can play an important role in achieving better water management. This chapter summarises the main policy problems that were identified by the project team in consultation with stakeholders and identifies a set of policy instruments that can help to respond and overcome those problems.

Policy instruments are thereby classified in regulation and incentice based instruments. Each type can then encompass monetary and non – monetary instruments, incentive based instruments may also include market instruments. This gives the following classification:

(1) Group 1: regulation based, monetary.
(2) Group 2: regulation based, non monetary.
(3) Group 3: incentive based, monetary.
(4) Group 4: incetive based, market.
(5) Group 5: incentive based, non – monetary.

3.1 Domestic sector

3.1.1 Lack of efficiency of water use devices

As shown in chapter 2.1 through replacing prevailing domestic water use appliances (toilets, showers, and washing machines) around 24% of water could be saved. Possible policy instruments that can increase the implementation of more water efficient household devices are listed in Tab. 3.

Tab. 3 Domestic sector – lack of efficiency of water use devices

Group1: regulation based, monetary instruments	No policy instruments could be identified
Group2: regulation based, non – monetary instruments	1. Ban water wasting devices. 2. Improve water use standards for water use devices
Group3: incentive based, traditional monetary instruments	1. Provide a subsidy (grant, non repayable) to households for adopting water saving devices. 2. Provide a subsidy (low interest loan, repayable) to households for adopting water saving devices. 3. Provide water saving devices for free to households
Group4: incentive based, market based monetary instruments	No policy instruments could be identified
Group5: incentive based, non – monetary instruments	1. Improve awareness of water saving devices by (increasing) promotion. 2. Promoting development of more water efficient devices

3.1.2 Too low water price

The current water price is 3.7 yuan/m^3 in Beijing and 3.4 yuan/m^3 in Tianjin which only a very small share of the average income. Compared to global best practices which show water is to be priced at its true cost including operational and maintenance costs, water price is too low to encourage people to save water. Several policy instruments can be used to tackle the problem of a too low water price as shown in Tab. 4.

Tab. 4 Domestic sector - too low water price

Group1: regulation based, monetary instruments	Increase water price
Group2: regulation based, non - monetary instruments	No policy instruments could be identified
Group3: incentive based, traditional monetary instruments	Establish stepped water pricing where people have to pay only for water exceeding certain basic needs
Group4: incentive based, market based monetary instruments	No policy instruments could be identified
Group5: incentive based, non - monetary instruments	Raise awareness for economic value of water

3.2 Industry—lack of adoption of technologies for clean production

The total water - saving and pollution reduction potential for the three studies industries in the Yongding River Basin is 3.41×10^8 m^3/a and 55,598 t COD/a, respectively. Compared to the total industrial water demand in the Yongding River Basin this could save around 95% of the current industrial water used. Policy instruments will be needed to utilize this potential. A list of possible instruments is shown in Tab. 5.

Tab. 5 Industry - lack of adoption of technlogies for clean production

Group1: regulation based, monetary instruments	1. Set a stricter fine for factories that do not comply with standards and strengthen enforcement. 2. Change charges for wastewater discharge so also the pollution load is charged
Group2: regulation based, non - monetary instruments	1. Increase standards for clean production that have to be fulfilled for wastewater treatment and reuse. 2. Create a pollution prohibition policy by factories (zero discharge) more specific
Group3: incentive based, traditional monetary instruments	1. Increase subsidy (grant, non - repayable) for implementation of clean technology. 2. Provide subsidy (low interest loan - repayable) that can be used by factories to implement clean technologies. 3. Provide tax benefits for companies adopting clean technologies
Group4: incentive based, market based monetary instruments	Water right trading
Group5: incentive based, non - monetary instruments	1. Increase awareness for clean technologies amongst industrial water users. 2. Increase awareness for clean technologies in industrial processes for the wider public

3.3 Agriculture

3.3.1 Lack of efficiency of irrigation technologies

Water – saving irrigation techniques in agricultural irrigation can not only reduce water wastage, increase crop yield and quality, but also reduce non – point source pollution in rural areas through water – saving methods in preventing the loss of chemical fertilizers and pesticides. The study has shown that around 40% ~ 60% of the agricultural water demand could be saved with state of the art irrigation technologies. Tab. 6 shows a list of policy instruments that can facilitate the implementation of such irrigation technologies.

Tab. 6 Agriculture - lack of efficiency of irrigation technologies

Group1: regulation based, monetary instruments	Set fines for applying non – efficient irrigation technologies
Group2: regulation based, non – monetary instruments	1. Prohibit practice of flood irrigation. 2. Increase coordination between local water resources department and the related departments (e. g. agricultural department). 3. Provide a regulation that encourages farmers to use treated wastewater for irrigation where feasible
Group3: incentive based, traditional monetary instruments	1. Subsidy (grant) to increase the investment of local government for water saving irrigation, including water saving projects and technologies. 2. Subsidy for low interest loans. 3. Exempt farmers who apply water saving irrigation technologies from water resources fee. 4. Provide financial support directly to farmers for adopting water saving irrigation technologies
Group4: incentive based, market based monetary instruments	No policy instruments could be identified
Group5: incentive based, non – monetary instruments	1. Raise awareness and capacity of farmers through establishing Water User Associations. 2. Improve awareness through propaganda for water saving irrigation technologies. 3. Strengthen the management of irrigation system so to make irrigation technologies more efficient. 4. Establish a rating system for the efficiency of different irrigation technologies

3.3.2 Too low water price

Agriculture is China's main water consuming sector and farmer can currently use water free of cost (apart from subsidised costs for pumping). As farmers receive water for irrigation without any costs, there are no incentives to save water.

Tab. 7 Lists the possible policy instruments to tackle this issue.

Tab. 7 Agriculture—too low water price

Group1: regulation based, monetary instruments	Levy charges for all irrigation water use
Group2: regulation based, non – monetary instruments	Allocate water among different users based on water rights
Group3: incentive based, traditional monetary instruments	Reduce electricity state subsidies for irrigation
Group4: incentive based, market based monetary instruments	1. Water right trading. 2. Base water price on water quota
Group5: incentive based, non – monetary instruments	1. Improve the awareness of economic value of water. 2. Farmer participation

4 Discussion

For – two sectors, industry and agriculture, policy workshops were conducted, where affected stakeholders and governmental representatives were invited. The policy problems and responses were then discussed with these stakeholders and their opinion on the instruments was then elicited (based on structured interviews and questionnaires).

4.1 Industry

In the group of the regulation based instruments, the creation of a pollution prohibition policy is ranked first with respect to the expected impact. All suggested policy instruments are related to a high effort for implementation and are believed to receive high acceptance from policy makers and public. The acceptance of industry as affected party is expected to be highest for the pollution prohibition policy and a charge for pollution loads. The acceptance for application of fines for the non – compliance with standards and the increased standards for clean production is estimated to be lowest.

All incentive based policy instruments are related to a high impact on industrial water use. In the opinion of the interviewed persons, the provision of subsidies needs least efforts to be implemented. Acceptance of public and policy makers is estimated to be high and very high, whereas acceptance of industry is not taken as granted for all measures: awareness raising campaigns for industry and public are not as well rated as monetary incentives.

4.2 Agriculture

The two policy instrument with the highest expected impact belong to Group 3 and are the provision of subsidies to increase investment of local governments, direct financial support for farmers, but also two regulation based policy instruments (set fines for applying non – efficient irrigation technologies and prohibition of flood irrigation) are associated with a high impact. There are some policy instrument to which respondents associated only low impact (> 50% low or very low impact): the increased coordination between local water resources departments, exemption of water resources fee and all instruments of Group 5 except the strengthening of management systems. All policy instruments are related to high or very high effort to be implemented.

In both case study areas, respondents identified the setting of fines for application of non – efficient irrigation technologies to be least accepted among farmers. In general, policy makers are expected to accept all policy instruments, whereas the public, especially in Yongxian, is not expected to welcome policy instruments that aim at increasing water saving in agriculture.

5 Conclusions

The study showed that structural measures need to be combined with non – structural measures to reach highest impact. It is difficult to find the right set of policy instruments to cater to expectations of all stakeholders. The affected parties, which were in the case of the survey farmers and industry, are expected to favour incentive based monetary policy instruments. The implementation of the policy instruments is the most difficult part and related to high efforts for all policy instruments.

Another approach would be to make use of additional, yet unused sources, such as rainwater harvesting. The promotion of the use of rainwater harvesting and storage is a possible solution as the drinking water demand of the human and livestock and the supplementary irrigation of the basic income security farmland can be covered with the available resources in Yongding River Basin.

Acknowledgements

The research leading to these results has received funding from the EU 6th Framework Programme FP6 – INCO under grant no 032397/MAI – TAI (Managing water scarcity: Intelligent tools and cooperative strategies).

References

Cobacho R, Arregui F, Gascón L, et al. Low – flow Devices in Spain: How Efficient are They in Fact? An Accurate Way of Calculation[J]. Water Science & Technology: Water Supply, 2004, 4(3):91 – 102.

Ministry of Housing and Urban – Rural Development of P. R. China. China Urban Construction Statistical Yearbook 2008[M]. Beijing: China Planning Press, 2009. (in Chinese)

Ministry of Land, Infrastructure, Transport and Tourism of Japan. Water Resources in Japan. http://www. mlit. go. jp/tochimizushigen/mizsei/water_resources/contents/current_state2. html.

Ministry of Water Resouces of P. R. China. China Water Resources Bulletin 2009[M]. Beijing: China Water Power Press, 2009. (in Chinese)

Townsend D. Going Green and the Benefits for Schools. http://www. firstelevenmagazine. co. uk/going – green – and – the – benefits – for – schools.

Wang X C, Jin P K. Water Shortage and Needs for Waste Water Reuse in the North China[J]. Water Science & Technology, 2006,53 (9): 35 – 44.

Wang W, Tang X, Huang S, et al. Ecological Restoration of Polluted Plain Rivers Within the Haihe River Basin in China[J]. Water Air Soil Pollute, 2010a, 21(1):341 – 357.

Wang L, Wang Z, Koike T, et al. The Assessment of Surface Water Resources for the Semi – arid the Yongding River Basin from 1956 to 2000 and the Impact of Land Use Change[J]. Hydrological Processes, 2010b, 24:1123 – 1132.

Wei Y, Miao H, Ouyang, Z. Environmental Water Requirements and Sustainable Water Resource Management in the Haihe River Basin of North China[J]. Int. J. Sust. Dev. World, 2008, 15 (2):113 – 121.

Willis R M, Stewart R A, Panuwatwanich K, et al. Alarming Visual Display Monitors Affecting Shower end Use Water and Energy Conservation in Australian Residential Households[J]. Resources, Conservation and Recycling, 2010,54: 1117 – 1127.

Willis R M, Stewart R A, Panuwatwanich K, et al. Quantifying the Influence of Environmental and Water Conservation Attitudes on Household End Use Water Consumption[J]. Environ Manage., 2011, 92(8):1996 – 2009.

Yang Y, Zhang X, Yu X, et al. Industrial Water Saving and Pollution Reduction Potential Analysis for Water – shortage Area in China: Taking the Hai River—the Yongding River Basin as Example, 2011.

Zhang Shiqiu, Deng Liangchun, Yue Peng, et al. Study on Water Tariff Reform and Income Impacts in China's Metropolitan Areas: The Case of Beijing[R]. Beijing: The World Bank Project Report, 2007.

Zhong Yuxiu, Li Peilei, Jiang Nan. Baseline Study of Selected River Basins in China. EU Project 032397 (INCO) Managing Water Scarcity: Intelligent Tools and Cooperative Strategies, 2007.

Rapid Discrimination of Earth – rock Dams' Stability under the Condition of Exceeding Standard Flood

Li Yun[1,2], *Wang Xiaogang*[1], *Xuan Guoxiang*[1,2], *Zhu Long*[1] and *Zeng Chenjun*[1]

1. Nanjing Hydraulic Research Institute, Nanjing, 210029, China
2. State Key Laboratory of Hydrology – Water Resources and Hydraulic Engineering, Nanjing, 210029, China

Abstract: It's very important for flood control decision to make rapid discrimination of dam breach. However, there is no efficient discriminant method for earth – rock dam's stability under the condition of exceeding standard flood because of the complex structure of dams and too many uncertain factors. In this paper, a discriminant model was established based on discriminant method for earth – rock dam. It considered the material characteristics of dam body. The results demonstrate that the model is qualitatively reasonable and quantitatively right. The importance of factors were ranked from big to small as follow: catchment area, incipient frictional velocity, dam height, dam width and dam length.
Key words: earth – rock dams, exceeding standard flood, discriminant analysis, incipient friction velocity

1 Introduction

The extreme weather events in China significantly increase because of the changes of global climate. The exceeding standard flood is more and more likely to occur suddenly, abnormally and unpredictable. It makes a big challenge to the dam safety. The dam – failure events recorded in our country are more than 3,500 and most of the dams are earth – rock dams. There are 11 events happened just in 2010. Looking for an efficient and rapid discriminant method for earth – rock dam failure under the condition of exceeding standard flood is of great significance to the promotion of the decisions and flood prevention abilities.

There is no direct methods towards the analysis of dam safety under the condition of exceeding standard flood. The limit equilibrium method is usually used to evaluate the stability of dams. which can provide the smallest safety coefficient after analysis of the dams under each operation condition. It can be used to evaluate the stability of dams. LiZongKun etc. analysis the coupling effects of seepage and stress fields in the dam. It can be seen that the interactions of seepage and stress in the dam can not be overlooked and the coupling effect will produce adverse effect to the stability of the dams. The earth dam slope stability are analyzed and discussed based on the shear strength reduction technique by Sun lili and LiZongLi. These methods are often difficult to apply when it involves the exceeding standard flood because of their complex calculation and too many parameters. The authors used the research achievement of barrier dam's stability for reference. A technique of discriminant analysis was applied to the analysis of dam stability under the condition of exceeding standard flood for the first time. It was proved it worked well. The effect of the catchment area, the dam height, dam width and dam length in the analysis are considered with the exception of material parameters of the dam. This paper attempts to consider material parameters of the dam in the discriminant analysis model so that the model's physical meaning is more clear and the accuracy rate of the model is better. The study shows that the discriminant analysis model which takes the material parameters into account is qualitatively reasonable and quantitatively right. The model can be used into discriminant analysis of the dams under the condition of exceeding standard flood very well.

2 The variable which on behalf of the dam materials' comprehensive erosion – resisting ability—the incipient friction velocity U_{*c}

The previous studies have shown that the incipient friction velocity U_{*c} can be chosen as the index of the dam materials' comprehensive erosion – resisting ability in the research of dam – break problems. The incipient friction velocity U_{*c} can be get from tests generally and the methods can be found in the references (Hong Qa – lin et al. , 2005,2006). Unfortunately, the data collected about the dam – break is lack of quantitative description of the dam materials' comprehensive ability at the moment. It's rather difficult to achieve the incipient friction velocity U_{*c}. Thus, an approximate method of analysis is used to get the parameter in this paper.

Xie Renzhi has drew the relationship figure between the scour coefficient and the storage capacity based on the statistics of 400 dams both at home and abroad. What's more, he also defined the soil coefficient. The dam soil were divided into 5 types and the corresponding soil coefficient can be seen in Tab. 1. The coefficients are empirical coefficient calculated by Xie Renzhi through the storage capacity, Erosion volume etc. These soil coefficient are inconvenient to use because they are difficult to acquire directly in practical terms. This is also one of the main reasons why U_{*c} was chosen instead of the soil coefficient in this paper.

There are a lot of test data between the incipient velocity and sediment grain size in literature (Qian Ning et al. , 2003). Alao, the relation curves are given in the literature. Through these data, the incipient friction velocity U_{*c} corresponding with the different sediment particle size can be got by using Eq (1) and Eq(2). The curve can be drawn as shown in Fig. 1. The range of grain size is wide, so it can nearly cover all the usual sediment particle size.

Tab. 1 The coefficients of soil property

Soil group	Compactedness	φ
1	very loose	12.5
2	a little loose	6.7
3	medium	3.65
4	a little dense	1.68
5	very dense	0.495

Note: To facilitate the application, a little change in the classification of "compactedness" was made.

$$\frac{U_c}{U_{*c}} = \frac{C}{\sqrt{g}} \tag{1}$$

U_{*c} is the incipient friction velocity in the equation:

$$C = 7.66\left(\frac{R}{K_s}\right)^{\frac{1}{B}} \tag{2}$$

where, Ksis the sediment's Nikuradse roughness, $K_s = D_{50}$ and $R = h$ in this paper.

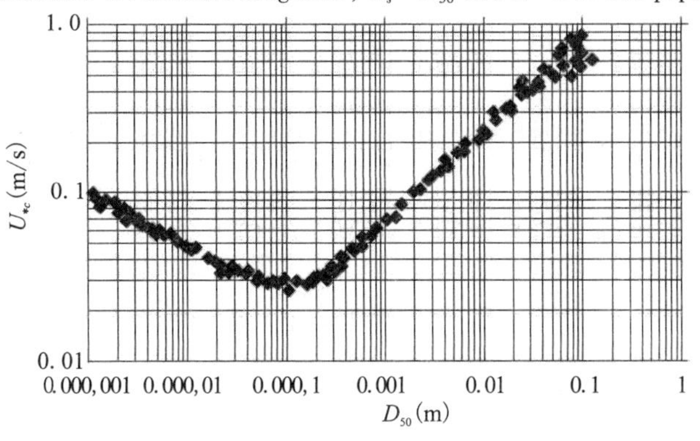

Fig. 1 The relation curve between D_{50} and U_{*c}

According to the Tab. 1 and Fig. 1, the incipient friction velocity U_{*c} goes bigger along with the increase of medium particle diameter D_{50} when the soil coefficient is bigger than 0.000,1 m. The bigger one is corresponding to the smaller soil coefficients. As it's difficult to find the quantitative relationship between soil coefficient and the incipient friction velocity U_{*c} directly. Here the D_{50} were divided into 5 parts according to the soil coefficient(take 0.000,1 ~0.1 m as the research object) and the incipient friction velocity can be got from fig. 1. The relationship can be established between the incipient friction velocity and soil coefficient, as it can be seen in Tab. 2.

Tab. 2 The relationship among φ, D_{50} and U_{*c}

φ	D_{50}	U_{*c}
$\varphi > 12.500$	0.000,1	0.030
$6.700 < \varphi < 12.500$	0.050	0.560
$3.650 < \varphi < 6.700$	0.075	0.680
$1.680 < \varphi < 3.650$	0.090	0.750
$\varphi < 1.680$	0.100	0.800

Through the transformation above, the soil coefficient can be replaced by the incipient friction velocity which can be achieved directly from tests or calculation.

3 The discriminant analysis method

The discriminant analysis is one of the multivariate statistical methods which can differentiate samples from different types. Its object of study is the training sample. That means the classification of the original data is known in advance. And then based on the original data a discriminant function can be sought. To judge the type, the original data should be plugged into the discriminant function. The purpose of the discriminant analysis is to seek for a model that can find the biggest difference between the two groups and classify the samples into different groups. Then, using the existing classification the ownership of the new samples can be determined.

3.1 The fundamental principles and calculation process of discriminant analysis

The discriminant analysis is order to seek a linear function composed of the original variables to maximize the ratio between inter - group and intra - group. The aim is to look for the most important one w_j of x_1, x_2, x_3, x_4, $\cdots$, x_j in the linear function which can make the difference of y between groups maximum relative to the difference in groups.

$$y = w_1x_1 + w_2x_2 + w_3x_3 + \cdots + w_jx_j \tag{3}$$

where, x is variable and the w is discriminant weight.

The basic concept of discriminant analysis can be illustrated by geometry graph (as shown in Fig. 2), supposing there were only two variables and they done not exist high correlation. There are two populations named 1 and 2. We want to separate the 1 and 2 by valuing the two variables x_1, x_2. The two populations have a partial coincident picture. However, they can be well separated by the two linear discriminant functions. Thus, to get better effect ,we can use the linear discriminant function 2 to discriminate the two populations instead of the original variables.

There are two discriminant analysis methods frequently used, the typical discriminant analysis and the Bayesian discriminant analysis. The basic principle and the calculation process are shown as follows.

3.1.1 The typical discriminant analysis

Suppose there are p factors in the group, we want to construct a likelihood function:

$$L(x_1, x_2, \cdots, x_p) = y = c_1x_1 + c_2x_2 + \cdots + c_px_p \tag{4}$$

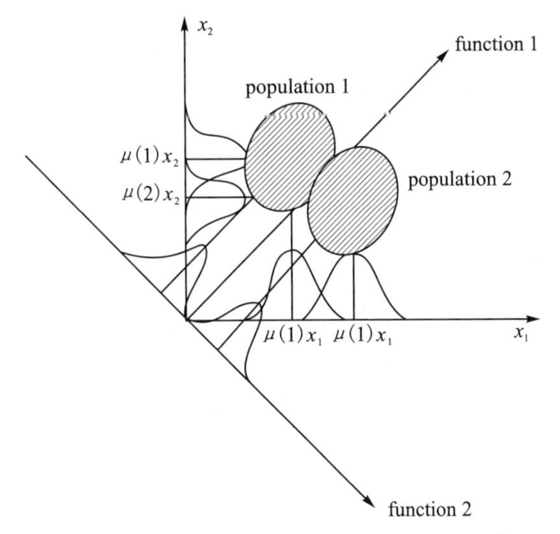

Fig. 2 The principle of discriminant analysis

where, the x_i Represents the i variable; c_i is the coefficient.

Suppose there are n_1 samples belong to the first group and n_2 samples belong to the second group, their matrix is as follows:

$$X^{(1)} = \begin{bmatrix} x_{11}^{(1)} & \cdots & x_{1n_1}^{(1)} \\ \vdots & \ddots & \vdots \\ x_{p1}^{(1)} & \cdots & x_{pn_1}^{(1)} \end{bmatrix} \tag{5}$$

$$X^{(2)} = \begin{bmatrix} x_{11}^{(2)} & \cdots & x_{1n_2}^{(2)} \\ \vdots & \ddots & \vdots \\ x_{p1}^{(2)} & \cdots & x_{pn_2}^{(2)} \end{bmatrix} \tag{6}$$

We can distinguish the two kinds of samples according to the $\boldsymbol{X}$ at the top right corner. A constant value y_0 should be defined first. Then we put the p factors into the function y. If $y > y_0$, we can regard the sample as the first class, otherwise, it belongs to the second class. Suppose we have achieved the discriminant functions and we put the data into the functions. Then:

$$y_j^{(1)} = c_1 x_{1j}^{(1)} + c_2 x_{2j}^{(1)} + \cdots + c_p x_{pj}^{(1)} \quad (j = 1,2,3,\cdots,n_1) \tag{7}$$

$$y_j^{(2)} = c_1 x_{1j}^{(2)} + c_2 x_{2j}^{(2)} + \cdots + c_p x_{pj}^{(2)} \quad (j = 1,2,3,\cdots,n_2) \tag{8}$$

suppose:

$$\overline{y^{(1)}} = \frac{1}{n_1} \sum_{j=1}^{n_1} y_j^{(1)} \tag{9}$$

$$\overline{y^{(2)}} = \frac{1}{n_2} \sum_{j=1}^{n_2} y_j^{(2)} \tag{10}$$

Thereupon, to the classification, it's obvious to require that:

(1) When the $Q = (\overline{y^{(1)}} - \overline{y^{(2)}})^2$ is bigger, the consequence is better;

(2) Make differences in the group as small as possible, in other words, make $\sum_{j=1}^{n_1} (y_j^{(1)} - \overline{y^{(1)}})^2$, and $\sum_{j=1}^{n_2} (y_j^{(2)} - \overline{y^{(2)}})^2$, as small as possible. Make:

$$F = \sum_{j=1}^{n_1} (y_j^{(1)} - \overline{y^{(1)}})^2 + \sum_{j=1}^{n_2} (y_j^{(2)} - \overline{y^{(2)}})^2 \tag{11}$$

According to the two standards above, we should make

$$I = \frac{Q}{F} \tag{12}$$

as big as possible。

Take logarithm on both sides of the equations above:

$$\ln I = \ln Q - \ln F \tag{13}$$

Obviously, the $\ln I$ will get its maximum value when the I get the maximum value . So if $c_i (i = 1, 2, 3, \cdots, p)$ meet the extrem conditions, it should make:

$$\frac{\partial \ln I}{\partial c_i} = \frac{\partial \ln Q}{\partial c_i} - \frac{\partial \ln F}{\partial c_i} = 0 \tag{14}$$

Then:

$$\frac{1}{Q}\frac{\partial Q}{\partial c_i} - \frac{1}{F}\frac{\partial F}{\partial c_1} = 0 \tag{15}$$

So:

$$\frac{\partial Q}{\partial c_1} = \frac{Q}{F}\frac{\partial F}{\partial c_i} = I\frac{\partial F}{\partial c_i} \tag{16}$$

Thus some conclusions can be derived from the formula $c_e (e = 1, 2, \cdots, p)$ the solution of the equation:

$$\sum_{e=1}^{p} c_e l_{ke} = t_k (k = 1, 2, \cdots, p) \tag{17}$$

Make the I to its maximum value. So the linear discriminant function can be got:

$$y = c_1 x_1 + c_2 x_2 + \cdots + c_p x_p \tag{18}$$

The criterion is as follows:

Determine y_0 first, usually:

$$y_0 = \frac{n_1 \overline{y^{(1)}} + n_2 \overline{y^{(2)}}}{n_1 + n_2} \tag{19}$$

We always assume, $\overline{y^{(1)}} > y_0 > \overline{y^{(2)}}$ so when a sample data is put into the equation: The sample belongs to the first class if $y > y_0$, otherwise, it belongs to the second class. In the practical calculation in this research, we put the sample data into the equation. Then , it can be discriminated according to the relationship between the results and 0.

3.1.2 The Bayesian discriminant analysis

The basic idea of the Bayesian is using the known prior probability to infer the posteriori probability. The posteriori probability and the wrong rate of each sample are calculate to differentiate the sample's classification by the biggest posteriori probability and made the expected loss to the minimum.

Suppose $q_i(x)$ is the prior probability of the g populations, And the corresponding probability density function is:

$$f_i(x) = (2\pi)^{-k/2} |V_i|^{-1/2} \exp\left\{\frac{1}{2}(X - \overline{X})' V_i^{-1} (X - \overline{X})\right\} \tag{20}$$

The V_i is the covariance matrix of the population i and k is the number of variables.

According to the Bayesian probability formula, it can be obtained that the posterior probability is as follows:

$$p(x \epsilon i \mid x \epsilon t) = \frac{q_i f_i(x)}{\sum_{i=1}^{g} q_i f_i(x)}, i = 1, 2, \cdots, g \tag{21}$$

Now, we define $y_i(x) = \ln [q_i f_i(x)]$, then the posterior probability can be described as follows:

$$p(x\epsilon i \mid x\epsilon t) = \frac{\exp[y_i(x)]}{\sum_{i=1}^{g}\exp[y_i(x)]}, i = 1,2,\cdots,g \tag{22}$$

The discriminant functions can be simplified. Here is the formula for reduction:

$$y_1(x) = c_{0i} + c_{1i}x_1 + c_{2i}x_2 + \cdots + c_{pi}x_p + \ln q_i (i = 1,2,\cdots,g) \tag{23}$$

The criterion is as follows: Put the sample data into the discriminant functions $y_i(x)$, then we can get g data files. Comparing the data files with each other, we will get the maximal $y_i(x)$. Of course, we can regard the sample as this class. Their are two classes in the research of dam safety. Thus, we can confirm the category of the sample by compare $(y_1 - y_2)$ with 0.

3.2 The requirements of discriminatory analysis

The 26 data files of damaged dams collected are as follows: Qiaodun, Huangshandong, Gaoshuxia, Shijiagou, Ljiaju, Changmao, Shimantan, Zhangba, Tiefosi, Banqiao, Oros, Sandaohe, Heilongweng, Liujiadong, Mahe, Qianjin, Yanxigou, Liulihe, Dadichong, Hongwafang, Changchong, Shangmating, Jianguo, Guanmentian, Zhonghua, Xiangshan. Part of the dam parameters are added through investigation. In order to improve the effectiveness and integrity of the samples, there are 15 damaged dams and 11 safe dams under the condition of exceeding standard flood in the data.

To establish the discriminant function, the basic requirements of the sample data are as follows:

(1) The relationship between independent variable and the dependent variable can fit the linear assumption.

(2) The value of the dependent variable is independent, and must be determined in advance.

(3) The independent variable must subject to multiple normal distribution. Discriminant analysis is relatively robust when the sample variables are in violation of normal distribution hypothesis. It would not almost affect on the result as long as the distribution is not deviating intensively from the normal.

(4) All the independents have a equal variance and the covariance matrix are equal too. But discriminant analysis method is a "robust" statistical method, it can still be analyzed when the data is in violation of this rule.

(5) The independent variables can not exist among each other. The equation and variables's coefficients may be changed when the multiple linear exists, however, the discriminatory analysis is very robust.

The data can always meet the first two requirements automatically, so we test the other three using conditions here.

The K – S verification method is used for the normal distribution inspection (K – S check's original hypothesis: the data fits normal distribution), the results are shown in Tab. 3. From the Tab. 3, it is known that the parameters can meet the normal distribution better after transformation (the friction velocity approximately meet the requirements).

Tab. 3 The appraisal results of normality test by the way of K – S

Variable	Catchment area	Dam length	Dam width	Dam height	U_{*c}
The significant level of original parameters's twotailed test	<0.001	0.225	0.227	0.559	0.034
Transform Mode	logarithmic	logarithmic	logarithmic	logarithmic	exponential
The significant level of transformational parameters's twotailed test	0.972	0.859	0.930	0.999	0.049

Note: The one whose significant level is 0.05 or more accord with the normal distribution.

To judge whether there are multiple linear between independent variables, the correlation coefficient matrix between variables are checked firstly. The calculation results are shown in Tab. 4. Generally, it will be problems of linear when the correlation coefficient of the variables on the analysis is more than 0.9. And, it may have a common linear problem if the correlation coefficient of the variables on the analysis is only more than 0.8. The correlation coefficient are shown in Tab. 4, it is known that the correlation coefficient between variables are all under 0.8. it won't fail to work, so the data collected can be calculated.

Tab. 4 The correlation matrix of all variables

Variable	Dam length	Dam width	Dam height	Catchment area	U_{*c}
dam length	1.000	0.396	0.420	0.703	-0.057
dam width	0.396	1.000	0.778	0.572	-0.218
dam height	0.420	0.778	1.000	0.716	-0.225
catchment area	0.703	0.572	0.716	1.000	-0.098
U_{*c}	-0.057	-0.218	-0.225	-0.098	1.000

Note: All the variables are transformed.

3.3 Discrimination model and its quantitative correctness

Discrimination model were established by using typical discriminant analysis method and the Bayesian discriminant analysis. Typical discriminatory analysis model is:

$$Y = -0.65LX_1 - 2.403LX_2 + 3.966LX_3 + 1.58LX_4 - 1.395LX_5 - 8.57 \quad (24)$$

Bayesian discriminant analysis model is:

$$Y^* = -2.352LX_1 - 8.693LX_2 + 14.346LX_3 + 5.72LX_4 - 5.049LX_5 - 31.004 \quad (25)$$

where, LX_1 is dam length; LX_2 is dam width; LX_3 is dam height; LX_4 is catchment area, LX_5 is U_{*c} (Note: All the variables are transformed).

The discriminant result are shown in Fig. 3 and Fig. 4. As shown in the figures, the one whose discriminant score Y is greater than 0 is as the destroyed one, and the others belong to the unspoiled group . We use 0 to mark the destroyed ones and 1 to mark the others in the initial material. The results of the typical discriminant analysis and the Bayesian discriminant analysis are consistent as shown in the figures. The difference of the discrimination model is only on the variable coefficients. Discriminant matrix is shown in Tab. 5.

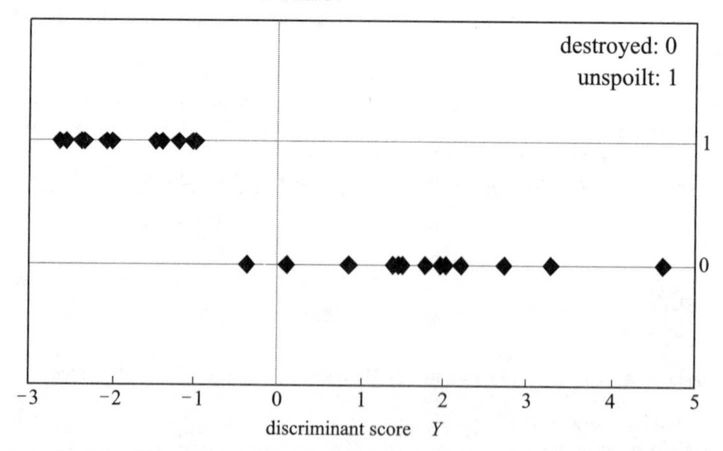

Fig. 3 The distribution of the results by typical discriminant analysis

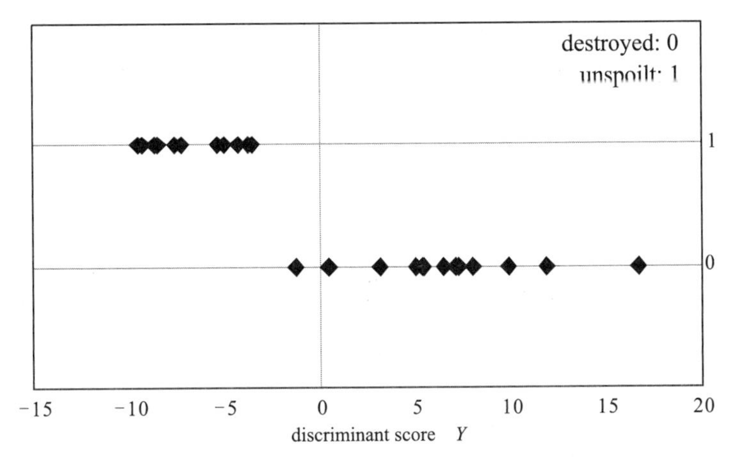

Fig. 4 The distribution of the results by Bayesian discriminant analysis

Tab. 5 The results of the discriminant analysis

		Discriminant analysis (both methods)		
		Discriminant result		Totality
		Destroyed	Not destroyed	
Initial materials	Destroyed	14	1	15
	Not destroyed	0	11	11
Totality		14	12	26

Total accuracy:96.15%
Accuracy of the group destroyed:93.33%
Accuracy of the group not destroyed:100.00%

In order to further test the stability and effectiveness, a cross - validation of the established model was done. It works as the following way: take out most of the samples given to rebuild model and use the rest of samples to forecast. Then, get the prediction error and record their quadratic sum. This process will be go on until all the samples are forecasted. According to the result the cross validation accuracy of the discriminant analysis model is 92.3%. It is clear that the discrimination model including material parameters of dam has a quantitative higher accuracy in this paper.

3.4 Discriminant variables' influence weight and the model's qualitative rationality

According to the discriminant analysis above, the influence weight can be got by comparing the coefficients of each variables. The bigger the absolute value of standardized discriminant function coefficients is, the more the influence weight is. The standardized discriminant function coefficients of the models are shown in Tab. 6 respectively.

From the Tab. 6, it is known that the catchment area is the most important factor that affect the stability of the dam; The second is U_{*c} which on behalf of the resistance of the dam materials. And in terms of the geometric parameters of the dam itself, the order of the importance influence is: dam height, dam width, dam length. It means that the material properties of the dam has a very important function in resisting the flood and is an indispensable factor in the research of the stability of the dam.

Tab. 6 The coefficient table of standardized discriminant function

Variable(Logarithm)	The standardized discriminant function coefficients
catchment area	1.064
the incipient friction velocity U_{*c}	-0.648
dam height	0.645
dam width	-0.573
dam length	-0.243

According to discriminant analysis principle, the positive standardized discriminant function coefficients and negative ones embody the negative and positive influences on the discrimination results respectively. It is known from the table that the coefficients of catchment area and dam height are both greater than 0. That means the stability of the dam will be reduced if the catchment area or dam height increases. This is because that the larger the catchment area, the greater the flood flow. The higher the dam, the higher the water head between the upstream and downstream of the reservoir. The coefficients of dam length, dam width and the incipient friction velocity are all less than 0. That means the stability of the dam will be enhanced if these parameters increase and this rule is obvious. The parameters' standardized discriminant function coefficients correctly reflects the effect on the stability of the dam. It also reflects the discriminant model established in this paper is qualitative reasonable.

4 Conclusions and Remarks

It's very important for flood control decision to make rapid discrimination of dam breach when the exceeding standard flood appear. There is no reliably rapid determination method available so far both at home and abroad. Because there are a large quantity of small reservoirs in our country and a lot of them are lack of hydrological data, a fast determination method based on the most basic parameters of the dam is presented in this paper. A model of discriminant is established which involve in the material properties of the dam. According to the results the model can be used for discrimination of earth - rock dam failure and it is qualitatively reasonable and quantitatively right. Additionally, the influence weight, which has important realistic meaning and practical value, can be got through the method.

Something should be noted, the incipient friction velocity of the samples in this paper are achieved through the transformation of the experience parameters "φ" in the reference(Xie Ren - zhi, 1993). So, it has a certain degree of uncertainty. The more accurate incipient friction velocity is to further improve the accuracy of the prediction model.

References

Peng Xuebin. Dam Stability Analysis Method and Application Example [J]. Science and Technology of West China, 2009, 8 (4): 49 - 50.

Chen Zuyu. Stability Analysis of Soil Slope—Principle · Method · Procedure [M]. Beijing: China WaterPower Press, 2003.

Li Zongkun, Wang Peng fei, Zhao Feng yao. Stability Analysis of Earth - rockf ill Dam Based on Fluid - solid Couping [J]. Journal of Zhengzhou University (Engineering Science), 2009, 30 (3): 44 - 47.

Sun Lili, Li Zongli. Slope Stability Analysis of the Earth Rockf ill Dam Based on the Strength Reduction Method [J]. Journal of Water Resources & Water Engineering, 2009, 20 (5): 113 - 116.

Shi Zhenming, Li Jianke, Lu Cunliang, et al. Research Status and Prospect of The Stability of Land – Slide Dam [J]. Journal of Engineering Geology, 2010,18 (5):657 – 663.
Tong Yuxiang. Quantitative Analysis for Stability of Landslide Dam [D]. Taiwan: National Central University, Graduate Institute of Applied Geology, 2008.
Chai Hejun, Liu Hanchao, Zhang Zhuoyuan, et al. Preliminarily Stability Analysis of Natural Rock – field Dam Resulting from Damming Landslide [J]. Geological Science and Technology Information, 2001,20 (1):77 – 81.
Nanjing Hydraulic Research Institute. the Research of the Sickness Danger Reservoir and Dam – failure Law [R]. Nanjing: Nanjing Hydraulic Research Institute, 2010.
Ermini L, Casagli N. Prediction of the Behavior of Landslide Dams Using a Geomorphological Dimensionless Index [J]. Earth Surface Processes and Landforms, 2003(28):31 – 47.
Korup O. Geomorphometric Characteristics of New Zealand Landslide Dams [J]. Engineering Geology, 2004(73):13 – 35.
Li Yun, Wang Xiaogang, Zhu Long, et al. Discriminant Analysis for Earth – rock Dams' Stability under the Condition of Exceeding Standard Flood [J]. Advances in Water Science.
Nanjing Hydraulic Research Institute. Dam – failure Tests and Simulation Technology Research [R]. Nanjing: Nanjing Hydraulic Research Institute, 2010.
Hong Dalin, Miu Guobin, Deng Dongsheng, et al. The Starting of Cohesive Soil and Its Application in Engineering [M]. Nanjing:Hohai University Press, 2005.
Hong Dalin, Miu Guobin, Deng Dongsheng, et al. Relation of Starting Shear Stress and Physical and Mechanical Indexes of Cohesive Undisturbed Soil [J]. Advances in Water Science, 2006,17(6):774 – 779.
Xie Renzhi. The Dam – failure Hydraulics [M]. Jinan: Shandong Science and Technology Press, 1993.
Qian Ning, Wan Zhaohui. Dynamics of Sediment Movement [M]. Beijing:Science Press, 2003.

Detection of Water Leaks in Foum El – Gherza Dam (Algeria)

N. Hocini and *A. S. Moulla*

Centre de Recherche Nucléaire d'Alger (CRNA), Alger – Gare, 16000, Algiers, Algeria

Abstract: The main objective of this work was detect water leakage combining classical and nuclear techniques (isotopic and radiotracer). Classical methods concerned the follow up of physico – chemical parameters (conductivity, temperature and chemical composition). Isotopic and radiotracer techniques concerned the isotopic composition (oxygen – 18 and tritium) and labelling of the reservoir (Fluorescent tracers), respectively. The investigation was performed by a research team from the 'Algiers Nuclear Research Centre' in collaboration with engineers from the 'National Agency for Dams'. The chemical and isotopic results have shown no influence of dam water on the surrounding aquifers. Dye tracing has shown a faster water circulation through complex pathways for the right bank as compared to the left one.

Key words: dam, leakage, detection, isotopes techniques, fluorescent tracers, conductivity, temperature

1 Introduction

This work was carried out within the framework of a Regional Co – operation AFRA programme supported by IAEA (RAF/8/028). This programme consists of the strengthening and development of scientific knowledge in African countries, mainly in the detection of a dam leakage and safety. The main objective of this work was to detect the origin of water leakage combining conventional, tracing and isotope techniques. Classical methods concerned the monitoring of changes in physico – chemical parameters (conductivity, temperature and chemical composition). Isotopic and tracing techniques concerned the determination of the isotopic composition of on – site available different water bodies (oxygen18 and tritium) and the labelling of the reservoir (Rhodamine – Wt fluorescent tracer) respectively.

2 Description of the study area

Foum El – Gherza dam is located at 18 km east of Biskra province in south – eastern part of Algeria (Fig. 1). Its water is collected mainly for irrigation purposes.

The dam model project was designed in 1946 by the Algerian Hydraulics Laboratory (Neyrpic). The completion of the construction phase was in 1952 and first operation immediately showed leaks at the downstream part of the dam. Since then, leakage continued and the maximum water loss (20.7×10^6 m^3) was recorded from 1981 to 1982.

The dam regulates about 13×10^6 m^3 of water conveyed by Wadi El – Abiod (Fig. 2) ephemeral river and tributaries during a whole hydrological cycle for a catchment of 1,300 km^2.

2.1 Geological and hydrogeological settings

The massif where the dam is founded is composed of a relatively thick fissured karstic Maestrichtian limestone laying over a Campanian marl stratum (Fig. 3).

Three main aquifers are present in the investigated region. These are from the shallowest to the deepest the following:

(1) The alluvial phreatic aquifer: it is contained in the alluvial deposits and is recharged by precipitation and infiltration from the riverbed and from irrigation channels.

(2) The Miopliocene sands and the Senonian – Eocene carbonates aquifers. They are deeper and are both still artesian at some locations.

Fig. 1 Map showing location of dam site

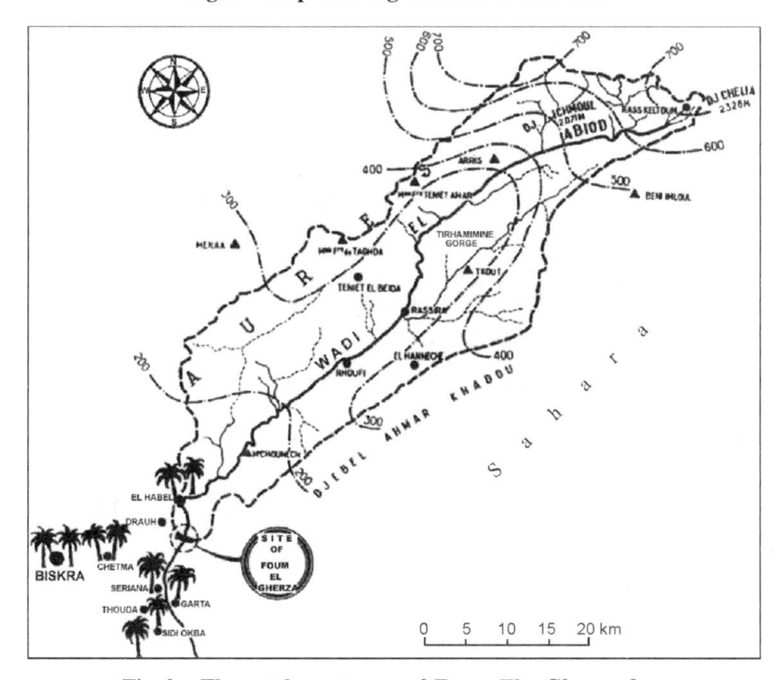

Fig. 2 The catchment area of Foum El – Gherza dam

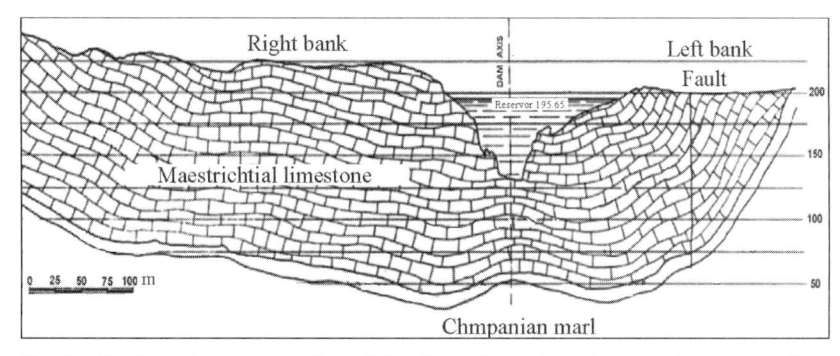

Fig. 3 Geological cross – section of the dam site and geology of site surroundings

2.2 History of the leakage phenomenon

The first filling and operation of the reservoir started in 1952 upon completion of the construction which was then resumed between 1954 and 1957 by the reinforcement of the hydraulic works and the injection of a grouting curtain. Just after dam filling, leaks started to appear at the immediate downstream of the dam (1.6×10^6 m^3 in 1952/1953, and 2.0×10^6 m^3 for the next two years,…). The maximum value was observed for season 1981/1982 during which not less than 20.7×10^6 m^3 were recorded (Fig. 4). Due to lack of precipitation, the leakage rate started to fall down and during summer 1994 (June 24th) no more water was present in the reservoir.

Seepage takes place both at the left and the right banks. The leaks at the left bank are visible and their flow rate is rather low. They are collected within a small irrigation channel which follows the riverbed towards the irrigated areas. On the contrary, right bank leaks flow via a two – row network of drains and are directly collected within the irrigation gallery.

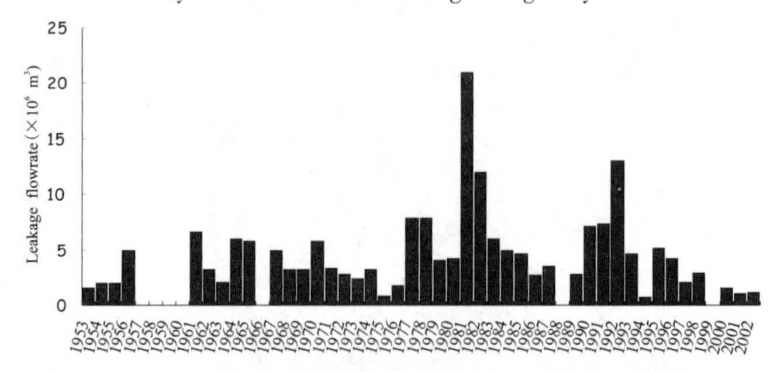

Fig. 4 History of the leakage at Foum El – Gherza since first filling of the reservoir

3 Experimental work

One evaluation mission and two field trips were carried out (Fig. 5). During the first field campaign, samplings for all water bodies that are present within the immediate vicinity of the dam were effected. In addition, conductivity and temperature profiles were recorded for the accessible piezometers on both banks and for some points in the lake itself.

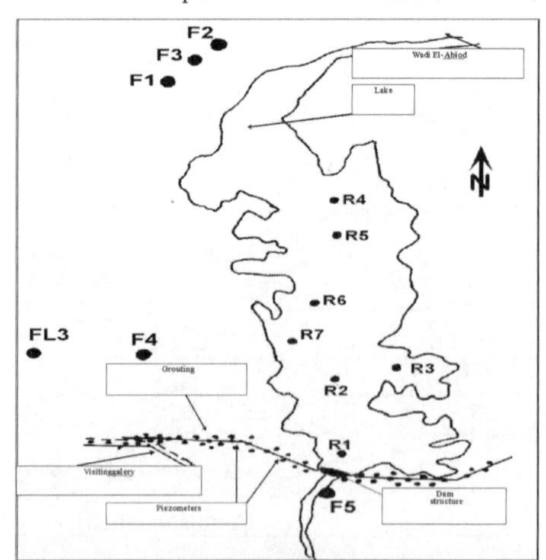

Fig. 5 Schematic location of sampling sites during field campaigns

The first in situ observations have shown the existence of lateral infiltrations through the massif. Land collapse and rockfall were also noticed. Excavations and large cracks were brought to sight by the decrease of water level in the lake (6.5×10^6 m^3 at that time). It was even possible to hear water flowing through the carbonate fractures on the left bank.

During the second field mission and besides recording profiles similarly as during the first field trip, tracer experiments using Rhodamine – WT were achieved. Making use of such a tool, an estimation of the total flowrate at the outlet of the irrigation gallery was performed.

Moreover, the reservoir water was also labelled in the vicinity of the banks for the sake of interconnection experiment purposes. The volume of water in the lake was about 5.9×10^6 m^3.

4 Results and conclusions

The achievements and the results gathered from the field campaigns that has been effected allowed us to identify the problems affecting this dam through the overall observation of the features of the physical medium (geology) where it has been built.

The results obtained from temperature and conductivity profiles that were drawn for the probe accessible piezometers have shown the presence of very complex vertical and horizontal flows as depicted in Fig. 6(a) and Fig. 6(b). This could be due to the geological characteristics of the site.

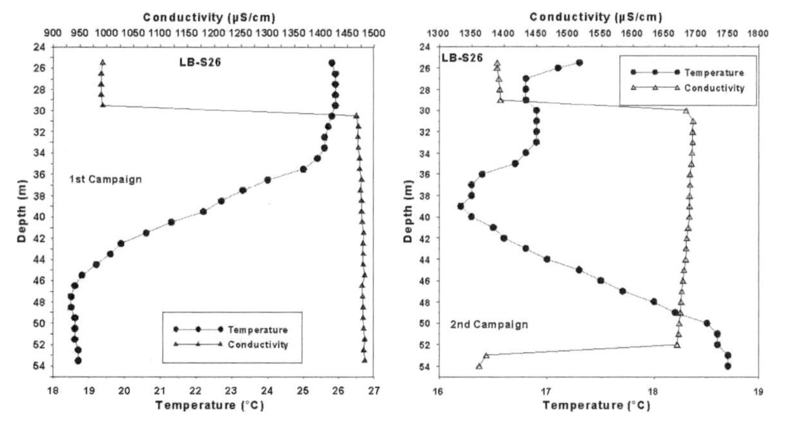

(a) Comparison of left bank S – 26 piezometer EC& T profiles of the two campaigns

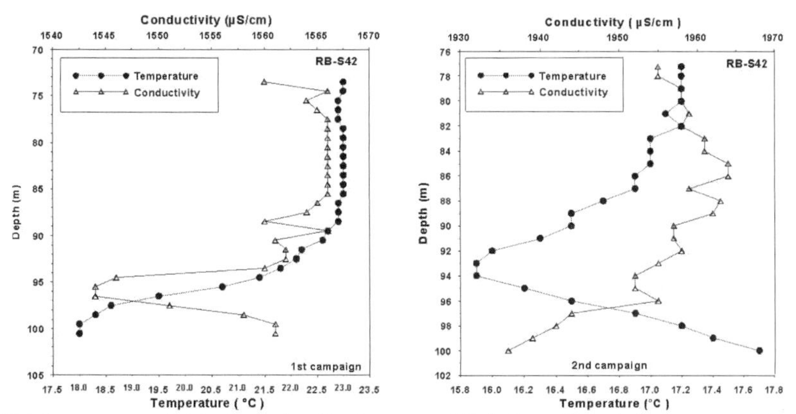

(b) Comparison of right bank S – 42 piezometer EC & T profiles of the two campaigns

Fig. 6

With regard to the chemical composition, a Piper diagram (Fig. 7) showed that there is no relationship between lake water and groundwater that is occurring in the immediate vicinity of the reservoir. This was further confirmed by the isotopic results through oxygen - 18 and tritium contents as summarized in Tab. 1.

An interconnection experiment using Rhodamine - WT fluorescent tracer was performed afterwards. It consisted of labelling the reservoir water at a distance of 2 m from the shores. The monitoring of tracer arrival at the downstream springs showed that Rhodamine was detected respectively after two days at the right bank and after one week at the left bank, since injection started (Fig. 8).

The investigation described in this paper leaded us to the conclusion that the implementation of such a pilot study and its associated preliminary findings seems to be satisfactory. However, according to the complexity of the geological site, more experiments need to be performed in order to better understand and better address the leakage phenomena.

Tab. 1 Isotopic composition of some samples

Sample	Tritium(T. U.)	$\delta^{18}O$(‰)
Borehole F1	2.1	-7.3
Borehole F2	1.3	-7.5
Borehole F3	< 0.4	-7.5
Borehole FL3	1.4	-7.9
Borehole F4	2.0	-7.4
Borehole F4 bis	1.7	-7.5
Borehole F5	< 0.4	-7.4
Reservoir	9.7	-0.2

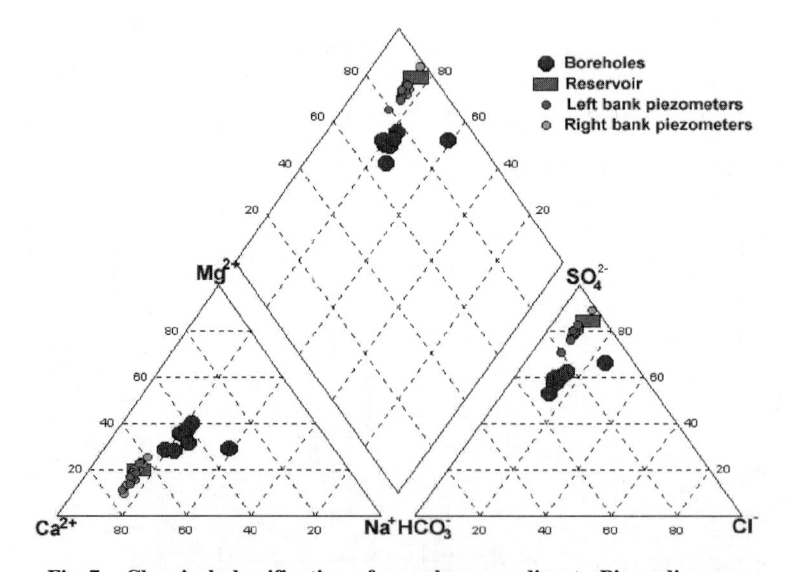

Fig. 7 Chemical classification of samples according to Piper diagram

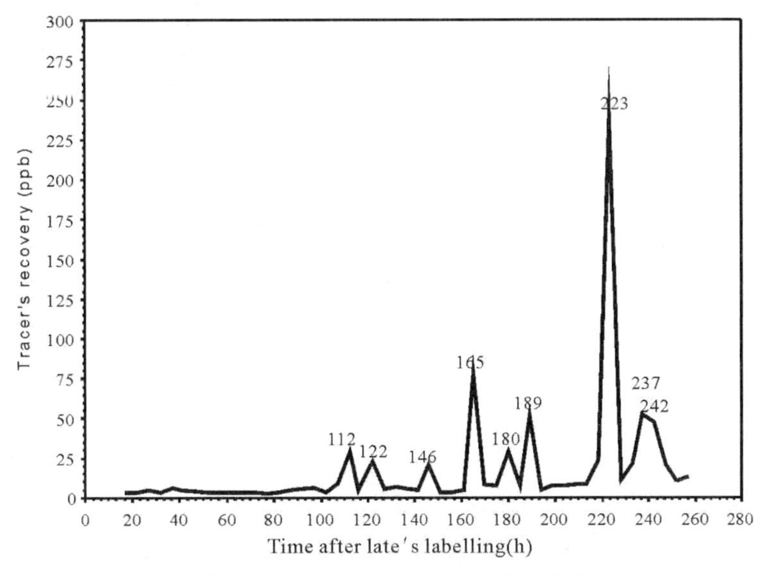

Fig. 8 Tracer recovery as a function of time

Acknowledgements

The above investigation was carried out within the framework of IAEA – AFRA – RAF/8/028 regional cooperation project. The authors are very grateful to the staff and colleagues of the 'Agence Nationale des Barrages' for their fruitful co – operation and logistic assistance during field missions. All analyses were performed at the Applied Hydrology and Sedimentology Department, Algiers Nuclear Research Centre whose staff contribution is gratefully acknowledged.

References

Plata Bedmar A.. Detection of Leakage From ReserVoirs and Lakes. Use of Artificiels Tracers in Hyrology[C]// Proceedings of an Advisory Group Meeting. AIEA – TECDOC – 601, May 1991.

Remini B., Hocini N., Moulla A. S.. La Problématique Des Fuites d'eau Dans le Barrage de Foum El – Gherza (Biskra). Premier Séminaire National sur le Développement des Zones Arides et Semi – arides. University of Djelfa, Algeria, (16 – 17 May 1999).

Remini B., Hocini N., Moulla A. S.. Water Leaks in Foum El – Gherza dam (Algeria), EIN – International No. 6, pp. 55 – 59, Décembre 2001.

Hocini N.. Etude des Fuites Dans les Barrages au Moyen des Techniques Isotopiques. Premières Journées d'Etudes sur les Applications des Techniques Nucléaires en Ressources Hydriques et en Agriculture, COMENA/CDTN, Alger 30/11 – 02/12/98.

Moulla A. S.. Détection du Cheminement des Fuites Dans Les Barrages et Autres Réservoirs Artificiels: Cas du Barrage de Foum El – Gherza. Seminar on Nuclear Techniques et Applications in Socio – economic Fields: Water Resources, held on the Fringe of the XIIIth AFRA Working Group Meeting of the African National Co – ordinators, Algiers Sheraton Club des Pins, 22 – 24/04/2002.

Plata Bedmar A., Araguas – araguas L.. Detection and Prevention of Leaks from Dams[M]. Ed: A. A. Balkema Publishers.

Study on Safeguard Measurement and Construction Optimization of Dike Section of the Yellow River

Zhou Li[1] and *Li Yan*[2]

1. Construction and Management Department, Yellow River Conservancy Commission (YRCC), Zhengzhou, 450003, China
2. Transportation School, Wuhan University of Technology, Wuhan, 430070, China

Abstract: The Yellow River dike is not only the main safeguard facility but also the important and indispensable routes for flood controlling. To defense the Yellow River flood is always attached great importance by our country and our party. Since P. R. China founded, the Yellow River dike has been widened and heightened for four times and the section of dikes changed correspondently. The roads on top of dikes underwent the structures of earth style, mortar, graval, pitch, et al. And top of dike underwent the structures of weir, concrete drainage (fieldworks), water barrier drainage, et al. the slope of dike underwent the planting of trees, grasses, tree cutting and maintaining trees of shoulders of dike, trees at front of river and trees on alluvial dike as well as trees in safeguard field, et al. It is necessary to study the structure of dike section and safeguard measurement to ensure safety of flood control and decrease the maintenance workload in order to maintain dike integrated and establish the abundant dike engineering in flood control, ecology, traffic and cultural contents. To regulate the management of dike and promote modern development of dike management and ensure safety of dike we studied and analyzed dike section and safe guard measurement in plan, design and construction as well as management by large amount of investigation of the Yellow River dike status and exiting problems, and we put forward concrete index and managing standard by terms of practice in order to offer technology support for the Yellow River dike construction and management.
Key words: dike section, optimization of structure, safeguard measurement, the Yellow River

1 Structure evolution of the Yellow River dike section

The Yellow River dike has been widened and heightened for four times since 1949. Because it formed on base of civilian levees in history, the section structure design was not reasonable and its quality was weak.

Under natural condition of complex, its status can change gradually. Inner dike body there exist much of hidden troubles such as caves, craves and loosen earth clod. These potential problems bring safety impact on our engineering and it needs us to research for safety measurement to ensure safety use. With implementation of dike design and standard, the dike section structure should be optimized accordingly in heightening dike. For example, dike top width was regulated from 6 ~ 8 m at foundation of our country to 10 ~ 12 m in 2004. The dike top road changed from gravel, sand or concrete to bitumen road in 1998.

Dike top drainage way has changed from weir block style to prefabrication concrete drainage or concrete blocking plank as concentrative form. Some adaptable drainage ways have been still in use and it is used to lower dike and small conflux area and with well planting vegetation. The drainage way of dike slope has changed from bricklaying to grass, 37 lime earth and prefabrication concrete or mould casting on site. There ever planted much of trees at fronting slope of river dike beforetime. By consulting experience of overseas in river management and analysis of detriment through digging trees in dike, YRCC clean out of slope trees overall in 1988 and only kept dike shoulder trees for viewing ecological effect. And therefore it underwent change of tree planting, trees cutting and dike shoulder trees keeping in dike slope. Flood controlling trees has planted since 1990s. To alleviate surfing erosion for dike slope YRCC workout flood controlling trees construction plan and executed it gradually. The main contents of this plan was to plant lower willow trees and higher willow trees in different width to defense flood at front slope in average dike section. The dike back alluvial area was created since P. R. China founded. The top height of dike back alluvial area was controlled in terms of river flood controlling water line and its width in 1985 was 30 ~ 50 m. Before 2003, the width of alluvial area was 50 ~ 80 m. Since 2004, the width was determined from 80 ~ 100 m after implementation of standard dike construction. Adaptable trees were planted after top alluvial area reinforced. Defense trees are planted in safety guard area at back of dike after borderline determined.

2 Dike section structure status and problems

Since floods of the three big rivers in 1998, our country made much investment in water facilities. After reformation of water system in 2006, the finance can be guaranteed in engineering maintenance and conception of the Yellow River engineering management also has been updated. To unify dike design standard and regulate section structure and reinforce management YRCC has issued many engineering plans and design standards. But due to long dike line and numerous engineering sites as well as limitation of other affection, the Yellow River dike sections are not in uniform and non – regulated yet. This has seriously influence on whole dike appearance and at same time it added workload of maintenance.

2.1 Problems of dike top roads

Dikes line is about 2,000 km in length. It includes the Yellow River dike, the Qin River dike, the Daqinghe River dike, the Weihe River dike, flood detaining area dike et al. If Ratings of dike buildings is different, the dike top road structure design is different.

The obvious problems existing are: firstly the Yellow River dike top of 1,370 km has not been all hardened yet, traffic under conditions of urgency can not be guaranteed. Secondly, parts of dike top roads which were built with concrete are damaged and spoiled badly owing to not adapting road sinking. Thirdly, dike top roads are lower than dike shoulders after being widened and heightened, resulting in un – fluent drainage and badly water accumulated and thereby cause seepage failure. Fourthly, part of road to dike top erodes dike shoulders and dike slope and structure of alluvial area, thus has much influence on dike integrated and its safety. Meanwhile, the motor way of the third rating can be used in only 10 years or so in terms of criterion of motor way design. At present, parts of dike top roads have attained to or surpassed the design limitation and are damaged badly, whereas, the rebuilding items of roads is difficult to implement on schedule due to lacking effective renewing mechanism of reformation and the complex construction procedures.

2.2 Problems of drainage

There are two kinds of drainage of dikes, one is concentrative way and another is dispersive. The drainage canals may be divided into three types which are brick, concrete and earth made. There existing much of damage to drainage canals every year, and what caused these problems are as follows.

2.2.1 Un – reasonable layout of drainage

Firstly, it is un – reasonable layout of drainage. The dike drainage canals was lain along dike shoulders and its structures are built with prefabricated slot ware and it lacks of consideration in jointing sectional drainage canals and thus results in un – fluent drainage and water accumulation in hardened top ways after rain. It is easy to form earth cave and dike road damage after rain soaking and crushing down by heave vehicles. Secondly, there does not form the network in sectional and lengthways drainage canals. For example, there are not top drainage canals in some dike sections, crossly it is set with sectional drainage ones. In some dike sections, there lays dike top drainage canals concentratively that do not joint with drainages in alluvial area, even they didn't set drainage canals at all. This results in no way for safeguard area drainage.

2.2.2 Materials and their structures are not reasonable

For example, drainage canals built with bricks were dug too deep in depth and large in section area. This easily causes stone to be loosening and sunk in dike shoulders. In winter, these canals are readily to be damaged because of freezing. The upper layer of bricks is easy to be loosened and damaged. The prefabricated concrete slot generally is made with pure concrete materials, transportation and installation may cause much spoil. There are many joints of drainage canals and slot process is difficult. The joint part of canals and earth body is not easy to fill out. The width and density of grass planting in dispersive drainage of dike shoulders can not attain to design demands. Thus affects the drainage effects.

2.2.3 Constructions are weak and damaged badly

Some drainage canals are too close near road shoulders (average not more than 1 m). During digging canals, it penetrates the harden road surface of lime earth layer and does harm to road base structure and accelerates damage of harden road surface, and consequently limits use life of road. There are no penetration proofing plastics at bottom of concrete drainage canals or processed with three – seven lime earth. So it can not guarantee quality and causes hidden problems inner dike.

2.2.4 To set lengthways drainage canals at dike top enhances workload of maintenance and affects engineering appearance

To build sectional drainage canals can decrease water eroded pit and meanwhile increase workload of maintenance. For example, drainage canals built with bricks and prefabricated constituent parts may cause spoiling in part and do damage to slot. There exist trashes, grass, litters and alluvial mud. The maintenance is large workload. Under condition of rain it easily produces penetration and craves and seepage et al.

2.3 Problems on ecological engineering measures

Flood controlling trees in average sections of dike are of shortage and its system is not improved. Part sections pursue planting benefit and lower willows and higher willows grow too fast. The tree top is higher than flood defense water level and does influence on surf defense effects.

Firstly, the problems in planting trees at dike shoulders are hidden troubles for flood control when trees are planted at both sides of dike shoulders because the tree roots in outside are located below flood water defense line. Other problem is that planting pine trees and white pines needs much maintenance and safeguard, at same time it affects traffics of vehicle. Problems of trees planting in alluvial area are that the saplings are too small in tree diameter (not more than 3 cm). Trees grow slowly and uniform in heights. Trees shortage and breach affects the overall ecological effect. Secondly, it deviates from function demands of flood controlling trees in alluvial area. Thirdly, it has influence on planting effect of flood controlling trees if part of obstacle buildings can not be cleaned out.

3 To optimize dike section structure and safeguard measures

To regulate dike management and ensure safety of its use and establish modern dike concentrated with ecology, culture and scenery and at same time reduce maintenance workload, in light of criterions such as dike engineering design regulations and dike engineering management design standard, we optimized dike sectional structure and study the safeguard measures from plan, design, construction and management. And we put forward concrete index of dike section structure. Combined with our work situation, we put fourth criterion of the Yellow River dike engineering planning design in order to offer support to the Yellow River dike construction and management.

3.1 Dike body structure

(1) Width of dike top should meet to needs of construction, management, and flood control. The width of dike top should be designed as 10 ~ 12 m.

(2) Height and width of dike top should keep criterion and height error should be within scope of −5 ~ +5 cm. And width error should be −10 ~ +10 cm. the dike shoulders line should be straight and arc of dam shoulders should be pliable and smooth and without concavity and convexity. The error within 5 m of concavity and convexity is not more than 5 cm.

(3) Gradient of fronting slope and back slope is 1∶3. Slope surface should be smooth and error is not more than 5 cm. At foot of dike should be flat and dike foot line is smooth. With 10 m of dike section, the error should not be more than 10 cm.

3.2 Structure of alluvial area in back of the dike

(1) At different dike sections, the top height of scope of alluvial area is respectively in 0 ~ 3 m in which the top of Huayuankou and Luokou is as same as dike top and its covering thickness is 0.5 m.

(2) The width of alluvial area is 100 m in principle. If it is difficult for immigrant, the alluvial area width is not less than 80 m. Top of alluvial area should be flat and be planted trees.

(3) The slope gradient of alluvial area is 1∶3 and slope surface is smooth, and dike foot line is clear. Within 10 m along slope section, the error is not less than 20 cm.

(4) To determine borderline outside of dike foot in terms of law and plant flood control trees.

3.3 Structures of front and back platform of dike

At the outside margins of platform should be constructed with weir, width and breadth is all 0.3 m. Out margin slope is 1∶3 and inner slope gradient is 1∶1. Every other 100 m, it should be set a grid levee and its width and breadth is all 0.3 m. Surface of platform should be flat and error within 10 m is not more than 5 cm. At top of platform are planted with trees.

3.4 Structure of dike road

(1) Design of dike road engineering should consult the third rating motor way and road surface should be paved with pitch in breadth of 6 m. And it should be set 2% of bidirectional cross slope in the section.

(2) At both side of road should be set block stone with same height of road surface. Between the block stone and dike shoulder earth weir is concreted with gravel or be planted with grass.

3.5 Drainage canals settings and its structure

(1) The drainage ways at top of dike may be set two styles that is concentrated and dispersive according to dike height and scale. When it is adopted as dispersive way, wide weirs should be set at dike shoulders with height of 0.15 m and top breadth of 0.8 ~ 1.0 m. The block stone should be buried with height, thickness and depth of 0.3 m × 0.1 m × 0.15 m. If it is adopted as dispersive ways, sward safety strip with breadth of 0.5 m should be set at both sides of dike shoulders.

(2) If it adopted as concentrated ways of drainage, the dike should be set drainage canals in lengthways and the distance is 100 m collocated with interlaced ways at fronting and back river. On average the space between drainage canals at slope of non − wrapped vulnerable sports or spur dikes are 50 m.

(3) Drainage canals of dike slope should be adopted with concrete structure and its section should be as inverted trapezoid. Energy dissipation pool should be set at foot of dike and the top of pool should be lower than ground 2 ~ 3 cm. Both side of drainage canals and their bottom should be

paved with earthwork fabric or 37 lime earth and combined with base tightly.

3.6 Auxiliary roads to dike

(1) The breadth of roads to dike jointed with local ways is 6 ~ 8 m. The average road breadth is 4 ~ 6 m and set 2% of bidirectional cross slope in both sides, and lengthways slope is not higher than 8%. Sward breadth of safeguard strip of auxiliary roads at both – side is not less than 0.3 m, bilateral slope gradient is 1:2 and at surface of slope is planted with trees.

(2) Auxiliary road should be hardened commonly and its length extends to borders of front and back dike.

(3) The auxiliary roads to dike should be kept integrated and be smooth without caves and deformity or spoiled dike body and sections of alluvial area.

3.7 Layout of flood controlling trees

(1) Breadth of flood controlling trees is 50 m above Gaocun and 30 m below Gaocun. Near to dike may be planted trees of higher willows and outside dike may be planted with lower willows. Higher and lower willows take each half of strip. Distance between two trees is 2 m × 2 m with higher willows and 1 m × 1 m with lower willows, the allocation may be lain by cinquefoil.

(2) It should be set the border stake at outside border of flood controlling trees. Inner side of border should be set lateral weir, breadth and height is all 0.3 m.

(3) The grown trees which have been as timber or its height surpasses top of dike should be cut intermittently. And it should be renewed after cutting. Flood trees strip should not be formed as gap and surviving rate of trees is not less than 95%.

3.8 Planting layout of adapting trees

(1) To guarantee the needs of trees materials in flood control it is mainly planted with poplar trees and willow trees in alluvial area. It may also be planted other kinds of trees combined with practice of local site and distance between two trees is 3 m × 4 m with allocation of cinquefoil.

(2) Planting trees should be lain out vertically with dike axis line for beauty and clean. The different trees should be alternatively planted in order to reduce insect pest and surviving rate of trees is not less than 95%.

3.9 Layout of dike shoulder trees planting

Both side of dike top is planted trees which is main kinds of scenery tress and distance between two trees is determined by tree type and distance to dike shoulder line is not less than 0.3 m. Trees on top of dike should be selected both bushes and arbors and with uniform tree distance and diameter in same kinds of trees. Surviving rate is not lower than 98%.

3.10 Layout of safeguard trees of dike

The strip land which safeguard for dike shoulder be flatted beforehand and then planting trees with willows and poplars etc. the trees distance is 2 m × 2 m on average. The outside border should be set border stake. Surviving rate is not less than 90%.

3.11 Sward

To nourish water and land for reducing damage by rain and conserving water and soil, it should be planted sward at both side of dike shoulders, slope of front and back river, two sides of roads to dike, slope of roads to dike, top of dam and dangerous dike sections as well as earth dam slope. Sward should be selected kudzu which is anti – drought or grass of adaptable kinds. The distance between two grass frusta is 0. 2 m with allocation of cinquefoil. Coverage rate is not less than 98%.

Sward strip band breadth at both sides of dike shoulders and roads to dike and two sides of dam shoulders should be not less than 0. 5 m without exeption.

4 Conclusions

In 2002, YRCC put forward the fourth standard dike construction and implemented since 2004. At present, standard dike construction has been finished for 1,371 km in length. Study on optimization of dike section and safeguard measurement offers the technology support to the Yellow River standard dike construction.

Constructing integrated Yellow River dike engineering concentrated with flood controlling line, traffic line for urgency and scenery line plays great roll in economic and social development and keep integrated engineering to ensure safety in flood controlling. Meanwhile, it promotes the Yellow River engineering management into a new stage in standardization and modernization.

Flow Analysis in Tidal Channel Connected with "UTSURO" (Tidal Reservoir)

H. Isshiki[1], *K. Sawai*[2], *Y. Ogawa*[2], *K. Akai*[3], *S. Takada*[4], *S. Bao*[3] and *S. Nagata*[1]

1. Saga University, Saga, Japan
2. Setsunan University, Osaka, Japan
3. NGO "UTSURO" Research group, Wakayama, Japan
4. Ministry of Land, Infrastructure, Transport and Tourism, Japan

Abstract: Flow discharge in a tidal river could increase in the case it is connected with an upstream UTSURO or reservoir. When the distance between the reservoir and the river mouth is short, water levels in the sea, river and reservoir change almost simultaneously with time, and the discharge in the river is approximately proportional to the area of the reservoir. On the contrary, when the distance is long, the phase lag increases and the wave property appears strongly, since the river and the reservoir constitute an oscillation system. We expect that the flow in the river increases much, and the effect such as the digging of the river bed and the prevention of sedimentation takes place. As examples of such phenomena, we introduce two cases. The first one is the mouth of Naka river in Ibaragi prefecture, Japan which is connected to Hinuma pond through Hinuma river. The second one is Huangpu river near Shanghai in China which is connected to Dingshan lake. An application of tidal effect to the control of Yellow river is also discussed. We carried out some numerical calculations using a simple mathematical model for these cases. Some interesting results are obtained.

Key words: tide, river, UTSURO, resonance, riverbed digging

1 Introduction

The tide can be used to increase the flow velocity. Since the strong flow digs the riverbed and discharges the sediment, the river becomes deeper or the water depth is maintained. This research aims to use tidal energy for controlling the depth of the river.

An example is found in Ibaragi prefecture, Japan. Naka and Kuji rivers flow almost parallel each other. There is a big difference between the water depth at the river mouth of Naka and Kuji rivers. The mean depth at the river mouth of Naka river is about 5 m, and that of Kuji river is less than 2 m. Kuji river has experienced the clogging at the river mouth. On the contrary, the clogging has never taken place in Naka river. There is a pond or a small lake named Hinuma pond near from Naka river, and the pond is connected to the mouth of Naka river through Hinuma river. The connection of a river mouth to a pond in case of Naka river may explain the big difference of the water depth in the two rivers. The pond may amplify the effect of tide and induce strong flow in the river mouth of Naka river, since the river and pond constitute an oscillation system. In 1993, Waki et al. has reported the effect of Hinuma pond on the river mouth of Naka river. According to the report, the riverbed digging as deep as 18 m, at the confluence of Hinuma and Naka rivers, is reported and discussed.

As another example, we discuss a case of Huangpu river in China. Shanghai port is located 30 km upstream from the river mouth of Huangpu river. Huangpu river is connected to Dingshan lake 100 km upstream from the river mouth. Dingshan lake generates tidal flow of more than 2,000 t/s in Huangpu river. The tidal flow seems to contribute very effectively for maintaining the water depth in Huangpu river.

We also discuss an application of riverbed digging by tide to the control of Yellow river in China. If the riverbed of the river mouth is dug, it will increase the river slope upstream, and floods may lower the height of the riverbed upstream.

2 Theory

2.1 Model

We use a very simple model. A channel consisting of waterways (or rivers) and reservoirs (ponds or lakes or UTSUROs) is shown in Fig. 1. We assume the waterways and reservoirs are rectangular parallelepiped. Let b_m, d_m and l_m, $m = 0, 1, \cdots, M-1$ be the breadth, depth and length of the m-th component of the channel.

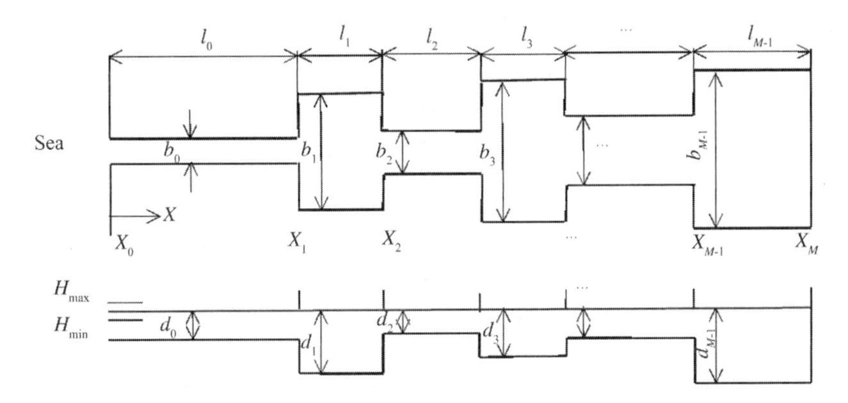

Fig. 1 A waterway and a reservoir

2.2 Equations of fluid motion

We take x -coordinates along the center plane of the waterways and reservoirs with the origin at the interface of the sea and channel, z -coordinates vertically upwards with $z = 0$ on the calm water surface and y -coordinates perpendicular to xz -plane. The x -coordinate of the m -th component of the channel satisfies , $x_m < x < x_{m+1}$, $m = 0, 1, \cdots, M-1$. One-dimensional motions are assumed in the channel.

We assume the hydrostatic pressure for pressure p:

$$p = \rho g(\xi - z) \quad , \tag{1}$$

where ρ, g and ξ are the density of water, gravitational acceleration and surface displacement. The water flow in the waterway and reservoir is assumed linear and one dimensional for simplicity. Let t, u and μ be time, flow velocity and riverbed friction coefficient, respectively. The equations of motion and continuity are given by

$$\frac{\partial u}{\partial t} = -g\frac{\partial \xi}{\partial x} - \frac{\mu}{\rho d}u \tag{2a}$$

$$\frac{\partial \xi}{\partial t} = -d\frac{\partial u}{\partial x} \tag{2b}$$

respectively, where μ, u is the friction force acting on water from the unit area of the riverbed and the water depth d is d_m in $x_m < x < x_{m+1}$, $m = 0, 1, \cdots, M-1$. If we eliminate u from Eq. (2), we obtain

$$\frac{\partial^2 \xi}{\partial t^2} = gd\frac{\partial^2 \xi}{\partial x^2} - \frac{\mu}{\rho d}\frac{\partial \xi}{\partial t}. \tag{3}$$

This equation may explain the effect of the riverbed friction qualitatively. The flow velocity u is obtained from Eq. (2a).

The boundary condition at $x = 0$ may be given approximately by

$$\xi = A_0 \cos\omega t \text{ at } x = 0, \tag{4a}$$

where A_0 and ω are the amplitude and circular frequency of the tide at sea. The boundary condition at $x = x_M$ is given by

$$\partial\xi/\partial x = 0 \text{ at } x = x_M. \tag{4b}$$

The intersection conditions at $x = x_m$, $m = 1, 2, \cdots, M-1$ are given by

$$[\xi]_{x=x_m-0} = [\xi]_{x=x_m+0} \tag{5a}$$

$$b_{m-1}d_{m-1}[u]_{x=x_m-0} = b_m d_m [u]_{x=x_m+0}. \tag{5b}$$

Eq. (5a) and Eq. (5b) refer to the continuities of pressure and flux at the intersection, respectively.

Let ω be the circular frequency of the tidal wave. The steady state may be written as

$$\xi(x,t) = \mathrm{Re}[Z(x)e^{i\omega t}] \tag{6a}$$

$$u(x,t) = \mathrm{Re}[U(x)e^{i\omega t}] \tag{6b}$$

The wave number k, wave period T and wave length λ are given by

$$k = \frac{2\pi}{\lambda} = \frac{2\pi}{T\sqrt{gd}} = \frac{\omega}{\sqrt{gd}} \tag{7}$$

$Z(x)$ is determined by solving the following boundary value problem:

$$\frac{d^2 Z}{dx^2} + \left(\frac{\omega^2}{gd} - i\frac{\mu\omega}{\rho g d^2}\right)Z = 0 \text{ in } x_m < x < x_{m+1}, m = 0, 1, \cdots, M-1, \tag{8a}$$

$$Z = A_0 \text{ at } x = 0 \tag{8b}$$

$$dZ/dx = 0 \text{ at } x = x_M, \tag{8c}$$

$$[Z]_{x=x_m-0} = [Z]_{x=x_m+0} \tag{8d}$$

$$\text{and } b_{m-1}d_{m-1}[U]_{x=x_m-0} = b_m d_m [U]_{x=x_m+0} \text{ at } x = x_m,\ m = 1, \cdots, M-1. \tag{8e}$$

From Eq. (2a), U is given by

$$U = \frac{ig}{\omega - i\mu/(\rho d)} \frac{dZ}{dx},\ x_m < x < x_{m+1},\ m = 0, 1, \cdots, M-1. \tag{9}$$

2.3 Solutions of fluid motion

We assume the solution of Eq. (8a) as

$$Z = e^{(\sigma + ik)x} \text{ in } x < x < x_{m+1},\ m = 0, 1, \cdots, M-1, \tag{10}$$

where σ and κ are real constants. The characteristic equation is given by

$$\sigma^2 + 2i\sigma\kappa - \kappa^2 + \frac{\omega^2}{gd} - i\frac{\mu\omega}{\rho g d^2} = 0. \tag{11}$$

From Eq. (11), we have

$$\sigma^2 - \kappa^2 = -\frac{\omega^2}{gd},\ 2\sigma\kappa = \frac{\mu\omega}{\rho g d^2}. \tag{12}$$

Solving Eq. (12), we obtain

$$\sigma = \pm\sqrt{-\frac{\omega^2}{2gd} + \frac{\omega}{2gd}\sqrt{\omega^2 + \frac{\mu^2}{\rho^2 d^2}}},\ \kappa = \pm\sqrt{\frac{\omega^2}{2gd} + \frac{\omega}{2gd}\sqrt{\omega^2 + \frac{\mu^2}{\rho^2 d^2}}}. \tag{13}$$

In the following, we define and as

$$\sigma = \sqrt{-\frac{\omega^2}{2gd} + \frac{\omega}{2gd}\sqrt{\omega^2 + \frac{\mu^2}{\rho^2 d^2}}},\ \kappa = \sqrt{\frac{\omega^2}{2gd} + \frac{\omega}{2gd}\sqrt{\omega^2 + \frac{\mu^2}{\rho^2 d^2}}}. \tag{14}$$

σ and κ satisfies

$$\sigma \to 0,\ \kappa \to K,\ \text{as } \mu \to 0. \tag{15}$$

Let(σ_m, κ_m) be (σ, κ) for $d = d_m$. The solution satisfying Eq. (8a) is given by

$$Z = \alpha_m e^{-(\sigma_m + i\kappa_m)(x - x_m)} + \beta_m e^{(\sigma_m + i\kappa_m)(x - x_m)},\ x_m < x < x_{m+1},\ m = 0, 1, 2\cdots,\ M-1. \tag{16}$$

If we substitute Eq. (16) into Eq. (9), we obtain

$$U=\frac{ig(\sigma_m+i\kappa_m)}{\omega-i\mu/(\rho d_m)}[-\alpha_m e^{-(\sigma_m+i\kappa_m)(x-x_m)}+\beta_m e^{(\sigma_m+i\kappa_m)(x-x_m)}] \tag{17}$$

Substituting Eq. (16) and Eq. (17) into Eq. (8b) and Eq. (8d) for Z – condition and Eq. (8c) and Eq. (8e) for U – condition, we obtain equations for the unknowns α_m and β_m, respectively:

$$\alpha_0+\beta_0=A_0, \tag{18a}$$

$$\alpha_{m-1}e^{-(\sigma_{m-1}+i\kappa_{m-1})(x_m-x_{m-1})}+\beta_{m-1}e^{(\sigma_{m-1}+i\kappa_{m-1})(x_m-x_{m-1}}=\alpha_m+\beta_m,\ m=1,2,\cdots,M-1, \tag{18b}$$

$$b_{m-1}d_{m-1}\frac{ig(\sigma_{m-1}+i\kappa_{m-1})}{\omega-i\mu/(\rho d_m)}[-\alpha_{m-1}e^{-(\sigma_{m-1}+i\kappa_{m-1})(x_m-x_{m-1})}+\beta_{m-1}e^{(\sigma_{m-1}+i\kappa_{m-1})(x_m-x_{m-1}}]$$
$$=b_m d_m\frac{ig(\sigma_m+i\kappa_m)}{\omega-i\mu/(\rho d_m)}[-\alpha_m+\beta_m],m=1,2,\cdots,M-1, \tag{18c}$$

$$\frac{ig(\sigma_{M-1}+i\kappa_{M-1})}{\omega-i\mu/(\rho d_{M-1})}[-\alpha_{M-1}e^{-(\sigma_{M-1}+i\kappa_{M-1})(x_M-x_{M-1})}+\beta_{M-1}e^{(\sigma_{M-1}+i\kappa_{M-1})(x_M-x_{M-1}}]=0 \tag{18d}$$

Since α_m and β_m are given by a linear combination of α_{m-1} and β_{m-1} from Eqs. (18b) and (18c), the above equations may be solved conveniently by using the transfer matrix method.

3 Numerical calculations

3.1 Combination of Hinuma river and Hinuma pond in Japan

A very interesting example of tidal effect on river flow at river mouth is found in Naka river, Ibaragi prefecture, Japan. Naka and Kuji rivers flow almost parallel each other. The water depth at the river mouth of Naka river is about 5 m in average. The deepest water depth is as big as 18 m. In the contrast, the mean water depth of Kuji river is less than 2 m. Kuji river has experienced the clogging. However, the clogging has never taken place in Naka river. There is a pond called Hinuma pond with area 9.35 km^2 near from Naka river, and the pond is connected to the mouth of Naka river through Hinuma river. Hinuma river and Hinuma pond constitutes an oscillation system. The resonant effect of the oscillation system in case of Naka river may explain the big difference of the water depth in Naka and Kuji rivers. The pond may amplify the effect of tide and induce strong flow at the confluence of Hinuma and Naka rivers. The confluence is located 1 km from the seashore.

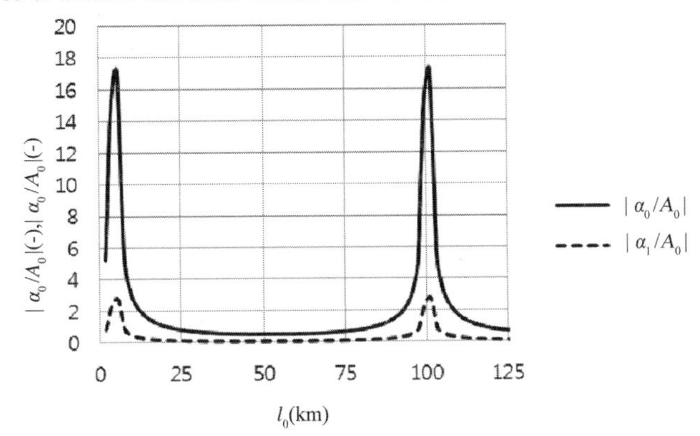

Fig. 2 Effect of waterway length on the resonance of Hinuma river and pond system in case of no riverbed friction

The dimensions of the waterway and reservoir are $M=2$, $b_0=50$ m, $d_0=2$ m, $l_0=10$ km, $b_1=2$ km, $d_1=2$ m and $l_1=4.7$ km. When the depth of water d_0 is 2 m and the riverbed friction is neglected, the wave length λ_0 is 191 km, since the wave speed is $\sqrt{gd_0}=4.43$ m/s and the wave period is 12 hr. The resonance occurs when the length of the waterway is 47.8 km ($=\lambda_0/4$), 143.4 km, ⋯ or 5.7 km, 101 km, ⋯ without or with the reservoir, respectively. In Fig. 2, we see more the effects of the waterway length on $|\alpha_0/A_0|$ and $|\alpha_1/A_0|$ in combination of Hinuma river and Hinuma pond. In this case, the resonance occurs at $l_0=5.7$ km. Since the actual length l_0 of Hinuma river is 10 km, the fluid motion may be influenced significantly by resonance. In Fig. 3, the amplitudes of surface displacement Z and flow velocity U are shown in case of $l_0=10$ km. The flow velocity is too big and unrealistic.

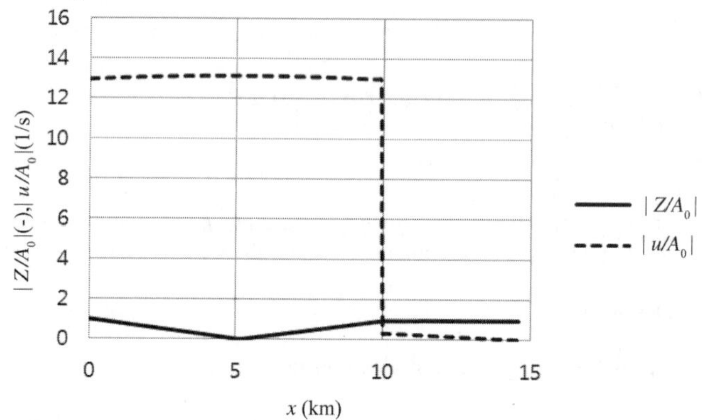

Fig. 3 Amplitudes of Surface displacement Z and flow velocity V in case of no riverbed friction

Since the water depth is very small in this case, the damping due to the riverbed friction may play a very important role. The effects of the riverbed friction on resonance are shown in Fig. 4 for $\mu=0$ Ns/m^3, 0.25 Ns/m^3. The resonance is much suppressed by introducing the riverbed friction. The effects on the amplitudes of the surface displacement $|Z/A_0|$ and flow velocity $|U/A_0|$ in case of $l_0=10$ km (actual length of Hinuma river) are shown in Fig. 5. If we assume $\mu=0.5$ Ns/m^3, the prediction of the surface displacement and flow velocity seems much improved. When the tidal difference at the river mouth of Naka river is 1.5 m, the present theory gives the tidal difference at Hinuma pond about 0.45 m. Since the observed tidal difference of Hinuma pond is about 0.5 m, the present theoretical prediction may be said realistic. So, in the following calculation, we assume 0.5 Ns/m^3 for the riverbed friction μ.

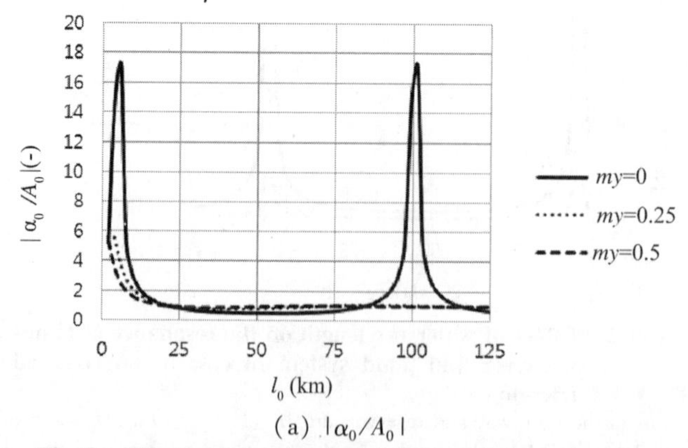

(a) $|\alpha_0/A_0|$

Fig. 4 Effects of the riverbed friction on the resonance

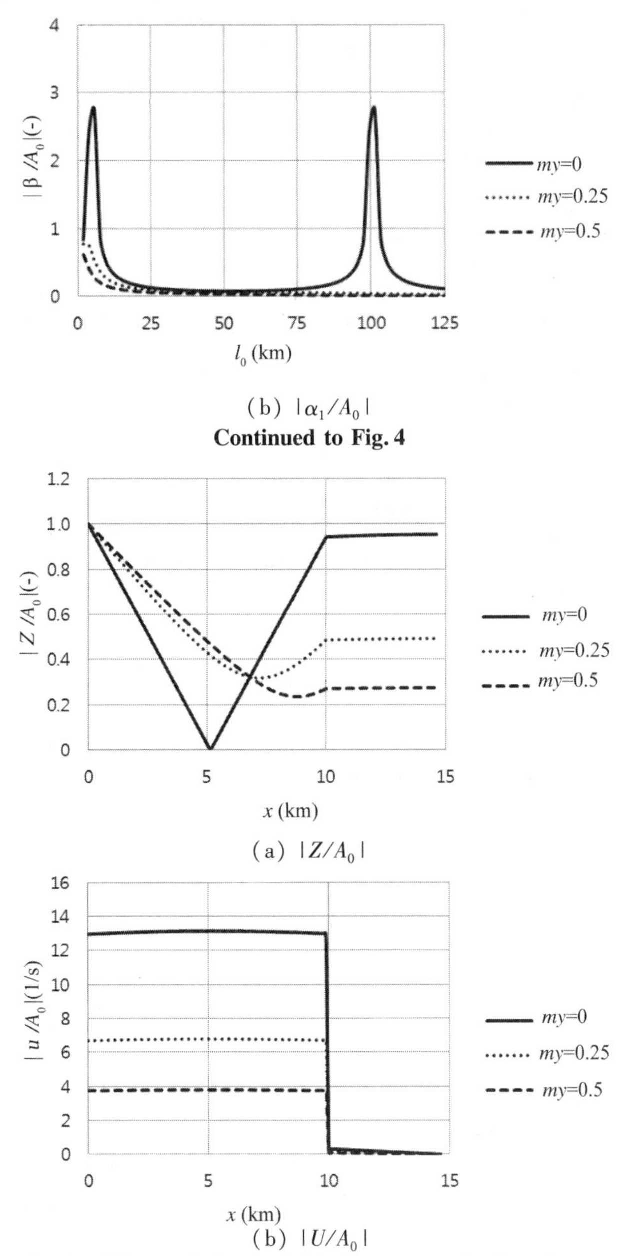

(b) $|\alpha_1/A_0|$

Continued to Fig. 4

(a) $|Z/A_0|$

(b) $|U/A_0|$

Fig. 5 Effects of riverbed friction on amplitudes of surface displacement and flow velocity

3.2 Combination of Huangpu River and Dingshan Lake in China

Shanghai port is located 30 km upstream from the river mouth of Huangpu River. The Huangpu River is connected to Dingshan Lake 100 km upstream from the river mouth. Big ships navigate through Huangpu River to Shanghai port, and many large barges pass through Huangpu River and Dingshan Lake to Tai Lake. These barges utilize the tidal flow for the navigation to save energy. The area of Dingshan Lake is 62 km^2 and generates tidal flow of more than 2,000 t/s in Huangpu River. The tidal flow seems to contribute very effectively for maintaining the water depth in Huangpu River.

In this example, we assume the length of Huangpu River is 100 km, and the breadth and depth of Huangpu River are given by

$$b(x) = 500 - (200/100{,}000)x,\ d(x) = 10 - (4/100{,}000)x,\ 0 \leqslant x \leqslant 100{,}000 \tag{19}$$

where the units of b, d and x are meters. The breadth, depth and length of Dingshan Lake are 10 km, 2 m and 6 km respectively. The real area of Dingshan Lake is 62 km. Since the water depth of Huangpu River is much bigger than that of Hinuma River, the effect of the riverbed friction may be much smaller. We studied the effects of the size and shape of the reservoir on the resonance numerically. The resonant effects seem small in this case. The area of the reservoir determines the resonant point. The effect of the shape is not so important.

The amplitudes of the surface displacement ξ and flow velocity U are shown in Fig. 6. When the area of the reservoir is 60 km^2, the flow becomes maximum at the intersection of Huangpu river and Dingshan Lake. The theory gives the tidal difference at Dingshan Lake of about 1.05 m for the tidal difference at the river mouth of Huangpu River of 3 m. The theoretical prediction is similar to the observed tidal difference of 0.7 m. The mean of the flow velocity amplitude of Huangpu River is 1.33 m/s under the same condition. Dingshan Lake is connected to Tai Lake by a river. This did not decrease the difference of the tidal difference at Dingshan Lake between the predicted and observed values. The predicted tidal difference at Tai Lake was almost zero.

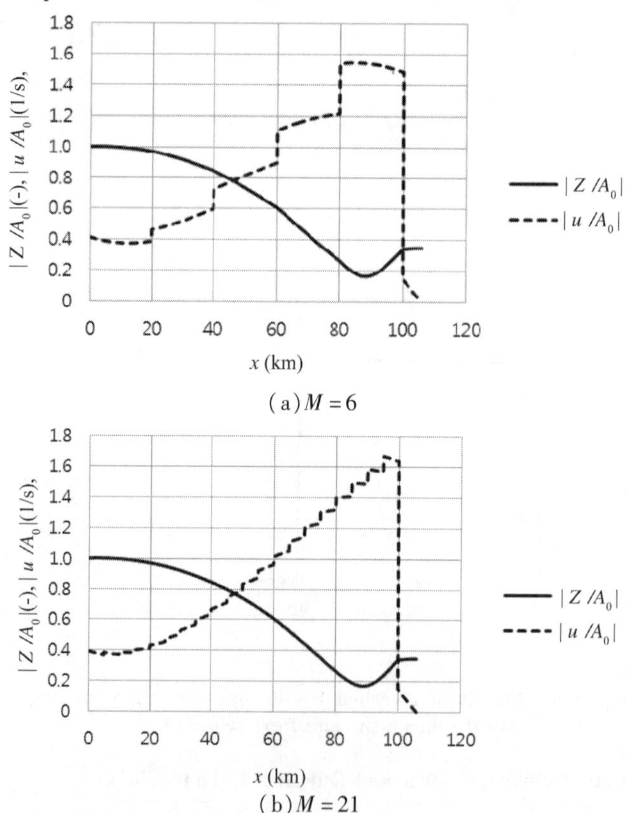

(a) $M = 6$

(b) $M = 21$

Fig. 6 The amplitudes of surface displacement Z and flow velocity U ($\mu = 0.5$ Ns/m^3)

3.3 Application to Control of Yellow River in China

In 3.1, we discussed an interesting example of tidal effect on riverbed digging at the mouth of Naka river in Japan. We may expect a similar result, if we combine Yellow River with an UTSURO or reservoir. There are old river trails of Yellow River near from Bohai bay. We can use one of them as a waterway to a reservoir build in the sea.

The dimensions of the waterway and reservoir are $M=2$, $b_0=2$ km, $d_0=15$ m, $l_0=130$ km, $b_1=10$ km, $d_1=15$m and $t_1=12$ km. When d_0 is 15 m and the riverbed friction is neglected, the wave length λ_0 of the waterway is 524 km. The resonance occurs when the length of the waterway is 131 km ($=\lambda_0/4$), 393 km, ⋯or 79 km, 340km,⋯ without or with the reservoir, respectively. We assume the riverbed friction μ as 0.5 Ns/m^3 in the following calculations. In Fig. 7, the effect of the waterway length on $|\alpha_0/A_0|$ and $|\alpha_1/A_0|$ or the resonance is shown, when b_1, d_1 and l_1 are 10 km, 15 m and 12 km respectively. We also conducted calculations on a series of reservoir size. The resonance may not be neglected in this case. The reservoir area determines the resonance as shown in Fig. 8 and Fig. 9.

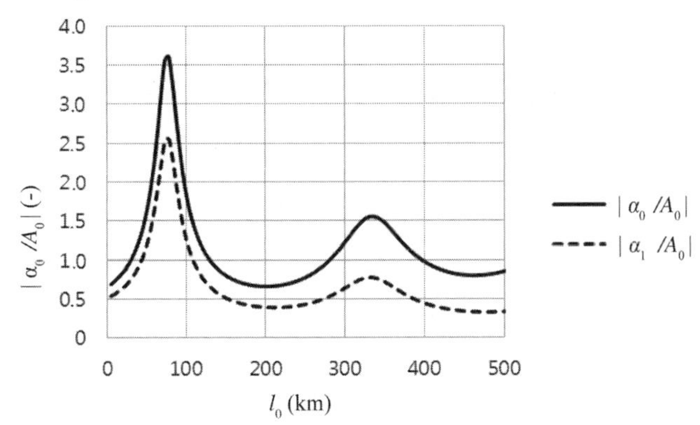

Fig. 7 Effects of waterway length on resonance

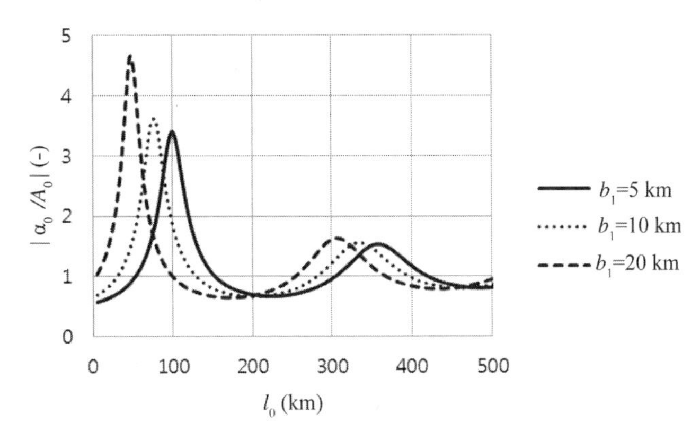

Fig. 8 Effects of reservoir breadth b_1 on the resonance

The area of the reservoir determines the amplitudes of the surface displacement and flow velocity as can be seen in Fig. 10 and Fig. 11. If the length l_0 of the waterway is 130 km, the flow becomes maximum at a point from 50 km upstream from the sea, when the area of the reservoir is

$b_1 \times l_1 = 120\ \text{km}^2$. If the area is smaller, the point becomes closer to the sea. If it is bigger, the point becomes farther to the sea. However, from Fig. 11, we know that the smaller reservoirs give faster velocity in this case, since the waterway length is closer to the resonant one. So, we must design the length of the waterway and the size of the reservoir carefully.

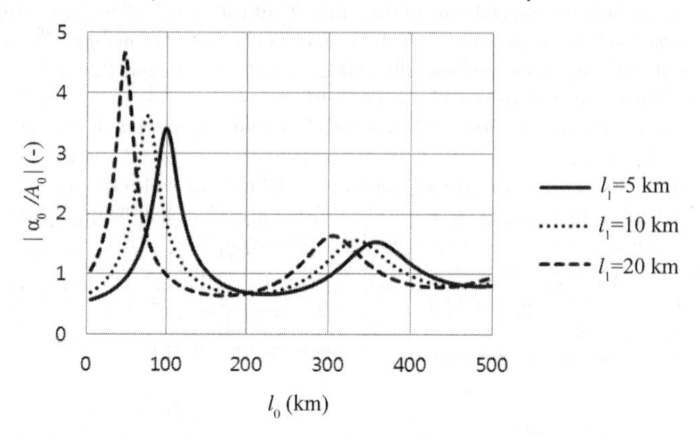

Fig. 9 Effects of reservoir length l_1 on the resonance

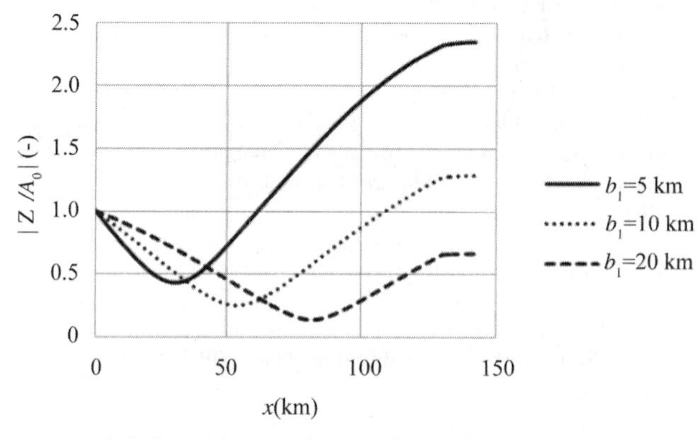

(a) $b_1 = 5$ km, 10 km, 20 km and $l_1 = 12$ km

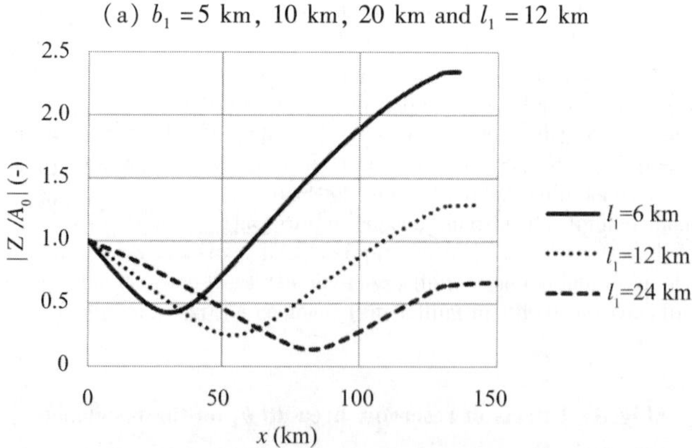

(b) $b_1 = 10$ km and $l_1 = 6$ m, 12 km, 24 km

Fig. 10 Effects of reservoir size and shape on the amplitudes of surface displacement

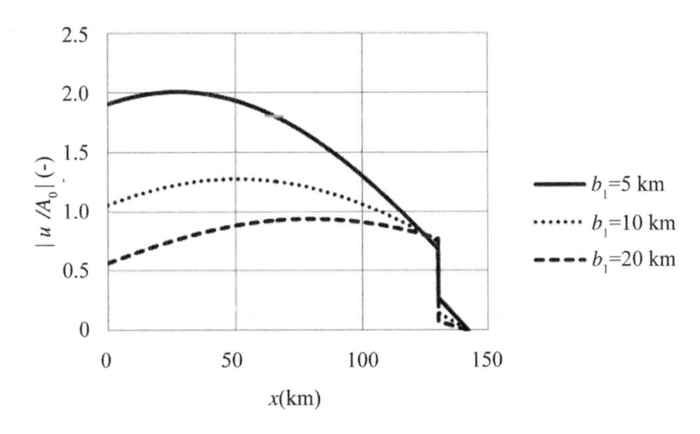

(a) b_1 = 5 km, 10 km, 20 km and l_1 = 12 km

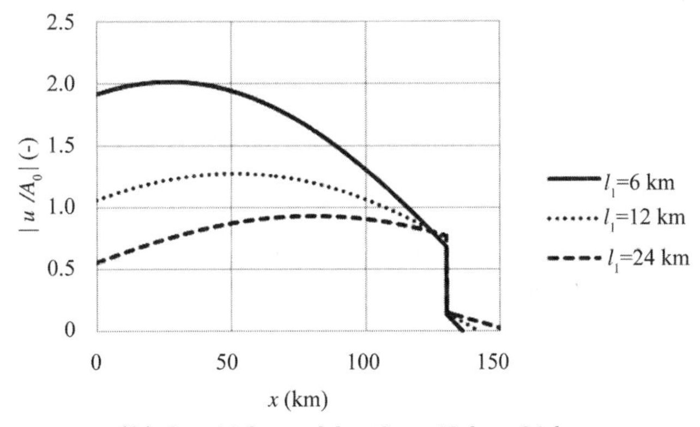

(b) b_1 = 10 km and l_1 = 6 m, 12 km, 24 km

Fig. 11 Effects of reservoir size and shape on the amplitudes of flow velocity

4 Conclusions

Flow discharge in a tidal river increases when it is connected with an upstream UTSURO or reservoir. In some cases, the flow velocity in the river is significantly increased because of the dynamical effects due to combination of the river and a reservoir. The area of the reservoir has a significant influence on the resonance of the river – reservoir system.

In the present paper, we discussed two examples. One is the riverbed digging of Naka river in Japan connected to Hinuma pond through Hinuma river, and the other is maintenance of water depth of Huangpu river in China connected to Dingshan lake. In the former case, the estimated average flow velocity of about 3 m/s is generated by tide because of resonance effect, when the tidal difference is 1.5 m at the river mouth of Naka river. In the latter case, the estimated mean flow velocity of about 1.33 m/s is induced by tide, when the tidal difference is 3 m at the river mouth of Huangpu river. It seems to contribute not only to the maintenance of water depth but also to the energy saving for navigation of large barges at Huangpu river. Application to the control of Yellow river is also discussed. The tidal energy utilization may find an interesting application to the control of not only the river mouth but also the whole river.

If the riverbed friction is neglected, the theory can't explain the above mentioned phenomena realistically. So, the riverbed friction is included in the present theory within the framework of a linear theory. The introduction of the friction improves significantly the appropriateness of the theoretical prediction especially in case of Naka river.

Acknowledgement

The authors express their sincere thanks to Prof. Emeritus K. Ashida of Kyoto Univ. for his continuous encouragements.

References

K. Akai, The UTSURO, ISBN 978 -4 -434 -12459 -4 (2008).

M. Waki, M. Fujisiro, T. Kawamura et al.. A study on tidal phenomenon of the Nakagawa river, 1993 PACON China Symposium(June, 1993).

H. Lamb. Hydrodynamics [M]. New York Cambridge University Press 1932.

Riverbank Collapse Causes and Characteristics of the Upper Yellow River with Wide Valley in Desert Area

Shu Anping, ***Li Fanghua*** and ***Duan Guosheng***

School of Environment, Beijing Normal University, Key Laboratory of Water and Sediment Sciences of Ministry of Education, Beijing, 100083, China

Abstract: Riverbed rise and suspended river occurred in the wide valley desert reach of the upper Yellow River. Riverbank collapse is one of the main sediment sources, which causes much attention. To understand riverbank collapse mechanism and evaluate sediment quantity brought by riverbank collapse, riverbank collapse causes and characteristics were analyzed in this study. Based on the particle size gradation tests, the study reaches were divided into two kinds. Type I was composed of sand from the surrounding desert while type II was made up of silt from sediment. Riverbank collapse causes were displayed from the view of both soil mechanism and hydraulics with respect to the two kinds of riverbanks. The factors of soil mechanism included physical properties (e. g. cohesion and friction) and morphology characteristics (e. g. height and slope). Water table, river stage and velocity were the main hydraulic factors. Wuhai and Dengkou were selected as the representative observation sites corresponding to the two kinds of riverbanks. By field observation and indoor experiments, their physical features and morphology characteristics related to riverbank collapse causes were demonstrated.

Key words: riverbank collapse, Cause, Characteristic, the Upper Yellow River, the wide valley reach

1 Introduction

Riverbank collapse is the result of the interaction between flow and riverbank. Flow scours riverbanks and takes bank materials away directly. On the other hand, riverbanks become steeper and immerse in the flow, both of which can decrease riverbank stability. Ability of riverbank to resist erosion and scour is subjected to physical properties such as soil structure and composition. Riverbank collapse is common all over the world. It always occurs in the lower Mississippi River in the United States and the Rhine in Europe. Riverbank collapse is widespread in the seven rivers in China.

The main factors affect slope stability are as follows: ①Slope morphology (e. g. slope height and angle, cross section. ② Physical properties (e. g. cohesion and friction). ③Soil structure. ④Water table and moisture content. ⑤Shake caused by earthquake et al. In addition to the above causes, hydraulic factors are critical to riverbank stability. Researches on riverbank collapse involved river stage, seepage, current, secondary flow, et al.

There is little study on riverbank collapse in the upper Yellow River. The wide valley reach is in the lower part of the upper Yellow River, where the interaction between flow and sediment is the most complicated and serious. In this area, ecology is fragile and sensitive to climate. Soil erosion intensity is mainly caused both by wind and water. Desertification and river channel evolution coexist here, which draw much attention of scholars who are interested in the interaction of desert and river. This area plays an important role in both economy and society with providing energy resources and foods for people in the northwest China. And it is also a main residence of the minority including Mongol nationality and Hui nationality. The suspended river in the upper river, is a threaten to the lower part, which alerts both political circles and scientific community. It is urgent to focus on how to keep the Yellow River develop securely, especially the wide valley reach. By field observation and indoor experiments, some data about the study reach have gained. Based on this, the causes and properties of riverbank collapse are analyzed in this paper, which are essential for the study on riverbank collapse mechanism and evaluation of the sediment percent caused by riverbank collapse.

2 Field observation and experiment

2.1 Study area

The wide valley reach is located at the lower part of the upper Yellow River, starting from Xiaheyan and ending at Toudaoguai (Ningxia – Inner Mongolia reach), with a total channel length of 990 km and a gradient of 0.000,25. It is characterized by low gradient, wandering channel, loose riverbed materials, high sediment load. A large area of alluvial plains (Yinchuan Plain and Hetao Plain) and deserts (Tenger Desert, Hedong Sandyland, Ulan Buh Desert and Hobq Desert) surround the reach. The study area is a typical region for the interaction of river and deserts. Both wind erosion and water erosion are more serious than those of the other regions around the world. This reach crosses several fault basins including Zhongwei basin, Zhongning basin, Yinchuan basin and Hetao basin. Zhongwei basin and Zhongning basin are between Xiaheyan and Qingtongxia. Yinchuan basin is from Qingtongxia to Dengou. Hetao basin starts from Dengkou and ends at Hekou, as shown in Fig. 1.

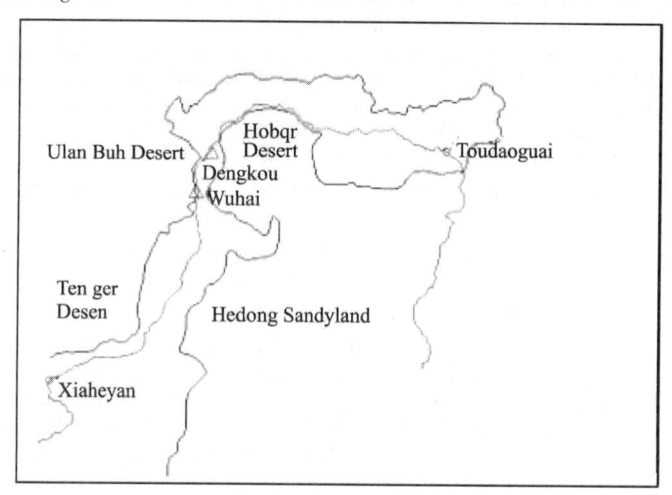

Fig. 1 Study area

This area has a warm temperate climate and semi – arid grasslands zone. The mean annual rainfall ranges from 150 mm to 363 mm, with a potential evapotranspiration of 1,000 mm to 2,000 mm. About 2/3 of the flow in the Ningxia – Inner Mongolia reach is from low latitude to higher latitude which is a critical reason for the disaster of ice floods.

2.2 Observation reaches

2.2.1 Riverbank collapse

The Ningxia – Inner Mongolia reaches are sorted into two kinds according to riverbank materials. Type I is composed of sand from the surrounding desert and type Ⅱ is made up of silt from sediment. Wuhai and Dengkou observation sites were selected to represent type I and type Ⅱ respectively. Both stripe collapse and arc – collapse are common in the two. Stripe collapse, a kind of riverbank collapse pouring into a river as a stripe under flow scour, is found in straight channel or slightly bent channel, as Shown in Fig. 2.

(a) (b)

Fig. 2 Stripe collapse

Arc - collapse caused by secondary flow in a bent channel is a kind of riverbank collapse as pyriform, as Shown in Fig. 3.

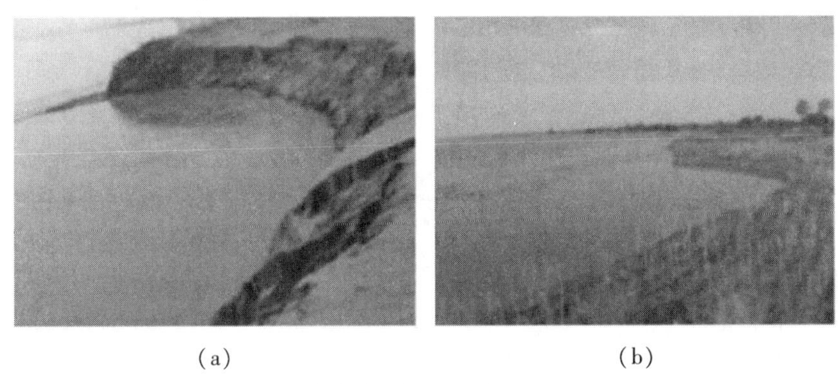

(a) (b)

Fig. 3 Arc - collapse

2.2.2 Particle size distribution

Riverbank materials of Wuhai and Dengkou are from Ulan Buh Desert and river sediment respectively. There were 260 samples tested for particle size distributions using laser particle sizer Microtrac S3500. Fig. 5 and Fig. 6 display the particle size gradation curves of the two kinds of riverbank. The median particle diameters are around 0. 18 mm in Wuhai and around 0. 03 mm in Dengkou as shown in Fig. 4. Most of the particles sizes are greater than 0. 062 mm in Wuhai and more than 80 percent of particles are smaller than 0. 062 mm in Dengkou.

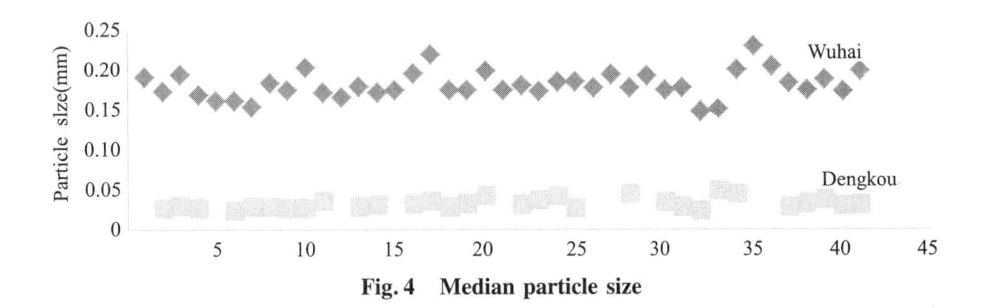

Fig. 4 Median particle size

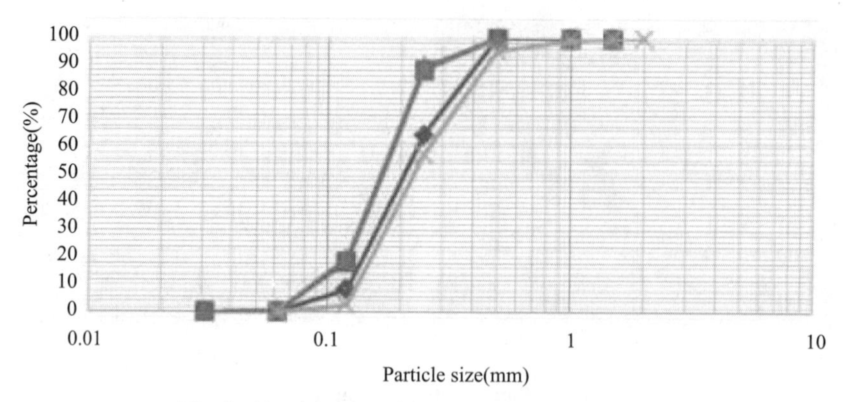

Fig. 5 Particle size distribution curve of Wuhai

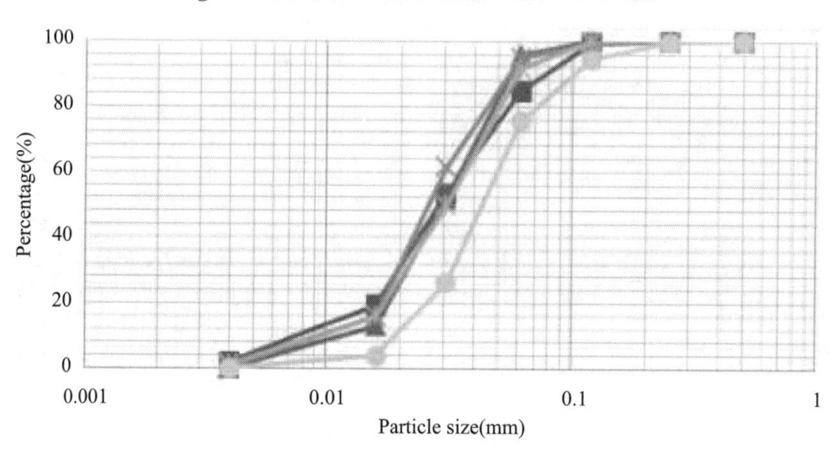

Fig. 6 Particle size distribution curve of Dengkou

3 Riverbank collapse causes and characteristics

3.1 Soil mechanism

3.1.1 Soil physical properties

Soil physical properties include soil composition, density, void ratio, moisture content, cohesion(c), friction (φ), et al. Cohesion and friction, the decisive factors of soil shear strength, are crucial to slope stability. The larger values of cohesion and friction result in higher slope stability.

The mean wet densities of Wuhai and Dengkou are 1.95 g/cm^3, 1.94 g/cm^3 and the dry values are 1.59 g/cm^3, 1.55 g/cm^3. Cohesion and friction are shown in Tab. 1. These data illustrate that the grain size is larger and cohesion is smaller in Wuhai than those in Dengkou. According to the grain size composition, riverbanks can be classified as non - cohesive riverbank (d_{50} > 0.1 mm) and cohesive riverbank (d_{50} < 0.10 mm). The different types of riverbank correspond to different mechanical characteristics and incipient motion. For cohesive riverbank, there is cohesive force between particles addition to thrust, uplift force and effective gravity in non - cohesive riverbanks. The grains start moving by individual in non - cohesive riverbanks and by bulk in cohesive riverbanks.

Tab. 1 C, φ

		Wuhai			Dengkou	
φ(°)	27.01	29.17	24.16	14.67	9.51	14.51
C	8.24	6.27	9.93	29.75	28.60	33.86

3.1.2 Riverbank morphologies

Riverbank morphologies include vertical morphology (e.g. slope and height) and longitude morphology (e.g. curvature). A steeper and higher slope leads to a higher risk of riverbank collapse caused by gravity and hydraulic power. The curvature of a channel has an effect on riverbank collapse by acting on flow structure.

Fig. 7 shows the channel morphology of Wuhai. Ulan Buh Desert supplies plenty of riverbank material. The riverbank heights and slopes vary with wind and sand tunes.

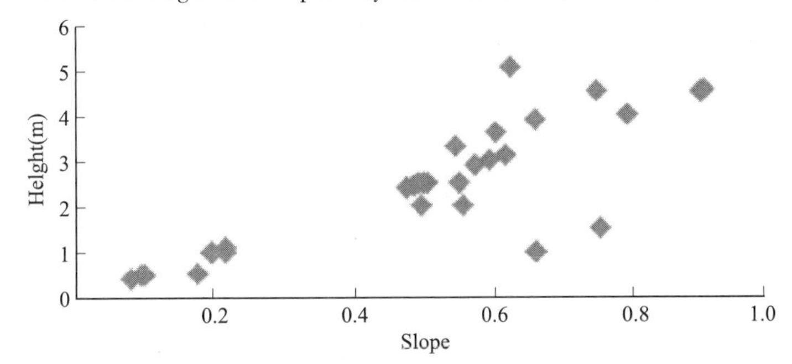

Fig. 7 Riverbank morphology of Wuhai

The riverbank is vertical to riverbed in most cross sections as shown in Fig. 8 in Dengkou. The riverbank heights vary from 0.8 m to 2.6 m. The land use type in this area is farmland where human activities are frequent.

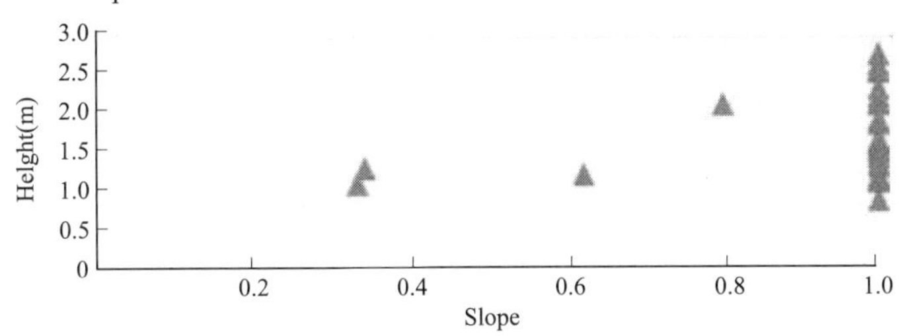

Fig. 8 Riverbank morphology of Dengkou

3.2 Hydraulic characteristics

3.2.1 Water table

Changing discharge in flood season and dry period affects water table directly. Water level acts on riverbank stability in the following ways. Firstly, water table can change slope gravity, shear strength, confine pressure and pore water pressure. Secondly, some materials dissolved into water make the soil composition and structure changed. Thirdly, water tables maybe lead to surface slide.

3.2.2 Flow shear force

Discharge affects velocity and shear stress. Flow regime affects distribution of flow shear stress and sediment carrying capacity by acting on turbulent fluctuation. Secondary flow transports sediments in the lateral inward direction building up a rhythmic sequence of forced (point) bars and pools respectively at the inner and outer bends of a train of meanders. The establishment of a bar - pool pattern then gives rise to a further "topographically induced" component of the secondary flow, which drives an additional contribution to sediment transport and further modifies the bed topography. Bank - collapse occurs seriously in the downstream of curve vertex because of the second flow.

4 Conclusions and discussions

(1) The Ningxia - Inner Mongolia reaches are divided into two kinds according to riverbank materials. Type I is composed of sand and riverbank morphologies are various and type II is made up of silt and riverbank morphologies are uniform.

(2) Riverbank collapse is a result of gravity and hydraulic forces. In this study, root reinforcement is neglected for scarce plants in type I and shallow root crop in type II. Soil physical properties include soil composition, density, void ratio, moisture content, cohesion (c), friction (φ), et al. Hydraulic properties include water table, discharge, velocity and flow regime.

(3) By field observation and experiments, quantitative achievements have been made with respect of riverbank morphologies and materials. Research on hydraulic force needs further study to quantitate.

References

Yu M H, Duan W Z, Dou S T. Riverbank Mechanism[C]// Decade Anniversary of Flood Fighting in1998 Proceeding, 2008.

Zhang X N, Jiang C F, Ying Q, et al. Review of Research on Bank Collapse in Natural Rivers [J]. Advances in Science and Technology of Water Resources, 2008, 28(3): 80 - 83.

Chen X Z. Edaphic Mechanics and Substructure[M]. Beijing: Tsinghua University Press, 2006.

Zhang X D, Bao L S, Li D Y, et al. Soil Mechanics[M]. Beijing: China Communication Press, 2011.

Wang Y G. Study on Mechanism of Bank Failure in the Alluvial River[M]. Beijing: China Institute of Water Resources and Hydropower Research Doctoral Dissertation, 2003.

Zheng Y R, Tang X S. Stability Analysis of Slopes under Drawdown Condition of Reservoirs[J]. Chinese Journal of Geotechnical Engineering, 2007, 29(8): 1115 - 1121.

Xie L F. Study on Slope Stability under Seepage Action[D]. Nanjing: Nanjing Hydropower Research Institute, 2009.

Qian N, Zhang R, Zhou Z D. River Evolvement Theory[M]. Beijing: Science Press, 1989.

Zhang R J. River Sediment Dynamics[M]. Beijing: China WaterPower Press, 1998.

Zhang L D. Study on Starting Velocity of Cohesive Sediment [J]. Journal of Hydrodynamics, 2000, 15(1): 82 - 87.

Wang D W, Yu M H, Liu X F. Scouring Mechanism and Simulation Methods of Alluvial Rivers [J]. Engineering Journal of Wuhan University, 2008, 41(4): 14 - 19.

Thorne C R. River Width Adjustment. I: Process and Mechanisms[J]. Journal of Hydraulic Engineering, ASCE, 1998, 124(9): 881 - 902.

Shao X J, Wang X K. Introduction to River Mechanics[M]. Beijing: Tsinghua University Press, 2005.

Wang Y G, Kuang F G. Study on Bank Erosion and its Influence on Bank Collapse[J]. Journal of China Institute of Water Resources and Hydropower Research, 2005, 3(4): 251 - 257.

Yao Z Y, Ta W Q, Jia X P, et al. Bank Erosion and Accretion along the Ningxia—Inner Mongolia Reaches of the Yellow River from 1958 to 2008[J]. Geomorphology, 2010, 127: 99 - 106.

Xia J Q, Wang G Q, Zhang H W, et al. A Review of the Research on Lateral Widenning Mechanisms of Alluvial Rivers and Their Simulation Approaches [J]. Journal of Sediment Research, 2001(6): 71 - 78.

Development and Application of a Decision Support System for River Basin Management in the Netherlands

Sandra Junier

Delft University of Technology

Abstract: River Basin Management requires balancing of the different functions of water in the basin. The effect of various changes in water management practices is often hard to determine and choosing between different options is therefore very difficult. Decision makers can be served by software systems that calculate the outcome of measures and compare the value of the options available.

This paper is concerned with the development process and the application of the instrument the Water Framework Directive Explorer (WFDE) in the Netherlands. The WFDE is a decision support system (DSS) that has been specifically developed to support the implementation of the Water Framework Directive by exploring the effects of different measures to improve water quality and ecology. Many parties were involved in the development process as the envisaged users were both national and regional water managers.

During the development it became apparent that tensions existed between the expectations and needs of the different stakeholders and also between what they wanted and the available expertise. A number of these tensions appear to correspond with tensions in River Basin Management in general. A main issue in the development process proved to be the rules describing the relation between changes in the physical environment and the targeted changes in ecology and water quality. As these relations were not established at the time of the development of the DSS, a parallel process of developing these rules was started. This has stimulated a great deal of discussion between members of different disciplines as well as within communities.

Key words: RBM, DSS, WFDE, the Netherlands

1 Introduction

River Basin Management (RBM) is an approach to water management that views upstream and downstream effects, water quality and water quantity as well as water and adjacent land – use in a basin in an integrated manner (Mitchell 1990; Mostert 1998; Moss 2004; Cash et al. 2006). New questions for RBM arose with the Water Framework Directive (2000/60/EC; WFD). They related to both institutional matters such as the interplay between the different authorities involved (Moss 2004; Junier et al. 2011) as technical, such as what is a good status of a water body, how to measure and monitor this status, and what measures lead to achieving this status.

Developing RBM policy is, like environmental policy development in general, "a complex process, which mixes legal requirements with issues of technical feasibility, scientific knowledge and socio-economic aspects, and which requires intensive multi-stakeholder consultation" (Quevauviller et al. 2005 p203). When the WFD was being developed, technical experts involved were well aware that the knowledge required to implement it, was not yet available (Lagacé et al. 2008). In the Netherlands a lack of expertise existed concerning ecological quality elements and for example the impact of measures (Raadgever et al. 2009).

Modelling instruments and decision support systems (DSS) can be important tools for either water management and/or planning processes in water management (Rekolainen et al. 2004; Mysiak et al. 2005; Gourbesville 2008) as a means of providing and assessing information on water management problems. They are also viewed as valuable for participation processes (Welp 2001; Bots et al. 2011). DSSs are defined as "interactive, computer-based systems, which help decision makers use data and models to solve unstructured problems" (Gorry and Morton 1971, quoted in Turban and Aronson 2001, p. 13). A DSS provides expertise in the form of models and/or databases that can be accessed through an interface. DSS can support the analysis of the present state or provide predictions for the future (Gourbesville 2008), or a combination of the two.

Developing a decision support system that supports WFD implementation is complicated. De Kok and Wind (2003) state that the use of DSS in water management remains limited in spite of their potential and analyse the complications of developing a DSS in general. Gourbesville (2008) specifically describes a number of problems related to developing a single DSS at a European, national or basin scale to support the planning process leading up to River Basin Management Plans as required by WFD. Summarised they are: incorporating all domains, problems and possible measures, for all different stages of the planning process, making it too large to develop, use and maintain; data-availability, the difference in data from different domains and incorporating local characteristics as each basin is different; achieving scientific robustness, validity and transparency; concern that system may become an institution that may hindering scientific debate and development, may reduce willingness scientists to contribute; a single system may dominate in such a way that other products will be hard to develop and market (Gourbesville 2008, p319 – 320).

Borowski and Hare (2005) studied a group of water managers and researchers from different European countries, comparing their requirements for DSSs. They concluded that water managers prefer DSSs that fill a specific gap; are simple, data-rich, spatial, hydrological instruments that can be used for specific tasks; and support decision making, but not to provide a decision.

Besides these aspects of content, the development process itself and the way trust is, or is not, established through this process are main drivers for success. When experts themselves or their expertise are not trusted their expertise will not be used. Wynne has showed that trust in expertise can be eroded when, for instance, uncertainty is not communicated or when standardisation leaves no room for individual variations (Wynne 1992).

To investigate if the development of a DDS for WFD in practice demonstrates what was found in the literature, this paper discusses the decision support system (DSS) called "WFDE" (WFDE) that was actually designed to support the WFD planning process at (sub) basin level. By enabling an exploration of the effects of possible measures and presenting them in a visually attractive way it was thought suitable to support discussion among policy makers and the politically responsible decision makers. The version of the instrument discussed here, is no longer available. A completely new release will be available as of Autumn 2012.

RBM is by nature interdisciplinary. The WFDE recognizes that, as it is an instrument that incorporates the knowledge of a number of different disciplines, such as hydrology, hydraulics, ecology, water quality, modelling and so on. These disciplines have different accepted practices and theories; bringing them together requires finding the points of connection and agreement.

The WFDE is in the terms of Power (2002, in Gourbesville 2008) a knowledge – and model-driven DSS. It was to disclose existing knowledge, in the form of rules that steer the model, and make it available in a simple to use format. Technically speaking there are three vital elements for the success of such a decision support instrument: the quality of the instrument itself, the knowledge rules that steer the model and the data available for the processing. However, as described above, many more factors may determine the outcome.

This research was performed in 2009 and 2010 and consisted of a literature study involving: policy documents on the implementation of the WFD in the Netherlands; scientific papers on the same; documents on the development of the WFDE, such as project plans, newsletters, minutes of user groups and steering group, project presentations etc. The study of de development of the new, completely overhauled, version of WFDE is on-going and will be reported on in later papers. Furthermore, a total of 14 people were interviewed: developers (5), users (4) and funders (4) of the WFDE, as well as civil servants involved in the implementation of the WFD (3). Some interviewees belong to more than one category. The (open) interviews took one to two hours. The interviewees were sent the minutes of the interview and were asked to comment on both the minutes and the draft case study report. This paper presents only part of the actual research. The full case study report can be downloaded from: http://www.actoranalysis.net/i-five/.

This paper starts with a description of the development process: who were involved, what was the objective and what was the result of the process. Secondly it describes the actual use. To explain the difference between the intended and actual use a detailed analysis of the tensions that arose in the development of the WFDE is provided in the next section. This will be followed by the concluding remarks.

2 Development of the instrument

May 2004 the proposal to develop the WFDE was filed at the funding agency "Leven met Water". At first the support for the development was low. Many at authorities concerned, both national and regional were sceptic, because they thought it would be too difficult. The DG Water, Rijkswaterstaat and the waterboards were the prospected users. Eventually, the initiators at RIZA, a research institute belonging to Rijkswaterstaat, and Delft Hydraulics, an independent research institute, found enough support. Early 2005, they began the development and testing of a first prototype. A regional division of Rijkswaterstaat and a waterboard participated in the consortium and hosted the pilot – study, so pilots were performed at a sub – basin and a local level. Alterra and the Delft University of Technology had a minor part in it, as well as two consultancies (Witteveen + Bos and Royal HasKoning). The start of the project was mainly funded by "Leven met Water", a national funding programme for the stimulation of knowledge development on water management issues. The participating water management organisations and Delft Hydraulics contributed in kind and STOWA, the research institute funded by the waterboards co-funded the project. At the end of 2006 the grant from "Leven met Water" ended. From 2007, DG Water, together with STOWA financed the continued development of the Explorer. Rijkswaterstaat and Waterschapshuis funded the maintenance.

The development of the WFDE can be described in a number of phases (in Tab. 1). In the first phase a first prototype was developed. The second phase saw the elaboration of the prototype to the basic version of the full instrument, including all water types and all measures. The third phase was concerned with implementing the Explorer. Implementation gradually shifted to a more evaluation oriented mode, phase four, that lead to phase five: the redesign and redevelopment of the WFDE. In the table below a brief description of the five phases is presented.

The official English title of the project was "WFDE, a planning kit to support policy development and communication on ecological objectives of (sub) river basins for the Water Framework Directive" (Consortium development WFDE 2006). At the start of the development, the objective was to develop the instrument in such a way that it could support the decision making process by visualising the effects different proposed measures would have on the water quality elements required by WFD. The instrument was intended to be fast enough to be used in interactive sessions. In this way it would also be able to support communication between different stakeholders. The parties involved would be able to use the graphs and maps available in the WFDE and quickly compare the effects of different measures.

It was also expected that use of a shared body of knowledge, materialised in the instrument, would contribute to a standardisation of the WFD policy planning process in the Netherlands. If all water management organisations would use the same methodology and the same terminology, this would enhance the transparency and the efficiency of the decision making process (Consortium development WFDE 2005). Additionally, an instrument such as this would help justifying to the European Commission how objectives were determined, and measures chosen, which the national water management organisations required.

The WFDE was designed to support policy makers, meaning those who develop the elements of the proposed river basin management plans. It would be the connection between the technical experts and the policy makers as well as the connection between policy makers and the politically responsible decision makers. It would specifically support WFD implementation, as can be seen by the use of the Ecological Quality Ratio (EQR) to describe the ecological state. The interface allowed users to see the status of water bodies on maps of the area. The status was shown by using the colour schemes belonging to the EQR method. Users could choose measures from a list, apply them to a water body, and see whether this measure improves the score of the water body. The instrument did not claim to predict the effect exactly, but it did roughly quantify the effect to indicate whether a measure would have effect and provide a rough estimate of the costs involved. In this way it was possible to select the most promising measures which could then be studied (not by means of the Explorer) in more detail to determine exactly how and where to implement them.

Tab. 1 Phases in development WFDE

	When	Phase	Activities
1	Early 2005 to August 2005	Exploration	First prototype, one water body type, eight measures. One pilot.
2	October 2005 to March 2007	Elaboration	Extension of prototype to include all types of water bodies and more than forty measures. Four pilots in the Netherlands, one in Belgium.
3	March 2007 to Late 2008	Implementation	Release basic version, release patches, improvements, courses, information meetings
4	Late 2008 to January 2010	Evaluation and improvement	Internal evaluation, evaluation visits to all waterboards. Evaluation report (Reeze and Vlieger 2009). Development of new ecological knowledge rules. Development of vision documents. Discussion of redesign.
5	January 2010-now	Redesign	Start new design process and new programming, aim to have prototype end of 2010 and pilots in 2011.

The instrument consists of a calculation core, a knowledge data base and an area specific data base. The first two are largely standardised. The knowledge rules relate factors, such as flow speed, sedimentation and nutrients to measures such as changes in hydro - morphology or point-source pollution to outcomes that are translated in terms of the WFD. Only minor changes in parameter values are possible. The third part consists of a structured database that is basically empty and needs to be filled by the user (or someone hired by the user) with area specific data to make a complete model that is valid for the area. This model is a vital part as well: it determines the boundaries and properties of the system; the quality of the model and the data included in it determine to a significant extent the quality of the outcome. The instrument could be used on different spatial scales depending on the model that the users developed. The software was free. To avoid confusion in this document the WFDE is called a (knowledge) instrument or DSS and the specific model (water system schematisation) on which it performs its calculations is "the application".

Before the WFD was formulated, several software solutions existed to help water managers analyse the current chemical and hydrodynamic state of surface waters, but none covered the whole range of aspects required for the implementation of the WFD. Especially for ecological quality elements no adequate instrument existed. The WFDE would be the first instrument to deal with all ecological quality elements and also the first to integrate hydrodynamics, chemistry and biology to determine the effects of measures on the chemical and ecological status. In the chemical balances, point source and diffuse pollution were represented. The instrument was planned to include costs aspects as well, but this functionality was developed in a limited manner.

The developers of the WFDE seem to have known the article by Borowski and Hare (2005) cited in the introduction. The WFDE was intended as a simple to use, spatial instrument, incorporating a large body of knowledge that would be used for the specific task of assessing potential measures for WFD. The Explorer would support decision making, by showing the potential effects of measures. It would provide options, not a single decision, and so it complies with the requirements the water managers in the article formulated. Users were involved from the beginning through the pilot projects, and meetings were organised with various groups of potential users. The project team showed in the early reports to be aware of the difficulties of making decision support tools that are actually implemented in water management practice. They expressed their intentions to avoid the common pitfalls, such as insufficiently engaging with stakeholders.

Generally speaking, WFDE supported defining the chemical and ecological present state of water bodies, setting objectives and choosing measures to achieve these objectives, so suitable in different stages of the planning process. It focussed on the parameters relevant for WFD, including effects of industry and agriculture on water as well as measures in those fields. All water bodies types and over 40 possible measures were included. It was not designed for economic analysis, but did provide a rough estimate of costs. It could be used on different scales, incorporating local characteristics, depending on the application developed. Presentations on the Explorer suggested that this is "one tool for all questions". Not all requirements that Gourbesville (2008) described for a single DSS were met, but certainly a considerable amount. So did the problems he predicted arise? The next sections will shed light on that.

3 Use of the WFDE

The WFDE started as a decision support system for water management organisations, to support communication between different stakeholders and between technical experts, policy makers and decision makers on setting objectives and identifying possible measures, however, it was not used as such. Some waterboards decided from the start not to use the WFDE at all, but used other tools, complemented with expert judgement. About three quarters of the waterboards did start developing their own application using the instrument. Most of them only used it to a limited extent, some not at all, as will be explained in this and the next sections. Furthermore, it was used by Rijkswaterstaat, mainly the Waterdienst (the central research and policy support institute) and some regional offices.

A few waterboards used the instrument to answer specific questions. A notable example is the Brabantse Delta. As one of the pilot studies, they developed a model using the Explorer to answer the question whether the Volkerrak – Zoommeer, a former sea – arm turned into a fresh water lake, would benefit from a return to a salt – water state as it now faces eutrophication and algae blooms in summer. The instrument provided interesting insights in the causes for the algae bloom in the lake, the sources of the nutrients and the management consequences of keeping it fresh or turning it into a salt lake.

The Waterdienst used the Explorer to support the development of the Management Plan for the State waters, which is one of the plans that form the basis of the RBMPs. They studied chemical balances and the potential effect of measures on chemical quality. These results were used in the assessment of the programmes of measures for state waters. To a lesser extent a few waterboards used the Explorer for chemical analysis.

In the end Rijkswaterstaat and some waterboards used the tool to determine the current state of the surface water in their area. Some modelled the effect of possible measures (comparable to scenario analysis) and used the results when writing their advice to the board. As far as known, no organisation used the instrument for communication between policy makers and decision makers or for interaction with other stakeholders in water management.

All in all the actual use made of the instrument in the development of the RBMPs has been limited and was not in accordance with the original goals. The ecological functionalities were hardly used at all. In summary this was due to:

(1) Timing of the development process in relation to the timing of the process of the implementation of the WFD.

(2) Lack of trust in the instrument due to unreliable results and missing ecological rules.

(3) Nnot recognising the role of experts like ecologists, hydrologists etc.

(4) The instrument being insufficiently geared to the users' needs .

(5) Insufficient user – friendliness.

The next section will discuss the tensions that these reasons resulted from.

4 Tensions in the development of the WFDE

A number of reasons can be provided for the difference between the intended and the actual use of the instrument. They result from the tensions in the development process that shaped the actions and discussions of the consortium. These tensions were hard to resolve; in many cases a precarious balance was all that could be achieved as will be shown in this section.

4.1 Who are the users: decision makers of technical experts?

The users that the consortium envisaged were the waterboards and Rijkswaterstaat, so the national and the regional water managers. The users within those authorities would be the policy makers in cooperation with technical experts of different disciplines, as well as through them the political decision - makers in the WFD process. It would be suitable for use in group sessions where representatives of different stakeholders would discuss the measures to be taken. The early documents also mention the possibility that other stakeholders like provinces and municipalities could use it. The development process shows a shift in the perceived users and the perceived use, which originated from the actual users.

The first users were the water management organisations that hosted the pilot studies. From autumn 2006 more waterboards got involved. As specific expertise was required, the development and testing of the application was in the hands of a diverse group of experts: modellers, hydrologists, water quality experts (chemical quality) and ecologists. No policy makers handled the instrument at that time. These experts wanted to understand and adapt the working of the instrument. They challenged the results of the WFDE. Some results were way off in a, to them, incomprehensible way. As long as that was so, they would not advise policy makers to use it, let alone in interactive meetings with decision makers.

These first users asked for changes in the instrument: more transparency in the knowledge database, new functionalities and higher reliability of results. As these changes were implemented, the tool moved in another direction. It became slower and more complicated to use, so less useful for policy makers and decision makers. It became a tool for experts, but as it had been designed for non-experts the user interface did not cater well for experts. Getting the data in was laborious and time-consuming. In the communications on the WFDE this shift in users was not visible, creating expectations that could not be met.

The different uses policy makers and specialists have for a decision support tool is easily explained by their different roles in the WFD implementation process. Policy makers will set up a framework and will ask specialists from various fields to supply specific information to fill in the framework. Sets of measures needed to be developed and assessed for effectiveness, in a process going from rough outlines of present state of water bodies, to problem analysis, to specific measures. Policy makers depend on specialists for the detailed analysis and will not use a tool such as the WFDE until the specialists assure them that the quality of the results is sufficiently reliable. This does not mean they have no interest in using it.

The developers in the consortium did study the role the instrument could have in the general process of WFD implementation, but not the internal process. They did not think sufficiently about those who would have to prepare the instrument before any policy maker would touch it. They had not realised that these people would not only have an important preparatory role for the decision making process, but also had a professional opinion about the quality of the information the instrument supplies and should supply. To address the needs of users, it is necessary to go beyond the need of the organisation in general and study the process within the organisation to find the needs of all the actual users.

4.2 General exploration or detailed analysis?

The WFDE was developed for a general exploration of options, matching the needs of the policy makers and decisions makers. It would give an indication of promising measures and an indication of the kind of improvement the measures would provide. After that general view on a basin or sub-basin scale, other tools would have to be used to see what these measures would actually do locally and what the best place would be to implement them. This was feasible for Rijkswaterstaat that controls the large waters. However, for the waterboards the modelling needed to be more detailed. They wanted to analyse their own waterbodies to assess bottle-necks before exploring which measures would be effective in solving them.

Although the communication concerning the WFDE stressed the fact that the instrument could only give general indications, exploring the effects of measures, and advises the use of other tools for detailed analysis, users frequently did try to use it for detailed analysis. This was partly due to a lack of other tools to do the analysis with. A developer stated that users projected un-realistic expectations on the instrument. If the instrument could not perform the functions the user required, they perceived it as not functioning. The use of the instrument caused frustrations with both users and developers. Users were dissatisfied with the instrument and developers felt they were unjustly held responsible for something they did not promise to deliver.

4.3 Modelling on a (sub-) basin or water body scale?

Another important issue was working from a basin model and down – scaling to make sub – basin model and then local models at (sub-) water body level, or start with the local ones and up – scaling and integrating them to create models on higher levels. An expert from the national level expressed his concern that if you develop a model from the local level up, you will never get a "balanced model", because each waterboard will want to do things differently. Furthermore, he assumed that the developers would never be able to get all the waterboards to join in, so the model would be left with gaps.

On the other hand, in a meeting on April 14th 2009, organised for a group of water quality and ecology specialists from waterboards, they clearly stated that a model scaled down from (sub-) basin level would not serve their purpose. Many ecologists feel that ecology needs to be viewed at water body level and preferably even at a lower scale. Sub-basin level, as the WFDE was intended to use as a basis, is useless in their view, if it is not based on specific local information. Local a – biotic and biotic conditions shape ecology and for this reason can be different even within a water body. The uncertainties of working from coarse to fine in ecology were seen as unacceptable.

They could develop a model at a finer scale, but the knowledge rules did not apply well to that scale, producing unreliable results. Furthermore, they would need detailed data to develop a model like that and even if they did have access to such data, it was an enormously labour-intensive procedure to import the data.

The developers proposed a country – wide application for analysis on a national and river basin scale, and regional applications on a scale suited to the region. This seems to accommodate all parties, however, it implicitly meant working from a national down to a regional scale. Rijkswaterstaat benefitted from this approach, this was what they needed.

4.4 Ambition versus realism?

Throughout the development of the WFDE a great deal of ambition was shown. The instrument would be innovative, of real use in the decision making process and fill a gap in the available tools. The presentations that were given at various moments, the information to the steering committee, or the RBO, breathe optimism about fulfilling these ambitions. Several interviewees mentioned that expectations were high, too high even. Others remarked that you need to be ambitious and optimistic to get funding and also to get potential users interested.

Another related aspect is that users may project their own ideas on the instrument and be disappointed if they are proven wrong. As said before, individual users did so. Similarly, some influential people, such as national policy makers and representatives from the funding agency, suggested publicly that the WFDE was the tool for WFD implementation, expressing their need for a harmonising tool as well as further increasing the level of expectations.

During the development more issues were added to the lists of issues to be included, such as water bed properties, more information on cost - effectiveness, more detailed analysis, and more reporting tools. The level of ambition increased while the first users in pilots and the early adopters were still struggling to get the basics working. Mid 2007, the developers admitted in a user meeting that not all they had wanted to achieve had been realised, but continued development would solve the issues. Optimism still reigned. In August 2007, the steering group decided to promote the use of the WFDE abroad and the steering group members would take action to get acknowledgement by the ministries involved that it was to be the Dutch standard tool to support the implementation of WFD.

The development process of the WFDE was performed in a pressure cooker. In the first years each year they had to prove that making a single DSS for WFD was realistic. Next year they needed the full working prototype to keep their funding. When support was at its heights, the pressure was there to deliver the product as soon as possible. Water boards started using the instrument before it was fully finished but were disheartened by the number of patches. In 2008, the optimism gradually changed into scepticism. Some said the instrument did not work and would never work. The height of the expectations contributed to the depth of the disappointment.

4.5 What expertise to include?

The intention was to use existing expertise and make this available to everyone through the WFDE. However, the first set of knowledge rules for ecology was incomplete and those that were available did not perform well; understanding why was not easy because the rules were not transparent. This led to distrust among the users, which was especially unfortunate as many ecologists were not keen on modelling to start with. They seemed to share the idea that it is impossible to generalise local ecological knowledge to other regions. Ecology depends on a large amount of factors and the relations between them are insufficiently known to be able to simplify them into rules that can be transferred to other places. On many ecological aspects, too little systematically collected data exists to warrant statistical analysis leading to plausible rules. The required monitoring will solve the lack of data, but budgets for monitoring are tight and the discussion on the best way to go about monitoring continues. Another problem is that certain water types are too rare to base any statistics on.

After they had been confronted with the distrust of potential users, the consortium decided to search for more expertise in the field of ecology in other research institutes and at the waterboards, hoping that by incorporating their expertise the instrument would be found acceptable. However, a generally accepted body of knowledge did not exist. Experts at waterboards knew the processes relevant for their own waterbodies, but in a too subtle way to translate them to general rules. The knowledge gathered at waterboards was area - specific; data was not collected on the same parameters or not in the same way. The developers eventually used a set of new knowledge rules modified from rules developed by a neural network method for another modelling instrument that was used developed to evaluate the proposed programmes of measures for the (sub) river basins that were submitted to the Ministry. These were based on data collected for assessing the present state of water bodies and were formulated in the form of decision trees, therefore sufficiently transparent. However, again they were not undisputed. The critique concentrated on the limited data - sets the rules were derived from, the purely statistical basis as opposed to a causal basis and the strict and discontinuous pathways of the decision trees. In any case, they came too late to be of use for the planning process, so the impression that the ecological modelling in WFDE was insufficiently valid persisted.

5 Concluding remarks

Although the developers of the WFDE did take into account the results of that Borowski and Hare (2005) study they did not manage to develop a DSS that was implemented in the river basin planning process. Users were to a certain extent involved and the project team was aware of the common pitfalls. Even so, they could not avoid that many users, both the intended and the actual users, were disappointed in the WFDE. This is also a matter of money: understanding the policy process the DSS is designed to support and engaging the potential users in the development process take time and therefore money. As budgets for development were low and uncertain, priority was given to the technical development that was the core business and produces the visible output needed for funding to be continued.

During the development of WFDE it became apparent that tensions existed between the needs and expectations of the different stakeholders. The development of the Explorer is an example of the importance to know who you are working for and what their specific needs are. The actual users in this process redirected the development from general to more detailed and from a communication to an analytical tool. Another major tension existed between what stakeholders (and developers) wanted and the expertise that was available. Many users concluded that with this level of expertise available in the instrument, the DSS was not trustworthy enough to be used. Furthermore, time pressure frustrated the development. To be of any use the instrument had to be finished before the analysis of the waterbodies and the development of the programme of measures, but being forced to deliver before the development was completed, was a recipe for disappointment.

The tensions concerning ambition and the use of expertise had the biggest influence on the trust users had in the instrument. Management of expectations is not an easy thing, though. Create high expectations and disappointments can be expected; create no expectations and no – one will be interested. Still, it is something that needs careful attention and may require professional support that is not commonly available in a team of developers. Engaging the community of ecologists is again not easy. They are not a homogenous group, but you cannot engage them all individually. Contacting through their own canals: professional organisations, journals, conferences and so on, appear to be the most effective way. The knowledge base is absolutely vital for a DSS such as this. In this case the knowledge was disputed, resulting in distrust. No accepted body of knowledge existed to enable ecological modelling at the start of the development of the DSS and no accepted knowledge could be developed during the process. The developers appear to have underestimated the effect this would have on the acceptance of the instrument as a whole.

Two tensions concern the content of the instrument: modelling on a (sub-) basin or water body scale and general explorations or detailed analysis. Interestingly, these tensions relate to a number of general tensions in RBM. Both have to do with Integrated River Basin Management requiring a broad overview over the entire basin to see problems and solutions in a broad context, while at the same time wanting to have participatory decision – making leading to tailor made solutions supported by the local people. As difficult it is to balance these two in RBM as a whole, as difficult it turned out to be in developing a DSS to support River Basin Management.

The WFDE was developed to supply expertise to support WFD implementation, but generally speaking it did not. It had little impact on the actual decision making process. On the other hand, it did stimulate discussion of what expertise was needed for WFD implementation, what was available, the quality of available expertise and the need to develop more. It also demonstrated that there is a need for such an instrument. Many specialists would like to have a tool for analysis of the system and the effect of measures on water quality and ecology. Therefore, developers and funders of the WFDE continue to support development of the instrument. They expect it to be the important instrument in the next period of RBMP development. With many lessons learned they had a promising starting point for the development of the New WFDE, that will be released Autumn 2012.

Acknowledgements

The research was funded within the framework of the IWRM-NET Funding Initiative, funding for this case being provided by the research department Waterdienst of the Ministry of Infrastructure and the Environment (then Ministry of Transport, Public Works, and Water management). Furthermore, the author would like to thank all the interviewees for sharing their experiences and providing documents.

References

Borowski I, M. Hare. Exploring the Gap Between Water Managers and Researchers: Difficulties of Model-based Tools to Support Practical Water Management[J]. Water Resource Management, 2005(21): 1049 - 1074.

Bots P, Bijlsma R, et al. Supporting the Constructive Use of Existing Hydrological Models in Partcipatory Settings: a Set of "Rules of the Game"[J]. Ecology and Society, 2011,16(2).

Cash D W, Adger W, et al. Scale and Cross-Scale Dynamics: Governance and Information in a Multilevel World[J]. Ecology and Society, 2006,11(2).

Consortium development WFD Explorer. KRW - Verkenner: Blokkendoos voor de implementatie van de Kader richtlijn Water. Hoofdrapport, 2005.

Consortium development WFD Explorer. Projectplan 'KRW - Verkenner Fase 2', versie 3, 9 Januari 2006.

De Kok, J. L. and H. G. Wind. Design and Application of Decision - support Systems for Integrated Water Management: Lessons to be Learnt[J]. Physics and Chemistry of the Earth, Parts A/B/C, 2003,28(14 - 15): 571 - 578.

Gorry, G. A. and M. S. S. Morton. A Framework for Management Information System. Sloan Management Review, 1971,13(1): 55 - 70.

Gourbesville, P. Integrated River Basin Management, ICT and DSS: Challenges and Needs. Physics and Chemistry of the Earth, 2008(33): 312 - 321.

Junier S., I. Borowski, et al. Implementing the Water Framework Directive: Lessons for the Second Planning Cycle[C]// The Water Framework Directive: Action Programmes and Adaptation To Climate Change. P. Quevauviller, U. Borchers, K. C. Thompson and T. Simonart. Cambridge, RSC Publishing, 2011: 80 - 96.

Lagacé, E., J. Holmes, et al. Science-policy guidelines as a benchmark: making the European Water Framework Directive. Area, 2008, 40(4): 421 - 434.

Mitchell, B. Integrated Water Management; International Experiences and Perspectives[M]. London, Belhaven Press, 1990.

Moss, T. The Governance of Land Use in River Basins: Prospects for Overcoming Problems of Institutional Interplay with the EU Water Framework Directive[J]. Land Use Policy, 2004 (21): 85 - 94.

Mostert, E. River Basin Management in the European Union[J]. European Water Management, 1998,1(3): 26 - 35.

Mysiak, J., C. Giupponi, et al. Towards the Development of a Decision Support System for Water Resource Management[J]. Environmental Modelling and Software, 2005(20): 203 - 214.

Quevauviller, P., P. Balabanis, et al. Science policy integration needs in support of the implementation of the EU Water Framework Directive[J]. Environmental Science and Policy, 2005(8): 203 - 211.

Raadgever, G. T., A. A. H. Smit, et al. Omgaan Met Onzekerheden Bij de Implementatie Van De Kaderrichtlijn Water. Deelonderzoek 2. Utrecht, Universiteit van Utrecht.

Reeze, A. J. G, B. d. Vlieger. KRW Verkenner ecologie: 1. Verbeterpunten En Verdere Ontwikkeling. Apeldoorn, Arcadis.

Rekolainen, S., J. KÄämäri, et al. A Conceptual Framework for Identifying the Need and Role of Models in the Implementation of the Water Framework Directive[J]. International Journal of River Basin Management, 2004,1(4): 1 - 6.

Turban, E, J. E. Aronson. Decision Support Systems and Intelligent Systems. Upper Saddle River, New Yersey, Prentice Hall.

Welp, M. The Use of Decision Support Tools in Participatory River Basin Management[J]. Physics and Chemistry of the Earth, 2001(26):7 - 8.

Wynne, B. Misunderstood Misunderstandings: Social Identities and Public Uptake of Science[J]. Public Understanding of Science, 1992,1(3): 281 - 304.

Design Research on Reusable Non – Rescue Submerged Groins in the Lower Yellow River Training Works

Liu Yun[1], *Xi Jiang*[2] and *Zhao Dachuang*[1]

1. Engineering Bureau, Yellow River Henan Bureau, Zhengzhou, 450003, China;
2. Yellow River Reconnaissance Design and Research Limited Company, Zhengzhou, 450003, China

Abstract: To control river regime changes, systematic river training works have been built on the lower Yellow River. However, with the operation of the Xiaolangdi Reservoir, great changes have taken place in the incoming water and sediment of the lower reaches, and the control effect of the river training works on the river regime has weakened. On the basis of an analysis of continued construction of control works, deformed river regime training and emergency handling of submerged groins in the past on the lower Yellow River, this paper presents a general idea of building reusable non – rescue submerged groins. Based on comparison among different types of submerged groins that have been built on the lower Yellow River with respect to construction technology, self – safety and emergency handling of the works, investment and removability, the paper suggests adopting assembled specially – reinforced prestressed reinforced concrete pipe pile groin as the basic groin type of the reusable non – rescue submerged groins. According to the need of the lower Yellow River training, the paper studies proper plain layout and design discharge of reusable non – rescue submerged groins. The design study can provide a theoretical basis for further spreading and application of the reusable non – rescue submerged groins.

Key words: the lower Yellow River, non – rescue, submerged groin, river training works

1 Introduction

The river training works on the lower Yellow River are designed and arranged in accordance with median flood. Viewed from the main stream line at large flood and median flood occurred in the lower Yellow River in recent years, the river training works have played a part in controlling the main stream and protecting floodplains and villages, and have prevented the adverse situation that the main stream directly lash the groins of the Yellow River and threaten the groin safety. However, in recent years, since the situation of incoming water and sediment in the lower Yellow River has changed, there have occurred a number of new problems in the river regime change in the lower Yellow River, which requires that there be some adjustment to the past way of river training to make the works construction suitable for the new training environment as far as possible.

The reusable non – rescue submerged groin is a new structure type that has been advanced just to solve the present river regime problems. The so – called reusable non – rescue submerged groin specially refers to a kind of buttress that can be removed quickly and completely when they are not needed, that can keep stable without emergency tackling under design working condition, that can direct flow under design discharge and have little effect on flood discharging under median and large flood condition and that allows water to flow across the crest.

2 Advancing of the structure type of reusable non – rescue submerged groins

2.1 Continued construction of river training works

The river training works on the lower Yellow River are designed and arranged in accordance with median flood. Since distribution of river training works on the lower Yellow River has primarily been completed, the proposed river training works are mostly upstream or downstream extensions of the existing works. Among them, most downstream extending buttresses are positioned in the straight flow directing sections, whose aim is to send the flow to the opposite bank to prevent adverse river regime. While actually, some works are such positioned that they have gone beyond the works scope that has been arranged according to training scheme for median flood, and some have even been positioned in the reserved flood discharging river width. To reduce the possible adverse effect of these buttresses on large flood and water – sediment exchange between channel and floodplain, some buttresses have been changed from the traditional groin type to submerged groin type, and the design crest elevations are somewhat lower than the other kinds of buttresses. Since the characteristics of channel sections where different works are located are not the same completely, the layout type, structure adopted and even the standard adopted for design of the downstream extending buttresses are not the same completely either. To bring into better play the river training works, we must select reasonable dam types and optimize works layout, which has become the first problem to be solved for further river training works construction.

2.2 Abnormal river regime

In recent years, especially since the Xiaolangdi Reservoir was put into operation in 2000, with increased industrial and agricultural water consumption and strengthened reservoir detention and storage, in the lower Yellow River, median floods of over 4,000 m^3/s have occurred in somewhat less frequency, while floods of small discharge (about 1,000 m^3/s) have occurred more number of times and lasted for a long time. The main stream at small discharge magnitude kept wandering within the channel with planned flood discharging width of 2.0 ~ 2.5 km, and abnormal river regime developed continuously in some reaches. Under the circumstances, not only dangerous situations in small flow increased and the main stream was not controlled well, but also, in case of large flood, transverse river or rolling river are more probable, seriously threatening the groin safety of the lower reaches.

2.3 Submerged groin rescue in the past

As a structure type of extension of training works in the straight sections, submerged groins have been used in the works of Mazhuang, Wuzhuang and Shunhejie. The submerged groin refers to a kind of dams with scour protection on the crest that allow certain quantity of flood to overflow. The Tiexie submerged groins adopt the structure with wire mesh narrowing, filled with ripraps and wrapped with wire meshes (referred to as "wire mesh dam" hereinafter). The Mazhuang submerged groins adopt the structure of long sack foot mattress, riprapped dam foundation and stone mesh slope protection (referred to as "long sack foot mattress dam" later) in the first 100 m, and used long sack blanket foot mattress and sand – filled sack dam body in the extended 100 m. The Shunhejie submerged groins are built in the structure of earth – filled blanket foot mattress, long sack narrowing and stonemesh slope and foot protection.

Though these works have played a good role in directing flow, with the actual operation, there have occurred a number of problems. Since the works cut into the main stream, they are vulnerable to scouring, and because they are more likely to be overflowed than other training works, this kind of works are easy to be in danger. In 2003, dangerous situation occurred in Shunhejie downstream extended submerged groins: the main stream bypassed the works and directly lashed against the Dagong Works. The Dagong Works upper floodplain and bank were severely scoured and the works had to be extended upstream, resulting in passive rescue and increased project investment. Therefore, it is necessary to find out submerged groin types suitable for the characteristics of the lower Yellow River that can keep stable, can control river regime and are economically rational.

Therefore, for coping with continuously occurring abnormal river regime in some reaches and problems in works construction in the straight sections of the planned works, to develop a reusable and non – rescue submerged groin type will help further development of river training works construction in the lower Yellow River.

3 Selection of reusable and non – rescue submerged groin types

A comparison was made among the above – mentioned types of submerged groins already built and other non – rescue types already built on the lower Yellow River including reinforced concrete pile dams and spile dams (see Tab. 1) to select dam types that are economical, safe, well adaptable and suitable for the characteristics of current river training.

In Tab. 1, comparison was made in 4 aspects: construction technology, works safety and rescue in flood control, project investment and removability. It can be see from Tab. 1 that each of the four submerged groins has its own advantages and disadvantages. For satisfying the requirement of being reusable, needing no rescue and being submerged groin types, each of the four types has its limitations. The above – mentioned comparison result considered, it is decided that a dam of assembled, reinforcement – specially – designed prestressed concrete pile structure be adopted as the basic reusable, non – rescue submerged groin type.

Tab. 1 Comparison of submerged groin structures

	Wire Mesh	Sack Foot Mattress	Reinforced Concrete Pile	Spile
Construction	Easy	Difficult in deep water	Difficult in water	Difficult in water
Works safety and rescue	Safe, easy to rescue	Safe, but not easy to rescue	Non – rescue	Non – rescue
Project investment	Low	Low	High	High
Removability	The upper part is easy to remove, but the lower part is difficult, some removed part can be reused	Difficult, and the removed part can not be reused	Can not be removed	Can not be removed

4 Plain layout of reusable, non – rescue submerged groins

4.1 General layout of works

The aim of further carrying out river training in the lower Yellow River is to further control river regime and direct the main stream. At the same time, it is required that the effect on water – sediment exchange between channel and floodplain be reduce as far as possible in works construction.

Since the distribution of training works in the lower Yellow River has been completed, and the distribution of bend section works have also been completed primarily, the main aim of the currently planned works is to further control river regime in cooperation with the existing works. On account of this situation, the plain layout of the planned works is arranged as shown in Fig. 1.

Since the newly planned works will be located in the straight section, and their function is to further stabilize the main stream so that the main stream may be directly sent to the next works smoothly, and this kind of works, after completion, will narrowing flood discharging width, their crest elevation should be lowered as far as possible to reduce the effect on flood discharging. Therefore, their vertical cross – section should be arranged as shown in Fig. 2.

Since the planned non – rescue submerged groins are all located in the straight sections, if they are arranged perpendicular to the flow, it will be unfavorable for main stream control, and moreover, the main stream may go through the crotch at small flood. Therefore, the reusable non – rescue groin should be arranged as parallel groin type along the planned diversion line.

4.2 Selection of submerged groin works' plain layout schemes

The prestressed reinforced concrete pile will be adopted for the submerged groins. As a kind of pile groin, its permeability and impermeability and permeable rates at different permeable situations are

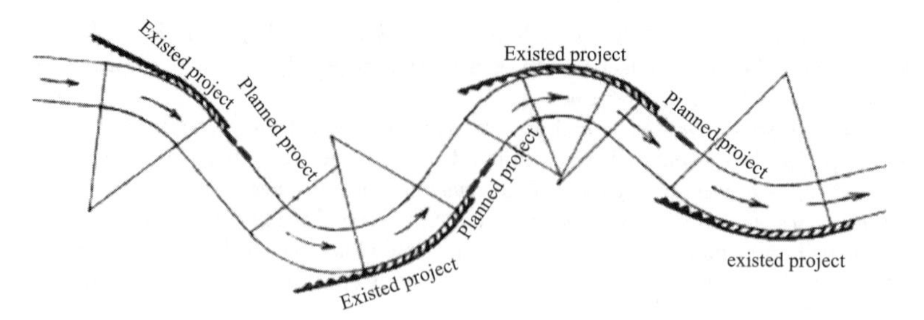

Fig. 1 Plain layout of submerged groins

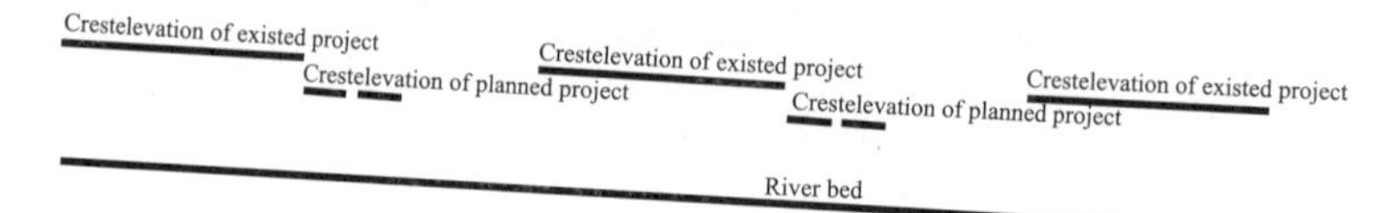

Fig. 2 Vertical layout of submerged groins

studied respectively.

On the plain the pile groins can be arranged as permeable and impermeable.

The impermeable pile groin usually adopts sheet pile wall (see Fig. 3). The wall body is usually made up of precast reinforced concrete sheet piles. When reuse is intended, precast prestressed and reinforced concrete sheet piles should be used, whose body is usually rectangular.

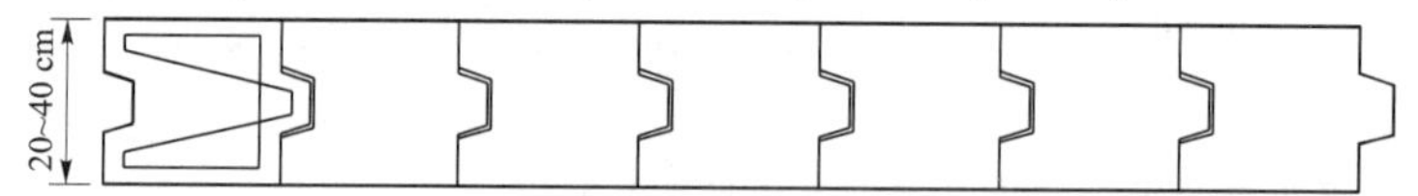

Fig. 3 Plain view of reinforced concrete sheet pile wall

The permeable pile groin usually adopts round reinforced concrete caisson piles. The caisson piles currently widely used on the lower Yellow River usually have central distance of 1.1 m, clear spacing of 0.3 m and diameter of 0.8 m with permeable rate of 33%, see Fig. 4.

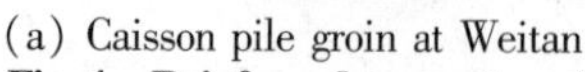

(a) Caisson pile groin at Weitan (b) Caisson pile groin at Zhangwangzhuang

Fig. 4 Reinforced concrete caisson pile groins on the ower Yellow River

According to the comparison between impermeable and permeable pile groins, and based on the following reasons, it is decided to use the permeable pile groin type.

(1) Permeable groins have been built in Works of Zhangwangzhuang, Andong and Weitan, and

according to the flow directing effect of Andong works after nearing the flow and the model experiment with Weitan Works as the prototype carried out by the Institute of Hydraulic Research, Yellow River Conservancy Commission (YRCC), the permeable pile groins can play a good role in directing flow as the traditional groins and have good engineering adaptability.

(2) According to the model experiment, under the same working condition, the scour hole in front of the pile of the impermeable pile groin is deeper than that of the permeable pile groin, and calculated based on this, the total pile length needed is smaller than that of the impermeable groin.

(3) According to model experiment and theoretical analysis, the permeable pile groin has a consecutive hole formed in front of and behind the pile, and part of the earth behind the pile is washed out, therefore, its stress is better than that of the impermeable groin, and both the flexural torque and shear of the pile groin are smaller than those of the impermeable groin.

(4) On comparison of permeable pile groin and permeable pile groin, according to the layout of permeable pile groin by the YRCC, i. e. its central distance of 1.1 m, clear spacing of 0.3 m and diameter of 0.8 m with its total length reduction of about 10%, the permeable pile groin adopted would be of good economic benefits.

4.3 Recommended plain layout of reusable non – rescue submerged groin works

Since the reusable non – rescue submerged groins can be removed when the works are no longer needed, the single – pile section should not be large. At present, the precast reinforced concrete pipe pile has the largest diameter of 0.5 m, therefore, it is decided that the plain layout of the reusable non – rescue submerged groins be arranged as follows:

The permeable groin is adopted, arranged along the straight flow directing section of the diversion line, with single pile outer diameter of 0.5 m, clear spacing of 0.3 m and permeable rate of 37.5%.

Factors such as removability and non – rescue, etc. considered, it is recommended that the prestressed reinforced concrete pipe pile be adopted for the submerged groin, whose cross – section is annular. This structure type can adapt to different flow conditions and save concrete material; adopting prestress technology can prevent cracks while the piles are being moved, installed and removed; the pipe piles can be prefabricated in the factory so that works quality can be guaranteed.

5 Determination of design discharge for reusable, non – rescue submerged groins

Design discharge is the most important parameter in the design of river training works, and in the design of river training, works' design water level, design river width of the works reaches and river bend factors are all directly connected with design discharge.

5.1 Factors to consider while determining design discharge

Reusable non – rescue submerged groins are mainly constructed in the straight sections of the planned bends and in temporary works built to deal with emergency. In works design, except for the factors considered for normal river training works design, the following restrictive factors should be considered.

5.1.1 Requirement of channel discharging capacity

At present, river training of the lower Yellow River is aiming at median discharge. In design of river training works, an important design parameter is flood discharging river width. The so – called flood discharging river width is, after river training has been carried out according to the plan, the vertical distance from the end of a training project to the straight line connecting the end of the training project upstream and the head of the training project downstream. In the design of river training works on the lower Yellow River, the flood discharging channel width usually ranges from 2.0 km to 2.5 km.

Within the flood discharging width, water retaining structures are not allow to be built so that the river bed may be scoured severely and cut down for flood going through during flood periods. While the reusable non – rescue submerged groins are constructed within the flood discharging width mainly due to the need to control river regime change, and in this case, the effect of newly - built works on large flood discharging should be considered as far as possible in works design.

5.1.2 Requirement of water - sediment exchange between floodplain and channel

In recent years, due to decreased incoming water from the upper reaches and strengthening of reservoir regulation capacities on the upper reaches, chances of dangerous situation have decreased greatly, and hence the number of times of the flood going over the floodplains. Due to decreased overbank flood, water - sediment exchange between channel and floodplain has been reduced, causing the main channel of the lower Yellow River to continuously silt and rise, going gradually higher than the floodplain and forming the adverse condition of secondary suspended river. Whether the design discharge of the river training works is rational or not and whether the elevation of the works groin crest compared to that of the floodplain is high or low, those also have direct effect on the chance of water - sediment exchange between floodplain and channel. Therefore, the effect on water - sediment exchange between floodplain and channel is also an important factor in further carrying out river training and determining design discharge of works rationally.

5.2 Determination of design discharge

At present, in the river training scheme in accordance with median discharge on the lower Yellow River, the design discharge adopted in river training works design is 4,000 m^3/s. The design discharge is determined mainly according to the bankful discharge. On considering the environment of reusable non - rescue submerged groins works and the conditions of future incoming water and sediment, it is not suitable to still choose this discharge as the design discharge.

While determining design discharge for submerged groins, on the one hand, the requirement of river training should be given full consideration, on the other hand, the effect of submerged groin construction on flood discharging and water - sediment exchange between floodplain and channel should be reduced to the lowest possible level. Therefore, in determination of design discharge, the above - mentioned two requirements should be both considered. The design discharge is derived with two calculation methods.

5.2.1 Sediment runoff method

The sediment runoff W at different discharge magnitudes is calculated respectively. The river width for training used in the calculation is derived from the formula $B = kQ^{0.5}$, the river width at median flood is derived by averaging the river widths at observed bankful discharge at Huayuankou from 1986 to 2005, and the average river width is 1,678 m. The Engelund - Hansen sediment transport capacity formula revised by Zhang Yuanfeng et al. according to the observed data from the lower Yellow River is used as the sediment transport capacity formula. The revised Engelund - Hansen sediment transport capacity formula can be simplified to $S = Bmv^{4.6}$, where, s is sediment load, m is a coefficient and v is average flow velocity. In calculation of sediment runoff at different discharge magnitudes, the occurring frequencies of different magnitudes are derived from the observed data at Huayuankou Hydrological Station from 2001 to 2005. Therefore, $W = TPS$, where, P is the occurring frequency of different discharge magnitudes, T is the total time of a year with second as the unit. For convenience of display, the sediment runoff at different discharges are compared with that at the discharge of 800 m^3/s, producing the relation between Q and W/W_{800} under training conditions for different discharges, see Fig. 5. It can be seen from Fig. 5 that, since the discharge of 800 m^3/s has a relatively long duration and its discharge is also relatively large, at schemes for different training discharges, their actual sediment runoffs are larger than those at other discharge magnitudes. It should be noted that the Fig dose not show the ratio of the sediment runoff at different discharge magnitudes over 1,600 m^3/s to that of 800 m^3/s. According to calculation, after exceeding 800 m^3/s, though the discharge magnitudes last for a shorter time, their sediment runoffs will still be higher than that of 800 m^3/s, and will form another peak value at the discharge of 3,500 m^3/s.

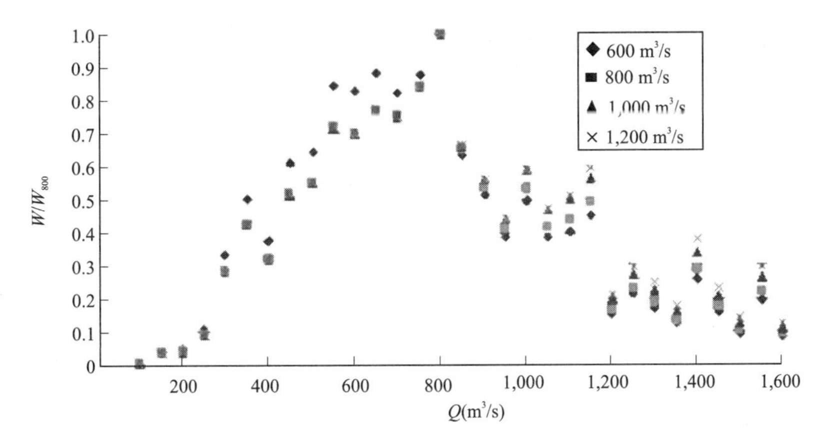

Fig. 5 $Q \sim W/W_{800}$ **relation under schemes for different training discharges**

5.2.2 Bankful discharge method

For the lower Yellow River, even in the reaches where the main stream is well controlled, the main stream still wanders within a certain scope, and even during a flood event, the main stream may swiftly swing due to change of discharge. Fig. 6 shows the changes of large cross – sections in Huayuankou channel of the lower Yellow River in recent years. It can be seen from Fig. 6 that the channel of the lower Yellow River is still a wide and shallow channel. Though, due to water and sediment regulation and clear water discharging after the Xiaolangdi Reservoir is put into operation, bankful discharges increased to some extent, the swinging property of the channel has not changed completely, and, at small discharges, the main stream is sometimes on the left and sometimes on the right without a steady flow route and cross – section, and its swinging amplitude still exceeds 3,000 m.

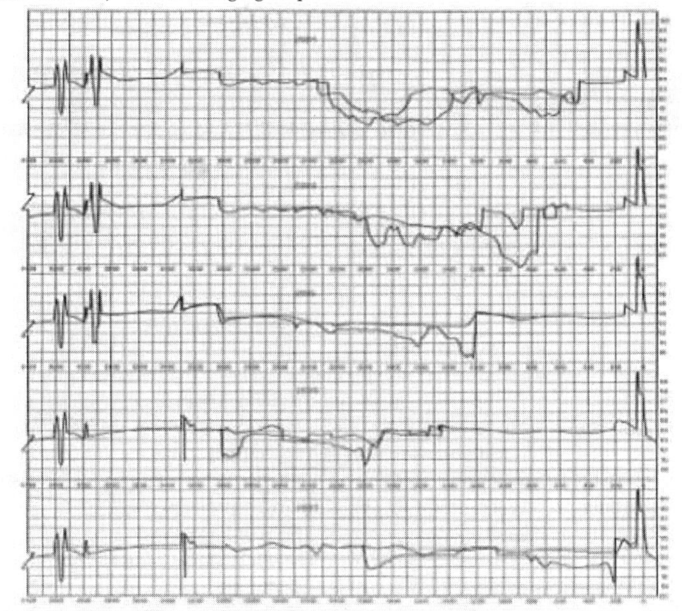

Fig. 6 Changes of cross – sections in the Huayuankou channel from 1997 to 2004

It can also be see from Fig. 6 that there usually exists a smaller channel within the main channel, that is to say, during low water period, the main stream will concentrate in this main channel, which will be referred to as "low water channel" hereinafter. When the water level is higher than the low water channel, the water will spread quickly over the whole main channel. Under this condition, the river will become wide, shallow and dispersing, the main stream is slow and weak of sediment carrying capacity, with great changes of scouring and silting, and the river regime is very likely to change greatly. Therefore, it is important to study the bankful discharge of the low water channel and to control the low water channel for control and stabilization of the river regime during low water period and for design of rational submerged groins.

To help analyze channel changes of the lower Yellow River, YRCC have been arranging channel cross - section measurement before and after flood period each year. According to the observed cross - sections, the typical channel cross - sections since 1996 are taken as the basis for calculation of the low water channel. The main channel has roughness of 0. 015 and bed gradient of 0.000,19, and the bankful discharges corresponded to by the low water channel estimated as uniform flow with the observed cross - sections are given in Tab. 2. The bankful discharges of the low water channel averages 820 m^3/s.

Tab. 2 Calculation of bankful discharges of low water channel

Year	1997		2001		2003		2004		Average
	Before Flood Period	After Flood Period	Before Flood Period	After Flood Period	Before Flood Period	After Flood Period	Before Flood Period	After Flood Period	
Bankful Discharge	960	500	310	620	1,637	456	1,030	1,010	820

6 Experimental project

In May 2009, a 50 - m - long reusable non - rescue submerged groin is built near the Nanguotou Control Works on the lower Yellow River, and a pipe removal experiment is carried out smoothly. During the water and sediment regulation period in 2009 when the flood discharge exceeded 4,000 m^3/s, the floodplain at north of the experimental project was continuously caved in by the river water, while the experimental project neared the flow in time and successfully stood the test of flood impact.

7 Conclusions and suggestions

According to the change of actual incoming water and sediment in the lower Yellow River and the forecast of future water and sediment conditions in the lower Yellow River, and with the real problems faced by the river training of the lower Yellow River considered, it is proposed that the reusable non - rescue submerged groin is an ideal river training works type. According to the experimental projects, the works can primarily attain the design goal, solving well the problems of controlling river regime at small flow and avoiding the effect on large and medium floods.

The research of application of the reusable non - rescue submerged groin is a special industrial research project for public good of the MWR. Since the pilot projects were carried out on the floodplain and experiment has not implemented in water, whether the works can be constructed smoothly under different working conditions has yet to be verified. Meanwhile, due to small scale of the pilot projects, its overall benefit has to be further verified. In future work, on the one hand, the main design parameters should be further improved and optimized, and on the other hand, further observation of the pilot projects has great significance for the project popularization.

References

Hu Yisan. Flood Control of the Yellow River[M]. Yellow River Conservancy Press, 1996.

Li Yongqiang, Chen Shoulun, Liu Yun. Discussion on Composite River Training Scheme for Improving the Sediment Transport Capacity of the Lower Yellow River[J]. Hydro. Power Journal, 2011(3).

Yao Wenyi, Wang Puqing, Chang Wenhua. Effect of Water Retarding and Silting of the Perameable Pile Groin and the Scour Process at the Pile[J]. Sediment Research, 2003 (2).

Qi Pu, Sun Zanying, Liu Bin, et al. Research of Both Bank Training Scheme for the Swinging Channel of the Lower Yellow River[J]. Journal of Hydraulic Engineering, 2003(5).

Geophysical Exploration of a Buried River Valley for Groundwater in Wukro, Tigray Region

Paulos Beyene

Ministry of Science and Technology P. O. Box 2490 Addis Ababa, Ethiopia

Abstract: The study was conducted with the aim of delineating a buried river channel with good ground water potential on the area and evaluating the usefulness of the two geophysical methods in such geological situation. The geophysical method used was electrical resistivity (vertical electrical sounding and profiling) and gravity survey to locate favorable areas for ground water development by delineating the extent and trend of buried channels. A total area of 1,000 m × 800 m = 0.8 (m^2) was covered by the method. A base line of 1 km long is laid out parallel to the buried channel, 5 lines each 1km long are cut perpendicular to the base line, with 200 m profile interval and readings were taken every 20 m. The gravity survey consisted of 246 stations carried out on six lines each 1 km long at an average line spacing of 200 m, and a station interval of 25 m along the profiles. The buried channel is characterized by undulations with shallows and deeps. The geophysical results were well correlated with drill holes and the local geology of the area. Low values of apparent resistivity and low gravity correlated with high yielding of the well and relatively high values of resistivity correspond to the sandstone formation. Based on the results, drilling was conducted on the area. The three drill holes intersected the buried channel streams and each well in the area are yielding 20 ~ 25 L/s. Since buried stream channels usually contain water saturated sand and/or gravel which are good aquifers and potential sources of ground water which acts as conduits for water flow. The depth to the water table is 10 m from the surface. Maximum thickness of sand and gravel could be intercepted. The use of electrical resistivity and gravity method can be taken for ground water development program in the future to help and locate more appropriate locations for drilling. Because of the presence of similar geological conditions the methods are successfully used to locate buried channel aquifers as potential sources of ground water.

Key words: buried river channel, aquifers, streams, yield

1 Introduction

Ground water is the earth's largest accessible store of freshwater that sustains stream flow during period of dry season and is a major source of water in semi – arid and arid areas. It forms an integral part of the hydrological cycle and is the only viable source of water supply to human use and development.

Wukro area is covered by sedimentary rocks and unconsolidated materials ranging in age from Mesozoic to recent. Sandstone, limestone and shale are found dominantly covering the area in different proportions. Sandstone and limestone are the main source of ground water in the area and the search for ground water starts with an investigation of unconsolidated sediments.

The unconsolidated deposits of high ground water resource potential exist located mainly on a narrow and long ancient river channel. Buried river channels usually contain sand and/or gravel which are good aquifers. However as the buried channels have no surface expression, techniques that would assist to locate and delineate them are of utmost importance in the sitting of high yielding, more successful boreholes and also in better understanding of the hydrogeology of buried channel streams. Infact, modern drainage follows the old drainage, and generally flows from North to South West in a very large area in broad flat channels within undulating landscapes.

A number of geophysical exploration techniques are available which enables and insight to be obtained rapidly in the nature of water bearing layers. These include: Electrical method, Electromagnetic, Seismic, Gravity, Magnetic, Gravity, soil temperature and Geophysical borehole Logging. Of the several geophysical techniques, electrical resistivity and gravity methods are useful in providing supplemental information for location of burial sites (e. g. , buried channel streams, trenches, their depths and boundaries) and Hydrogeologic conditions (e. g. , depth to water table or water bearing zones, depth to bedrock, and thickness of soil, etc). The resistivity of rocks is primarily a function of porosity, salinity, and temperature of the pore fluid because the conducting medium is mainly water. The importance of electrical method is more because of complexity of the buried channel stream (D. S Parasins, 1971). The method has been found to be cost effective, time saving and simple.

It is important that well sites are chosen principally on hydrological grounds. One of the most essential steps for ground water investigation is collection of available data that expresses the surrounding ground water conditions, but in most cases of our country acquision of well data is not adopted. The electrical resistivity, gravity method and the corresponding well data should be related so as to produce a model that can best explain the situation In this regard, electrical method and gravity methods was conducted in Wukro tannery, located about 5 km south West of Wukro town, in the eastern zone of Tigray regional state to delineate the extent and trend of a buried channel stream and also to evaluate the usefulness of the methods in such geological situation. Accordingly, the field data was acquired during the period 10 March to 8 April 1996.

2 Geology of the Area

The study area, Wukro, is located in the eastern zone of Tigray Regional state. Geographically, it is located between longitude 39°30′ and 14°N 45′E and latitude 13°47′ to 14°N. It has an area of about 0.36 km^2.

Fig. 1 shows location map of the study area.

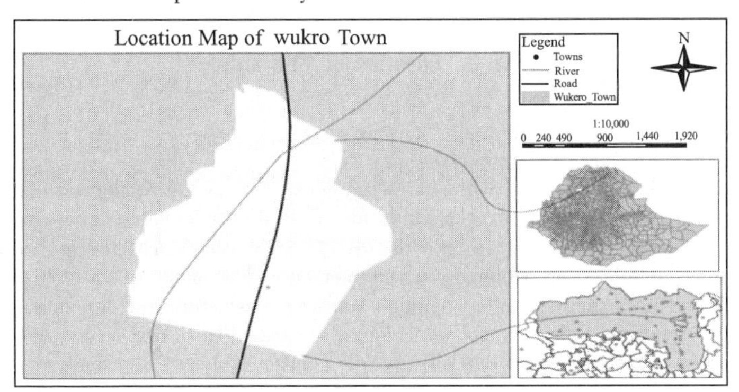

Fig. 1

The study area is a flat lying sedimentary environment. The main lithological units are limestone, sandstone, alluvium and shale. The limestone is outcropping all the survey area and which is mostly covered by transported/residual soil. It is apparently jointed and fractured on the surface. It has variegated layers in terms of texture and composition, but big cavities and karst holes were not observed. It is dissected by Genfel River. The sandstone covers the northern and north western parts of Wukro town. Especially it forms a big hill at the western part. The shale is usually found intercalated with limestone on the survey area. In one of the river cut, buried river mouths were observed where stream channels cut during an earlier time were drowned or buried by latter sediments. The geophysical survey was aimed to follow and delineate the pattern and extent of the buried channel sites on the channel. Because buried stream channels usually contains water saturated sand and/or gravel which are good aquifers and potential sources of ground water which provide as conduits for water flow in the area.

3 Geophysical methods for ground water investigation

The most commonly geophysical Techniques applied for ground water investigations are: Electrical Resistivity, Magnetic, Electromagnetic, Seismic, Gravity, Soil temperature and Borehole logging. The choice of a particular method is governed by the nature of the terrain time and cost consideration. The resistivities of rocks are primarily a function porosity, salinity, and temperature of the pore fluids because the conducting medium is mainly water (J. S Sumner, 1975).

Some of the geophysical investigations that can be done by the electrical resistivity method for ground water studies are: Correlating lithology and drawing geophysical sections: Bed rock profile for subsurface studies: Contact of geological formations: Delineating buried channel streams and Water quality in shallow aquifers and ground water pollution as in oil field brine pollution, pollution by irrigation waters and pollution by sea water intrusion, which cause change in electrical conductivity.

The VES points are located in the following coordinates: Line 400N/00, Line 400N/300E, Line 400N/400W, Line 00/00, Line 00/300E, Line 00/400W, Line 400S/00, Line 400S/300E, and Line 400S/400W (Fig. 2).

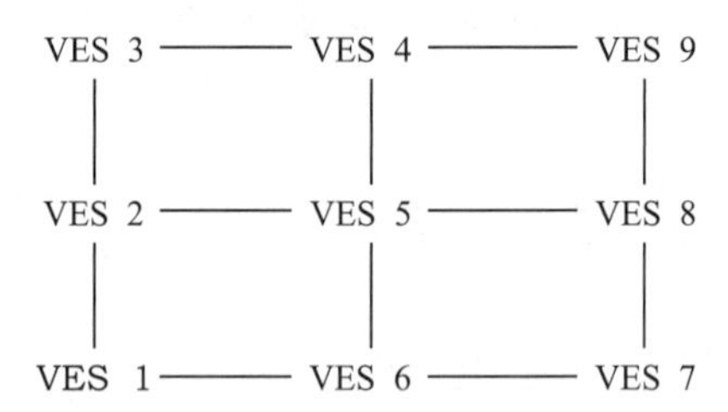

Fig. 2 Location of VES sites

3.1 The gravity survey method

The gravity method is involved with the measurement of variation in the gravitational field of the earth. Changes in the density of rocks within the earth's crust produce variations in g, the acceleration due to gravity, the unit of which is the gal ($1 \text{ gal} = 1 \text{ cm/s}^2$). The extremely small variations in this acceleration are measured by gravity meters.

The gravity method can be used in defining subsurface structure and lithology. The only requirement is that there be a significant lateral density contrast. In buried river valleys, there is a variation thickness of unconsolidated deposits overlying denser bedrock, and these commonly appear as gravity lows on gravity maps or profiles. Thus, the method offers information on the location of a buried channel and about its width and depth.

The accuracy of the gravity data depends on the accuracy of the gravity observations and the accuracy of other quantities, namely station location, elevation and density of the near surface rocks that must be used to reduce the gravity observations to a form suitable for interpretation. In order to achieve high accuracy in gravity, survey the positions and elevations of the stations have to be determined by precise leveling.

3.1.1 Instrumentation and field procedure

The instrument used for the gravity survey was a Lactose and Romberg model G-780 gravity meter. Model G gravity meters have a range of over 7,000 mgals, a reading accuracy of + 0.01 mgal, and a drift rate of less than 1 mgal pet month. The gravity data werc collected with reference to a single base station located on line 00/ station 00. A 20m station spacing was used, and at each station the gravimeter was leveled and counter/dial readings taken. The base station was repeatedly occupied almost every hour for correcting the drift and tidal effects.

3.1.2 Data processing and presentation

As a first step in data processing the counter/dial readings of the gravimeter were converted into milligal values using the conversion table given in the instrument manual. The East – West distance of the stations from the base were also determined, for the purpose of latitude corrections, and the errors are estimated to be less than + 5 m. Standard data reduction procedures (Grant and West, 1965) which consist of drift correction, latitude correction, free air and Bouguer Correction were applied to the field data with respect to the base station, Due to the flat topography of the area terrain effects are negligible and thus has not been applied.

A density of 2.0 g/cm^3 was used for the Bouguer Correction because densities of near surface samples measured that ranges from 1.8 g/cm^3 to 2.2 g/cm^3 with an average of about 2.0. The estimated error in the Bouguer gravity values that was caused by a positional error of 5 m in the north – south position (0.005 mgal), an error of + 3crn in the station elevations (0.009 mgal) and an error of less than 0.06 mgals in the gravity observations due to drift totally amounts to less than 0.07 mgals.

The resulting Bouguer anomaly map is shown in Fig. 3, which is the superimposition of large scale (regional) and local (residual) effects. As it is the local anomalies that are of prime interest in groundwater investigations the regional field has to be separated from the residual anomalies.

After considering the nature of the Bouguer gravity, a first order polynomial was found to be appropriate to represent the regional trend; and is presented here in Fig. 4. The regional trend, the constants of which are determined by a least – squares procedure, was then removed from the Bouguer anomaly map using the GRIDTRND program (Geosoft, 1991); and the residual gravity map shown in Fig. 5 was obtained.

The Bouguer, Regional and Residual gravity maps are presented at a scale of 1:10,000 with a contour interval of 0.2 mgals.

3.1.3 Results and interpretation

The Bouguer gravity map (Fig. 3) which is a result of the combined effect of local and large scale masses, is highly influenced by the regional trend which dips to the northeast with a gradient in the range of 0.2 to 0.7 mgal/100 m. The presence of local anomalies is also indicated at places by the "nosing" and closing of the Bouguer gravity contours as at line 00 between stations 200 W to 300 W.

Fig. 3

The regional gravity map (Fig. 4) is represented by a first order polynomial fitted to the Bouguer gravity map, and it is certainly due to sources of greater depth and lateral extent that are of interest to ground water resources. Although, the proper definition of the regional is not possible due to the limited size of the survey area, the steep gradient (0.4 mgal/100 m) possibly indicates a structure trending in the NW – SE direction.

The residual gravity map, (Fig. 5) is of utmost interest from the view point of delineating the buried river channel. The prominent feature observed on the map is the distinct gravity low which

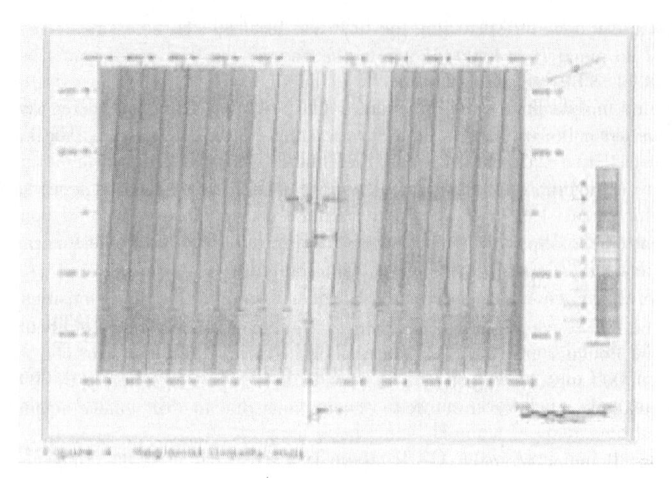

Fig. 4

spans the survey area from the northwest to the Southeast. This broad and continuous gravity low clearly shows the suspected NW – SE trending buried river channel.

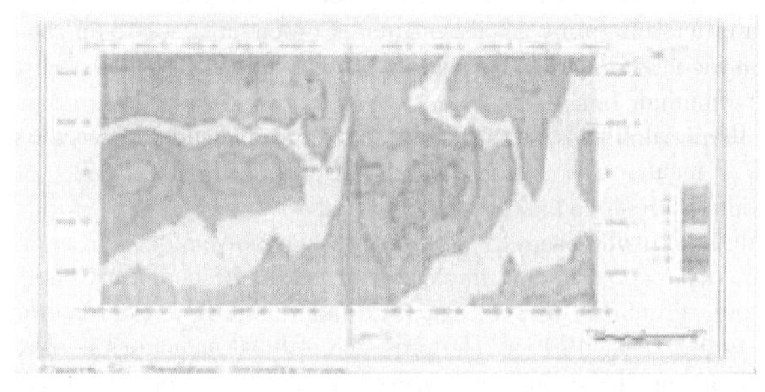

Fig. 5

The gravity low appears to be wide in the northern (line 400N/0 – 500W) and southern (line 400 S/00 – 150E) parts of the grid, but it is due to these areas being located at the bends of the buried channel.

A small gravity low is also observed at the southeastern of the area on line 525 S east of station 300 E. This gravity low would probably join the one located west of it had the survey been continued further south with the implication that it is in fact the response of the same buried channel that swings to the east to form a continuous meander.

The wavy nature of the gravity low implies that the buried channel is spatially complex and that it has a meandering channel with high degree of sinuosity. The residual gravity map also shows a series of closed gravity lows, with amplitudes as low as –0.5 mgal, as the one located between lines 0 to 400 S and stations 0 to 250 E. This may suggest the presence of a series of shallows and deeps along the course of the channel and probably indicates the undulating nature of the buried channel.

A positive anomaly of high amplitude located in the northern part of the area on line 00, west of station 100W may be associated with localized shallow bed rock where as the elongated gravity high observed all over the lines between stations 200 E to 500 E and with decreasing amplitude to the south may be interpreted as being due to a subsurface ridge.

3.2 Electrical method

Two methods of investigation are generally employed in the electrical resistivity method of traversing.

3.2.1 Vertical Electrical Sounding (VES)

Vertical Electrical Sounding is used form mapping of the horizontal or low dipping boundaries; practice is to use a system of expanding electrodes. Increase in the current electrode separation necessarily increases the depth of current penetration; therefore the variations in resistivity correspond to the change in sub surface characteristics in a vertical plane.

Ground water survey starts from data collection, interpretation of geomorphic and geologic features on available maps, aerial photographs and landsat imagery followed by field reconnaissance and further map interpretation. The survey plan of Vertical Electrical Sounding is going to be decided after the hydro geological conditions of the investigated area. In any field study, before setting up sounding apparatus, the following points were considered to minimize common interpretation errors. Relatively flat topography: Spreading performed along the inferred/interpreted geological structure to the direction of sloping: checking of survey apparatus and avoiding of buried surface material such as cables, pipes, and high voltage eclectic lines.

For the Vertical Electrical Sounding (VES) survey the schlumberger array was used with a maximum electrode separation AB/2 of 500 m. 9 VES sites were surveyed by this array along 5 profiles. These profiles run mainly from North to the South direction to cut the expected structures and creeks. The apparent resistivity values for each VES's as shown in the Tab. 1.

Tab. 1 Layer resistivity, thickness, depth and RMS error in %

VES stations	Layers	Resistivity (Ω · m)	Thickness (m)	Depth (m)	RMS error
		20	2	2	
		4.4	7	9	
1.	5	10.9	11	20	3.7%
		2.3	27	47	
		489	—	—	
		49.9	3	3	
		4.4	5	8	
2.	5	99.7	10	18	3.95%
		4.2	14	32	
		476	—	—	
		30.0	4	4	
		8.0	8	12	
3.	5	45.6	15	27	3%
		4.5	45	52	
		520	—	—	
		123.3	4	4	
		4.0	6	10	
5.	5	33.4	10	20	3.8%
		4.5	12.5	32.5	
		234	—	—	
		53	3.5	3.5	
		3.4	6	9.5	
8.	5	32	9	18.5	2.7%
		4.5	15	33.5	
		567	—	—	
		47.9	4.4	4.4	
		9.1	7	11.4	
9	5	104	9	20.4	3%
		4.4	25	45.4	
		432	—	—	

During the presentation, I will display all the nine VES's using Oasis Montaj software. (This time difficult to copy the curves with out the software.)

Based on the interpretation of VES data, a geoelectric sections were produced along Line 1N, 2N, 3N and along VES 2, VES 5 and VES 8 traverses. Quantitative interpretations were done on the VES data, yield a geoelectric sections. The geoelectric sections, the profiling curves and VES results are interpreted and well correlated with each other. Since the geoelectric section made along Line 400 N, 00 and 400 SN have no significant contribution for the mapping of the buried channel stream, the sections neither are nor prepared for presentation.

Based on the interpretations, the geoelectric section made along VES 2, VES 5 and VES 8 (Fig. 6) traverses shows five lithological units. The first unit which is mapped as dry top soil has a thickness of 3 ~ 4 m with a resistivity range of 49.9(Ω · m) to 123.3(Ω · m). The second unit, which is characterized by relatively low resistivity of 3.4(Ω · m) to 4.4(Ω · m) and ranging in thickness from 3.4 ~ 4.4 m is due to water saturated clay soil. The third unit, which is characterized by 32 ~ 99.7(Ω · m) and ranging in thickness from 9 to 10 is due to the response of the limestone unit. The fourth unit, which is characterized by low resistivity value of 4.2(Ω · m) to 4.5 (Ω · m) and ranging in thickness from 12.5 m to 15 m is due to buried channel stream. This unit, being composed of medium to coarse sand and fine gravel, forms the most important water bearing unit in the area. The bottom unit, which a resistivity value ranges from 234(Ω · m) to 567 (Ω · m) and infinite thickness is due to sandstone bedrock formation.

3.2.2 Profiling method

The electrode separation is kept constant for different values and the center of the electrode spread is moved from one station to another station to have the same constant electrode separations. The method helps in mapping vertical or steeply dipping beds, lithological contacts, faults, etc. Profiling was carried out along 5 parallel lines and a resistivity contour map of the area showing isoresistivity lines was prepared. This indicates that the areas of high/low resistivity were useful in identifying aquifer formations.

For the resistivity profiling survey a three pole array with electrode separations of $a = 40$ m, 80 m and 120 m was used and readings were taken every 20 m interval. The three electrode separations were conducted on three lines corresponding to three depth levels chosen to show responses at near surface, intermediate and deeper depths of investigation for the selected array.

The resistivity profiling curves at $a = 40$ m of Fig. 6, 7 and 8 show that, the low resistivity zone is found between 40 W to 360 W on Line 400 N, 00 to 250 E on Line 00 and 00 to 150 E on Line 400 S.

The river channel has a maximum width at line 400 N and minimum at line 400 S. As we go from north to south, its width becomes narrow. The resistivity values caused by the buried channel stream appear to be lower than the common resistivity of rocks. This condition indicates that, buried river channel usually contains water saturated sand and/or gravel which are good aquifers and potential sources of ground water which provides conduit for water flow. Most of the area that is characterized by intermediate value of resistivity that ranges from 100 ~ 300 (Ω · m) is the response of the combined effect of the thinly intercalated shale with limestone and high values of resistivity corresponds to the sandstone unit.

Similarly, the resistivity profiling curves at $a = 80$ m (intermediate level) and $a = 120$ m at deeper level will be displayed during the presentation as a section map using Oasis Montaj software.

3.2.3 Data processing and presentation

For the resistivity profiling, the current "I" of the transmitter and the voltage V_p of the receiver together with an appropriate geometric factor are used for calculating the apparent resistivity of each station. Then the data are entered in to the computer and processed in our case using Excel to produce the apparent resistivity profiling curves.

For the VES survey, the apparent resistivities are calculated and plotted on a log - log paper at the spot for the purpose of controlling the measurements, and then these date are entered into the computer for calculating the resistivity and thickness of each layer, using the RESIST software

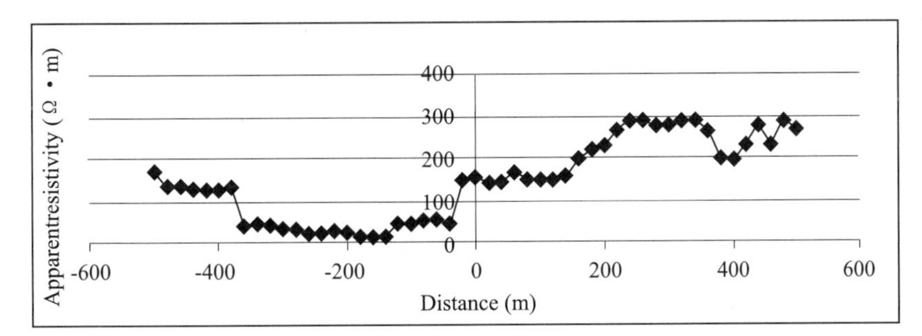

Fig. 6 Apparent resistivity curve along line 400 N(a =40 m)

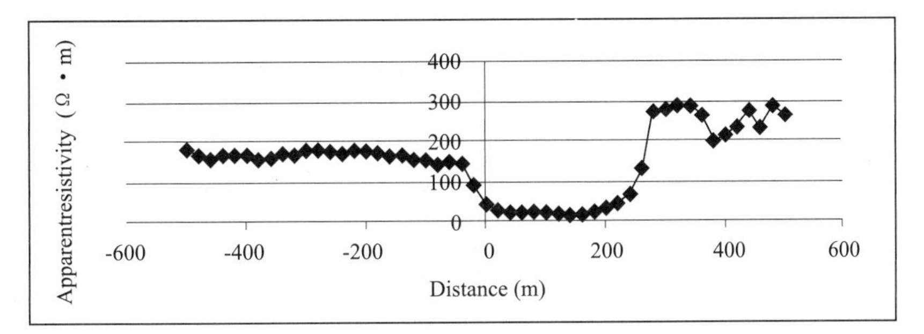

Fig. 7 Apparent resistivity curve along line 00 N(a =40 m)

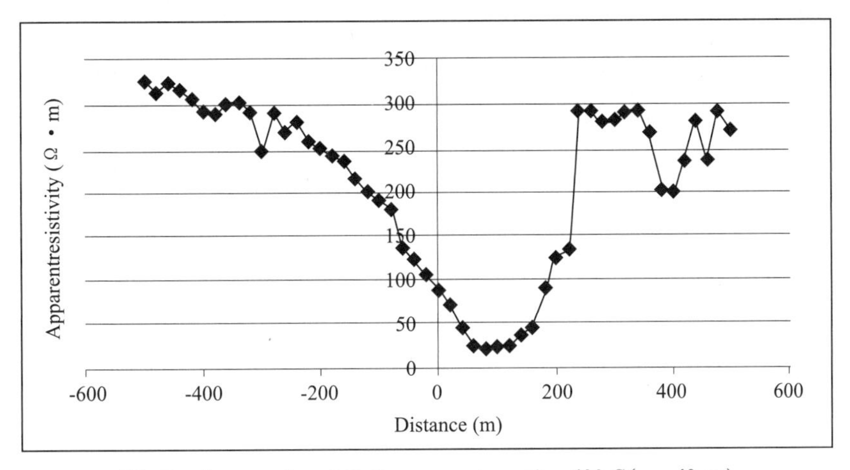

Fig. 8 Apparent resistivity curve along line 400 S(a =40 m)

package. Then, interpretation of the results is made. Based on geophysical survey, three drill sites were recommended and a fourth one was also allocated in case if the three wells give less yields than required. Accordingly, drilling was conducted on the area; the three drill holes intersected the buried channel streams. The aquifers are found to be confined and the wells in the area are yielding 20 L/s to 25 L/s.

The geophysical results are well correlated with drill holes and the local geology of the area. Low values of apparent resistivity correlated with high yielding of the well and high values of resistivity correspond to the sandstone unit. The pattern of the buried channel stream shifts towards south east, as we go from Line 400 N to 400 S and its width also decreases. The geophysical results obtained from the VES, profiling and gravity surveys have showed the effectiveness of the method to map the different lithologies, the buried channel stream and depth of the bed rock.

4 Conclusions and Recommendations

The electrical and gravity surveys carried out in Wukro area has been found to be useful for delineating the buried channel stream. The geophysical results obtained from the VES, profiling and gravity survey have showed the effectiveness of the method to map the different lithologies, the buried channel stream and depth of the bed rock.

Of the several Geophysical Techniques, Electrical resistivity and gravity method which provides valuable information necessary for the selection of appropriated drill sites. The resistivity method was delineated the limestone/shale, sandstone formations and the extent, and pattern of the buried stream channel. The gravity survey shows that the buried channel has a meandering pattern with a high degree of sinuosity. It also exhibits undulations (shallows and deeps) along its course. This is evident from the presence of a series of high amplitude and closed gravity lows within the wide zone as on line 3 N between stations 260 E to 300 E The most favorable locations to site boreholes are within the closed gravity lows which probably correspond to the deeper parts of the buried channel and therefore are likely to contain thick succession of sand and gravel deposits.

Based on the geophysical results, three drill sites were recommended and a fourth one was also allocated in case if the three wells give less yields than required. Accordingly, drilling was conducted on the area, the three drill holes intersected the buried channel streams and Artesian wells were produced during drilling. The aquifers are found to be confined and the three wells in the area are yielding 20 ~ 25 L/s.

This study has provided information on the depth to the ground water and thickness of the auriferous buried stream channel in sedimentary environment, Wukro, Tigray region. This information is going to be relevant to the development of an effective water scheme for the area and possibly beyond other areas with the same environment.

The two methods can be taken for ground water development program in the future to help and locate more appropriate locations for drilling. Because of the presence of similar geological conditions the electrical and gravity method are successfully used to locate buried channel aquifers as potential sources of ground water.

References

Grant F S, West G F. Interpretation Theory in Applied Geophysics. Mc Graw - Hill, New York, 1965:583.

Parasins D S. Principles of Applied Geophysics, Chapman and Hall, London. Geosoft Inc. 1971.

GRIDTREND: A Program to Remove a Polynomial Trend From a Grid[M]. Geosoft Inc., Toronto, Canada, 1991.

Sumner J S. Principles of Resistivity for Geophysical Exploration. 1975.

Discussion on Construction and Management of Dam System Project in Small Watersheds in Sediment - Rich Area of the Loess Plateau

Ma Anli, *Dong Yawei* and *Chen Guirong*

Upper and Middle Yellow River Bureau, YRCC, Xi'an, 710021, China

Abstract: Yellow River Conservancy Commission (YRCC) has approved of dam system construction in 83 small watersheds in the sediment - rich area suffering from serious water loss and soil erosion on the Loess Plateau. In order to deeply summarize the experience of warping dam construction on the Loess Plateau, to popularize their successful mode of control and mechanism of management and operation, and to promote the smooth project construction, 12 small watersheds which have favor conditions for gully control are selected to develop demonstration dam system construction. These 12 selected small watersheds are Jingyanggou in Datong County of Qinghai Province, Chenggouhe in Anding District and Chengxichuan in Huanxian County of Gansu Province, Niejiahe in Xiji County of Ningxia Autonomous Region, Xiheidai in Jungar Banner and Fansiyao in Qingshuihe County of Inner Mongolia, Yuanping in Hengshan County, Mazhuang in Baota District, and Yulingou in Mizhi County of Shaanxi Province, Shuerliang in Hequ County, and Chakou in Yonghe County of Shanxi Province, and Yanwahe in Jiyuan City of Henan Province. Based on the demonstration items such as mode of construction, practice of the 'three systems' reform and the disclosure system for project construction, mode of project operation and management, technical application, and construction of demonstration dam system, the layout optimization and key technology for dam system construction in different types of areas is discussed, and the experience of 'three systems' reform and the disclosure system for warping dam construction is summarized to promote standard and scientific warping dam construction on the Loess Plateau. Meanwhile, the experience of project management and maintenance and reform of property right system is popularized in order to provide a complete set of effective management system for large - scale construction of warping dams on the Loess Plateau.

Key words: dam system, demonstration project, three systems, disclosure, management and protection

1 Project profile

In 2005, Yellow River Conservancy Commission (YRCC) began to construct demonstration dam systems in 12 small watersheds which were selected from 83 small watersheds requiring dam system construction. Through innovating and perfecting the management system of warping dam construction projects, the successful experience of layout, construction management, and operational mechanism, etc. is summarized. Under the principle of 'taking local actual conditions into consideration' and 'abiding by the law of nature and economy, the projects take the demonstration effect of dam system construction in small watersheds depending on application of high and new techniques and continuous s&t innovation of water and soil conservation technologies. The project construction sets up the example for warping dam construction on the Loess Plateau, and accelerates the overall development of local dam construction.

Demonstration of dam system construction and management in small watershed include the following five items, which are, the first, mode of dam system construction in small watershed, the second, practice of the 'three systems' reform and the disclosure system for project construction, the third, mode of project operation and management, the fourth, technical application, and the fifth, construction of demonstration dam system.

The dam system projects of the 12 small watersheds, such as Jingyanggou in Datong County of Qinghai Province, etc. mainly distribute over the sediment – rich area in the middle reaches of the Yellow River and involve with the administrative divisions of 12 counties (banners and cities) in seven provinces (autonomous regions) covering the Loess Plateau. These watersheds mainly distribute over the loess hilly and gullied area and the rocky mountain area. The Yulingou in Mizhi County and Yuanping in Hengshan County of Shaanxi Province, Fansiyao in Qingshuihe County and Xiheidai in Jungar Banner of Inner Mongolia, and Shuerliang in Hequ County of Shanxi Province belong to the first sub – region of the loess hilly and gullied area; Mazhuang in Baota District of Shaanxi Province and Chakou in Yonghe County of Shanxi Province belong to the second sub – region of the loess hilly and gullied area; Niejiahe in Xiji County of Ningxia Autonomous Region belongs to the third sub – region of the loess hilly and gullied area; Jingyanggou in Datong County of Qinghai Province belongs to the fourth sub – region of the loess hilly and gullied area; Chenggouhe in Anding District and Chengxichuan in Huanxian County of Gansu Province belong to the fifth sub – region of the loess hilly and gullied area; and Yanwahe in Jiyuan City of Henan Province belongs to the rocky mountain area.

These 12 small watersheds have the total land area of 970.3 km^2, 914.3 km^2 of which is suffering from water loss and soil erosion, amounting to 94.2% of the total area. Having fragmented land surface, sparse vegetation, and centralized rainstorm, these small watersheds suffer from serious water loss and soil erosion. Water erosion is dominant in the watersheds, with alternation of water erosion and wind erosion. There is a large area suffering from water loss and soil erosion, about 90% of which has the erosion modulus greater than 5,000 $t/(km^2 \cdot a)$. Local water loss and soil erosion is complex and diversified, involving the types of water erosion, wind erosion, and gravity erosion, and the forms of surface erosion, gully erosion, splash erosion, blowing dust, collapse, and sliding, etc. Serious water loss and soil erosion makes the soil fertility becomes low, which seriously affects local agricultural production and induces poverty of local people and deterioration of local ecological environment. Along with the growth of population, unreasonable development and utilization of water and land resources are increasing day by day, which further aggravates water loss and soil erosion.

2 Mode of dam system construction in small watershed

Based on the actual conditions of theses 12 small watersheds, such as area and shape of watershed, landform of gully, water resources conditions, degree of slope treatment, and the status of local economic and social development, dam system construction can be divided into three types according to the layout of dam system. The first is the layout giving prominence to dam land utilization; the second is the layout relying on key dam for comprehensive utilization of water and sediment resources; and the third is the layout laying stress upon fully utilization of water resources.

2.1 The Layout Giving Prominence to Dam Land Utilization

The following 6 small watersheds have the layout giving prominence to dam land utilization, i. e. Yuanping in Hengshan County, Mazhuang in Baota District, and Yulingou in Mizhi County of Shaanxi Province, Shuerliang in Hequ County of Shanxi Province, Xiheidai in Jungar Banner and Fansiyao in Qingshuihe County of Inner Mongolia. These small watersheds has the common characteristic that, after many years' comprehensive control, such a layout of dam system has take shape that large dams serve for controlling flood and sediment, and warping dams in branch gullies retain sediments for agricultural production. However, because of lacking key dams in gullies and imperfect project layout, many dams have already been silted up after many years' service. In such small watersheds dam construction mainly focuses on reinforcing the existing warping dams, constructing auxiliary facilities, and building more key dams to retain sediment for agricultural production.

2.2 The layout relying on key dam for comprehensive utilization of water and sediment resources

The following 4 small watersheds have the layout relying on the key dam for comprehensive utilization of water and sediment resources, i. e. Jingyanggou in Datong County of Qinghai Province, Chenggouhe in Anding District and Chengxichuan in Huanxian County of Gansu Province, and Chakou in Yonghe County of Shanxi Province. These small watersheds have the common characteristic that, after many years' comprehensive control, slope treatment for erosion control has become perfect but gully control remains week. These small watersheds mainly relies on warping dam groups which fail to form integrated control system owing to absence of unified planning, insufficient quantity of key dams, and scattered layout of existing key dams. After many years' service, many medium and small warping dams have been seriously damaged by flood and can no longer effectively control flood and sediment. In these small watersheds, construction of demonstration dam system has the theme of 'constructing key project for gully erosion control'. Key dams are built at proper places in the trunk channels and branch gullies to control flood and sediment. Then, production dams are built downstream the key dams, and small reserviors are built in the places with water resources. Key dams can protect the downstream production dams and small reserviors, and dam land and terrace in valleys can be irrigated by the water retained by key dams. By such a way, high utilization rate and yield rate of dam land are ensured by fully utilizing water and sediment resources.

2.3 The layout laying stress upon fully utilization of water resources

The are 2 small watersheds having the layout laying stress upon fully utilization of water resources, i. e. Yanwahe in Jiyuan City of Henan Province and Niejiahe in Xiji County of Ningxia Autonomous Region. These small watersheds have the common characteristic that there is permanent flow in river channels and local management is mostly featured by reasonable development and utilization of water resources. Construction of dam systems in these watersheds adopts the mode of 'flood retention in the upper reaches and flood storage in the lower reaches depending on combination of dams and reservoirs'. Dams are built in the main gullies of the lower reaches for flood retention, which becomes the water source for irrigation of terrace in valleys and for comprehensive water resources utilization. Sediment arresters are built in the main gullies of the upper reaches to reduce the sediment running into reservoirs, so as to prolong the service life of reserviors. And moreover, more medium and small - size warping dams are built in the branch gullies.

3 Reform of "Three Systems" Disclosure System for Project Construction and Mode of Project Operation and Management

The structure of construction and management is perfected in accordance with the actual conditions of the 12 small watersheds. Responsibility system of project legal person and supervision system of construction are carried out all - around and bidding system is actively put into practice for key dam projects, and moreover, the disclosure system for project construction is put into trail use. Under the principle of 'quality of construction utmost', contract management and fund management are strengthened to ensure the smooth construction of demonstration dam system in the small watersheds.

3.1 Reform of "Three Systems"

3.1.1 Carrying out responsibility system of project legal person

Different forms of responsibility system of project legal person are taken for the dam system construction in the 12 small watersheds. Department of Engineering Construction for Water and Soil Conservation, subordinated to Upper and Middle Reaches Yellow River Bureau, acts as the project legal person of Yuanping Small Watershed and Mazhuang Small Watershed in Shaanxi Province. The project legal persons of the other 10 small watersheds are water conservancy bureaus (stations) of the counties (banners and cities) of the project locations. The project legal persons are in charge of whole project construction, operation and management, and responsible for quality of construction, construction progress, and fund management. In addition, the project legal persons are responsible for the competent authorities of the projects. The measure makes the engineering construction standardized and scientific, and ensures the project construction implemented smoothly.

3.1.2 Carrying out supervision system of construction

A small watershed is taken as a unit, and qualified supervisors are selected for overall control of quality, progress and investment of projects. The key dam construction project in Yanwahe Small Watershed of Henan Province adopts the mode of site supervision for overall control of materials, process and quality of project construction. The key dam construction projects in other 11 small watersheds adopt the mode combining petrol inspection with site supervision. Site supervision is required for construction of the vital parts and concealed works of key dams and medium - size warping dams. Quality and schedule of project construction are ensured by taking different forms of supervision.

3.1.3 Promoting project bidding system

Based on the government incremental investment, the dam system construction projects in these small watersheds adopt the project bidding system by issuing bidding announcements on media such as newspapers and internet. The projects are subject to 'open, fair, and just' bidding after examining the achievement, qualification and credit of the candidates. Construction undertakers are chosen by means of open bid, bid invitation and bid negotiation. In order to ensure the quality of construction and inspire the workers, lifelong responsibility system is applied to the construction undertakers. The 'three - in - one' quality inspection system is adopted for construction quality control by the procedures of primary inspection, re - inspection and final inspection. The measure improved the engineering quality and construction progress.

3.2 Disclosure system for project construction

In order to strengthen public supervision, invite public participation, and propagandize the effect of dam system construction, disclosure system for project construction is adopted for dam system construction in the 12 small watersheds for trail use. Project information is publicized by issuing announcements and setting up billboards, introducing the items such as project name, scale, investment, contractor, construction undertaker, supervisor, fund and labor required, and benefits for local people, etc. Sign boards are built on the top of the key dams after completion of project construction. Application of disclosure system enables local people to know more and well of the projects and reduces the activities violating regulations in some extent.

The Disclosure System was conducive to the cadres and the masses further understand the meaning of the dam construction, to coordinate the contradiction between contradiction occupying the land and people, to promote the reform of property right system of warping dam, to better promote the construction effect and the role model of dam system.

3.3 Mode of project operation and management

Large quantity of warping dams needs to be built in many places on the Loess Plateau, involving with various types of water loss and soil erosion and different status of economic and social development. Many feasible modes of project operation and management have been found during warping dam construction in different places and the positive operation mechanism which integrates 'construction, management, and application' and unifies 'responsibility, rights, and benefits' is set up in accordance with local conditions. Centering on reform of property right system, the dam system construction in the 12 small watersheds follows the principle of 'the contractor shall be responsible for project operation and management and will be the beneficiary'. The right of management and the right of use are transferred by means of handover to villages, contracting, leasing and auction, etc. so as to determine who would be responsible for project operation and management against flood. Meanwhile, in order to ensure the safe operation of projects, communication networks between the management personnel and the three levels of county, township, and village are set up for flood prevention, and the dam system projects are accepted as the county - level key flood - prevention projects and brought under unified control. The measure built the benign operation management mechanism that construction, management and use in combination, and responsibility, power, benefit in unified.

4 Application of new techniques

According to the conditions for dam system construction in the 12 small watersheds and the objectives of demonstration, new techniques and '3S' technology are applied to dam system construction in some small watershed, by such means to make the dam systems scientifically advanced and give full play to project benefits. The application of new technology improves the technical level of monitoring of soil and water conservation in small watershed of Loess plateau.

4.1 New technique for dam construction

4.1.1 Method of dam construction

Sluicing siltation method has been applied to warping dam construction in Niejiahe Small Watershed for a long time, from which outstanding achievement has been made and rich experience of construction has been acquired, involving process of some technical data such as the ridge width, filling rate, and slurry concentration, as well as the mode and materials of water discharge. Many years' operation shows that the sluicing - siltation method overcomes the technical roadblock that it is difficult to treat the joint sections of dam bottom, bank slope and water release structure. Especially, application of this method in the collapsible loess region can make the compactness of dam body much higher than that of the earth dam built by the rolling method. It also avoids the collapsible problems, and makes the dam body joint better with the banks, and moreover, investment can be saved because sluicing siltation method requires less construction machinery.

During the demonstration dam system construction in Niejiahe Small Watershed, this method is applied to body filling of all the key dams and medium and small - size warping dams built in the gullies having permanent flow or assured water sources satisfying the conditions required by sluicing siltation.

4.1.2 Discharge

Discharge rate and discharge performance are the key factors ensuring the safety during sluicing siltation earth dam construction. Various and diversified modes, measures and materials have been adopted for water discharge in the past many years. Presently, porous corrugated pipe is widely accepted and technically mature because of its high discharge performance, simple and easy construction procedures and low cost. It will be used for the construction of sluicing siltation earth dams in the future and will be further optimized to achieve the best performance. However, the consolidation of hydraulic - fill dam body in the later period is not quite clear, so samples are taken from the in - service dam bodies for test. Five dams which have been in service for 1 ~ 5 years are selected for slurry concentration and consolidation test.

4.2 New Technique and New Material Test for Water Release Structures

It is planned to adopt new - type PVC pipe, instead of traditional reinforced concrete pipe, as the water release structure of the medium and small - size warping dams in Niejiahe Small Watershed for test. To use PVC pipe for water release structure has the following advantages.

(1) Good water tight, corrosion resistance, and hydraulic performance, light in weight, easy

transport, low constriction cost, and long service life.

(2) Long pipe course, small quantity of socket pipe joint with gum ring that is easy to be installed, safe and reliable.

(3) While in the same length, PVC pipe is much lighter than reinforced concrete pipe, which makes installation so easy that only hand haulage and adhesive joining are required.

(4) The smooth, acid – alkali – resistant pipe surface needs no anticorrosion treatment. PVC pipes can be directly buried in corrosive soil.

(5) Having smooth inner wall, low roughness coefficient (generally $n = 0.009$), and small flow friction, PVC pipe has the delivery capacity 30% higher than that of concrete pipe and reinforced concrete pipe in the same size. It is characterized by low cost and long service life. Some concrete or reinforced concrete discharge pipes will be replaced by PVC pipes.

Because of light in weight, cost saving, rapid installation and easy factory fabrication, plastic materials can be adopted as water release structure of the medium and small – size warping dams for water and soil conservation. However, there is little successful experience referable at present. Especially, plastic materials have unique characteristics in aspects of covering depth, stress strain, seepage prevention, water sealing, and connection of inlet and outlet. Therefore, it of great significance to investigate the technical indicators of such plastic pipes by making special studies and tests and further know well of the key techniques required by construction and installation.

4.3 Application RS Technology

Remote sensing is an important technical approach for rapidly acquiring the information of land use and water and soil conservation measures in the study area or monitoring area.

Depending on the Spot satellite image of August 2005 of Niejiahe Small Watershed and the QuickBird satellite image of July 2005 of Mazhuang Small Watershed, the status of land use and water and soil conservation of these two small watersheds are monitored by integrating computerized interpretation with on – site verification.

What the Shuierliang Small Watershed adopts is the all – band and multi – band digital image data acquired by PRISM (panchromatic remote – sensing instrument for stereo mapping) of ALOS (advanced land observing satellite) and AVNIR – 2 (advanced visible and near infrared radiometer type – 2) of Japan. The data are of high cost performance ratio.

4.4 Application of GPS and GIS Technology

Based on earlier GPS coordinate data of the dam sites, the remote sensing photomap of warping dam distribution in Shuierliang Small Watershed is made by RS image interpretation with the assistance of DEM information. Using MAPGIS software, accurate positions of dam sites and the names of inhabited places are marked on the photomap processed by color composite and fusion. The photomap clearly shows the distribution of the warping dams built in Shuerliang Small Watershed in recent years, as well as the oriented texture formed by topographic inequality and the relative positions of surface features such as inhabited places and roads.

GPS/GIS – based technology is applied to dynamic monitoring of sediment deposition of warping dams. No. 3 key dam and No. 4 medium dam at Yangsizui are monitored, both of which are in the gully located on the northwest of Chuanjiawa Village in Shuerliang Small Watershed. The gully runs from northeast to southwest,

Following the principles for layout of GPS surveying network and selection of control points, and taking the actual situation of the surveying zone into consideration, 4 control points that are easy to be preserved and accessed are selected after field reconnaissance, and steel nails for GPS surveying are buried as the marks at the control points.

GPS control network is established adopting GPS static surveying. Based on the control points of GPS control network, the elevation of sedimentation level before and after rainfall is measured using RTK technique, and DEM is generated from the measured data by interpolation using ArcGIS. Then the volume of yearly sediment deposition is figured out by calculating the elevation difference before and after sedimentation.

Using ArcGIS, the elevation data of sedimentation level is processed by interpolation to generate the sedimentation level before and after sedimentation. Finally the yearly volume of sediment deposition of No. 3 key dam and No. 4 medium dam at Yangsizui is worked out from elevation conversion.

5 Dam system demonstration

Demonstration dam system in small watersheds is an innovative mode of dam construction. Dam systems are built in small watersheds aiming to control water loss and soil erosion, improve local ecological environment, promote social and economic development, improve the production condition and living condition of the people in the mountain area, and improve the flood control performance of gullies as well. Depending on scientific argumentation and reasonable layout, dam systems are put into service for retaining and reducing sediment, building farmland by silt deposit, controlling flood, and promoting 'returning cultivated lands to pastures and woods'. From the scale point of view, the model construction of dam system of the small watershed is the largest in the management of small watershed in the Loess Plateau, and the project is of great significance.

5.1 Comprehensive control of gully and slope

'Comprehensive control of gully and slope' features largely in dam system construction in Chakou Small Watershed. It integrates control of coterminous gullies with slope treatment for erosion control, aiming to provide all - in - one and systematic control for a gully with a certain size, thereby constructing excellent project for water and soil conservation. 'Comprehensive control of gully and slope' follows the idea of 'integrated construction of farmland, forests, roads, and trenches' and 'comprehensive treatment of plateau surface, slopes, gullies and rivers'. Terrace fields are built on gentle slopes by using machines, and trees and grass are planted on steep slopes. A number of cross dikes are built at gully bottom to make it terraced. Land is built by leveling up river shoals and flood hazard is reduced by dredging the network of floodway channels. In order to build high - quality basic farmlands, dam fields and bench terraces are built together with roads. Depending on scientific planning and reasonable layout and following the principle of 'planting trees, grass, and fruit trees only at proper places', a unique mode of development comes into being, that is, planting walnut trees on hilltop and pine and cypress on hillside, and trapping sediment and storing water at the foot of hill for fish farming and water foul breeding. This is a new mode of development which improves the ecological environment of a gully and changes the appearance of the watershed. Presently, comprehensive control of gullies and slopes in the small watershed is carried out depending on construction of dam system and giving play to the functions of warping dams. Attention is paid to developing bottom lands in gullies by both silting and leveling - up. In summary the detailed measures are controlling flood by dam systems consisting of key dams and medium and small warping dams, leveling up the gullies covered by dam systems, promoting terracing of gullies and slopes, and constructing dam lands with dependable and high yield.

5.2 Connecting up dams, water cellars and ponds

Focusing on warping dam construction and high - efficient utilization of water, 'connecting up dams, water cellars and ponds' is to deliver the water retained by warping dams to the ponds and water cellars in farmlands by pumping stations and small - size pumping facilities, thereby forming an integrated network. In details, it requires to build warping dams in gullies for water impoundment and to build ponds and water cellars in farmlands. The rainwater coming in the flood season (July to September) would be retained in the warping dams built in the gullies. Then the water is delivered to the ponds in villages for further clarification and finally the clarified water is distributed to the water cellar of each household for irrigation in case of water shortage. This is a way which rearranges the time sequence of water utilization, and employs ponds and water cellars for water storage just as what the dams and reserviors do. It reduces evaporation and infiltration and improves the utilization rate of water resources. Moreover, it reduces the workload of the key dams in the flood season.

Niejiahe Small Watershed is the first one that simultaneously develops dam system construction for water resources utilization together with construction of pilot test site. The method of ' connecting up dams, water cellars and ponds' is adopted for comprehensive control of Niejiahe Small Watershed. Presently there are 8 key dams, 3 small water – storage dams, and 1,052 hm^2 of high – standard basic farmland. Besides routine control, Water Conservancy Bureau of Xiji County invested 700,000 yuan for construction of 6 water ponds and 372 water cellars. Water is delivered up to the hills and brings 100 hm^2 of irrigated land under cultivation. The presents of Niejiahe Village used to rely on dry land, and 90% of peasant households were short of food in 8 months of a year. Now the village has 20 hm^2 of irrigated land and the grain yield per capita comes to 570 km. There are 60 hm^2 of alfalfa under cultivation, and every household rears goats. Animal raising contributes 300 yuan to income per capita of peasants this year.

The mode of ' connecting dams, water cellars and ponds' and system construction in Niejiahe Small Watershed show us the ' bridge' between water and soil conservation on the Loess Plateau and income increase of local peasants. It also proves that the dam system project in Niejiahe Small Watershed is an excellent long – standing project with high quality and satisfactory benefits. It is the ' protector' of the watershed and ' granary' of local people. Moreover, it improves local ecological environment and helps local people cast off poverty and becomes well off.

5.3 Joint development of crop farming and fish farming

Monitoring in the period from 2006 to 2010 shows that dam fields have not been built up in some small watersheds because of the small quantity of rainfall, especially in Yanwahe Small Watershed, which is a small watershed located in the mountain area of Jiyuan City, Henan Province. However, an effective mode is developed in this small watershed in accordance with local actual conditions, that is, to effectively utilize and protect water and soil resources, to jointly develop crop farming and fish farming, and to build basic farmland having dependable and high yield, which greatly improves local ecological environment, living and production conditions, and economic benefits of local people.

5.3.1 Yield increase by irrigation

Dam system construction plays a good role in resisting drought and ensuring stable yield in dry years, and transforms some dry lands into irrigated lands. Four key dams, i. e. Fotanggou, Wanggou, Liugou, and Jiangzhuangbeigou, supply irrigation water for 45.6 hm^2 of farmland; three medium warping dams, i. e. Xigou, Zhaolaozhuang – 4, Qiushugou, supply irrigation water for 13.56 hm^2 of farmland; and 10 small – size warping dams, i. e. Yangzhuang – 1, Yangzhuang – 2, Chenhuzhuang – 1, Chenhuzhuang – 3, Houduangou, etc., supply irrigation water for 37.1 hm^2 of farmland. Taking 2010, which was a normal year, as example, investigations show that the yield of the irrigated farmland came to 54.87 kg/hm^2, but the yield of the non – irrigated farmland was only 39.67 kg/hm^2. It can be seen that the yield increase was 15.2 kg/hm^2, with the total increase of 329 t.

5.3.2 Fish farming

In 2010, the total aquiculture area of 5 key dams in Fotanggou, Wanggou, Liugou, Jiangzhuangbeigou, and Zaoshuzhuang and the medium dam in Jiucaipo came to 15.75 hm^2 with mixed culture of carp, silver carp and crucian. The total aquiculture area of 2 medium dams and 2 small – size dams in Xigou and Zhaolaozhuang came to 1.40 hm^2 with culture of carp. Investigations show that the per unit area yield was 1,950 kg/hm^2 and the fishery production came to 33.5 t. The gross output value of fish farming amounted to 334,580 yuan in case of the market price 10 yuan/kg.

Dam system construction in Yanwahe Small Watershed brings about great ecological benefits and economic benefits and inspires the enthusiasm of local people, and the people around, for warping dam construction. 13 small - size warping dams and 5 lift - irrigation facilities have been built up in the watershed by self - raised fund. Moreover, 5 medium dams and 11 small - size dams have also been built in the neighboring small watersheds by self - raised fund.

6 Conclusions

Being an important engineering measure for water and soil conservation and basic farmland construction on the Loess Plateau, warping dam plays a great role in effectively controlling water loss and soil erosion, reducing the sediment emptying into the Yellow River, improving utilization rate of water resources, promoting industrial restructuring in rural areas, and implementing 'returning cultivated lands to pastures and woods' and 'closing off hillside for afforestation'. The ecological environment of the small watersheds is obviously improved after project construction, basically realizing 'keeping the runoff on hills when there is light rain and keeping the runoff in gullies in case of flood'. Such a phenomenon can be seen that dam construction not only keeps gullies under control, but also effectively controls water loss and soil erosion and helps local people cast off poverty and become well off. Warping dam construction is a rapid approach for helping the people in the poor area. It acts as a manmade 'green factory' for the peasants in the poor area. It can be said that a dam means a 'bridge', a 'granary' or an 'oasis'. The successful experience of applying 'three systems' reform, contract system and disclosure system in warping dam construction is summarized based on project construction and management. Dam systems become technically advanced depending on application of new techniques, new materials, and RS, GPS, and GIS technologies. Related policies and successful experience of project management and maintenance and reform of property right system are discussed. A new mechanism is discovered, that is, rolling development with 'clear division of the right and responsibilities for construction and maintenance', and 'proper benefits and clear property rights for developers'. It can be popularized in similar dam system projects to promote warping dam construction in small watersheds and provide a complete set of effective management system for large - scale construction of warping dams on the Loess Plateau.

References

Du Feng, Zhang Shengli. Discuss on Warping Dam Construction in the Hilly and Gullied Area of the Loess Plateau [J]. Science & Technology Information, 2011 (04) :108.

Cao Quanyi, Gao Xiaoping, Ma Chunlin. Practice and Feasibility Analysis of Key Dam Construction in the Third Sub - region of Loess Plateau [J]. Yellow River, 2005,27 (4): 27 - 28.

Bi Cifen, Zheng Xinmin, Li Xin. Effect of Warping Dam Construction on Regulation of Water Environment of the Loess Plateau [J]. Yellow River, 2009, 31 (11): 85 - 86.

Li Min. Effect of Warping Dam on Control of Water Loss and Soil Erosion in the Middle Reaches of the Yellow River [J]. Yellow River, 2003, 25 (12): 25 - 26.

A New Proposal on Control and Water Purification of Yellow River by "UTSURO" (Tidal Pond)

Kazuaki Akai[1], *Kazuo Ashida*[2], *Kenji Sawai*[3], *Kuang Shangfu*[4], *Feng Jinting*[1], *Hiroshi Isshiki*[5], *Shen Jianhua*[6], *Shoichi Takada*[7], *Li Zegang*[1], *Wang Gang*[1] and *Sheng Genming*[8]

1. "UTSURO" Research Group
2. Kyoto Univ. (Prof. Emerius), Kyoto, Japan
3. Setsunan Univ., Osaka, Japan
4. IWHR, Beijing, China
5. Saga Univ., Saga, Japan
6. CTI Engineering Co. Ltd., Tokyo, Japan
7. Ministry of Land, Infrastructure, Transport and Tourism, Japan
8. Shanghai, Investigation, Design & Research Institute, Shanghai, 200000, China

Abstract: In order to meet the three major themes and four targets on Yellow River by Chinese government, we propose a new technology for flood control and water purification of Yellow River using "UTSURO". Originally, "UTSURO" comes from Chinese philosophy. It means vanity. A water region enclosed by an embankment in a sea with tide is called "UTSURO". If a part of the embankment is opened, a strong tidal flow may be generated every time when the tide changes. The tidal current digs the riverbed at the river mouth and makes the riverbed lower than the sea water level. It increases the river slope of the river mouth. Hence, the flow energy due to the gravity lowers the riverbed at the upper reaches, until the stability slope is reached. We also make a proposal on water purification of Yellow River using "UTSURO".

Key words: Yellow River, tide, UTSURO, riverbed digging, flood control

1 Introduction

"UTSURO" was invented in early 1980s by K. Akai, the first author of the present paper. Many studies have been conducted, and many proposals have been made between Japan and China after 1985. In the present paper, we are going to make a new proposal on flood control and water purification of Yellow River using "UTSURO".

Yellow River discharges a vast amount of earth and sand every year. The earth and sand accumulate at the river mouth, make the river length longer and reduce the river slope. Then, it makes the tractive force weaker, the riverbed higher and the discharge capacity of flood smaller. It finally causes dike break.

Chinese government has instructed solution of three major themes on Yellow River: ① Prevention of flood; ② Solution of shortage of water resource; ③ Improvement of ecosystem deterioration. And it is promoting four targets of the flood control: ① The embankment does not break; ② The riverbed does not become higher; ③ The shortage or cutting - off of the river stream does not occur; ④ The water quality does not deteriorate beyond a standard.

In order to meet the three themes and four targets of Chinese government, we propose a technology for flood control and water purification of Yellow River using "UTSURO".

2 A technology to lower riverbed by increasing river slope and stabilize river stream

2.1 A fundamental idea

The tractive force of a river is given by

$$\tau = \gamma RI \tag{1}$$

where, τ, γ, R and I are the tractive force, specific gravity of water, hydraulic mean depth and water surface gradient, respectively. Hence, in order to strengthen the tractive force τ, we should make I bigger and R deeper.

So, in order to maintain the depth of the river stream, we utilize the flow energy due to the gravity and the tidal energy due to the tide at the upper and lower reaches of Yellow River, respectively.

2.2 A technology to utilize tidal energy

2.2.1 A technology to generate tidal current

As a method to convert tidal energy into tidal current, Akai invented "a tidal current generator using UTSURO" and filed a patent in 1987 (Japanese patent: 2726817). He also invented "a system to control and utilize river water by a tidal current generator using UTSURO" and filed patents in 2003 (International patent: WO2004/090235AI; Chinese patent: 608805). This technology is used to reduce the burden of the gravitational energy. And the tidal energy is used for effective use of mud possible by separating mud and water from the vast amount of earth and sand discharged from Yellow River.

Originally, "UTSURO" comes from Chinese philosophy. It means vanity. A water region enclosed by an embankment in a sea with tide is called "UTSURO". If a part of the embankment is opened, a strong tidal current may be generated every time when the tide changes. In the present paper, we call this "UTSURO A".

The tidal current digs the riverbed at the river mouth and makes the riverbed lower than the sea water level (Fig. 1). It increases the riverbed slope. Hence, the flow energy due to the gravity lowers the riverbed at the upper reaches, until the stability slope is reached.

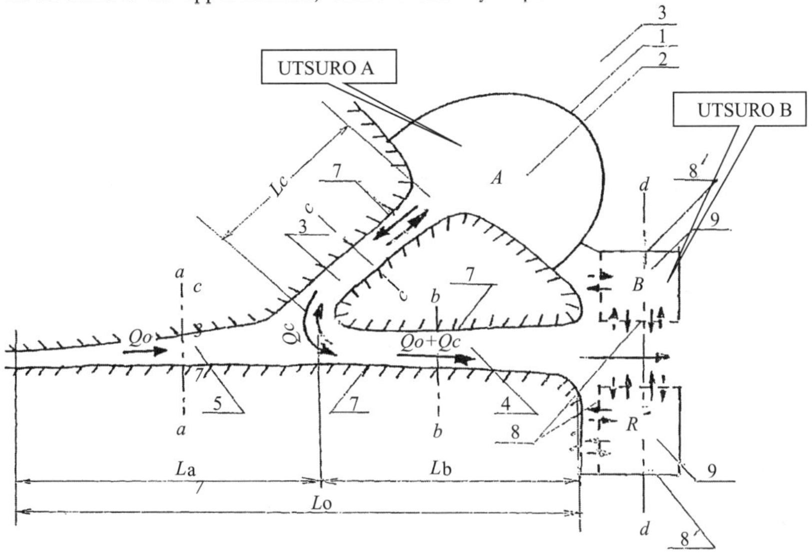

Fig. 1 Tidal current generator by an UTSURO A

2.2.2 A technology to purify water and reclaim land

A water region enclosed by an embankment for water purification is called "UTSURO B" in the present paper. An UTSURO B is used to purify water and to reclaim land or tideland by using mud in muddy water.

"UTSURO B" in the present paper was invented by Akai in 1979 and a patent was filed for "a water purifying system using UTSURO" (Japanese patent: 1806954). A paper on land reclamation using an UTSURO B was read at the Oceanographic Society of Japan in 1987. In 1981, Akai proposed the technology using "UTSURO B" for the land reclamation of Shanghai No. 2 International Airport. The land reclamation was completed as the second term construction recently.

3 Examples of dredging for maintenance of riverbed using tidal current

There are many phenomena in nature which are caused by a principle of tidal current generator of an UTSURO A. Very strong tidal currents at Kitan and Akashi straits in Osaka bay and very strong vortices at Naruto strait are examples of these phenomena. In the examples, Seto inland sea corresponds to an UTSURO A.

In 1989, Prof. Emeritus K. Ashida of Kyoto university, the second author of the present paper, said: "There are two similar rivers in Ibaragi Pref., Japan flowing into Pacific ocean, namely Kuji and Naka rivers. Kuji river suffers river mouth clogging. However, clogging has never occurred in Naka river. In case of Naka river, a pond called Hinuma is connected to Naka river via Hinuma river. Hinuma pond may play a role of an UTSURO A by Akai and the tidal current induced by Hinuma pond at low tide may dig the riverbed of Naka river mouth and prevent clogging. In 1993, Waki et al. reported the influence of Hinuma pond on the river mouth of Naka river. They clarified that the maximum water depth at the confluence of Naka and Hinuma rivers reaches 18 m.

There is a Dingshan lake with area of 62 km^2 at the upper reaches of Huangpu river in China. The tidal difference of Dingshan lake is 70 cm. Although the water surface gradient is 0%, 50×10^6 t of water flows in and out Huangpu river at every tide and generates a strong tidal current more than 2,000 t/s in average. The tidal current digs the riverbed and maintains the water depth of more than 10m. Hence, Dingshan lake is a huge UTBSURO A or tidal current generator. A theoretical study was conducted by Isshiki et al., recently.

We apply the above - mentioned UTSURO technologies to Yellow River and try to contribute to the solution of the three themes and four target aimed by Chinese government.

4 A concrete proposal

In order to develop the function of UTSURO fully, the following fundamental items should be considered.

Fundamental items to realize flood control of Yellow River and dredging for maintenance of the river mouth by applying "UTSURO" technologies

①No flowing of mud water from Yellow River to the UTSURO A If the mud water of Yellow River flows into the UTSURO A, the function of the UTSURO A is deteriorated. In order to prevent this situation, we must make the length of the tide conducting waterway long enough and satisfy $V_t < L$, where L is the length of the waterway and V and t are the flow velocity and duration of the tidal current.

②Discharge capacity at the river mouth

We make the size of the UTSURO big enough and generate the tidal current corresponding to the flood discharge always. The discharge capacity is then secured.

③Building a big port with big water depth at the river mouth

We should use the tide conducting waterway with length 100 km as a distribution basis for such as oil.

④Utilization of a vast amount of discharged earth and sand

We should make muddy water from a vast amount of earth and sand dug by tidal current and convey the muddy water to the river mouth. We arrange UTSURO B's according to a plan and conduct the muddy water into UTSURO B's. The muddy water separates into water and mud in UTSRO B's. The mud is used to reclaim land and tidal flat in UTSURO B's. The mud is also used to construct and reinforce an UTSURO A.

⑤Utilization of the above – mentioned system as a huge water purification system

An UTSURO B purifies water by sedimentation in the UTSURO B and oxidation through pebbles in the embankment of the UTSURO B. For the moment, we expect the water purification in an UTSURO B. In future, advanced ability of water purification and sunlight – hole effect in a closed water region of an UTSURO A would be utilized as leisure and fishery resources.

4.1 A proposal in 1990

In 1990, Akai made a design of Yellow River flood control system using UTSURO's as shown in Fig. 2. An old trail of Yellow River was used as a tide conducting waterway. An UTSURO A and UTSURO B were placed at Bohai bay and the river mouth at the time, respectively.

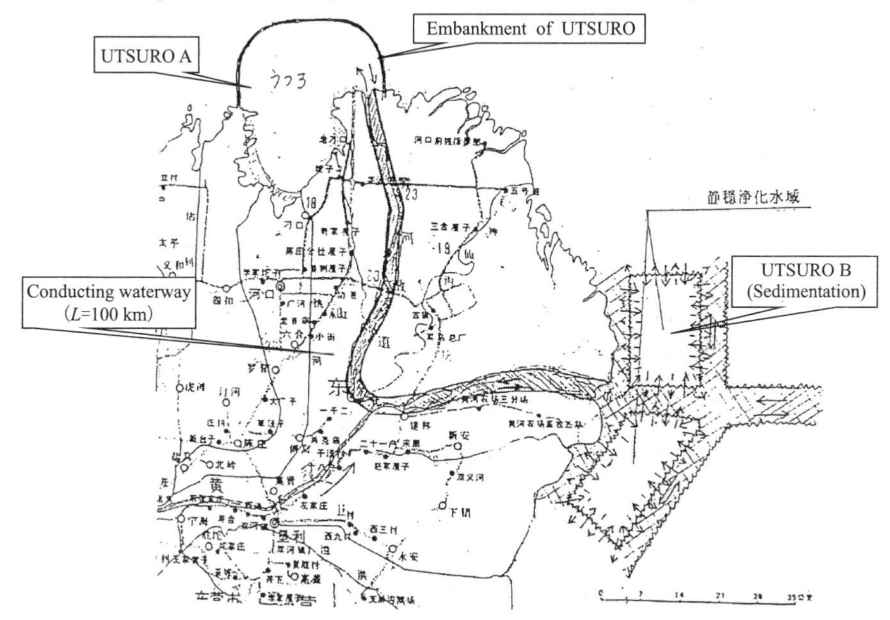

Tidal difference: 2 m, Area of UTSURO: 400 km^2, Length of waterway: 100 km, Cross section of waterway (breadth: 3 km, water depth: 10 m)

Fig. 2 Flood control and dredging for maintenance plan proposed in 1990 for Yellow River applying tidal current generator UTSURO A

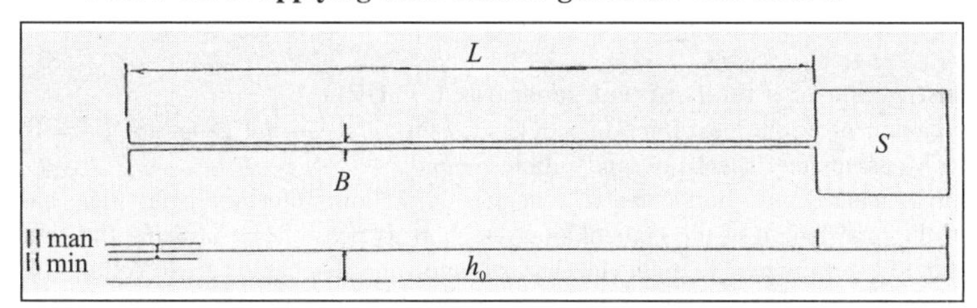

Fig. 3 Simulation model for tidal current generation

A simple model was made and a simulation was conducted. When the tidal difference in the UTSURO A was 40 cm, the flow velocity change was ± 50 cm/s and the flux change was ±200,000 m^3/s.

4.2 A new proposal of a tidal current generator (UTSURO A)

An UTSURO A was planned in Bohai bay in 1990. However, the water depth of Bohai bay is deep, namely 15 ~ 20 m, and waves and coastal current are also big. So, we now consider to change the location to Caizhou bay.

4.2.1 Location of a tidal current generator or UTSURO A

In addition to the above - mentioned fundamental items for the realization of the UTSURO system, we must satisfy the following conditions for the location of an UTSURO A used for the tidal current generator.

Conditions for the location of an UTSURO A

① We must select a place at the river mouth of Yellow River where the water depth is comparatively small and we can build the embankment of the UTSURO A easily.

② The tidal current and wave are small.

③ The water region is undeveloped now.

④ Materials for the construction of the UTSURO A are obtained easily.

⑤ A vast amount of earth and sand discharged from Yellow River can be used effectively in future.

The maximum water depth at the river mouth of Yellow River is about 20 m in Bohai bay which locates in the north, but less than 11m in Caizhou bay which locates in the east. The wave height due to north wind is also considered big in the north.

The recent geographical features tells us that Caizhou bay is more appropriate for the location of the UTSURO A.

4.2.2 Scale of an UTSURO A or the tidal current generator

Since the tide conducting waterway constitutes a part of the river mouth of Yellow River, we must generate tidal current more than the flood discharge of Yellow River.

The tidal difference at sea: 2.0 m

Flood discharge of Yellow River: 22,000 m^3/s

Scale of the UTSURO A: 1,200 km^2

Yellow River discharges a vast amount of earth and sand every year. We must construct an UTSURO A for the tidal current generation which is safe and fitted to the geographical features. We used Google aerial photographs to investigate it.

Flood control system using the UTSURO A is one of the methods to solve the three major themes and to realize the four targets of Yellow River flood control. The location of the UTSURO A or the tidal current generator is the most important in the technology of "Flood control and dredging for maintenance of the river mouth of Yellow River using a tidal current generator UTSURO A".

4.2.3 Construction of a tidal current generator UTSURO A

The embankment is the most important device of the tidal current generator UTSURO A. It is firmly closed by using steel sheet pile and rubble mound.

The embankment should be located at a region of shallow water and mild wave and wind. It should make the total length of the embankment as short as possible by utilizing the shore.

Furthermore, we utilize a vast amount of earth and sand discharged from Yellow River to reinforce the embankment. The embankment is reinforced by mud sedimentation with width of 3 ~ 5 km. After the completion, we should place a tide embankment for safety (Fig. 4).

A tidal current generator UTSURO A closes a water region with tide. In order to deal with flood discharge of Yellow River (22,000 m^3/s), a tidal current with flux of more than the flood discharge must be generated.

So, it is necessary to construct an UTSURO A with area of more than 1,200 km^2. The closure

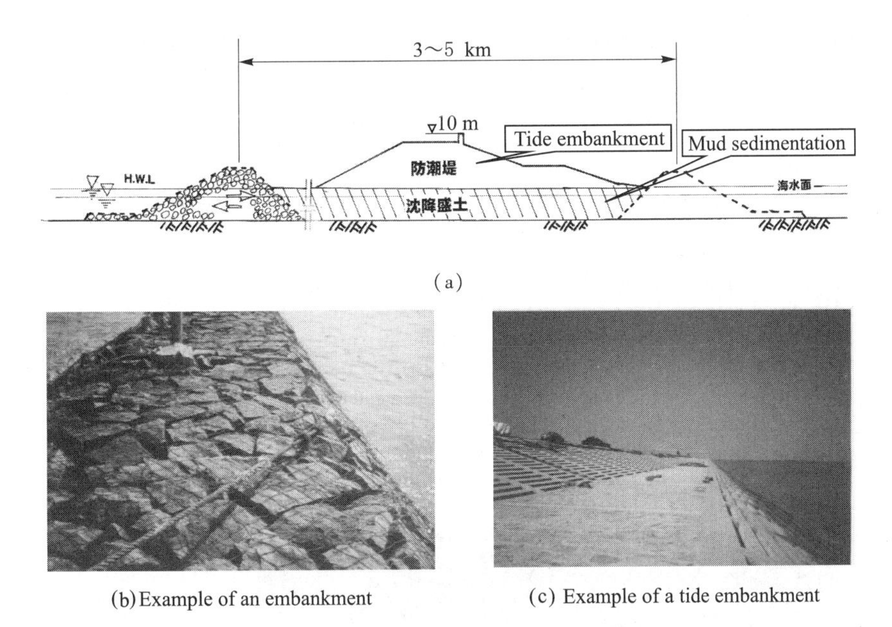

(a)

(b) Example of an embankment

(c) Example of a tide embankment

Fig. 4 Embankment of a tidal current generator UTSURO A

requires a very difficult construction. An UTSURO A consists of Rubble mound bank, Steel sheet pile bank, Separation levee and tide embankment is.

4.2.4 Structure of a tidal current generator UTSURO A

Structure of a new tidal current generator UTSURO A is shown in Fig. 5. Fig. 5(a) shows the present state of the river mouth of Yellow River. Fig. 5(b) shows an embankment of UTSURO A. The embankment of UTSURO A is built by sedimentation of earth and sand discharged from Yellow River in the past and present. First, we build closing levees on both sides of the embankment. The closing levees may be built with steel sheet pile in deep places and with rubble in shallow places.

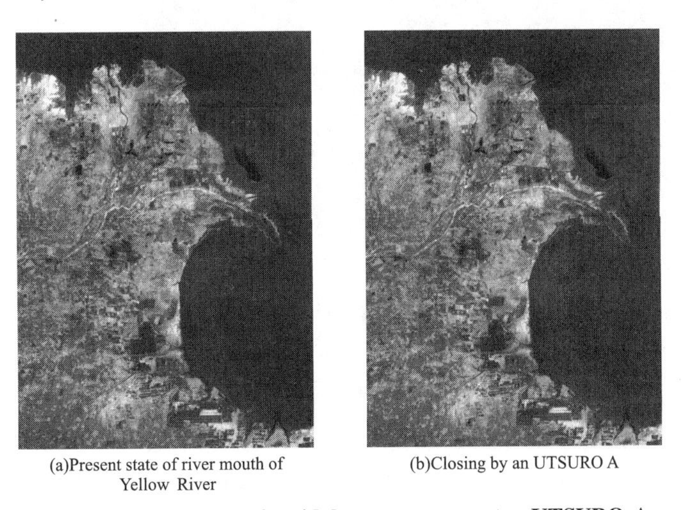

(a) Present state of river mouth of Yellow River

(b) Closing by an UTSURO A

Fig. 5 Construction of a tidal current generator UTSURO A

Furthermore, separation levees are built across the closing levees. The separation levees prevent water flow along the closing levees. Muddy water dug outside the closing levees by tidal current is brought to the water region where the embankment of UTSURO is constructed. The digging takes place in a waterway between one of the levees and land, where a rather strong tidal current is induced. The earth and sand is conveyed along the waterway by tidal current and into water region planned for the embankment of the UTSURO at high tide. There is no significant water flows there because of the separation levees. So, the earth and sand in muddy water sediment there, and the embankment of the UTSURO is constructed gradually.

The scale of UTSURO A is summarized as follows:

①Area of tidal current generator or UTSURO A: 1,200 km^2

②Length of Yellow River governed by tidal energy: about 50 km

③Length of pollution prevention waterway: about 80 km

④Area of water purifier by sedimentation or UTSURO B: about 500 km^2

⑤Advanced water purification system using closed water region: 0.5B. m^3/day

⑥Storage capacity of brackish water: 5B. m^3

⑦Huge waterway or harbor, landing place and leisure facility: about 270 km

Fig. 6 shows a conception of the finished system for the control and water purification of Yellow River. The UTSURO A is installed at Caizhou bay.

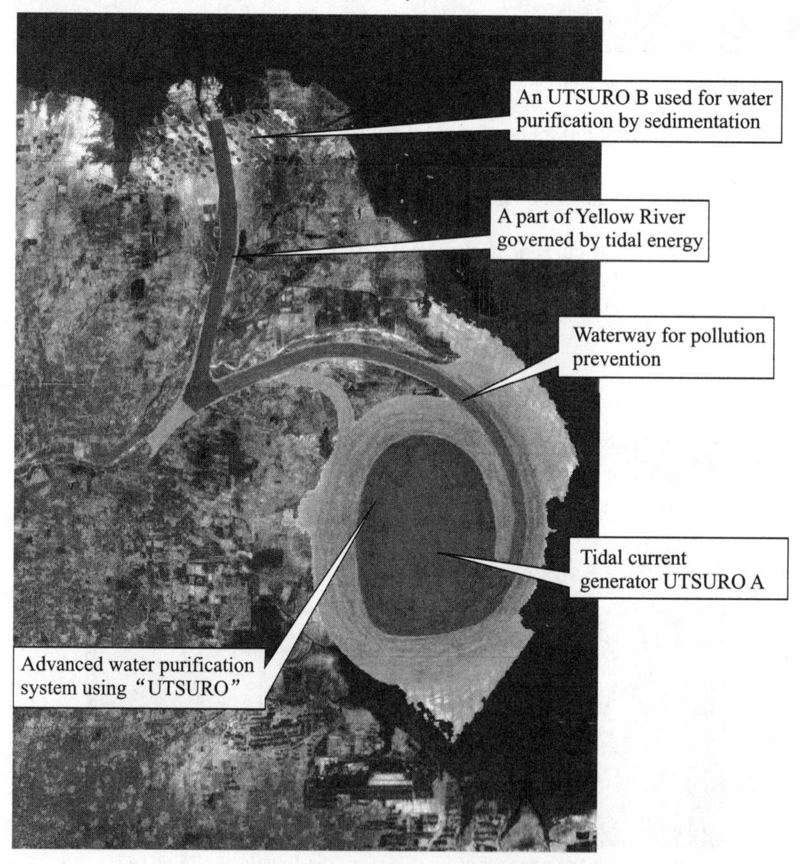

Fig. 6 A new proposal for a tidal current generator by "UTSURO"

The above-mentioned digging makes the water depth of the waterway deeper. Namely, the

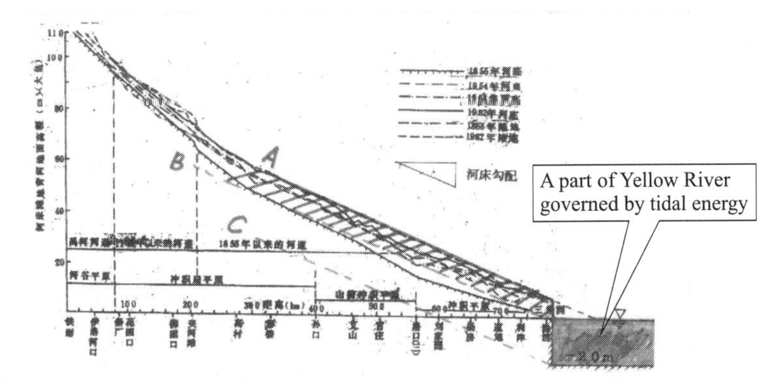

Fig. 7 Riverbed digging by an UTSURO A

riverbed of the main stream of Yellow River becomes lower, and the length of Yellow River governed by gravity is shortened. Then, the riverbed digging at the upper reaches is induced, until the stability slope is attained as shown in Fig. 7. Hence, the riverbed at the upper reaches is lowered.

5 Conclusions

We discussed flood control and water purification of Yellow River using "UTSURO". We showed the possibility to shorten the length of the river governed by gravity, to lower the riverbed by strengthening the tractive force, to control flood and to stabilize the river stream by utilizing tidal energy, a kind of natural cnergy. Furthermore, we can convey a vast amount of discharged earth and sand safely to the river mouth in the form of muddy water, sediment the muddy water without any running cost as planned and utilize the vast amount of mud as the construction material as planned. The mud will be utilized effectively as a resource and material for reclamation of land and tidal flat as planned.

Furthermore, in order to stabilize the river stream at the upper reaches of the River, the breadth of the stream should be narrowed, and the Hydraulic mean depth R should be increased. We should investigate both a technology to prevent the cutting – off of the river stream and a system for continuous water purification and riverbed digging at the same time.

Lastly, the authors thank sincerely those who have helped the authors in developing the technology of "UTSURO".

References

K Akai, K Ashida, K Sawai, et al. An Idea of Estuary Sedimentation Control and Land Reclamation by Tidal Dynamics ("Marine Hollow") — Considering Estuaries of Yangtze River, Yellow River and Haihe river[J]. Advances in Hydro – Science and Engineering, 1995(3).

K Akai (translated into Chinese by Y. Feng and J. Feng, proofread by H. Wang). Advanced Purification System Using UTSURO [M]. Zhengzhou: Yellow River Conservancy Press, 2009.

K. Akai (translated into Chinese by Y. Feng and J. Feng). A Proposal to Control of Huang River using UTSURO[M]. Zhengzhou: Yellow River Conservancy Press, 2010.

S Ueda, R Tsuda, K Akai, et al. Idea of Ocean Development in China – Land Reclamation –, Spring Meeting of Oceanographic Society of Japan, (1987) in Japanese.

M Waki, M Fujisiro, T Kawamura, et al. A Study on Tidal Phenomenon of the Nakagawa River, 1993 PACON China Symposium(June, 1993).

H Isshiki, K Sawai, Y Ogawa, et al. Flow Analysis in Tidal Channel Connected with "UTSURO" (Tidal Reservoir). The 5th International Yellow River Forum, 2012.

Research of Reach in Xiaolangdi Reservoir Area for De – siltation Construction with Pulsed Jet

Wang Puqing, ***Xie Yibing*** and ***Chen Hongwei***

Yellow River Institute of Hydraulic Research, Zhengzhou, 450003, China

Abstract: Xiaolangdi Reservoir is a gorge type reservoir in mountainous area with a high gradient of reservoir bottom and numerous branches along both sides, whose reservoir capacity of main stream accounts for about 60% of the total. After the reservoir ran into impoundment and operation, the high sediments flux into the reservoir from upstream results in sand bar at the terminal of backwater following their siltation. The location of the silted delta has prominent influence on the application of the reservoir capacity and the formation of tributary sand barrier bar. In the perspective of water transfer and sediments transfer, it is expected that the sediments into the reservoir can be deposited in front of the dam gates, which is in favor of the removal of sediments out of the reservoir and its adjustment, control and management. According to the siltation pattern and the bed composition of the reservoir area in post – flood season of 2009, two reaches are planned to be selected initially as test reaches: ①a reach for routine work 20 ~ 60 km far away from dam area; ② a reach for water transfer and sediments transfer work 60 ~ 100 km from the dam area.
Key words: pulsed jet, density current, reach for de – siltation, selection

1 Introduction of the topic

In order to sustainably maintain the effective reservoir capacity of Xiaolangdi Reservoir, it is of vital necessity to take manpower measures based on its siltation characteristics to create manpower density currents in its reservoir area and to remove the silted sediments upstream the upper gradient section of its delta sediments. And also it is of much importance to drain sediments as much as possible out of the reservoir or transfer as much as possible to before – the – dam area and drain them out of the reservoir during the annual coupled water and sediments regulation operation before the flood season. Additionally, so is it very important to further strengthen the acquaintance on the motion characteristics of Yellow River water and sediments, so as to provide guidance for the Yellow River regulation afterwards. The selection of reach for de – siltation construction is a key point in the formulation of construction scheme for pulsed jet de – siltation as well as the focus of this research, thus in need of a systematical research.

2 Principles of sediments drainage by creating density current with pulsed jet

2.1 Basic characteristics of pulsed jet

As a matter of fact, pulsed jet is a kind of intermittent pressurized high – speed current, and when the pulsed jet is impacting the medium, water hammer phenomenon will occur; based on momentum theorem, the reduction of liquid velocity and momentum will inevitably lead to the increase of pressure, and the corresponding pressure increment is just the water hammer pressure p_H, which is a kind of pressure step and transmits in the liquid medium at velocity of sound a ; for open water conservancy impacting system, water hammer pressure is generated when the jet impacts the target surface.

$$p_H = \frac{\rho a u}{1 + (\rho/\rho_s a_s)} \tag{1}$$

In the equation, u stands for velocity of jet, ρ stands for liquid density, a stands for velocity of sound in water, ρ_s stands for density of target and as stands for velocity of sound in target material.

Research findings of numerous experiments have proved that pulsed jet has bigger impact on the target compared with usual currents, thus providing an effective and feasible technological approach for people to break materials of high hardness and brittleness.

2.2 Basic scheme of sediment drainage out of reservoir

In a reservoir of a sandy river, pulsed jet devices (Fig. 1) are utilized to extract the clean water in the reservoir through the circulatory system and to generate high – pressure pulsed jet, which will impact the siltation bed surface on the reservoir bottom, and the sediments that is impacted, raised and suspended have certain momentum. However, this is not enough, passages or channels that can push the sediments towards before – the – dam area are yet to be provided. Since the impact force of the high – pressure pulsed jet is extremely large, it is assumed that mechanical power measures can be taken advantage of to form continuous and stable "deep grooves" on the proposed construction lines, in which way the muddy water caused by the pulsed jet can be provided with underwater movement passage of a certain depth and width. Manpower density current, after generated, can drain the sediments along the "passage for sediment drainage" for density current generated previously to the sediment deposition area before the dam, and through continuous operation, "move" of sediments in local reservoir area can be realized. This is the basic scheme of creating manpower density current with pulsed jet.

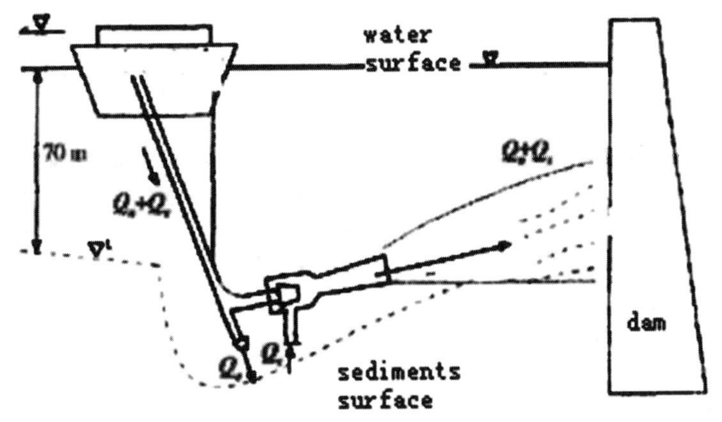

Fig. 1 Schema on working principle of pulsed jet device

3 Siltation & distribution characteristics of sediments in reservoir area

3.1 Longitudinal variation

Different reaches represent different siltation characteristics. To be specific, the river bed of reservoir area that is 55 km upstream or closer to the dam (basically lying on section HH35) continues to rise due to siltation. The backwater fluctuation area which is 55 ~ 110 km far away from the dam features close relation between the erosion & siltation variation and the operation mode of the reservoir, while the reaches more than 110 km far away from the dam have been basically balanced in erosion and siltation and been minor in variation of sectional pattern, thus belonging to relatively stable reaches.

Effected by its rising water level and the increase in amount of sediments flux into it, the siltation of the main stream of Xiaolangdi Reservoir, since the Reservoir being closed and brought into impoundment and operation in October, 1999 (till November, 2000), has represented a pattern of delta, and the vertex of the delta is about 70km far away from the dam. Since then, the pattern of the delta and the location of the vertex have varied and moved with the operation status of the reservoir's water level, and the overall tendency shows gradual downstream movement. Fig. 2 shows the siltation pattern of the main stream over the years. It can be seen that delta surface has always witnessed a substantial rise due to siltation. Compared with the post - flood season of 2002, the period from May to October of 2003 has witnessed a most substantial rise due to siltation in the position HH41, where the thalweg point got a rise of more than 40 m. At the same time, the change of erosion and siltation of the reaches in fluctuating backwater area is relatively complicated, which have undergone a continuous fall due to erosion in 2003, 2004 and 2005, and a minor overall change in other years of both erosion and siltation.

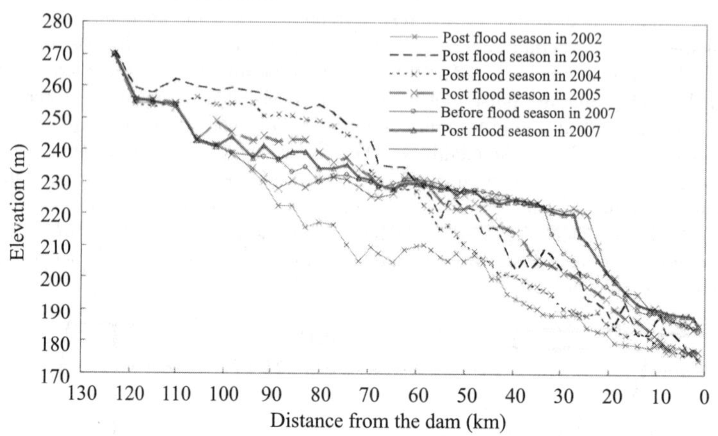

Fig. 2 Overlapping longitudinal sections of sediments siltation in reservoir area

The downward extension of the siltation delta surface and the siltation rise of the bottom in the reservoir area can exert a retardation effect on the continuous operation of the density current, thus directly influencing the sediments drainage out of the reservoir.

3.2 Lateral variation

The sectional erosion and siltation of the reservoir area varies with the sectional pattern and the hydraulic factors. The reservoir section (0 ~ 55 km) before the dam and in the density current siltation is located in the density current siltation reach, which shows an overall tendency of parallel uplift. For instance, it is shown in Fig. 3 Overlapping erosion and siltation variation of section HH1 that the period of the largest siltation amplitude is during the early stage of impoundment, namely from May, 2000 to October, 2000, during which the largest siltation thickness is nearly 15 m, and that till the post flood season of 2007, the river bed surface got an accumulated rise of 37.5 m.

In the fluctuating backwater area (55 ~ 110 km far away from the dam), the location of the main stream is unstable due to the fluctuation of the water level. Section HH35 (58 km far away from the dam) is located in the widest part of the reservoir area. After Xiaolangdi Reservoir was brought into impoundment, the accumulated uplift thickness of the siltation in post flood season of 2004 was 30 m, and ever since 2005, the bed surface has been basically in a state of balance (Fig. 4). The basic characteristics of the horizontal variation of sections in this area are siltation in early stage (2000 ~ 2004) and tendency of balance between erosion and siltation in later stage.

The valley in the upper section of the reservoir area is narrow and the amplitude adjustment of river bed erosion and siltation is relatively large. Section HH55 (100 km far away from the dam) is located in the reservoir's terminal of backwater and has a narrow and deep sectional pattern, and by contrast, its amplitude of variation in erosion and siltation is smaller. In other years, section HH33 witnessed minor erosion or balance between erosion and siltation. Characteristics of vertical variation are represented in the horizontal fluctuation of the sectional talweg point (as shown in Fig. 5), which got a relative large amplitude in both 2004 and 2008.

In a word, after Xiaolangdi Reservoir ran into impoundment and operation, the cross section of the reservoir area has shown an overall tendency of gradual siltation, which is specifically embodied in the parallel uplift of near dam area, the earlier – stage siltation in intermediate section of reservoir area and the later – stage equilibrium of crosion and siltation, as well as the horizontal swing of the sectional thalweg point in the upper section of the reservoir area.

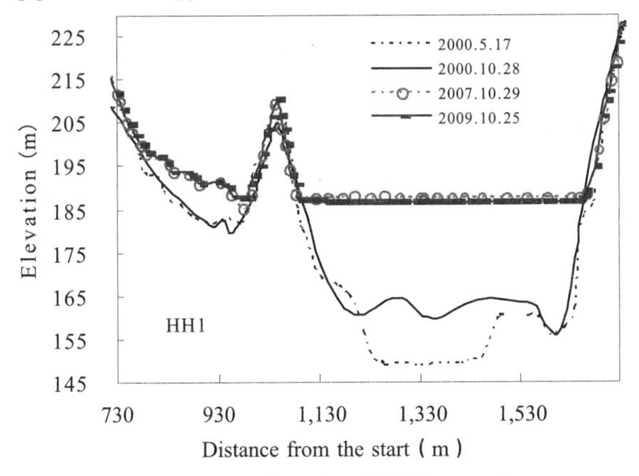

Fig. 3 Variation of section HH01 over the years

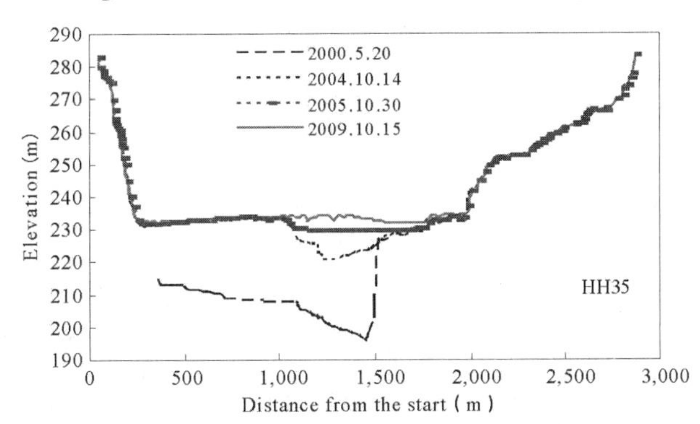

Fig. 4 Variation of section HH35 over the years

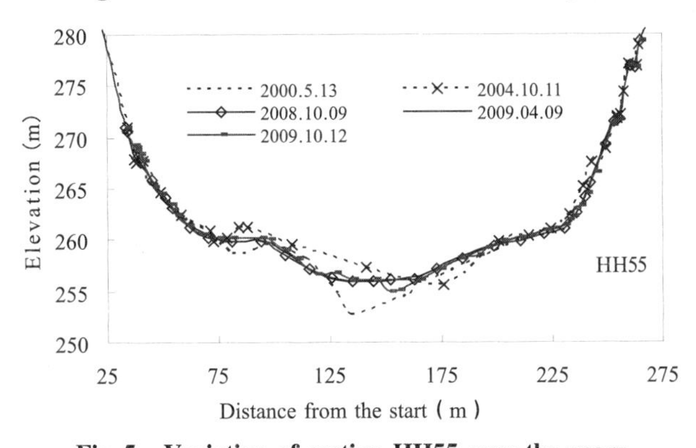

Fig. 5 Variation of section HH55 over the years

3.3 Composition variation of silted sediments

After the reservoir ran into impoundment and operation, the reservoir area 40 km far away from Xiaolangdi dam has got a relative large depth of water and is mainly fine sediments siltation areas; the tail of reservoir 100 km far away from the dam has been generally small in depth of water and is mainly coarse sediments area. Between the two areas is the transition area of fine sediments and coarse sediments. In the early stage of impoundment, the river bed composition of the reservoir generally tends to be coarse, the median diameter of river bed sediments in HH18 ~ HH28 reach ranges from 0.1 mm to 0.15 mm, and that of other reaches is above 0.04 mm (Fig. 6).

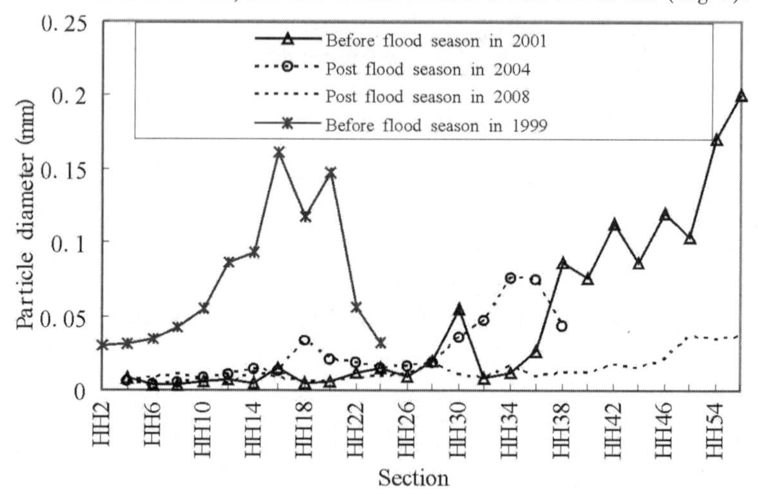

Fig. 6 Variation of average median particle diameter of bed sediments along the reservoir in Xiaolangdi Reservoir area

Between section HH54 and HH42 (115.13 ~ 74.38 km far away from the dam), the median particle diameter d50 of the sediments reduces suddenly from 0.195 mm (HH54) to 0.0186 mm (HH42) along the reservoir, and the weight percentage of fine sediments ($d < 0.025$ mm) also increases suddenly from 11.7% to 67.6%. In the siltation reach before the dam and downstream section HH36 (about 60 km far away from the dam), the median particle diameter of the sediments basically remains below 0.008 mm, and the sediments percentage of fine particles ($d < 0.025$ mm) is unexceptionally above 90%.

4 Selection of reach for de – siltation

4.1 Principles of selection

The final purpose of creating manpower density current with pulsed jet is to create manpower density current in Xiaolangdi Reservoir with the water impounded to remove the silted sediments upstream the upper gradient section of its delta sediments, to transfer them to the funnel area before the dam by taking advantage of the reservoir density current's motion features and to drain as many sediments as possible or directly drain the sediments out of the reservoir. Therefore, the selection of the reach shall conform to the following conditions apart from not effecting flood prevention, water supply for irrigation, power generation, safe production, etc.: longitudinal raised reaches that resist the motion of water and sediments; edge reaches out of the influence of impact during water transfer and sediments transfer processes; reaches in need of de – siltation where manpower density current can be easily generated.

4.2 Selection of reach for de – siltation

Xiaolangdi Reservoir is a gorge type reservoir in mountainous area with a high gradient of reservoir bottom and numerous branches along both sides, whose reservoir capacity of main stream accounts for about 60% of the total. After the reservoir ran into impoundment and operation, the sediments flux into the reservoir from upstream shape into silted delta at the terminal of backwater following their siltation. The location of the siltation delta has prominent influence on the application of the reservoir capacity and the formation of tributary sand barriers. From the point of water transfer and sediments transfer, it is expected that the sediments into the reservoir can be deposited on the near – dam reaches, which is in favor of the removal of sediments out of the reservoir and its adjustment, control and management.

In a certain sense, the development process of the reservoir area delta has indicated the evolution of the reservoir siltation. According to the siltation pattern and the bed composition of the reservoir in post – flood season of 2009, two reaches are planned to be selected initially as test reaches:

(1) A reach for routine work 20 ~ 60 km far away from dam area.

This reach is the sediment siltation delta reach of the reservoir area, and the location of the delta vertex has been undergoing continuous movement towards before – the – dam area since the reservoir was brought into impoundment and operation. The longitudinal section during 2000 ~ 2008 is shown in Fig. 1. Despite the annual water transfer and sediments transfer that leads to its fall due to erosion, its delta surface's overall tendency of developing towards before the dam hasn't shown any substantial changes. If this reach is selected for operation, it will help to improve the longitudinal section pattern of the reservoir area, to reduce the running resistance of the density current, to increase the sediments transferring capacity of the channel and to enlarge the sediment drainage amount during the process of water transfer and sediments transfer.

(2) A reach for water transfer and sediments transfer work 60 ~ 100 km far away from the dam area. In the period of water transfer and sediments transfer, the area before the dam features a low water level, a heavy flow rate and a complicated fluid state, thus the operating ships are uneasy to control, which can exert negative effects on safe production. The reach 60 ~ 100 km far away from the dam is a good choice to conduct de – siltation operation since it can been seen through the variation of talweg point that its elevation of river bed has remained high during 2001 ~ 2004 and has still been 2 ~ 12 m higher than that before impoundment.

5 Summary

In a certain sense, the development process of the reservoir area delta has indicated the evolution of the reservoir siltation. According to the siltation pattern of the reservoir in post – flood season of 2009, two reaches are planned to be selected initially as test reaches: ①a reach, 20 ~ 60 km far away from dam area, for routine work. ②a reach, 60 ~ 100 km from the dam area, for water transfer and sediments transfer work; Due to various reasons, the de – siltation construction with pulsed jet remains in the stage of research, and there is a large amount of work to do to put it into practice. The reaches for de – siltation put forward in this thesis is just an initial outline, which needs constant summarizing, perfection and improvement in actual engineering application.

Reference

Ning Feng, Gao Chuanchang, Wang Weishu. Research of Manpower Density Flow in Xiaolangdi Reservoir Based on Pulsed Jet[J]. South – to – North Water Transfer and Water Science & Technology, 2007,5(2):76 – 76.

Preliminary Study on the Regulation Scheme for the Flood Plain Areas in the Lower Reach of the Yellow River

Cui Meng*, *Liu Shengyun*, *Han Xia*, *Liu Hongzhen* and *Wan Zhanwei

Yellow river Engineer Consulting Co. Ltd, Zhengzhou, 450003, China

Abstract: The flood plain areas in lower reach of the Yellow River play an important role in the flood control and sediment reduction system in the lower reach of the Yellow River , as an important flood diversion and silt deposition area, are also the hometown of nearly 1.9×10^6 people. On the premise that flood control safety of the lower reach of the Yellow River is guaranteed, it is essential to improve the relations between floodplain regulation and the production and life of the masses and to rationally determine the regulation scheme for the flood plain areas of the in the lower reach of the Yellow River. Based on the multi-faceted analysis of the impact on downstream flood control, river erosion and deposition, economic indicators and operation of projects and possible social problems, the "gradual abolishing production dykes and applying the whole flood plain" can better play the role of flood detention and silt deposition, is also in favor of flood management and scheduling, can create the conditions for a win-win situation for the downstream river management as well as production and development of the masses in flood plain areas. Therefore, is taken as a recommended scheme for governance of flood plain areas in the in the lower reach of the Yellow River.

Key words: in the lower reach of the Yellow River, flood plain areas, gradually abolishing production dykes, utilizing the whole flood plain areas

The Yellow River downstream flows through the North China Plain from Baihe mountain area in Mengjin town of Henan Province into Bohai Sea in Kenli town of Shandong Province, with a total length of 878 km. The flood plain areas between the dykes in the lower reach of the Yellow River occupy an area of 3,154.0 km^2 (of which Fengqiu inverted irrigation area is 407 km^2 in area). Until the end of 2007, the flood plain areas (including Fengqiu inverted irrigation area) in the lower reach of the Yellow River have a resident population of 1,895,200 (of which the population of Henan Province and Shandong Province is 1,246,500 and 648,700 respectively).

1 Requirements on floodplain regulation based on flood prevention and sedimentation reduction and safety development of the flood plain areas in the lower reach of the Yellow River

The flood plain areas in the lower reach of the Yellow River, as an important flood diversion and silt deposition area, are of strategic significance in dealing with flood disaster and sediment in the Yellow River, and play an important role in maintaining the stability of the river course and guaranteeing flood control safety in the lower reach of the Yellow River. As a deluge takes place in the lower reach of the Yellow River, the flood plain areas in the lower reach of the Yellow River is a large natural detention basin. According to nine analyses of floods at Huayuankou station with peak discharge over 8,000 m^3/s, the average flood peak deduction rate in the reach from Huayuankou to Sunkou of the Yellow River is 24.37% and the average flood peak deduction rate of three comparatively large floods (happened in 1954, 1958, and 1982) is 35.02%. The channel storage volumes (large floods happened in 1958 and 1982) in the reach from Huayuankou to Sunkou are 2.589×10^9 m^3 and 2.454×10^9 m^3 respectively, which are equal to the total storage capacity of Gu town and Luhun reservoir. The flood plain areas in the lower reach of the Yellow River have obvious effects on flood detention and greatly relieve the pressure of flood protection in narrow reaches of the in the lower reach of the Yellow River. The flood plain areas in the lower reach of the Yellow

River play a role in both flood detention and silt deposition. According to the statistics of measured data, 9.202×10^9 t of sediment in all was deposited between June, 1950 and October, 1998 in the lower reach of the Yellow River, of which 6.37×10^9 t (contributing to 69.1% of the total amount of deposition of the whole section) was deposited in the flood plain areas. In the reach from Huayuankou to Sunkou, the total amount of deposition in the flood plain areas of this reach is 5.519×10^9 t, making up 86.7% of the total amount of deposition of the whole in the lower reach of the Yellow River. The flood plain areas perform well in silt deposition. In recent years, changes in the relation between runoff and sediment of the lower reaches and a great number of production dykes, built by the masses to prevent small water flow, cause the reduction of silt deposition in the flood plain areas, severe channel deposition and shrinkage of normal channel, and rapid formation and development of secondary suspended rivers.

The flood plain areas in the lower reach of the Yellow River are the hometown of nearly 1.9×10^6 people. Before the mid-1980s, the bankfull discharge of the in the lower reach of the Yellow River was about 6,000 m^3/s while nowadays it is only about 4,000 m^3/s. The change causes the plain areas to be attacked by medium and small floods more frequently. People not only have difficulties in developing production, but also can not guarantee their life and property security. Under the construction of socialist harmonious society, people strive to realize the people-oriented concept. Thus, on the premise that flood control safety of the in the lower reach of the Yellow River is guaranteed, it is essential to improve the relations between floodplain regulation and the production and life of the masses and to rationally determine the regulation scheme for the flood plain areas of the in the lower reach of the Yellow River.

2 Regulation scheme for the flood Plain areas

In recent years, based on the estimation of future water and sediment conditions and researches on the downstream channel regulation, the strategy for regulating downstream channel takes shape, which upholds "Stabilizing Main Channel, Regulating Water and Sediment, Consolidating Embankments at Wide Sections, and Policy Compensation". Based on this strategy, various points of view concerning the regulation of the flood plain areas and representative plans can be summarized. By consolidating embankments at wide sections, Emphases should be put on gradually abolishing production dykes and establishing the schemes for utilizing the whole flood plain areas, substandard production dyke scheme, and regional utilization scheme. Considering the system of flood prevention and sedimentation reduction within the strategy has basically been formed, the following analyses only focus on the establishment of these schemes and control measures on corresponding flood plain areas.

2.1 Scheme for gradually abolishing production dykes and utilizing the whole flood plain areas (Scheme one)

Scheme for gradually abolishing production dykes and utilizing the whole flood plain areas is designed to abolish production dykes in a planned and step-by-step way and to fully fulfill flood detention and silt deposition of the flood plain areas in the lower reach of the Yellow River.

In 1950s, no clear boundary exists between channels and flood plain areas. Frequent flow and sediment flux exchange between channels and flood plain areas results in sediment relatively evenly occupying the cross section. After 1960, production dykes have been built on a large scale. Since 1986, less inflow of the Yellow River causes production dykes to start to move towards newly-formed floodplain. The constructions of production dykes exert a great influence on floods. In this case, normal floods basically can not move out of the river channel; sediment is deposited within the channels; and conveyance capacity has been reduced rapidly. Relevant analyses indicate that the volume of annual average sediment deposition in downstream channels reaches 23.2×10^6 t from 1985 to the year that Xiaolangdi Reservoir is put into operation. Only 66×10^6 t of sediment is deposited in the flood plain areas, making up 28.4% of the total amount of deposition of the whole section. Compared with the volume of the 1950s, the volume of sediment deposited in the flood plain areas is reduced nearly 49%.

The comparative analysis has been conducted to study the impacts of the scheme for breaching production dykes and the scheme for abolishing production dykes on scouring and silting in the lower reach of the Yellow River. It takes runoff-and series calculated in ten years as example. Compared with the scheme for breaching production dykes, the implementation of the scheme for abolishing production dykes will increase 3.74×10^8 t of sediment deposited in downstream channels. Longitudinal distribution proportions of various reaches are slight changed while horizontal distribution proportions of between flood plain and channel undergo great changes. The proportion of sedimentation on beaches in the scheme for abolishing production dykes increases from 54.2% of the scheme for breaching production dykes to 70.1%. Compared with the scheme for breaching production dykes, the volume of sedimentation in the main channels has been generally reduced and its thickness of sedimentation from upper sections to lower sections decreases about 0.4 m by conducting the scheme for abolishing production dykes. The thickness of sedimentation in the flood plain areas increases relatively. Thus, the implementation of the scheme for abolishing production dykes can increase the volume of sedimentation in whole reach and the flood plain areas; decrease the volume of sedimentation and the thickness of sedimentation in the main channels; relatively increase the elevation difference between floodplain and channel; and favor river course flood passing and river flood control.

In the scheme for abolishing production dykes, the flood plain areas in the lower reach of the Yellow River can naturally undertake the task of flood detention and silt deposition. The scheme intends to carry out the regulation on secondary suspended rivers by developing the controls of river embankments and ditches and conducting the large-scale mechanical warping in the reaches from Dongbatou to Taochengpu. The scheme also adopt three methods, respectively organizing immigrants (0.35×10^6) to live in the back riverbank, building bedding cushion locally to help residents (0.841×10^6) settle down and temporary withdrawal of local residents (0.422×10^6), to guarantee the life and major property security of the masses in the flood plain areas. It also plans to stimulate post-flood economic recovery in the flood plain areas by reasonably compensating the flood plain area submerged by flood. Therefore, safety construction of the flood plain areas should be accelerated and regulations on the implementation of compensation policies on submergence in flood plain areas should be formulated as soon as possible. Based on this, production dyke will be gradually abolished and the flood plain areas can resume its function of flood detention and silt deposition.

2.2 Substandard production dyke scheme (Scheme two)

Substandard production dyke scheme is to adjust and transform the existing production dykes and to form the dykes up to certain standards. In case of floods, the flood control safety can be ensured in the flood plain areas with dykes below such standards while with dykes exceeding such standards the flood plain areas should fulfill flood detention and silt deposition.

The defense flow of substandard production dykes should not be lower than the estimated bankfull discharge. Nowadays, in the reaches above Taochengpu of the in the lower reach of the Yellow River the regulation discharge adopts 4,000 m^3/s. The forecast suggests that the bankfull discharge of the in the lower reach of the Yellow River will be 3,887 m^3/s ~ 4,169 m^3/s from the year (2020) that the Guxian Reservoir is completed to 2080. It means that normal channel of the in the lower reach of the Yellow River maintains a conveyance capacity of about 4,000 m^3/s. Thus, the defense standard of substandard production dykes should not be lower than 4,000 m^3/s. According to the statistic analysis of field data collected from Huayuankou station, the number of days when various discharges occur within flood season should be studied. At Huayuankou station, the discharge below 3,000 m^3/s (daily mean) has a relatively high probability of occurence. There is a large difference between the number of the days with discharge above 4,000 m^3/s and that of the days with discharge above 5,000 m^3/s. the number of the days with discharge above 5,000 m^3/s decreases. Substandard dykes only aim to moderately reduce the probability of overbank flooding in downstream flood plain areas and inundation loss of cultivated land. Large floods, especially hyper-concentrated floods, still need the application of flood plain. The dykes should not be built in a high standard. As substandard dykes are built by heightening and consolidating existing production dykes with relatively poor quality, the higher the defence standards are, the higher the probability of dyke breakdowns is. Based on the comprehensive analyses, the defense standard of substandard dykes is 5,000 m^3/s.

Domestic experts' opinions, recent research findings concerning river course regulation and results obtained from Criteria for Dividing Channel and Floodplain proposed and developed by Yellow River Committee have been taken into a comprehensive consideration. Combined with heightened and widened existing production dykes, the layout of substandard dykes and dyke lines will be accomplished. In the wide reaches above Taochengpu, the average width between dykes is about 4 km. In the reaches above Beijing—Guangzhou Railway Bridge, the average width is 5.7 km. In the reaches from Beijing-Guangzhou Railway Bridge to Dongbatou, the average width is between 3 km and 6 km. in the reaches from Dongbatou to Gaocun, the average width is between 2.4 km and 5.6 km. in the reaches from Gaocun to Taochengpu, the average width is between 1.1 km and 3.2 km. In the reaches below Taochengpu, the average width is about 1 km. The overall length of the planning dyke line is 807.4 km, among which newly-built dyke is 102.4 km in length and heightened and consolidated existing production dykes or control projects have a length of 705 km.

In this scheme, substandard production dykes can bear the floods below 5,000 m^3/s. As floods exceeding this standard take place, the flood plain areas still fulfill the function of flood detention and silt deposition. Thus, under this scheme, the regulation on secondary suspended river and safety construction of residents in the flood plain areas still need to be developed. The control measures are the same as those of scheme one. If floods above 5,000 m^3/s occur, compensation policies on submergence in flood plain areas should be implemented based on flood damage.

2.3 Regional utilization scheme (Scheme three)

Combined with existing production dykes, regional utilization scheme divides the flood plain areas into a number of flood diversion and silt deposition areas by dykes. As floods of certain standards or hyper-concentrated floods take place, these areas can achieve flood diversion and silt deposition. Thus, this scheme can reduce inundated areas and inundation losses of the flood plain areas and achieve the goal of "sacrificing the part, protect the whole".

2.3.1 The layout of regions

In the reaches from Beijing—Guangzhou Railway Bridge to Taochengpu, there are 31 natural flood plains, among which the area of 10 natural flood plains exceeds 30 km^2, with a total area of 1,416 km^2. Within these natural flood plains, Yuanyang flood plain, Yuanyang—Fengqiu flood plain, Kaifeng flood plain, Changhuan flood plain, Lankao—Dongming flood plain and Puyang—Xicheng flood plain, etc, have relatively larger areas which all exceed 100 km^2. There are 21 natural flood plains with an area of below 30 km^2, with a total area of 363.3 km^2.

According to the rule of "dividing flood, storing flood, retreating flood, keeping secure", with the consideration of factors, such as the population of each natural flood plain, its storage capacity after the completion of flood detention and silt deposition areas, etc., border dykes will be built for the natural flood plain with an area of above 30 km^2 to form flood detention and silt deposition areas. In the relative small flood plains with an area of below 30 km^2, due to a small population and cultivated land, building border dykes will occupy cultivated land. Thus, there is no need to enclose flood detention and silt deposition areas with border dykes.

Yuanyang flood plain, Yuanyang—Fengqiu flood plain, Chuanghuan flood plain, Kaifeng flood plain and other flood plains have relatively large longitudinal gradient ratios. For the purpose of reducing inundation losses as much as possible and meanwhile achieving relatively large storage capacity, lattice dykes will be built within these four flood detention and silt deposition areas to divide them into two small zones based on this division, in the reaches from Beijing—Guangzhou railway bridge to Taochengpu, natural flood plains can be divided into fourteen flood detention and silt deposition areas in all, with a total area of 1,416.1 km^2 (see Tab. 1).

Tab. 1

Num.		The name of flood plain areas	Town (city) involved	Area(km^2)
The left bank	1	Yuanyang flood plain I	Yuanyang town	**92.7**
	2	Yuanyang flood plain II	Yuanyang town	**190.6**
	3	Yuanyang—Fengqiu flood plain I	Yuanyang town, Fengqiu town	**86**
	4	Yuanyang—Fengqiu flood plain II	Yuanyang town, Fengqiu town	**38.5**
	5	Changhuan flood plain I	Changhuan town	**175**
	6	Changhuan flood plain II	Changhuan town, Puyang town	**127.6**
	7	Xicheng flood plain	Puyang town	**126.4**
	8	Luji flood plain	Fan town	**61.8**
	9	Qinghe flood plain	Taiqian town	**74.9**
The right bank	1	Zhengzhou flood plain	Zhongmou town, Kaifeng city	**80.3**
	2	Kaifeng flood plain I	Kaifeng city, Kaifeng town	**60.1**
	3	Kaifeng flood plain II	Kaifeng town, Lankao town	**76.7**
	4	Lankao—Dongming flood plain	Lankao town, Dongming town	**184.2**
	5	Juancheng—Zuoying flood plain	Juancheng town, Yuncheng town	**41.5**
Total				**1416.1**

2.3.2 Principles of application

In terms of the terrain and geography of each flood detention and silt deposition areas, the water level at the end of each area should not exceed the design water level of bedding cushion's safety construction in flood plain areas (at Huayuankou station, peak discharge of 20-year frequency is 12,370 m^3/s.). After partition, border dykes narrow the width of passing flow and raise the water level. Based on the analyses, the water level that the discharge of 8,000 m^3/s corresponds to is close to the design water level of bedding cushion. Thus, based on this discharge, the water level of each area will be controlled. In the wide reaches form Beijing—Guangzhou Railway Bridge to Taochengpu, every flood detention and silt deposition areas have a maximum flood diversion volume of 1.67×10^9 m^3. Beside flood detention and silt deposition areas, small flood plains in wide reaches also fulfill the task of flood detention, with the detention volume of about 1.032×10^9 m^3.

As the discharge of floods exceeds 4,000 m^3/s, 5,000 m^3/s, 6,000 m^3/s, 7,000 m^3/s and 8,000 m^3/s, the volumes of floods of different frequencies are analyzed and calculated at all stations of the in the lower reach of the Yellow River. As floods of different frequencies exceed different flood volumes, envelope curve of major floods happened in the upper and lower reaches of the Yellow River will be adopted. As floods above 4,000 m^3/s of 5-year frequency, 3-year frequency, 2-year frequency at Huayuankou station, the flood volumes will be 4.745×10^9 m^3, 2.89×10^9 m^3, 1.47×10^9 m^3.

The relationship between the volumes of floods with different frequencies at the cross-section of Huayuankou and the volumes of flood diversion in each flood detention areas are analyzed. Even if within wide reaches below 8,000 m^3/s, both fourteen flood detention and silt deposition zones and small beaches capable of flood detention fulfill flood diversion together, their volume of flood storage and detention can not fully deal with floods of 3-year frequency, with a volume of above 4,000 m^3/s. Thus, by means of regulation and control, when floods of 2-year frequency occur, flood diversion at an appropriate time can decrease the flood peak discharge of the narrow reaches below Taochengpu to 4,000 m^3/s. No flood-diversion with full insurance is adopted to deal with floods of 3 ~ 5 year frequency. As for floods of 5 year frequency and above, the whole flood plain areas will be applied to fulfill flood diversion.

2.3.3 Regulation and control measures

The design standard of border dykes should follow the standard that fight against floods with a discharge of 8,000 m^3/s. Border dykes will be built based on the research findings on the Criteria for Dividing Flood Plain and Channels and the current control projects and heightened and consolidated production dykes. The border dyke of fourteen zones and Changping flood plain has a total length of 497 km, of which newly-built border dyke is 148.8 km in length and the heightened linked dyke of control projects and the reformed existing production dyke have a length of 348.3 km.

In this scheme, border dykes can fight against floods below 8,000 m^3/s. As floods above 8,000 m^3/s take place, the flood plain areas still need to play a role in flood detention and silt deposition. Thus, as the scheme is implemented, the regulation on secondary suspended rivers and safety construction of residents still need to be paid attention to. The control measures are the same as those of scheme one. In terms of different utilization of flood plain areas, compensation policies on inundation of flood plain areas should be implemented in the flood plain areas participating in flood diversion according to their flood damage.

3 Schemes comparison

Judging from the flood control function, all these three schemes can play a role in flood passage and detention in the flood plain areas, but in the long run, when the production dykes or enclosure dykes are constructed, the downstream channel will be narrowed and the flood level be elevated at the same flow, a certain impact will be put on flood detention and clipping conducted by flood plain areas when the flood over engineering standards takes place and the risk will be increased to the levee because of substandard dykes in danger. Therefore, the scheme one is superior to the other two, taking into consideration the safety of the in the lower reach of the Yellow River.

In light of effects on river scouring and deposition, compared with scheme for breaching production dykes, the scheme for abolishing production dykes will increase siltation in the flood plain areas, decrease siltation in volume and thickness in major channels and comparatively magnify height difference between flood plain and channels, which is in favor of flood passage, even though siltation in the whole river is increased. For regional utilization scheme, research results shows that after the application of the regional utilization, the proportion of tender flood plain siltation increases, height difference between flood plain and channels decreases, traverse gradient ratio of beach face increases, "secondary perched river" is intensified, "the possibility" is increased for "yokogawa" "oblique river" " flood along dike" come into being and the threats are added to dike projects.

Taking into account the treatment effect and economic indicator, by building low-standard production dykes and enclosure dykes, the scheme two and scheme three can both cut down submergence loss in the flood plain areas and relatively enhance benefits guided by a certain standard, while the cost of constructing production dykes and enclosure dykes is increased. From the long term, due to the construction of a low-standard production dykes and enclosure dykes, the scheme two and scheme three will further weaken free exchange between water and sediment in the flood plain areas and make the relatively larger traverse gradient ratio of beach face harder to be changed and the control of "secondary perched river" more arduous. These three schemes should be subject to safety construction in the flood plain areas with the same measures on resettlement and scale; three schemes are required to implement flooding compensation policies in the flood plain, but compared to the scheme one, scheme two and scheme three have slightly lowered compensation scope and possibility. With the assistance of analyzing benefits and investment differences generated by individual schemes, the net benefit of incremental investment by scheme two is below zero in comparison with the scheme one and the scheme three is also below zero compared to the scheme one. Therefore, the scheme one is much better.

Considering scheduling decision-making and the impact of the operation and management, because of difficulties in accurate flood forecasting, scheme two and three will affect the correct command decision-making, make the flood prevention work more difficult and management tasks and degrees of difficulty will be greatly increased. As for the scheme three, it is more complicated to decide to utilize different the flood plain areas to fulfill flood diversion and silt deposition concerning floods of different standards.

From the perspective of possibly-arising social problems, the low-standard dykes or regional enclosure dykes are mostly constructed in the wide reaches of Henan Province. When the project is completed, in case of small and medium-sized floods, the flood plain areas in the wide reaches fail to fulfill flood detention and peak clipping. As small and medium-sized floods enter the narrow reaches, its flood peak flow and flood volume increase. As floods with the same flow level enter the narrow reaches, their inundation losses increase which is unacceptable for the masses living in the flood plain areas of the narrow reaches and may lead to certain social issues.

Therefore, the gradual abolition of the production dykes and the application scheme of the whole flood plain areas can better play the role of flood detention and silt deposition, and is also in favor of flood management and scheduling. Through the regulation on "secondary perched river", the safety construction will be improved and compensation policies be implemented in the flood plain areas. Combined with the construction of flood prevention projects and the implementation of the river-governing measures of "Stabilizing Main Channel, Regulating Water and Sediment, Consolidating Embankments at Wide Sections, and policy compensation", this scheme as a recommended program can create the conditions for a win-win situation for the downstream river management as well as production and development of the masses in flood plain areas.

4 Conclusions

The flood plain areas of the in the lower reach of the Yellow River play an important role in the governance of the in the lower reach of the Yellow River. The "abolishing production dykes and applying the whole flood plain" is taken as a recommended scheme for governance of flood plain areas in the in the lower reach of the Yellow River, based on the multifaceted analysis of the impact on downstream flood control, river erosion and deposition, economic indicators and operation of projects and possible social problems. However, the reality in the flood plain areas shows that the premise of abolishing the production dykes is that the safety and compensation issues must be resolved for masses in such areas. To abolish the production dykes is still more difficult in present situations in which security construction is lagged behind and people's living standards are very low. Therefore, security infrastructure construction must be accelerated in such areas with perfect supporting facility. Compensation policies must be implemented for inundated flood plain and water conservancy infrastructure construction must be strengthened in the flood plain areas. The state has issued preferential economic and taxation policies to promote the transfer of land, adjustment of planting structures, the development of intensive farming and modern agriculture and immigration and employment of masses in other places. Through the comprehensive management of the flood plain areas, an overall solution to the security and development issues for the flood plain areas will be put forward and the production dykes will be gradually abolished.

References

Comprehensive Management Plan for the Flood Plain Areas of the In the Lower Reach of the Yellow River [R]. Zhengzhou: Yellow River Engineering Consulting Co., Ltd., 2009.

Study on Problems of Production Dykes[R]. Zhengzhou: Yellow River Water Resources Research Institute, 2006.

Study on Division Methods for Flood Plain and Channel of the In the Lower Reach of the Yellow River [R]. Zhengzhou: Yellow River Water Resources Research Institute, 2005.
Solid Model and Experimental Study on Effects of Flood Detention and Desilting by Regional Utilization Scheme in the Flood Plain of In the Lower Reach of the Yellow River [R]. Zhengzhou: Yellow River Water Resources Research Institute, 2008 .

Initial Analysis on Structure System of River Basin Water Resources Management

Zhang Yankun, *Liu Shuhua*, *Ding Yuanfang*, *Liu Yuanyuan*, *Feng Yan* and *Xue Mei*

Songliao River Water Resources Commission of the Ministry of Water Resources, Changchun, 130021, China

Abstract: Based on the definition of the conception, connotation and principle of the river basin water resources management, then its target and content are defined. Based on the actual situation of the water management of the Songliao River Basin, according to the authorization and regulation of water resources management of the river basin management organization from the national laws and regulations and the "institution, function and staff" determined scheme, the management of the river basin water resources are divided into macro – management and micro – management on the task and content. The macro – management mainly includes the scheme making works of the basin planning, water allocation, control indicators of basin water resources, etc. and the operation works of water resources regulation and monitoring, etc. The micro – management mainly includes the approval works of the water project planning consent, the assessment of water – draw and utilization in construction projects, water drawing permit, consent of sewage outlet to the river, etc. and the supervision works of the water using planning, water function area supervising, etc. The complete architecture of river basin water resources management is constituted by the interaction of various aspects and different types of works.

Key words: river basin, water resources, management, structure system

1 Overview of river basin water resources management

The water resources have own distinctive characteristics of taking basin as basic unit to form a complete relationship of revenue and expenditure. The basin is a compound system with hierarchical structure and global function. The basin hydrological cycle is not only the resources base of the society economic development and the controlling factors of ecological environment, but also the focus of water and ecology issues. Therefore, the unified management of water resources based on the basin unit has become an international generally accepted scientific principle. Strengthening the management and coordinating contradiction among provinces and ensuring the reasonable water resources allocation and the water supply in the important water deficient area are important contents of water resources management.

1.1 Conception, connotation and principle of river basin water resources management

1.1.1 Conception of river basin water resources management

The water resources management is the related organization, coordination, supervision and scheduling of the water resources development, utilization, conservation and protection by using the administrative, legal, economic, technological and educational means. The water resources management coordinates relationship between water resources and society economic development, formulates and implements the related water resources management regulation and law, handles water utilization contradiction among areas and departments, supervises and restricts the unreasonable development of water resources and the harmful act of water source, makes reasonable programs for water resources allocation, proposes and implements the optimal regulation scheme of water resources for all water users with consideration the benefits of all parties, monitors the changes of water inflow and water quality and adopts the implements measures, and so on.

According to the water resources characteristics, the water resources management is summarized as follows: the equal attentions on development, utilization and protection of water resources, the unified management of water quantity and quality, the comprehensive management and unified regulation for surface water and groundwater, the benefits maximization for society, economic and environment, the formulation of principles and policies for the water resources management, the great attention and strengthen for the hydrological information and forecasting, etc.

The river basin is a natural catchment unit from riverhead to estuary. Taking the different reaches, different riverbanks, main stream and tributaries, water quality and quantity, surface water and groundwater, management, development and protection as a complete basin system, the river basin water resources management is the process of coordination and unified regulation based on the river basin with promoting beneficial and abolishing harmful.

1.1.2 Connotation of river basin water resources management

The connotation of the river basin water resources management mainly include the five aspects of subject, object, task, mode and characteristic.

(1) The subject of management, is to determine the administrator. The administrator of water resources is its owner. *The Water Law of the People' s Republic of China* clearly regulates "water resources are owned by the State.", so are owned by all people. The State is the subject of management, using the water management rights to realizes sustainable utilization of water resources.

(2) The object of management, is to determinen who is managed. The object of management has the integrity and is the protected and developed target. The integrity of water resources is embodied on the natural hydrologic characteristics. The basin water resources as a natural hydrological unit are a relatively independent entity, which is the object of management.

(3) The task of management, is to determine the content. The conception shows that the water resources management have four tasks, including the development, utilization, conservation and protection. The administrator realizes the sustainable utilization target by finishing the four tasks. The protection is the precondition of development and utilization. The purpose of the protection is the development and utilization. The purpose of development is the utilization, but development is restricted by protection. The utilization affects development and protection in turn. The utilization is restricted by conservation, while the conservation is to better utilization and protection, but also reduce the development dependence. These tasks form variety of complex management process by the mutual fusion, interaction, interaction and mutual restraint. The complexity is impossible to estimate by mathematical permutation and combination method, which forms in the deeper and broader category of space - time, psychology and behaviour.

(4) The mode of management, is to determine how to manage. The aim of water resources management is realized through some organization forms, management system and the corresponding management mechanisms. The substance of water resources management is to build some management system and management mechanism, to coordinate the relative optimization match between factors, to reach a certain optimization aim through some methods, measures and technologies. The management substance is the movement of management object toward the intended target by the manager under the restriction of management mechanism.

(5) The characteristic of management, is to determine the integrity.

The typical characteristic of water resources management is integrity, manifested as a natural hydrologic unit. It means the water resources management must take the basin as the centre. Out of this center is bound to destroy the characteristics of the integrity. The protection, conservation, development and utilization are also as a whole. The management aim - sustainable utilization can not be achieved without any one. That is, the four must be uniform management. In the past years, the one of the water resources management problems in China is paying attention to development and utilization, but neglecting protection and conservation, which destroyed the integrity of the tasks. The managers should also have the integrity. Another problem of the water resources management problems in China is that the water is managed by many departments and the policies are made by many departments. The cause of the problem is neglecting the integrity of administrator. Simultaneously, the administrator, the task and the object are a whole. The study of water management problems based on the integrity is the only way to realize sustainable utilization.

1.1.3 Principle of river basin water resources management

Because the water resources are an irreplaceable important natural resources and the important part of the basin environment elements, the river basin water resources management is the core of basin comprehensive management and coordination. The unreasonable development of river water resources of any one area or department in river basin, such as excessively water intake and water source pollution, will affect other water users, then affect the industry, agriculture, urban construction, people's living and ecological environment, etc. Therefore, the river basin water resources management must proceed from the whole and the global of basin, consider parties benefits of the different reaches, the different riverbanks, mountains and plains, urban and villages, industry, agriculture, transport service, fishery, tourism, etc, ensure the optimal balance of the river basin ecological system. The economic benefit, society benefit and environmental benefit are comprehensive considered to reduce the errors. So the basis principle to follow of river basin water resources management has two aspects:

(1) The regional water resources management should obey the basin water resources management. That means in the water resources system, the basin water resources management has priority to the regional management of the basin, and the regional water resources management should obey the unified coordinative and management of the basin.

(2) The special water resources management of department should be in obedience to the basin water resources comprehensive management. The water resources managements of various industries should be brought into the basin water resources management system and be subordinated to the basin water resources management.

1.2 Target and content of river basin water resources management

1.2.1 Target of river basin water resources management

The determination of the water resources management target should adapt the local social economic development target and the ecological environmental target, which not only consider the resources conditions, but also consider the economic capacity.

The determination of water resources management target belongs to the decision making category. The ultimate target of water resources management is to gain the maximum society economic benefit and the best ecological environment by the limited water resources. In other words, the target is to meet the water demand of the society economic development by the minimum input. No benefit management is ineffective management.

Based on the conception and target given above, the basic target of river basin water resources management should include four aspects as follows:

(1) Reasonable development of the local basin water resources (It includes: electricity generation, irrigation, shipping, fishery, water supply, tourism, etc.).

(2) Coordinating relationship between society economic development and water resources utilization, handling correctly the contradictions of water utilization among areas and departments, meeting maximally the reasonable increasing water demand for all the areas and departments in river basin.

(3) Supervising and restricting the unreasonable development and utilization of water resources and the pollution and harmful act of water sources, controlling the trend of water pollution, strengthening the protection of water resources, implementing the management of important equally water quantity and water quality and the integration of resources and environment.

(4) Reasonably allocating the limited basin water resources based on the overall planning, monitoring and scheduling the large hydropower projects to ensure the normal operation of the important rivers in the basin.

1.2.2 Content of river basin water resources management

According to the specific basin characteristics, the content of river basin water resources management includes five aspects, as follow:

(1) Making policy for water resources management. In order to manage water resources, a set of policies should be made to meet the social economic development in different periods, such as, overall planning and comprehensive utilization policy, investment allocation and resettlement policy, collection policy of water fees and water resources fees, policies of water resources protection and water pollution control, etc.

(2) Making the water resources development planning and implementation progressively. The programming and implementation of reasonable water resources development planning is one of the basic conditions for the effective water resources management. The works include the general arrangements for the development, utilization, conservation and protection of water resources, such as water source planning, water supply planning, water saving planning, water quality planning, etc.

(3) Allocation and scheduling of water quantity. According to the principle of comprehensive utilization with consideration to different reaches, riverbanks and departments, the allocation and scheduling planning of water quantity is made. For the arid basin with the water deficient and the dry years with water shortage, the domestic water should be put to the first place and be met limited. At the same time, the development of the industry and agriculture with large water consumption should be restricted. The water use planning and water saving should be carried out to alleviate the contradiction between water supply and demand. The unified management and joint scheduling of the surface water and groundwater should be put into effect to improve water utilization.

(4) Controlling and protection of water quality. With the increasing of water use in industry, urban and living, the large quantities of untreated sewage and substandard wastewater discharge into the rivers. The water pollution reduces the available water resources and even causes social disasters. The administrative measures should be adopted to supervise and control the sewage volume of industrial and mining enterprises and to promote the sewage treatment equipment placement. The economic measures should be adopted, including the sewage charges, over standard fine and pollution accident compensation, which could ensure the water quality of water supply.

(5) River hydrological regime and forecasting. After the implementation of the river multi - objective development, there are more and more buildings along rivers and so are management department. The importance of the river hydrological regime are increasingly apparent. In order to do the water resources management better, ensure reservoir operation safely and improve economic benefit, the hydrological observation and the river hydrological regime and forecasting work must be strengthened.

2 Management statuses of Songliao River Basin water resources

2.1 Basin management organizations

2.1.1 Songliao River Water Resources Commission of the Ministry of Water Resources

The Songliao Water Resources Commission is built in Changchun, Jilin Province, in 1982, which is approved by the National Iinstitutions Compilation Commission. It is the dispatched institutions of the Ministry of Water Resources at the scope of the Songhua River Basin, Liao River Basin, international boundary rivers (lakes) basin of Northeast China and river running into sea alone in Northeast China. Representing the Ministry of Water Resources, the commission exercises the water administrative responsibilities within the river basin and is the institutions with the administrative functions.

Since the foundation of the Songliao Water Resources Commission, following the aim of water control corporately, practicality and devotion, servicing for Songliao River Basin and striving for best, the commission plays an important role in many works of the Songliao River Basin, such as the water resources planning, flood control and drought relief, water management, water conservation, soil and water conservation, engineering construction and management of hydro projects, etc.

The 15 composed departments of the Songliao Water Resources Commission are as follows: ①offices; ②water administration and safety supervision department (water administration supervision corps); ③water resources department; ④ planning and programming department; ⑤financial department; ⑥personnel department; ⑦international rivers and technology department; ⑧construction and management department (Songliao basin branch station of the hydraulic engineering quality supervision central station of the Ministry of Water Resources); ⑨soil and water conservation department; ⑩flood control and drought relief offices; ⑪supervision department; ⑫audit department; ⑬ retired workers management department; ⑭ commission of Communist Party; ⑮Songliao commission of China agro – forestry water trade union.

The 8 subsidiary institutions are as follows: ①Chaersen reservoir management agency; ②hydrology bureau (information center); ③the center of river basin planning and policy research; ④Resettlement and Development Center; ⑤monitoring center station of soil and water conservation of Songliao River Basin; ⑥construction and management station of hydro projects; ⑦integrated services centre; ⑧integrated management center.

The only separated institution is the Water Resources Protection Bureau of Songliao River Basin.

2.1.2 Management range of the Songliao Water Resources Commission

The management range of the Songliao Water Resources Commission includes the Songhua River Basin, Liao River Basin, international boundary rivers (lakes) basin of Northeast China and rivers running into sea alone in Northeast China. Its administrative region includes the Liaoning province, Jilin province, Heilongjiang province, eastern four cities of inner mongolia autonomous region and part of Chengde city, Hebei Province.

The Songliao Basin is encircled by mountains on its western, northern and eastern side, and borders on the Bohai Sea and the Yellow Sea are on its southern side. In the south – central region of the basin lies the broad Liaohe Plain and Songnen Plain, and in its northeast region there exists the Sanjiang Plain. The gross area of the basin is 1.24×10^6 km^2 square km. The Songliao Basin is located in the upper – air prevailing westerlies of the northern latitude, and has more characteristics of the westerlies weather and climate. In the northeast region, there has a typical feature of the continental climate. Therefore, it belongs to the temperate continental monsoon climate zone. The winter is freezing cold and very long whilst the summer is humid and rainy. Part of the region belongs to cold temperate climate. The total quantity of the water resources of the Songliao Basin is 1.99×10^{11} m^3, including 1.704×10^{11} m^3 of surface water and 6.81×10^{10} m^3 of groundwater. The non – repeated amount of surface water and groundwater is 2.86×10^{10} m^3. The Songliao Basin is an important industrial, agricultural and forestry base in our country.

The Songhua River and Liaohe River both rank among the 7 largest rivers of our country. The Songhua River has two sources. The northern source is called Nen River, which originates from Yilehuli Mountain in Daxing, anling. And the southern one is the cecond Songhua River, which originates from the Tianchi of Changbai Mountain. The two sources converge at Sanchahe and then become one river, which is called the Songhua River. The Songhua River flows eastwards and pours into the Heilongjiang River in Tongjiang. Liaohe River originates from Guangtou Mountain in Qilaotu Mountains. Its headwaters are Laoha River. Laoha River flows northwards and pours into Xilamulun River in Hailiutu. Here it becomes the West Liaohe River. The West Liaohe River then flows eastwards and doesn't turn south until it passes through Zhengjiatun. It flows into the East Liaohe River in Fudedian. After that it is called Liaohe River. Then Liaohe River flows southwestwards and is divided into two branches in Liujianfang. One branch is called Shuangtaizi River, and it runs westwards and then flows into Raoyanghe River in Panshan. After that, it pours into the Bohai Sea. The other is called the Outside Liaohe River, and it runs southwards and then flows into Hunhe River and Taizi River in Sanchahe. Here it becomes Daliaohe River. Finally it poured into the Bohai Sea in Yingkou. In the basin there are more than 60 rivers which run into the sea all alone, and the seas include Japan Sea, Bohai Sea and Yellow Sea.

On our country's frontier in the northeast of China Region boundary rivers are distributed along the west, north and east sides of this region. They pass through the famous Daxing, anling, Xiaoxing, anling, Wandashan Mountains and Changbai Mountains. Those boundary rivers include Argun River, the main stream of the Heilong River, Wusuli River, Suifen River, Tumen River and the Yalu River, etc. From administrative division of view, they are subordinate to Inner Mongolia Autonomous Region, Heilongjiang province, Jilin province and Liaoning province. The area of the international rivers' basin accounts for nearly 1/3 of the Songliao Basin's area. On Chinese side, the area of the basin is 405.6 thousand square kilometers. The total quantity of the water resources here is one third larger than that of the Songliao Basin. The length of the reaches of boundary rivers on the frontier is up to 5,232.8 km, which is two thirds longer than the total length of all the boundary rivers of our country. Since the soil and water loss is quite serious in our country and the international water affairs are increasingly becoming more and more, the administration of international rivers in the Songliao Basin gradually becomes the key point of the administration of water resources in the Songliao Basin.

2.2 The basin administrative organization responsibility for water administration and law enforcement endowed with by laws and the relevant rules and regulations

The current valid national laws, rules, regulations and "institution, function and staff" determined scheme have stipulated detailed responsibilities and authority of the basin administrative organization and ensured the organization legal status to administer water resources in the basin.

2.2.1 The basin management organization management responsibility endowed with by Water Law

Major laws that related to water resources management by the Songliao Water Resources Commission are *The Water Law of the People's Republic of China*, *Law of the People's Republic of China on Prevention and Control of Water Pollution and Flood Control Law of the People's Republic of China*. Now the relevant provisions of the *Water Law of the People's Republic of China* are briefly introduced as follows.

One particular feature of *Water Law* is to establish a management system which combines the river basin management with the administrative region management. Under the first provision of Article 12 of *Water Law*, the state implements an administrative system on water resources which combines the management of the basin with that of the administrative region. The innovation in systems will exert an enormous effect on strengthening the management of water resources in the basin, solving the problem of stripe - and - block partition in water resources management fundamentally, allocating water resources properly, developing the comprehensive efficiency of water resources and promoting the coordinating and healthy development between economy and society in the basin. Besides, under the third provision of Article 12, the basin administration founded by the water administration department of the State Council in important rivers and lakes, exercises the responsibility, stipulated in laws and administrative regulations and endowed with by the water administration department of the State Council, of management and supervision on water resources within its jurisdiction. This provision has solved the problems, which have existed for many years, of the basin administration legal status and the responsibility it should have possessed in itself. According to the preliminary statistics, in the new *Water Law*, there are ten articles involving the responsibility of basin water resources management, which mainly aim at basin planning, the construction of water engineering, allocation of water volume, sewage outlet to the river, water drawing permit, water resources fee and water saving, etc.

2.2.2 The main responsibilities of water resources management by the Songliao Water Resources Commission endowed with by "institution, function and staff" determined scheme

Under *The Provision of the Songliao Water Resources Commission's Main Responsibilities, Organization Setting and Personnel Posts*, the main responsibilities of water resources administration by the Songliao Water Resources Commission are as follows.

(1) Being responsible for ensuring rational development and utilization of water resources. The Songliao Water Resources Commission, entrusted by the Ministry of Water Resources, organizes and draws up both the basin's comprehensive planning as regards the basin and those trans - provincial (including autonomous regions and municipality cities) rivers and lakes in the basin and the relevant professional and special planning, and supervise the planning being carried out. The

commission maps out policies and regulations regarding the basin water conservancy and organizes and promotes the preliminary work about the basin controlled project, the important trans - provincial (including autonomous regions and municipality cities) water conservancy project and the central water conservancy project. According to the authorization, the commission is responsible for the relevant planning, the inspection and approval for the central water conservancy project and the compliant inspection of the relevant water engineering project. The commission inspects and verifies the local water conservancy project of large or medium size technically. The commission is also responsible for presenting the annual investment plans for the central water conservancy project in the basin, the preliminary work about water conservancy and directly subordinated infrastructure project, and for organizing and carrying them out. The commission organizes and directs the later assessment which is relevant to the planning and construction of water conservancy in the basin.

(2) Being responsible for the river basin water resources management and supervision coordination of living, production and ecology water use of water basin. The Songliao Water Resources Commission organizes and carries out water resources investigation and assessment work and the basin hydropower resources investigation and assessment work by regulations. According to the regulations and authorizations, the commission organizes formulation of inter - provincial water allocation plan and annual river basin water resources scheduling plan and water resources regulation plan under drought emergency situation and organizes implementation. The commission organizes and carries out basin water drawing permit quantity control, basin water drawing permit and water resources assessment systems, carries out water resources scheduling of basin and basin main water project according to the regulations.

(3) Being responsible for water resources protection work. The Songliao Water Resources Commission organizes and formulates river basin water resources protection plan, organizes the formulation of provinces (including autonomous regions and municipality cities) water function regionalization of rivers and lakes and its supervision the implementation, checks the water pollutant carrying capacity, proposes the advices of the total limit pollutant discharge. The commission is responsible for the review and permission of pollution discharge outlets to rivers setting based on the authorization and for rivers water quality monitoring of provincial boundary water body, important water function area and important pollution discharge outlets. The commission guides and coordinates the protection of the drinking water sources, the development, utilization and protection of groundwater, and guides the work related to the local water - saving and the water - saving society construction.

(4) Guiding basin hydrology work. In accordance with the rules and authorizations, the Songliao Water Resources Commission is responsible for monitoring of hydrology and water resources and construction and management work of hydrological station network and for water quantity and quality monitoring work of the main part of river, directly managed river, lakes and reservoir.

2.2.3 The main responsibilities of water resources administration by the Songliao Water Resources Commission endowed with by the relevant administrative regulations and departmental rules

Over years, based on the water resources management regulations by basin management organization endowed with by *The Water Law of the People's Republic of China* and "institution, function and staff" determined scheme. The State Council and the Ministry of Water Resources have issued a series of administrative regulations and departmental rules, mainly include:

(1) State Council Decision on Administrative License for Necessary Administrative Approval Projects.

(2) Management Approach of Water Project Construction Planning Consent System.

(3) Management regulations of Water Drawing Permit and Collection of Water Resources Fees.

(4) Management Approach of Water Drawing Permit.

(5) Management Approach of Water Resources Fee Collection and Use.

(6) Management Approach of assessment of water - draw and utilization in construction pro-

jects.

(7) Management Approach of Water Function Region.

(8) Supervision and Management Approach of Pollution Outlet to the River.

(9) Hydrology regulation.

(10) Implementation Approach of Water Administrative License.

The responsibility, endowed with by *The Water Law of the People's Republic of China* and "institution, function and staff" determined scheme, has the characteristics of concretization and operation by these laws and regulations. The method and process of the river basin water resources management has been further standardized.

3 Analysis on structure system of Songliao River Basin water resources management

Summing up the work experience and management experience for many years, the author thinks that the core of river basin water resources management is dealing with the relationship of two aspects. One is dealing with water utilization relationship between human society and natural ecological environment in basin. The other is dealing with water utilization relationship of economy society among the administrative areas in basin. Dealing with the two aspects goodly would realize water harmonious utilization. Based on the basin management organization legal allocation and management responsibilities, the river basin management mechanism plays the role of dealing with the relationship between the two aspects. To play the role of river spokesperson successfully, there must be a clear understanding for the structure system of river basin water resources management. Then, how to construct the structure system of river basin water resources management, in my opinion, is that the structure system should be constructed according to specific management tasks and work contents of river basin management mechanism endowed with by the laws, regulations and "institution, function and staff" determined scheme. Seeing from the tasks and contents of river basin management organization, thc water resources management structure should be divided into two aspects: one is macro – management and the other is the micro – management.

3.1 Macro – management of river basin water resources

3.1.1 Conception and target

The macro – management of the river basin water resources management is the unified management of water resources taking the basin as unit with the target of correctly treating the water use relationship between human society and the natural ecological environment and among administrative areas for the water resources problems related to the whole basin. The management includes the development, utilization, conservation and protection of water resources. The management system should be built to meet the characteristics of the natural river basin and the multi functional unification of water resources, which could realize the optimal allocation and the maximum comprehensive benefits of water resources, ensure and promote the social and economic sustainable development of river basin.

3.1.2 Content of macro – management

The macro – management of water resources could be divided into two categories: one is the establishment of the technical scheme, such as river basin planning, river water allocation, the development, utilization and drainage of river basin water resources (including the amount control of water – draw and utilization, the control of water efficiency and the amount control of pollution discharge); the other is operational work, such as river basin water resources regulation, river basin water resources monitoring. The macro – management of river basin water resources is dependent on these two categories, neither is dispensable.

From the responsibilities and actual work situation of the river basin management origination, the river basin water resources management starts with the basin planning, that is to say the planning is the foundation of the river basin water resources management. The river water allocation is to solve intensive competition of water utilization among the inter – provincial (including autonomous regions and municipality cities) and the deterioration of the river ecosystem in the basin, which basically follows the ideas of planning. For example, the Yellow River water resources allocation in 1987 is earlier implemented in the modern Chinese. The Songliao Water Resources Commission has carried out the pilot work for the water allocation of Dalinghe River and Huolinhe River in recent years. The amount control of water - draw and utilization, the control of water efficiency and the amount control of pollution discharge are put forward in the context of the most stringent water management system in recent years, which control mainly the process of water - draw, water utilization, water drainage in the economic society. There is no monitoring, the management is not exist.

A macro - management process of basin water resources should be like this: first of all, making clear the river basin water resources, water use of the river ecological environment as well as the water consumption index of different level years for each administrative area, which could be completed by planning, river water allocation or quantity control of water - draw and water utilization. Secondly, compiling annual water allocation plans, annual total quantity control index scheme of water - draw and use based on the annual runoff, water storage and water use. Thirdly, determining the runoff, water storage and water utilization by the monitoring system and compiling the monthly or a period of ten days or real - time scheduling, carrying out the actual water resources scheduling; the last, summarizing the situation of water - draw, water utilization and water drainage in the period of the water utilization or assessment, and then carrying out reward and punishment depended on the assessment indexes.

3.2 Micro - management of river basin water resources

3.2.1 Conception and target

The micro - management of river basin water resources is the supervision behavior and the control behavior for the specific water - draw project in the basin. The former is supervising the construction, water - draw, water utilization and water drainage; the later is controlling the adverse influences for the natural ecology environment and other legitimate people of water utilization during promoting economic and social development. In addition, this also is the behavior of considering and regulating legitimacy of water drawing project from the different angles of economy, environment, society, etc.

3.2.2 Content of micro - management

The micro - management of river basin water resources mainly includes two categories: one is the license and approval work, such as the water project planning consent, the assessment of water - draw and utilization in construction projects, water drawing permit, consent of sewage outlet to the river, etc. The other is the regulatory work, such as the water drawing permit, water function region, supervision and management of the pollution discharge outlets to rivers. License and approval work are the first step that the specific water - draw projects is put into the basin water resources management, are the determination process of the scale of water utilization and water drainage fundamentally for the management object, are the important stage of micro - management and are the beginning of the micro - management. The regulatory work is another important stage after license and approval work and is the continuation of micro - management. Though the management of water - draw, water utilization and water drainage for the water drawing project of the two works, the adverse influences of water drawing project on the natural ecology environment and other legitimate people of water utilization could be controlled.

The micro - management process of river basin water resources should be like this: firstly, before the construction of water drawing project, a series of application and approval work and related procedures should be developed, such as the water project planning consent, the assessment of water - draw and utilization in construction projects, water drawing permit, consent of sewage outlet to the river, etc. The scale of water - draw, water utilization and water drainage of water drawing project and the influences of water drawing project on the natural ecology environment and other legitimate people are determined scientifically and reasonably. Secondly, before the project construction, it should obtain the official authorization from management departments, issue the related license and stipulate the responsibilities. Thirdly, annual water – draw plans of water drawing project should be declared. The management departments appraise and approve all water – draw plans of water users in the river basin according to the condition of runoff, water storage and water utilization. Fourthly, the water – draw departments develop productive activities depend on the approved annual water – draw plans. Finally, the management departments should check the condition of water – draw and water drainage of water drawing projects during the water – draw period and evaluate the conditions after the water – draw period, which would be the references of approval for the water – draw plans in the next year.

3.3 Interrelationship between the two aspects

The macro – management and micro – management of river basin water resources have both connections and discrepancies.

The connections show that: the macro – management is realized by the accumulation of specific micro – management. For instance, the total amount control of water – draw and water utilization is realized by all specific water users in river basin, as well as pollution discharge. For micro – management, the macro – management is taken as the framework and boundary. For example, the approval of specific water drawing project is controlled in the range of total water – draw control index in the whole river basin. The macro – management and micro – management are both the indispensable and important links and contents for the river basin water resources management.

The discrepancies show that: firstly, there are differences for the management objects of river basin water resources. The objects of macro – management are the local water administrative departments with various levels, while the objects of micro – management are specific administrative counterpart. That is the specific water drawing project. Secondly, the management scope is different. The macro – management is the surface and line management and its work scope is on the whole river basin level and provincial section. The micro – management is the point management, and is the management of the specific water drawing projects within the scope of administrative rights of the watershed management organizations. Thirdly, the perfection degree of law system, depended on the both managements, is also different. Compared with the micro – management, the corresponding laws and regulations of the macro – management are far behind. For instance, the laws and regulations of total amount control, water allocation, water resources regulation, and so on, are lack of systematization and integrity.

4 Summary and forecast of Songliao River Basin Water Resources Management

According to the structural analysis of river basin water resources management, the summary and forecast of Songliao River Basin water resources management have been made combined with the developed water resources management work by the Songliao River Water Resources Commission in recent years.

4.1 Macro – management Work

(1) Developing the planning and other basic water resources management work.

Since 2000, the Songliao River Water Resources Commission has successively organized and developed the whole river basin planning including the planning and revision planning of the integrated Songliao River Basin water resource and the the comprehensive planning and the special research of cross – provincial rivers such as the Huolin River, Chuoer River, Nuomin River, Hun River and Lalin River, etc. Combined with related provinces, a number of professional and special planning have developed, which laid the foundation of planning for the river basin water resources management. Then the existing planning and the planning work of the sensitive rivers are promoted powerfully for the next step.

(2) Promoting the water allocation work in river basin.

Since 2004, the SongLiao River Water Resources Commission has developed a series of research and experimental works for water allocation, successively completed the compile work for the water allocation schemes of the Daling River and the Huolin river. The water allocation scheme of the Daling River has already obtained approval by the Ministry of Water Resources. At present, according to the new round of work arrangement from the Ministry of Water Resources, the water allocation works of the Nen River, the Second Songhua River, Lalin River, the East Liao river, etc, are been developed. The water allocation works of the Songhua River, Taoer River, Nuomin River, Yalu River, Mudanjiang River, Chuoer River, Liao Rive, the West Liao Rive and Liu River would be developed in the next stage.

(3) Obtaining the phase achievements for the river basin control management of total water - draw amount.

In order to strengthen the total amount control management for water drawing permit in Songliao River Basin, establish the harmonious water related orders, the Songliao River Water Resources Commission has completed the total control index scheme of water utilization in 2009 according to the requirements of the Ministry of Water Resources. And on this basis, the total control index scheme of water utilization has accomplished in recent. The total control index scheme of water utilization in Songliao River Basin, as the core technological scheme established by the "three red lines", will provide necessary foundation and basis to the water resources management and protection in river basin and the provincial detailed index scheme. The construction of the basin - wide water resources management system would be constructed and the legislative preparation of the Nen River regulations would be developed.

(4) Making sturdy steps for the construction of water - saving society.

The Songliao River Water Resources Commission regularly organizes the compilation works of the water - saving society construction planning in river basin, fully considers the water - saving plans and measures in making river basin comprehensive planning, strictly checks the construction project in the project establishment and review, meets the water resources condition of the economic layout and the industrial structure, limits strictly the project of high water consumption, plays the control effect of the red line in the water use efficiency and participate actively the experimental work for the construction of the water - saving society in river basin. In the next step, combined with the Ministry of Water Resources, the commission would continue to supervise, inspect, estimate and accept the water - saving society construction of the experimental cities and strength the evaluation and examination for the control index of the red line in the water use efficiency of river basin.

(5) Gaining the new development for water resources protection.

Combined with the actively coordinate and strengthen cooperation, the advantage of coordination of the protection leading group organization of Songliao river are strengthened. The local working force of the local water resources protection and the water pollution control are promoted. The guiding force for the protection of drinking water source are improved and the investigation of the important drinking water source are developed. *The implementing scheme of water resources protection and supervision for implementation of the most strict water resources management system in Songliao River Basin* is completed. The water quality model system of main stream of Songhua River is built and the dynamic supervision of the important water function areas of Songhua River basin would be developed. The works of water resources protection planning, water ecological protection and repairing and environmental assessment of the river basin comprehensive planning would be strengthen. In additions, the control indexes of limited line of pollutants capacity in water function area are checked and examinated.

(6) Developing active regulation of river basin water resources.

Recent years, the regulation of basin water resources gradually has become the main method for the river basin management. Based on the utilizable regulation of Nierji hydro junction and Chaersen reservoir, the Songliao River Water Resources Commission has already developed the water resources regulation of the Nenjiang River Basin, which has gained the significant benefits in economic, society and ecology. In the next stage, the scope of the water resources regulation would be expanded continually and the unified regulation research of water resources of Songhuajiang River Basin based on the joint regulation of Nierji hydro junction and Fengman reservoir would be developed.

(7) Strengthening the monitoring of the water quality and quantity and increasing the information publication.

The water resources information is the base of management. In recent years, the Songliao River Water Resources Commission has strengthened the monitoring of hydrological information, at the same time, expanded the depth and width of hydrological information. As a result, the status of the River Basin water resources can be mastered fatherly in space and time. The commission has highly emphasized the publication of water resources related information for the society and departments related to water, such as *the basin water resources bulletin*, *the groundwater bulletin*, *the river sediment bulletin*, *the water quality of province boundary buffers bulletin*, etc. In the next stage, according to the unified requirement of the Ministry of Water Resources and the capacity construction needs of the river basin water resources management, the commission will speed up the information construction of basin water resources management, organize and implement the system construction of Songliao river basin water resources management, which provides the technical support for the realization of river basin water resources macro – management.

4.2 Micro – management work

(1) The approval of the water project planning consent must be strict. The construction projects meets the requirement of related planning and the macro – guidance and the control constraint effect of the planning are fully used.

(2) The assessment work of water – draw and utilization in construction projects could be accomplished, and the report review is strict. Then these behaviors of the post – evaluation of project, violation investigation, supervision and management could be developed. In the next stage, the water resources assessment of the social economic and development planning, the urban general planning, the important industry development planning, etc, would obtain the explorition approval, which would promote the suitability between the basin social economic development and the carrying capacity of the water resources and the water environmental.

(3) Strengthening the standardized management of water drawing permit. Focusing on the planned water use, measurement and monitoring, the supervision and management of water drawing permit should be strengthened. The approval of the water drawing permit must be strict. The replacement and cleanness of the water drawing license are made well. According to the Unified deployment of the Ministry of Water Resources, the census work of water users in Songliao basin are made well. In the next stage, the installation and application of water measure equipment for water users and the construction of data collection terminal of the water resources management system should be strengthened.

(4) The supervision and management of the water function area have been strengthened comprehensively. The inspections of the water function area are developed positively and the related inspections system is built. The water quality dynamic supervision of the important water function area of the Songhua river basin is developed and the recheck of water quality supervision sections of buffers of province boundary of whole basin is completed. In the next stage, the water quality target assessment of water function area and the limited line management of pollutants capacity of water function area should be promoted.

(5) Gaining the new breakthrough for the supervision and management of sewage outlet to the river. *The detailed regulations of sewage outlet to the river supervision management method implemented by Songliao river water resources commission of MWR* (*try out*) is formulated to provide system guarantee for the consent approval of sewage outlet to the river with legal implementation. The consent approval of sewage outlet to the river is developed. The annual supervisory monitoring of pollution quantities for sewage outlet to the river in province boundary buffers in wet period, normal period and low water period is entirely completed. These sewage outlets to the river at the river section of many cities are checked. In the next stage, the water resources assessment and the supervision and management of sewage outlet to the river should be considered comprehensively and the quality of water drainage should be managed rigorously.

5 Conclusions

The construction of management structure system of river basin water resources is beneficial for the systematic cognition of river basin water resources and the formation of framework thinking, which could be laid the good foundation for the clear division of the basin responsibility, the accurate work orientation, the reasonable work target and the successful work development.

Cause of Relative Stabilization of Tongguan Elevation in Recent Years

Hou Suzhen, ***Wang Ping*** and ***Chu Weibin***

Yellow River Institute of Hydraulic Research, Zhengzhou, 450003, China

Abstract: Tongguan elevation imposes local erosion base level for lower Weihe River and Xiaobeiganliu reach of Yellow river. Since the operation of Sanmenxia Reservoir, Tongguan elevation has caused widespread concern, for which an extensive research is carried out. To lower the Tongguan elevation, the mode of operation of Sanmenxia Reservoir was adjusted over and again and numerous structural and non - structural measures are applied. Based on measured data, the evolution process of Tongguan elevation is analyzed and the driving factors for the change of Tongguan elevation are put forward from the prospective of incoming water and sediment conditions and boundary conditions. The causes of its relative stabilization since 2006 are discussed in such aspects as change of incoming water and sediment, adjustment of operational water level of Sanmenxia Reservoir, the experiment of using spring flood to scour Tongguan elevation. The study results are available for the reference for the depression of Tongguan elevation and the research of control measures.

Key words: Tongguan elevation, stabilization, Sanmenxia Reservoir

1 Background

Tongguan is located at the downstream of confluent area between Yellow River and Weihe River, 113.5 km away from the Sanmenxia Reservoir Dam, imposing local erosion base level for lower Weihe River. The rise and fall of Tongguan riverbed has an important impact on the channel scouring and deposition and the flood control of both lower Weihe River and Xiaobeiganliu reach. In a long run, the change of Tongguan elevation (the water level of 1,000 m^3/s) has been always particularly concerned.

Since the operation of Sanmenxia Reservoir in 1960, the serious sediment problem and early - stage's high water level resulted in the serious sedimentation in reservoir and the rise of Tongguan level. To slow up reservoir sedimentation, Sanmenxia Reservoir changed its role from water storage and sediment trapping to flood detention and sediment discharge in March 1962. From 1966 to 1971, Sanmenxia Reservoir underwent twice reconstruction to expand the drainage size, after that, the reservoir sedimentation was brought under control, partial storage was restored, Tongguan elevation fell down significantly. From Nov. 1973, Sanmenxia reservoir began to follow the operation of "storing clear and releasing muddy". In a long run after that, Tongguan elevation remained stably. Since 1986, accumulative sedimentation occurred in the reservoir below Tongguan, resulting in the again rise of Tongguan elevation reaching 328.28 m after the flooding in 1995, rising by 1.64 m compared to the early stage when following the "storing clear and releasing muddy" in Oct. 1973.

Concerning the operation of Sanmenxia reservoir and evolution of Tongguan elevation, many scholars have done extensive researches. In 2003 to 2005, Yellow River Institute of Hydraulic Research, Tsinghua University, China Institute of Water Resources and Hydropower Research and Shaanxi University of Technology jointly carried out a in - depth and systematic study and proposed the recovery objectives and control measures for Tongguan elevation.

To cut down Tongguan elevation, many measures were applied. For example, from 1996 to 2003, the project of using jet flow for dredging was implemented at Tongguan reaches. Since Nov. 2002, the highest water level of Sanmenxia Reservoir in non - flood season is restrict to 318 m. The Dongluwan cut - off engineering was implemented in 2003. The Xiaobeiganliu warping test was initiated in 2004. The experiment of using and optimizing spring flood to scour the Tongguan elevation has been carried out since 2006. This paper has a discussion on the cause for relative stabilization of Tongguan elevation since 2006.

2 Evolutionary process of Tongguan elevation

After the operation of Sanmenxia in water storage, Tongguan elevation experienced a cyclic process as rise – fall – rise, reaching 328. 28 m after the flooding season in 1995, but remained stably from 1996 to 2001. (shown in Fig. 1).

In June 22 to 26, 2002, Weihe River and Beiluo River saw a high – sediment concentration and low flow, of which the peak flow at Huaxian Station was 890 m^3/s, with maximum sediment concentration of 787 kg/m^3, the maximum sediment concentration at Zhuangtou Station of Beiluo River was 453 kg/m^3, with peak flow of only 344 m^3/s, in this way, a hyperconcentration of sediment but small flood process was formed at Tongguan Station, with peak flow of 1,510 m^3/s, and maximum sediment concentration of 312 kg/m^3. After the flooding, the Tongguan elevation on one occasion came up to 329. 14 m, a historic maximum; through the scouring during flood season, Tongguan elevation was 328. 78 m, still at a high level.

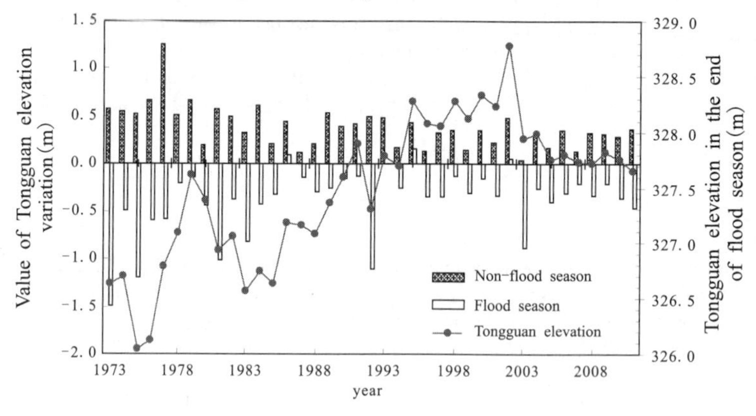

Fig. 1 Changes of Tongguan Elevation

In 2003 and 2005, owing to the impact of fall flood of Weihe River, Tongguan elevation fell down for a great extent. In 2003, there were 4 big flood in Weihe River, of which the largest peak flow at Huaxian Station was 3,540 m^3/s, that at Tongguan Station is 4,220 m^3/s, and the number of days with flow more than 3,000 m^3/s was as high as 13. Tongghuan elevation, due to the continued scouring in the floods, fell by 0. 71 m during flood season, fell by 0. 84 m in that year to 327. 94 m. During Sep. 29 to Oct. 11, 2005, Weihe River underwent a big flood when peak flow at Huaxian Station was 4,880 m^3/s, with maximum sediment concentration of 123 kg/m^3, the peak flow at Tongguan Station was 4,500 m^3/s, with corresponding sediment concentration of 37 kg/m^3. Such flood also resulted in a significant fall of Tongguan elevation down to 327. 75 m after the flooding season and restoring to the level in 1993 to 1994.

In 2006 to 2011, Tongguan elevation saw an average rise of 0. 30 m during non – flood season, but an average fall of 0. 315 m during flooding season, and a fall of 0. 10 m after 2011 flooding season compared to 2005, remaining stably. The remaining stably of Tongguan elevation, except for the impact of water – sediment conditions, attributes to both the control and operation of 318 m of Sanmenxia Reservoir during non – flood season and the spring flood scouring experiments.

3 Analysis on driving factors

Tongguan is located at the end of backwater of Sanmenxia Reservoir and the downstream of confluent area of Weihe River and main stream. For this reason, the rise and fall of Tongguan elevation is affected not only by water – sediment conditions, but also by the operation of reservoir and the boundary conditions of riverbed. The water – sediment conditions mainly reflect the strong or weak of stream dynamic conditions, the operation of reservoir has the impact on the upward sedimentation during non – flood season. In addition, the adjustment of scouring of reaches also affects Tongguan elevation.

The sensitivity analysis of factors shows that the fall or rise value of Tongguan elevation during flood season relates to stream energy of Tongguan reaches, scouring quantity during flood season of Tongguan reaches and the sedimentation quantity during non – flood season . Based on data from 1974 to 1996, empirical formula can be gotten as follows:

$$\Delta H_{1,000} = -7.314,5\gamma' WJ + 1.657,2\Delta W_{sX} - 0.276\Delta W_{sF} + 0.051 \quad (1)$$

where, W is for runoff in flood season, $\times 10^8$ m^3; γ' is the specific weight for muddy water, t/m^3; J is the average channel slope from Tongguan to Guduo; ΔW_{sX} is for sedimentation quantity in flood season, $\times 10^8$ m^3; ΔW_{sF} is for sedimentation quantity during non – flood season, $\times 10^8$ m^3; the unit of $\Delta H_{1,000}$ is m.

Fig. 2, taking 2009 and 2010 for an example, plots the relation between the change of Tongguan elevation and the coefficient of incoming sediment at common period and flood time during flood season. When the coefficient is less than 0.01 kg · s/m^6, the scouring occurs at Tongguan elevation, when more than 0.01 kg · s/m^6, sedimentation resulting in rise occurs at Tongguan elevation, and the rise value becomes increase with the increase of the coefficient. The riverbed conditions are different in different years, different with the change of coefficient of incoming sediment.

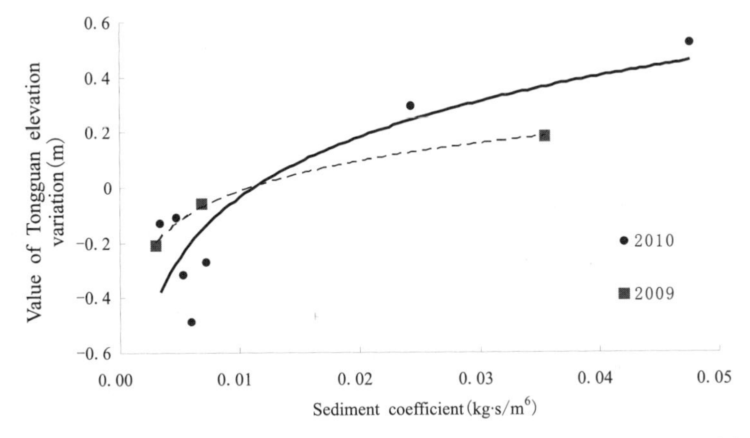

Fig. 2 Relation of change of Tongguan Elevation with the coefficient of incoming sediment during flood season

It can be seen that the change of Tongguan elevation during flood season has a large relativity with the runoff and boundary conditions; however, to similar boundary conditions, the change of Tongguan elevation has some correlativity with the coefficient of incoming sediment of stream.

4 Cause for stabilization in recent years

4.1 Change of water and sediment conditions

During 2006 ~2011, Tongguan Station had a average annual runoff of 23.47 $\times 10^9$ m^3, average annual sediment load of 1.846 $\times 10^8$ t. Compared with that in 1986 to 1995 when Tongguan elevation was in an accumulative rise, runoff reduced by 18%, sediment load reduced by 77%; if compared with that in 1996 to 2005, water flow increased by 16%, sediment load reduced by 65%; however, benefit from the dredging by jet stream at Tongguan reaches in 1996 to 2005 and the impact of the fall food in Weihe River in 2003 and 2005, Tongguan elevation reduced. Compared with annual average value, the water flow in 2006 to 2011 was not so much, with mean sediment concentration only of 7.86 kg/m^3, less than those in previous periods. In this way, the non-saturated sediment - laden flow increased the opportunity to scour the channel, but reduced the opportunity of sedimentation, and the Tongguan elevation remained at 327.7 m to 327.8 m, basically in a relative stabilization.

4.2 Impact by the distribution of sedimentation

Tongguan elevation reflects the scouring and deposition in riverbed. Related analysis shows that Tongguan elevation relates best (Fig. 3) to accumulative sedimentation between Huangyu36 - 41 (accumulative sedimentation is calculated from the flood season in 1973), the correlation coefficient thereof is 0.91. It can be seen that after the flood season, Tongguan elevation takes a linear relationship with the accumulative sedimentation quantity, becoming rise with the increase of it. After 2006, the sedimentation quantity in such reaches had a small change except due to small incoming water and sediment, also relating to the control operation of Sanmenxia Reservoir below 318 m in non - flood season and the experiment of using spring flood to scour Tongguan elevation since 2006.

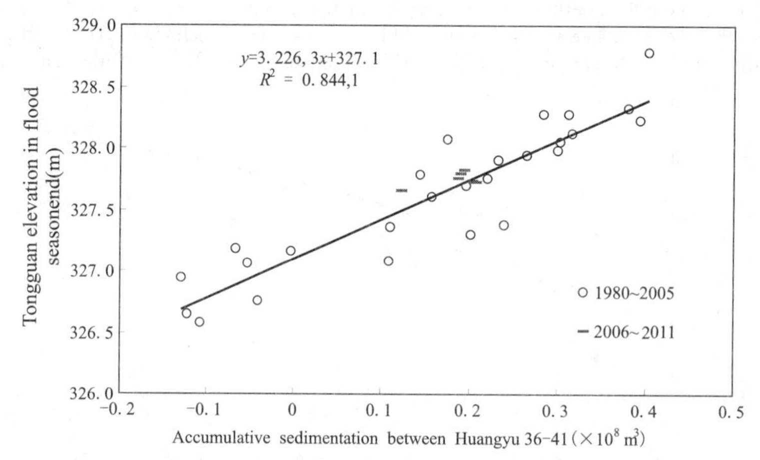

Fig. 3 Relation between Tongguan Elevation and the Accumulative Sedimentation between Huangyu 36 - 41

4.2.1 318 m control operation during non - flood season

Since the Sanmenxia reservoir had followed the "storing clear and releasing muddy" for control and operation in 1974, reaches of the reservoir below Tongguan featured sedimentation during non - flood season but scouring during flood season. With the drop of water level in operation during non - flood season, the sedimentation goes downward. During the initial stage following "storing clear and releasing muddy", the highest water level during non - flood season is controlled at 326 m or below, not more than 322 m since 1993, and not more than 318 m since Nov. 2002. After the highest water level during non - flood season is less than 318 m, the end of backwater is found around the crosssection Huangyu 34, so a long reaches at Tongguan or below is not directly affected by the water storage of reservoir. At the same time, the control operation at 318 m during non - flood season since 2003, the center of gravity of sedimentation of reservoir basin goes downward to Huangyu 31 or below, and reaches above Huangyu 33 is not affected by the reservoir operation, just in a natural evolution state (Fig. 4).

The adjustment of operational water level of reservoir turns the reaches of Huangyu 36 - 41

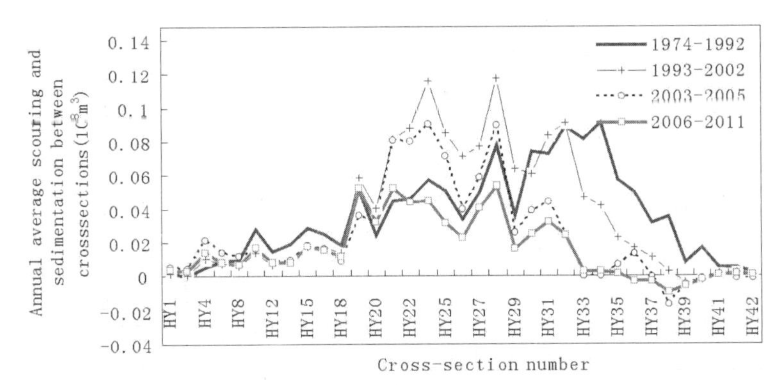

Fig. 4 Distribution of scouring and deposition below Tongguan during non – flood season

from the previous sedimentation to scouring. The low sediment flow during flood season produces sedimentation less than the scouring quantity during non – flood season. The accumulative scouring quantity in 2006 ~ 2011 was 3.35×10^8 m^3, which is conducive to the relative stabilization of Tongguan elevation.

4.2.2 Scouring experiment by spring flood

Since the scouring experiment by spring flood to lower Tongguan elevation in 2006, by the aid of adjusting the regulation mode of Wanjiazhai Reservoir during spring flood, optimizing the discharge flow, the Tongguan elevation is subject to scouring. The 6 years' experiment shows that through the optimized spring flood, peak flow at Tongguan Station reaches 2,500 m^3/s or above, the Tongguan elevation falls to some extent after the flood, ranging from 0.05 m to 0.20 m, average dropping by 0.11 m, an increase to some extent compared with the average dropping by 0.04 m in 1999 ~ 2005, the initial period after the operation of Wanjiazhai Reservoir when the average peak flow is 1,687 m^3/s.

The experiment not only directly cut down Tongguan elevation but also improved the distribution of scouring and deposition during non – flood season below Tongguan.

During spring flood, Sanmenxia Reservoir is in normal water storage operation during non – flood season, the reservoir basin is in a sedimentation situation, so the incoming sediment from main stream and Weihe River usually silts up in the reservoir below Tongguan. The scouring and deposition at the end of backwater or above is affected by the water – sediment conditions, the more the flow is, the more conducive to the scouring. The sedimentation distribution at the area affected by the backwater is affected by the initial water level of Sanmenxia Reservoir when the spring flood come in, the lower the initial water level is, the closer of center of gravity of sedimentation to the dam, and the more conducive to the reservoir's sediment removal by water lowering prior to and during flood season.

Fig. 5 shows the change of scouring and deposition of cross – section before and after spring flood. As viewed from it, scouring and sedimentation both occurred in cross – section Huangyu 26 below Tongguan. Scouring occurred at cross – sections above Huangyu 26, while sedimentation occurred at crosssections below Huangyu 22, of which sedimentation at Huangyu 19 section is the largest.

In this way, the sediment staying at the reaches above Huangyu 26 before spring flood goes downward to reaches below Huangyu 22 during spring flood, and sediment coming into reservoir basin during flood season almost all stay at the section below Huangyu 22 section, prone to be scoured out of the reservoir when operating at lower water level during flood season, which is conducive to reduce the accumulative sedimentation of reservoir basin and thus remaining the annual balance between scouring and sedimentation of reservoir.

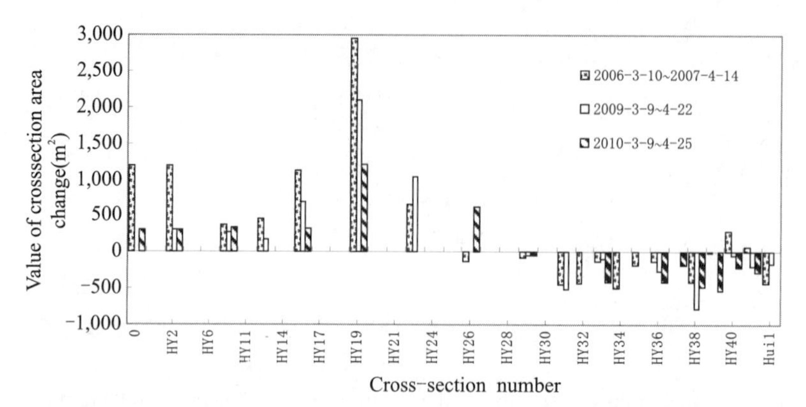

Fig. 5 Distribution of scouring and deposition below Tongguan during spring flood experiment

5 Conclusions

(1) Tongguan elevation, after the operation of Sanmenxia Reservoir, underwent an alternately change as rise - fall - rise. Under the 318 m control operation during non - flood season since 2003, through the flood scouring in 2003 and 2005, Tongguan elevation remained relatively stabilization in 2006 to 2011.

(2) Tongguan elevation change is not only affected by incoming water and sediment conditions and the operation of Sanmenxia reservoir, but also relates closely to the change of scouring and deposition, and takes a linear relationship with the accumulative sedimentation in reaches from Huangyu 36 to 41.

(3) In 2006 to 2011, Tongguan elevation remained relatively stable. Except for the significant reduction of sediment load, the 318 m control operation in non - flood season, the downward of center of gravity of sedimentation, the experiment of using spring flood to scour Tongguan elevation, improved the sedimentation distribution below Tongguan, effectively control the sedimentation of reaches from Huangyu 36 to 41, which played an important role in remaining the stabilization of Tongguan elevation.

References

Yang Qing'an, et al. Utilization and Study of Yellow River Sanmenxia Hydro - junction Project [M]. Zhengzhou: Henan People's Press, 1995.

Jiang Naiqian, et al. Study on Key Technologies regarding Dredging at Tongguan Reaches[M]. Zhengzhou: Yellow River Conservancy Press, 2004.

Yellow River Institute of Hydraulic Research (Huang Keji ZX - 2005 - 27 - 13). Summary of Dredging at Tongguan Reaches in 1996 to 2003[R]. October 2005.

Wang Ping, Jiang Naiqian, et al. Analysis of Scouring Deposition Effects in the Prototype Test of Sanmenxia Reservoir[J]. Yellow River, 2007 (7).

Lin Xiuzhi, et al. Analysis of Donglu Bay Tailoring during 2003 Flood Season[J]. Yellow River, 2005 (10).

Jiang Naiqian, Liu Bin, et al. Effect of Warping Test at Liulianbo Bottomland of Xiaobei Main Stream of Yellow River in 2004[J]. Yellow River, 2005 (7).

Hou Suzhen, Lin Xiuzhi, et al. Analysis of Test Using and Optimizing Spring Flood to Scoure Tongguan Elevation[J]. Journal of Sediment Research, 2008 (4).

Discussion on Measures for Floodplain Security Construction in the Lower Yellow River

Wang Hao[1] and *Qi Wenjing*[2]

1. Engineering Bureau, Yellow River Henan Bureau, Zhengzhou, 450003, China
2 Upper and Middle Yellow River Bureau, YRCC, Xi'an, 710021,China

Abstract: The floodplain on the lower Yellow River is the area for flood discharging and detention and sedimentation during the flood period, and is also the base of the people in the floodplain relying on for survival and development. At present, measures taken for floodplain security construction in the lower Yellow River mainly include on – the – spot shelter from flood, temporary evacuation and village relocation. Based on evaluation of the existing measures for floodplain security construction, and according to the requirement of social development and the actual problems faced with the floodplain harnessing in recent years, this paper presents a new measure for floodplain security construction, namely, constructing village platforms along the dikes in the area adjacent to the dikes. The flow direction of large flood should be considered as far as possible while arranging the direction of the village platforms along the dikes. Agricultural priority area and livestock farming priority area shall be positioned in order from the outside of village platforms to the flood discharge area of the river. At the same time, the proper areas within the livestock farming area and the flood discharging area of the river shall be reserved as ecological protection areas. This paper also makes economic evaluation of it, so as to provide a good thinking for further floodplain security construction in the future.

Key words: the lower Yellow River, floodplain, security construction

The floodplain on the lower Yellow River is the area for flood discharging and detention and sedimentation during the flood period, and is also the base of the people in the floodplain relying on for survival and development. With respect to safety against flood of the people in the floodplain and the special geographical situation of the floodplain, different measures are taken in flood control and harnessing of the lower Yellow River floodplain. For people's safety against flood, the main measures taken are village relocation, temporary evacuation and on – the – spot shelter; and for the problem of the secondary suspended river in the Yellow River floodplain, the main measures taken include silting up of along – dike trenches and string ditches.

1 Present situation and existing problems of floodplain security construction

1.1 Measures taken for floodplain security construction

The lower Yellow River floodplain is vast with a large number of population. To solve the problem of people's safety against flood, according to the present situation of the floodplain, natural geographical features of different reaches and flood characteristics, three schemes are taken for the lower Yellow River floodplain security construction.

1.1.1 Village relocation

To relocate the people to the back side of the dike so as to completely solve the problem of safety against flood of this part of the people. In principle, the villages "riding on the dike" within 1 km from the dikes, villages where the local governments and the masses are enthusiastic and that meet the requirement of relocation and the villages "falling into the river" shall all be relocated.

1.1.2 Temporary evacuation

The back up area of Fengqiu is less likely to be flooded, therefore the measure of temporary evacuation shall be taken, and the main task is to build roads and bridges connecting to the outside. For villages with the depth of flooding water less than 1m, the measure is to raise building platforms by the masses themselves as the masses customarily do, but due consideration shall be given to the measure of temporary evacuation.

1.1.3 On – the – spot shelter

Except the villages to be relocated or evacuated temporarily, the villages with the depth of flooding water on the floodplain larger than 1m or water depth within the villages large than 0.5 m shall build flood shelters for on – the – spot shelter.

1.2 Present situation of floodplain security construction

At present, the Yellow River floodplain security construction is mainly borne by the masses in terms of finance, with some allowance from the State. The finance channel of the State includes flood control funds, water damage relief, work relief and local funds of Henan Province. Since the great flood of the Changjiang River in 1998 (hereafter called "96.8" flood), the State has strengthened harnessing of major rivers, and the Yellow River floodplain security construction has reached an unprecedented scale. According to statistics, from 1998 to 2003, the State invested accumulated 424 million yuan in Henan Yellow River floodplain. After 2004, 133 million yuan was arranged with the ADB loan for village platform construction in the floodplain.

Since the "96.8" flood in the lower Yellow River was of high stage, some high floodplain areas that had not been flooded for years was submerged. After that, many villages in the floodplain began to move out and settled nearby on the back of the dikes. Up to now, altogether 20 villages with a population of 16.9 thousand in Henan Yellow River floodplain have been relocated.

Up to the implementation of the project of the floodplain harnessing financed with the ADB loan, 609 villages with 564.6 thousand people had had flood shelters, while 547 villages with 602.8 thousand people had not yet, accounting for 48.8% and 51.6% of the total floodplain population respectively. The existing flood shelters are mainly distributed over the reaches downstream of Dongbatou, of which the reaches between Dongbatou and Zhangzhuang have the highest proportion of shelters.

1.3 Problems in floodplain security construction

At present, with respect to the Yellow River security construction, there exist the following main problems. Firstly there is shortage of investment in flood shelter construction. The most recent investment was that invested in floodplain construction with the ADB loan in 2004. Secondly the flood shelters that have been built are of insufficient standard. Taking Henan Yellow River floodplain as an example, there are only 26 villages with 23.4 thousand people with platforms that are up to the standard, while there are 544 villages with 491.8 thousand people whose village platforms are not high enough. Besides, most village platforms have not been connected to form a certain scale. Thirdly, it is difficult to relocate villages. Due to problems such as finance, villagers' living habits, farmland capacity and re – division, et al. , village relocation is difficult to carry out. Fourthly, Roads for retreat are small in number and are of low standard.

At present, the general arrangement of the lower Yellow River floodplain security construction still adopts the overall settlement manner, that is, construction of village platforms as the main measure, supplemented with relocation of some villages, and temporary evacuation for low – risk areas.

2 Measure of building flood – avoidance connected platforms adjacent to dikes

2.1 General thinking for construction of flood – avoidance connected platforms adjacent to dikes

Based on the idea for security construction and development mode of the lower Yellow River floodplain presented by Liu Yun (Liu Yun and Li Yongqiang, 2012), we propose the following scheme for floodplain harnessing, namely, constructing village platforms along the dikes in the area adjacent to the dikes. The area of village platforms can be determined according to the number of population, long - term development as well as new countryside construction, with land for dike protection reserved. The flow direction of great flood should be considered as far as possible while arranging the direction of the village platforms along the dikes. Agricultural priority area and livestock farming priority area shall be positioned in order from the outside of village platforms to the flood discharging area of the river. The villages in those areas shall all be installed on the along - dike village platforms. At the same time, the proper areas within the livestock farming area and the flood discharging area of the river shall be reserved as ecological protection areas.

2.2 Design scheme for construction of flood - avoidance connected platforms adjacent to the dikes

With Xicheng Floodplain of Puyang as an example, comparison is made between schemes of building flood - avoidance connected platforms locally and building village platforms adjacent to the dikes. The Xicheng Floodplain corresponds to the left bank dike pile No. 66 + 100 ~ 105 + 000, with length of 39.0 km. Within the floodplain there were 123 villages and a population of 79.3 thousand (in 2005). The total area of the Floodplain is 135.50 km^2, of which 1.852×10^5 mu (1 mu = 666.67 m^2) is farmland, and 11.97 km^2 is occupied by villages. In the floodplain there are apple orchards, lotus ponds, fish ponds and vegetable greenhouses, with total area of 1.553×10^5 m^2.

With respect to construction of flood - avoidance village platforms, for the "rural" sector with protection of population less than 200 thousand, the flood control standard can be flood once in 10 ~ 20 years, the flood shelter in the floodplain is of the Grade IV, equal to buildings of Grade 4 (MWR, 1994). With the lower Yellow River flood process and special silting changes considered, flood once in 20 years is taken as flood control standard, and the corresponding design flood control discharge is that corresponding to the discharge of 12,370 m^3/s at Huayuankou. Concerning to the area of flood - avoidance connected platforms, and with current per capita homestead and living space of floodplain villages, per capita 80 m^2 shall be taken as standard for the top area of the flood - avoidance connected platforms (MWR, 2007).

Based on the above - mentioned standards, the Xicheng Floodplain of Puyang had a population of 79.3 thousand in the reference year (2005), and the natural population growth rate is assumed to be 7.8‰, and 50 design - level years later, the population will reach to 121.6 thousand, then 9.73 km^2 of land will be needed. While constructing connected platforms along the dikes, the width of the platform top needs to be 260 m. To facilitate the Yellow River dike management, and considering that the width of the flood - avoidance connected platforms should meet the requirement of settlement of floodplain population, the floodplain residential areas shall be positioned 100 m away from the dike root, and the area in between shall be filled up so as to be even with the village platforms.

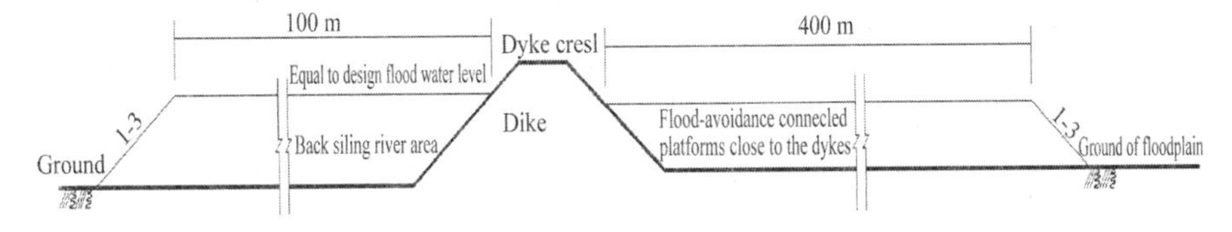

Fig. 1 Cross - section of flood - avoidance connected platforms along the dikes

2.3 Comparison with the original flood - avoidance village platform scheme

Under the existing overall planning, the standard of village platform construction is the same as it was before. Comparison will be made in several aspects.

2.3.1 Land occupation

The comparison is made only for some of the village platforms where villagers are installed. Since the flood – avoidance connected platforms are built near the dikes, the land occupied by side slopes would be somewhat reduced; this would reduce land occupation by side slopes of separate village platforms. With respect to total land occupation, total land occupation by flood – avoidance connected platforms would be 1.48×10^4 mu, 1.3×10^3 mu less than the 1.61×10^4 mu of land occupation by 20 dispersed large connected platforms. For needs of future urbanization and dike management, the connected platforms adjacent to the dikes would be widened to 400 m, then the land occupation would be 2.28×10^4 mu.

2.3.2 Macro investment comparison

If the flood – avoidance connected platforms adjacent to the dikes are just for resettlement, namely, they are not widened to 400 m, the construction will involve 6.92×10^7 m^3 of earthwork, which is somewhat larger than 5.37×10^7 m^3 of the original dispersed layout scheme. The reason for this is that there are many along – dike trenches and the terrain is low – lying in the Puyang reaches. According to overall calculation, without considering land occupation and resettlement compensation, the investment on the main work of near – dike layout will be 2,766,050 thousand yuan, large than 1,879,520 thousand yuan of dispersed layout by 886,530 thousand yuan. And if the platforms are widened to 400 m, the investment on the main work will be 3,039,620 thousand yuan.

Viewed in a simple way, dispersed layout is cheaper than near – dike layout in terms of total investment. On the one hand, the construction site is close to the main channel so that the unit price of earthwork is relatively low; on the other hand, since they are positioned on high floodplains at the original places, the construction quantities are relatively small. Yet, in actual, at present, the lower Yellow River is secondary suspended river to be harnessed, and the actual harnessing width will be over 500 m, and the filling – up level will be 0.5 m higher than local floodplains. From this point of view, building village platforms adjacent to the dikes will solve the problem of secondary suspended river at the same time. On the basis of the planned investment on the secondary suspended river harnessing, the total investment on the secondary suspended river harnessing and the dispersed layout construction will be larger than that on the construction of near – dike layout.

Secondly, in secondary suspended river harnessing, a lot of money will be spent on temporary land occupation, plus the temporary land occupation for dispersed layout, the total scale of temporary land occupation will be much larger than that of constructing flood – avoidance connected platforms adjacent to the dikes. Therefore, with Yellow River flood control and floodplain security collectively considered, the total cost of constructing flood – avoidance connected platforms adjacent to the dikes will be smaller than that of dispersed village platforms.

2.3.3 Control of large flood in the floodplains

With respect to the flood smaller than the design flood control standard of village platforms, during the flood period, the flood will not produce little effect on the masses of the floodplains. However, once the flood goes over the floodplains, since there exists along – dike trenches, the trenches will keep water for a long time after the flood, which will cause inconvenience to the masses going out. In case of flood larger than the design flood control standard of village platforms, during the flood period, since the dispersed village platforms are far away from the Yellow River dikes, the lives and property of the masses in the floodplains will be threatened, and it will be difficult to retreat temporarily. Comparatively speaking, for village platforms constructed adjacent to the dikes, on the one hand, proper defense may keep out somewhat large flood; on the other hand, the masses of the floodplains could retreat easily, and the lives and property of the masses will be protected well.

2.3.4 Effect of the scheme on flood control

With near – dike layout, during the large flood discharge period, the effect will mainly be restricted to reduction of the cross – sections, but the effect on the main stream direction and river regime change will be relatively small. With dispersed layout, according to the particular positions of dispersed layout, the effect on the flood routine on the floodplains will be much larger than that of the near – dike layout scheme. Secondly, the scheme of near – dike layout will eliminate the direct threat of flood discharge against dikes to the dikes, and guarantee the whole flood control system. Comparatively speaking, dispersed village platforms will speed up flow velocity of flood in the along – dike trenches, which will be disadvantageous to the dike safety.

3 Conclusions and Suggestions

3.1 Conclusions

According to the analysis above, the following conclusions can be made:

(1) With respect to construction of flood – avoidance connected platforms, it will be able to protect the masses in the floodplains against flood.

(2) Though the investment for construction of flood – avoidance connected platforms adjacent to the dykes is higher than that for construction of flood – avoidance villages platforms locally, construction of flood – avoidance connected platforms can play a role in the secondary suspended river training at the same time, therefore, its investment is less than that for construction of flood – avoidance village platforms in the floodplains in terms of the Yellow River flood control combined with floodplain security measures.

(3) Construction of flood – avoidance connected platforms has little effect on flood control in the floodplains.

(4) Construction of flood – avoidance connected platforms will free the floodplains so that they can be collectively used to fully develop ecological agriculture and ecological tourism and to further improve economic development in the floodplains.

3.2 Suggestions

This paper mainly made an analysis of the economic feasibility of the scheme and its advantage and disadvantage in terms of protection of the masses in the floodplains against flood. The effect on the scheme on the whole flood control system of the lower Yellow River needs to be further verified so as to provide further theoretical support for the scheme optimization. Besides, the presented scheme is a macroscopic one, and the comparison between schemes is mainly qualitative, while the quantitative comparison is relatively rough, and detailed comparison awaits further optimization of the scheme.

References

Liu Yun, Li Yongqiang. Discussion on the Security Construction and Development Model for the Floodplain in the Lower Yellow River[J]. China Water, 2012(6):30 – 32.

The Ministry of Water Resources of the People's Republic of China. GB 50201—94 Standard for Flood Control[S]. Beijing: China Planning Press, 1994.

The Ministry of Construction of the People's Republic of China. GB 50188—2007 Standard for Planning of Town [S]. Beijing: China Architecture and Building Press, 2007.

Preliminary Discussion on Construction Land of Flood Control Projects on the Lower Yellow River

Tan Hao, *Bao Ruchao* and *Pan Mingqiang*

Yellow River Engineer Consulting Co. , Ltd. , Zhengzhou, 450003, China

Abstract: Projects of reinforcing embankments by depositing and waterproof wall are mainly used on the Lower Yellow River. In this paper, the important effects of these measures in eliminating hidden danger of embankments, flood-fighting and improving eco-environment are analyzed. Through comparing these measures, it has a conclusion that reinforcing embankments by depositing takes an important role, and then the management mode of land expropriation and its important effects are discussed. Under the situation that land managements at the national and local increased and farmer' arable lands reduced, the land expropriation management mode about flood control constructions need innovating and improving. Departing from the requirements of the national land policy, economic and social development on the lower reaches of the Yellow River, the new land expropriation management mode is put forward that warping works are part of the flood control projects. In order to ensure the security of flood control projects, the manage department of the Yellow River expropriate the permanent occupation land of the projects of reinforcing embankments by depositing, at the same time, the silt body land also be used for farming. In the process of land expropriation, the farmers are paid fee of construction land compensation for agricultural production. The new land management mode is that, the water conservancy infrastructure is invested by the national, and the farmers acquire incomes from planting and the manage department of the Yellow River manage the construction land. The crop growth conditions on the lower reaches of the Yellow River, the reclamation design of silt body land, the tillage soil thickness and the soil composition are discussed in the paper. For the sake of long – term farming in silt body land to farmers, reasonable invest and feasible management mode are analyzed. This paper puts forward the opinion of management.

Key words: the Yellow River, problem of construction land, engineering construction, flood Control of the Lower Yellow River

1 The construction land problems of flood control projects in the Lower Yellow River

The lower reaches of the Yellow River is different from other rivers. The river has not been on the beach area for many years, many embankments hidden troubles are not easy to identify, once hit by a larger flood, the embankments will occur many dangers. During the "96 · 8" flood period, the Huayuankou peak flow was 7,600 m^3/s, there occurred 170 times danger accidents in the lower reaches of the Yellow River, seepage length 40,383 m, crack length 5,280 m, 8 piping, and 3 craters. From the geological exploration of the embankments, it can be found that the dry density of quite a lot levees is less than 1.5 t/m^3. The detection of embankments shows that there are many obvious hidden dangers in the levees, which are main loose body and cracks, and a small amount of holes. The most risks are 2 ~ 9 m below the top of the embankments, lying in the upper part of the levee. The Yellow River embankments are filled with sandy soil, even if built according to design requirements, it is still very difficult to control flood when huge flood is coming. The embankments basis is mostly sandy soil, mainly loam, sandy loam and silt. The formation structure is mostly double-layer structure or multi-layer structure. The Yellow River dike is made on the basis of the history of China dam, the foundation of which is not processed, so that it is of strong permeability. The Yellow River embankments had repeatedly burst flooding in the history, each burst blocking used a lot of soft materials such as straw willow, which further increased the foundation seepage strength. The lower reaches of the Yellow River dike reinforcement mainly need to solve the problem of dike burst. For the levee, it should not only meet the seepage stability requirements, but also eliminate levee damage caused by weak filling, poor soil and badger fox caves. Because of the complexity of the dike foundation on the lower reaches of the Yellow River, it is easy to concentrated seepage and seepage deformation. To prevent the dike foundation problems caused by soil flow, piping and other damages, it is suitable to adopt the dike reinforcing principle "cutting seepage on the before side, guiding seepage on the back side". Since the People's Yellow River harnessing, clay pipes, bamboo pipes, steel pipes, concrete pipe relief wells, cutoff wall, before after closure and reinforcing dike by depositing have been used to reinforce embankments. The measure of reinforcing embankments by depositing using the method fills up the low-lying ponds in the back side of the river, which enhances the stability of the levee, has a significant role to solve the vulnerability, seepage, piping and other dangers, and effectively improves embankments' flood defense capability. Because the sand of reinforcing embankments is taken from the river bed, and it can reinforce the embankments while reducing beach groove siltation combined with river regulation, therefore, it has always received strong support of the local government and the local people since 1970s. Compared to the measure of reinforcing embankments by depositing, the cutoff wall has the advantage not account for permanent land acquisition, however, the cutoff wall is hanging and the cutting seepage effect is limited. During 2000 ~ 2003, there are only 37 km cutoff walls built on the lower reaches of the Yellow River. With the "Recent key governance planning of the Yellow River Basin" and "Flood control planning of the Yellow River Basin" approved by the State Council, project of reinforcing embankments by depositing is planned as the mainly reinforcement measure, waterproof wall in reinforcement is used in hard implemented embankment sections.

At present, the management model of land expropriation about flood control projects is that the manage department of the Yellow River embankments expropriate the permanent occupied land of the flood control projects, the ownership and properties of land change. The collective land is changed into state-owned. Take the reinforcing embankments by depositing as examples, the silt body land is managed by the manage department of the Yellow River, and the top of silt body is planted with appropriate forest. The width of the embankments is 100 m. To relevant planning which needs nearly 500 km warping embankment sections, the land required enormous. In the context of per capita arable land is only 1.33 acres in the statistical part along the Yellow River in 28 counties and local economic construction is also a large number of land acquisition, land expropriation of flood control construction highlights the dilemma.

2 The land expropriation face more stringent policy constraints

According to the State Council ([1991] No. 74) called "large and medium-sized water conservancy and hydropower project construction land acquisition compensation and resettlement ordinance", the expropriation of arable land compensation fee, which is 3 ~ 4 times of the average annual output value per acre in the first three years that before arable land has been expropriated. The standard of the resettlement fee for each agricultural population, which is 2 ~ 3 times of the average annual output value per acre in the first three years that before arable land has been expropriated. The land expropriation of large flood control, irrigation and drainage engineering compensation standards can be lower than above land compensation standards. According to the regulations, the feasibility study of "Nine-Five" flood control project construction in the lower reaches of Yellow River used the standards that both land compensation and resettlement fees are 5 times of the average annual output value per acre in the first three years that before arable land has been expropriated. Implementation of the regulations not only provides important support of the land for the Yellow River flood control construction, but also provides a guarantee for the stability of the life of the masses on both sides of the Yellow River. With the rapid development of society economic, the demand of the land of economic construction is growing, and the state and local land management is becoming stringent. In 2004, the State Council issued "a decision on deepening reform of strict land management" ([2004] No. 28), which proposed that "the standard of land acquisition compensation and resettlement methods of large and medium-sized water conservancy and hydropower project construction are otherwise provided by the State Council" and "Provinces, autonomous regions, municipalities directly under the Central Government need to formulate and publish the uniform annual output value standard of the land expropriation of each city and county or district integrated land prices". Soon afterwards, on March 29, 2006, the new "large and medium-sized water conservancy and hydropower project construction land acquisition compensation and resettlement ordinance" is passed at the 130 executive meeting of the State Council. The Ordinance substantially increased the compensation standard, both the land expropriated compensation and resettlement subsidies of large and medium-sized water conservancy and hydropower project construction is 16 times of the average annual output value per acre in the first three years that before arable land has been expropriated. In accordance with the relevant provisions of the state, Henan and Shandong province each formulates relevant land expropriation policies.

3 The new model of land expropriation

Both temporary land expropriation and the only expropriation not turning land properties are two additional land expropriations which are different from the full land expropriation. Temporary land expropriation is limited to temporary project covers, and the deposit body is permanent works. Obviously, temporary land expropriation can not meet the requirements of engineering construction and management. With regard to the only expropriation not turning land properties, when the projects are completed, the land occupied by the permanent works is levied for the state-owned, with the nature of ownership changing and the manage department of the Yellow River owning the rights, at the same time, the deposit body land also be used in farming, and the land properties does not change. According to the "large and medium-sized water conservancy and hydropower project construction land acquisition compensation and resettlement ordinance" and relevant provisions, the farmers whose land is expropriated are paid full compensation for agricultural production. Because the only land expropriation not turning land properties does not change the nature of land, and not need the local government keeping the occupation-compensation balance, it do not have to pay tax on occupying farmland and farmland reclamation costs. From the point of view of the protection of the interests of farmers, it is better than the full land expropriation. The new model of land expropriation can not only protect the safety of flood control, protect the interests of farmers, but also strictly protect arable land. In the case that the land issue is obvious and part of the local land dispute has already affected the social stability, the method of only expropriation not turning land properties is undoubtedly a good way to mitigate land conflicts, and it provided innovative ideas for water conservancy and hydropower industry to crack the problem of land expropriation.

At present, the mode has been first attempted in construction in Guangdong and Guangxi province. The program that acres of orchards Wetland Reserve land has been expropriated using the mode of only expropriation not turning land properties has been approved in Guangdong Province, Guangxi Zhuang Autonomous Region government has approved 9 expropriation programs which used the mode, approving the land area of 505 hm^2. In the framework of existing laws and regulations, multi-industry groundbreaking attempt will provide an important accumulation of practice for the improvement of laws and regulations. Deposit area with the width of 100 m along the Yellow River embankments, compared to other flood control reinforcement measures, the particularity of the measure that reinforcing embankments with silt determines that it has congenital conditions to use the mode of the only land expropriation not turning land properties. In accordance with the strategic measures that silt build relatively "Underground River" in the Yellow River, the width of the reinforcing embankments with silt can be widened to 500 m. According to the mode of only expropriation not turning land properties, combined with the comprehensive management of flood plain on the lower reaches of the Yellow River, it provides ideas to solve the livelihoods and land issues of floodplain masses after occupied land was expropriated.

Combining the measure of reinforcing embankments by depositing with the new model of land expropriation, the land supply will increase. Along the outside of the lower reaches of the Yellow River embankments, there are a large number of ponds with long-term water, so that the land can not be cultivated. If taking the measure of reinforcing embankments by depositing and reclamation transformation, and using the land expropriation mode of only expropriation not turning land properties, farmland area will increase. According to estimation, newly arable land increased by about 10%.

Face to the dilemma that it is difficult to solve the land acquisition due to the situation that there are larger people but less land in the cities and counties along the Yellow River in Henan and Shandong Provinces, the land expropriation mode of only expropriation not turning land properties provides a guideline, and also it gets strong support of Henan and Shandong provinces government.

4 Feasibility analysis of the new mode of land expropriation

According to the "land reclamation Ordinance" that was promulgated in the 2011, the permanent occupation land damaged by water conservancy and other construction needs reclamation, but the deposit body is permanent structure, with a height of 6 ~ 7 m from the ground, the conditions of which is different from general reclaimed land and original farming land, so it is important to make the top of deposit area reach general farming conditions and guide farmers to have a long-term cultivation through reasonable management mode.

The areas on the lower reaches of the Yellow River is mainly growing wheat, corn, cotton, peanuts, potatoes and other crops. Taking the iconic substances organic matter that indicating soil fertility, and the Nitrogen, phosphorus and potassium content that reflecting the soil macronutrients on crop supply capacity as indexes, according to the soil survey and sampling analysis on reinforcing embankments with silt which has formed many years, in addition to the available potassium in relatively good condition, available phosphorus, available nitrogen and soil organic matter are all in poor condition, therefore, to the deposit area with rehabilitation and land acquisition using the only expropriation not turning land properties mode, it is possible to fill the cultivating layer at the top of the deposit area with original arable land. In the process of cleaning base of reinforcing embankments with silt, keeping piling up the original arable land, when the reinforcing embankments with silt construction is completed, the clear base soil is used as the tillage layer, in order to meet crop on organic matter, nitrogen, phosphorus, potassium and other nutrients needs. If the clear base soil in the construction is difficult to meet the needs of overburden layer, it can get tillage soil in adjacent deposit area according to the construction of arrangements. Due to the sandy soil of the deposit body lack of clay and soil organic matter, the soil structure and adsorption effect is very weak, the soil particle absorbing fertility and soil fertilizer conservation are very poor. It must arrange the water retention layer in the lower part of the sand layer.

The tillage layer is of good physical and chemical properties and high nutrient content under the influence of farming and cultivation measures, the thickness is about 20 cm or so, having a high degree of maturation, and also a main soil between the soil crop growth and soil management. The layer is mainly filled with clean base soil. The core soil is between the topsoil and subsoil, and that is the deep soil difficult to cover when cultivating, which is formed of materials to withstand topsoil leaching. Taking the root of crop widely grown on the lower reaches of the Yellow River as an index to analyze, the wheat is strong adaptable to soil, which roots mainly distribute within 50cm of soil, the maximum root depth is 1.2 m, but more than 80% of the root distribute at a concentration of 0 ~ 40 cm soil layer. Taking the moist soil layer depth as an index to analyze, the moist soil layer depth of wheat is 40 ~ 80 cm in growth period, and the moist soil layer depth of cotton is 40 ~ 70 cm. Analysis from the conditions of plant growth, The overall farming overburden thickness can be controlled by 80 cm, and comprehensively considering the reclamation investment, construction difficulty and appropriate soil material sources, farming soil layer thickness about 50 ~ 80 cm can meet the basic requirements of the crop.

For the convenience of farmers to grow and achieve earnings, the reclamation design should pay attention to facilities design of irrigation and drainage facilities, road and other auxiliary facilities of agriculture production. Combined with special circumstances such as flood control and farming requirements, the manage departments of Yellow River embankments manage the construction land and the farmers acquire planting income. According to the principle "who is benefited, who bears the cost", the cost of production is assumed by the farmers.

In order to strengthen the deposit area land management, considering the deposit flood control function and agricultural functions, the manage department of Yellow Rive embankments should sign an agreement with the local village committee, clear responsibility, farming tillage, crop types and related income distribution, and punishment measures. During the flood season and the subsequent construction, deposit land should first guarantee for flood control and engineering construction.

5 Conclusions

The new mode of land expropriation is a breakthrough compared with the existing land expropriation. In the case of guaranteeing flood control safety, it has great significance in safeguarding national 1.2×10^8 hm^2 farming land, reducing social contradictions, and safeguarding the interests of farmers. At the same time, it is good for the safety of flood control and the interest of the farmer to establish corresponding management system and strengthen management. The new mode of land expropriation needs to be perfected in the further.

Referenrces

Cheng Yi, Zhou Feng. Discussion on Object Survey and Compensation for Relocatees in Water Conservancy Project[J]. Yangtze River, 2010(23).

Chen Zhen. Analysis on and Selection of the Policies Concerning the Compensation for Land Expropriation and Requisition[J]. Natural Resource Economics of China, 2006(10).

Shen Xiaomin, Zhang Juan. The Comparison and Reference of Land Expropriation System about Chinese and Foreign[J]. Economic Research Guide, 2007(06).

Wang Shujuan. Discuss the Perfect of the System of Land Tequisition Compensation[J]. Public Administration & Law, 2006(07).

Chen Bochong, Hao Shouyi. Analyses of Land Acquisition Compensation[J]. China Rural Survey, 2004(06).

Wang Jingmei. Real Predicament and Way out of Land Expropriation System in Our Country[J]. Journal of Shenyang Institute of Engineering (Social Sciences), 2011(02).

Research on Prevention and Protection Technology of Soil and Water Conservation and its Adaption

Zhang Yongjiang[1], *Zhang Huiyao*[1] and *Zhai Ran*[2]

1. Soil and Water Conservation Supervision Bureau In Areas Bordering of Shanxi, Shaanxi and Inner Mongolia, Yulin, 719000, China
2. School of Earth Sciences and Engineering of Hohai University, Nanjing, 211100, China

Abstract: Though prevention and protection has been taken as the most important policy in water and soil conserve strategy, to implement the prevention and protection is particularly difficult in present utilization system of resources. In order to establish the conservation - awareness, implement the eco - environmental protection, management should be strengthened through strict law enforcement, policy guidance and economic levers.
Key words: soil and water conservation, prevention and protection, technology, research

In the Soil and Water Conservation Law of the People's Republic of China (hereafter called the Law of Soil and Water Conservation), prevention focused, protection first has been taken as the most important policy in water and soil conserve strategy. This progress is not only a tribute to Scientific Development View of harmonious between the man and nature in soil and water conservation, but also the experience of long - term soil and water conservation in China. Protection of vegetation cover has been taken as the first strategy in soil and water conservation planning. Lock the stable door after the horse has gone, rehabilitation of gullies and slopes has been taken place by the strategy of prevention and protection, such as reconverting farmland to forest (or grassland), ecological migration, natural vegetation protection.

In the intensive erosion center of the Loess Plateau, especially the main source of coarse sediment of the Yellow River, the task of implementing the strategic decisions about natural vegetation protection and ecological rehabilitation is very difficult.

Most of the main source of coarse sediment of the Yellow River located in the contiguous areas of Shanxi, Shaanxi and Inner Mongolia. Natural factor, geological structure and climate result in fragile ecological environment. Because of flimsy geological structure, very little but concentrated rainfall, which is the most important factors cause soil erosion, gullies criss - cross this area. Nature can rehabilitate the soil erosion cause by nature itself, thus build dynamic balance. However, the human races go against the natural law, human activities are so frequent that the balance has been broken, soil erosion become severe. Extracting materials from the earth and capital construction leading environmental destruction become the great agent of soil erosion. Natives still farming the slope land and Blown - sand soil, which yielded very little. The more they farm the poorer fields would be, those activities promoted the soil erosion. Study shows that in construction projects concentrated Ulan Moron River basin, added erosion 1.53 times controlled soil erosion. If we won't do better, soil erosion area and intensity will continue increasing. Not only rehabilitate, but also prevent.

1 Protect the natural vegetation

1.1 Focus on natural vegetation protection while drawing up the integrated rehabilitation plan in accordance with local conditions

Natural vegetation growing up within long - term nature evolution, accommodate itself to soil and rainfall, be harmonious with nature. Vegetations become lush when precipitation increase, to protect the surface soil better. If precipitation decrease, vegetation's growing slow down or even stop to save the soil water, can also protect the surface soil. This virtuous cycle could evolve within climate, and then protect the surface perfectly. Within the evolution power, grassland could evolve to forest without external interference. While drawing up the integrated rehabilitation plan in accordance with local conditions, instead of the policy that sow grass on which can grow grass, plant trees at where can grow forest, adopt the natural vegetation protection when eroded soil is less than allowable scope. In the past, prosperous weeds has been changed into fish scale pit to plant trees, which now be suspected of destroying the vegetation.

1.2 Avoid or protect natural vegetation as much as possible in construction project

Provisions specified under Article 24 of the Law of Soil and Water Conservation that the location or route of a construction project shall not be within a key area for prevention or management of soil and water erosion; Where this is not possible, the standards for prevention and management shall be raised and processes optimized to reduce surface disturbance and the scope of plantation damage and effectively control possible erosion of soil and water. As growing up within long - term nature evolution, Natural vegetation is valuable, once destroyed; recovery is hard or even impossible. In the source region of the Yellow River, there is a sort of grassland called stony - hill meadow steppe, on which meadow grows on sand and gravel, become a particular natural vegetation. Even the meadow be broken in very small scale, once the sand and gravel be exposed to the sunlight, its temperature can goes up to 60 ~ 70 ℃. The heat perished surrounding vegetation; the grassland goes like human beings' alopecia, and soon dies out. Natural vegetation, especially which grow in river source region or sources protection zones for domestic and drinking water, evolve for a long time, must be protected.

Road - works should be with the construction project. Of two evils, choose the less, instead of decreasing the cost, we should protect the natural vegetation. Under no circumstances should we seek temporary economic development at the expense of the environment and resources. Financial can no more be in effect when environment beyond retrieve.

2 Protect soil, vegetation, ecological environment resources with economic levers

The ecological compensation is a system designed to re - modify the development balance in order to avoid the ecological resources deployed retortion, which is one of the most important policy to implement economic regulation. As real public resources, soil, vegetation, ecological environment resources should be regulated and managed only by the government. With the ecological compensation system, users pay for fees that caused by environment protection and protectors benefit.

2.1 The system of utilizing resources with compensation should be established

Diseconomy in using resources become the impetus for overexploitation even prevents soil and water conversation. As requested, the volume of the fills should be equal to the volume of the cuts in one project, but balanced earthwork in project means higher cost, hence soil, vegetation, ecological environment resources cannot be effectively utilized. If relative organizations tax on natural resources so that cost for a nearby cut is higher than balanced earthwork. This measure could urge constructor to concern resources utilization, so that quantity of earthwork could be nearly halved, and soil, vegetation, ecological environment resources are protected effectively.

2.2 Government pays for ecological compensation

On important ecological function divisions, the sources of drinking water, the sources of rivers and lakes, intensive ecological environment disturbance should be restricted to remain the ecosystem, so that its accommodation could functions. In such divisions, government restores the natural ecology; make decision to protect the ecological environment, increase natural ecology restoration input, Reinforcing conservation district supervision. As a sort of compensation, local government and residents benefit from government input ecology protect project.

2.3 Ecological compensation between upstream and downstream in ecological function divisions

The protection projects in the upper reaches cost much. Residents upstream offer manpower, financial resources and materials for ecological environment protection from which the construction projects downstream benefit a lot. Meanwhile, ecological function divisions' economic is limited because of restricted in construction projects. To resolve these economic problems, government should find ways to solve the unbalance on the one hand, on the other hand, the resource user downstream should pay for the ecological protection upstream by ecological compensation ways. As one of compensation measure, soil and water resource users downstream pay the fees that called fee for soil and water development and utilization, which will be used for the ecology protection and rising the quality of life of the residents has downstream. In order to increase input for water conservation, water resource protection, returning land for farming to forests and grasslands, ecological migration, energy conservation and pollution reduction in the region upstream, users downstream can also invest the ecology management upstream directly to take part in the ecology protection.

2.4 Intensified collection of charges on soil and water conservation compensation in resources development

Provisions specified under the second paragraph of Article 32 of the new Law of Soil and Water Conservation that where a construction project or other production or construction activity conducted in a mountainous area, terrain, sand area or other area defined as vulnerable to soil and water erosion damaged the conservation facilities, geography and vegetation and the conservation function cannot be restored, compensations shall be paid and used exclusively for erosion prevention and management. It provides legal backing to intensified collection of charges on soil and water conservation compensation in resources development. Before this, collection of charges on soil and water conservation compensation in resources development base on the Rules for the Implementation of the Soil and Water Conservation Law. This method not only is at low standard so that can hardly modify the diseconomy in soil and vegetation utilization, but also has no reliable legal basis that be forbidden by many construction projection. To resolve this problem, Document Shaanxi Zheng Fa (2008) No. 54 had been issued by the provincial government of Shaanxi Province, in which there was a document called the Measures for the Administration of Collect and Use for the Soil and Water Erosion Compensation in the Exploitation of Coal, Oil and Natural Gas Resources, and became operational on January 1, 2009. According to Article 4 of the Measures, the calculating and levying standard is: coal 5 yuan/t in northern Shaanxi, 3 yuan/t in Central Shaanxi, 1 yuan/t in southern Shaanxi; crude oil 30 yuan/t, natural gas 0.008 yuan/t. The calculating and levying standard of soil and water compensation had been greatly augmented, so that mining enterprises and individuals shall pay for the soil and water erosion caused by their mining.

3 Implement the policy of ecosystem restoration

In prevented & protected areas and areas with fragile ecosystem, now residents convert their life style from farming to ecology tourism and rotation grazing. Which can only reduce the direct destroy. Especially in water conservation areas, human activities' indirect influence on ecological environment would cause new environmental pollution, which would influence the vegetation ecological restoration. Economic compensation should be given to residents to reduce their disturbing to the ecological environment, gradually implement ecological migration, and let human being leave these areas to help the ecological environment restoration.

3.1 Natural ecology restoration project

Natural ecology restoration project, use natural ecological balance and the natural ability of evolution, within proper protection, to restore the ecological system's function of climatic regulation, soil and water conservation in a certain area step by step.

Natural ecology restoration project should be implemented in some healthy ecosystem such as the three – river source area, the secondary growth area of the first level branches of the Yellow River, and the secondary growth area of the Ziwuling range in Mount Liupan. Though the vegetation in these areas is in good condition, disturbing from human activities bring them into degeneration, so that the intensity and area of soil erosion and desertification are increasing. To restore the ecological balance, the quickest way is implement natural ecology restoration project. Human activities in these areas should be forbidden, restore the ecological nature by rotation grazing, pasturing prohibition, reconverting farmland to forests and grasslands, ecological migration.

In some areas with fragile ecosystem, the population is thin, the precipitation is little, artificial rehabilitation is hard to implement and has lot to do with, lack of manpower, financial resources and materials hasn't been undertook. After ecological migration, the government could limit human disturbing, then implement natural ecology restoration project.

3.2 Reconvert farmland to forests and grasslands

The program to reconvert farmland to forests and grasslands has many success experience in our country。The project is making good progress from pilots been implemented in 1999. Notable successes have been attained, land greening is speeding up, vegetation is increasing, soil erosion intensity and hazards of sand storms has been lessened. This is a big contribution to the ecological environment protection and rehabilitation. Meanwhile, peasant farmers can receive direct subsidization from the reconvert program. Grain and cash for living expenses become the majority component of their incomes; those farmers receive the ecological compensation. In more recent years, a circular was issued by the Central Committee, to implement ecological environment protection and construction, which putting emphasis on Natural Forest Protection Project, afforestation of barren hills and wasteland suited to afforestation, Reconverting farmland to forests and grasslands on steep hillsides step by step. The most important principle of the reconverting program is, tailor measures to suit local conditions, specific guidance for different localities, seeking the truth out of facts, strive for real effect. Bring about harmony of economic returns and contribution to society and environmental protection. Plans should be carefully worked out to provide effective policy guidance for this work, and the wishes of farmers should be respected, efforts will be made to promote the advancement of science and technology; demonstration driving, Steadily Promote; provincial government in charge of all duty, system of target responsibility to governments at different levels. Consequently, the policy reconverting farmland to forests and grasslands will be standardized and normalized gradually in prevented & protected areas and areas with fragile ecosystem.

3.3 Ecological migration project

People live in areas where ecological environment is abominable, resources is meager, farmland is limited, economically backward, soil erosion is severe, their lifestyle make the ecological environment worse.

Ecological migration is that to protect the ecology in a certain area or to rehabilitation. In some region where environment is severe, ecology migration is to move resident to areas where can afford poverty alleviation.

In the migration from the area of the Three Gorges Project, except for migration of cities, towns and enterprise, there are some people which live below poverty line, or live in the area with fragile ecosystem or severe soil erosion, this migration is one sort of ecology migration.

In Ningxia Hui Autonomous Region, the government move the poor from remote, ecological unbalanced and drought areas to the nearby of pumping water irrigation project, highway and suburbs. Migrations work on the Basic Farmland, and their income increase Stably, Living Standards Further Improved. In areas with fragile ecosystem, implement ecological nature restoration within enclosure after migration.

4 Strengthen the degree of soil and water conservation law enforcement

The diseconomy in soil, water and vegetation utilization, reduce the ecological environment resources protecting consciousness of people. To restore the ecological environment resources protecting consciousness, we need to stand the concept of living in harmony with nature, scientific thought of development and the legal awareness of ecological environment protecting. We should standardize ecological environment protecting in the production and construction, and crack down on exploitations environmentally disastrous within application of the law. Ecological environment protecting in production and construction should be encouraged, so that ecological environment protecting and restoring could be compensated and approved by our society.

4.1 Strengthen the degree of soil and water conservation law censorship and enforcement

In the new Law of Soil and Water Conservation, the legal rules are clear about soil and vegetation resources protection. However, large - scale production and construction project phases only on the major project, soil, water and vegetation resources protecting project even be neglected in small - scale production and construction project. Hence, implement the new Law of Soil and Water Conservation, need the policy everyone should abide by the law, and those who violate the law must be dealt with standardize the exploitation. Nowadays, the examination and approval of the water and soil conservation programme, soil and water conservation censorship, and water and soil conservation facilities checking for acceptance are important ways to ensure that constructor implement the Simultaneous Policy.

4.2 Enforce the management, protecting and patrolling in ecology restoring area and prevented & protected areas

In the long - term soil and water conserving comprehensive harness practices, the operator summing up experience 30% control, 70% manage. Areas with fragile ecosystem or severe soil and water erosion were always troubled by poverty and backwardness; people there have no sense of ecological environment protection. Therefore, implement ecological migration to protect the soil, water and vegetation resources on the one hand, enforce the management, protecting and patrolling in the sources of rivers and lakes, the sources of drinking water, ecology restoring area and soil and water conserving comprehensive harness areas on the other hand. We should build up a patrol to stand the concept of environment protect; sign out the protected areas, promote environmental protection, investigate and punish the violation of environmental laws.

Soil and water prevent and protect policies are the best way to protect the area with fragile ecosystem, de - evolving ecologically sound area and scenic sites which we've long been proud of. These policies help the ecological environment to rehabilitate, restore its function of ecological nature rehabilitation. This is not only the shortcut to recover the ecological balance system, but also the only way.

References

Lin Minghua. Prediction Technique Research on Growing Soil Erosion in Construction Projects in the Middle Reaches of the Yellow River[J]. Bulletin of Soil and Water Conservation, 2006.

Zheng Xinmin. The Ecological Problem and Solution in the Source Regions of the Yellow River [J]. Yellow River, 2000.

Yang Shaolin. Ecosystem Restoration and Its Supporting Measures[J]. Soil and Water Conservation in China, 2004.

Zhang Yongjiang. Ecological Function Division in Soil and Water Conservation Plan in Sanchuan River Case[J]. Soil and Water Conservation Science and Technology in Shanxi, 2006.

Zhang Yongjiang. Compensating Mechanism of Soil and Water Resources[J]. Soil and Water Conservation in China, 2006.

Analysis of the Effectiveness of Prototype Test for Lowering Tongguan Elevation by Utilizing and Optimizing Spring Flood in 2007

Chang Wenhua, *Hou Suzhen*, *Lin Xiuzhi* and *Hu Tian*

Yellow River Institute of Hydraulic Research, zhengzhou, 450003, China

Abstract: the paper has introduced the processes and results of the prototype test for lowering Tongguan elevation by utilizing and optimizing spring flood in 2007. Some achievements were gotten, such as the water level of Tongguan (6) at the discharge of 1,000 m^3/s decreased for 0.05 m, about 3.47×10^6 t sediment in Xiaobeiganliu reach was scoured, about 14.45×10^6 t sediment in Wanjiazhai Reservoir was scoured, the topset bed of the delta war scoured and pushed forward for 5.5 km. Comparative analysis against the scouring effects of Tongguan elevation in 2007 with that in 2006 has been made in three aspects include the water level of Tongguan(6) at the discharge of 1,000 m^3/s, the average bed elevation and the area of cross section. The results show that the test in 2007 gets good effectness. The causes for slight lowering of Tongguan elevation in 2007 include great increasing of roughness factor in Tongguan reach, great reduction of discharge and slight lowering of water level by backwater produced. It also points out that the siltation changes in Tongguan, with the water level at the discharge of 1,000 m^3/s in Tongguan (6) as one of the representations for Tongguan elevation, sometimes cannot thoroughly reflect the variation in bed siltation.

Key words: spring flood, prototype test, Tongguan elevation, peak discharge, sediment concentration

1 Foreword

Tongguan section lies in the lower reaches of the confluence area between Yellow River and Wei River, and the lifting of Tongguan bed is of great importance to the flood prevention issues in the Lower Wei River. For this reason, the changes in Tongguan elevation always attract attentions of people. Since the lifting of Tongguan elevation is worked by multiple factors comprehensively, then only adoption of multiple measures can lower the elevation of Tongguan. In recent years, the Yellow River Conservancy Commission has taken many measures to lower the elevation of Tongguan, such as "Adjust the Operational Mode of Sanmenxia Reservoir", "Desilt Tongguan Reach of Yellow River" and "Curve and Cutoff of Dongluwan". In addition, it also studies the measures employed for lowering Tongguan elevation such as "Remediation for Estuary of Wei River" and "North Luo River Diversion and Direct Flowing into Yellow River". The "Test for lowering Tongguan Elevation by Utilizing and Optimizing Spring Flood" is a new measure taken to lower Tongguan elevation by virtue of scouring processes of floods.

Spring flood is a flood process produced by the breakup of ices in Ningxia – Inner Mongolia reaches. With big flood peak and flood volume, it possesses certain effects on lowering Tongguan elevation by scouring. Tongguan elevation has been lowered by 0.19 m during the spring flood season in 1987 to 1998. Upon the operation of Wanjiazhai Reservoir in 1998, impound and cut off the flood peak in spring flood season to reduce the flood peak and flood volumes discharged to Tongguan and convert the former scouring lowering of Tongguan elevation to basically scouring free or slightly siltation. With respect to these facts, the Yellow River Conservancy Commission has conducted "Prototype Test for Lowering Tongguan Elevation by Utilizing and Optimizing Spring Flood Scouring" and has lodged the test purpose as "Utilize and Optimize Spring Flood Processes, Lower Tongguan elevation, Improve the Sediment Silting Conditions in Wanjiazhai Reservoir and Sanmenxia Reservoir, Further Deepen the Recognition to the Water – sediment Moving Rules in Yellow River". The test is a success and has reached the anticipated targets.

The "Prototype Test for Lowering Tongguan Elevation by Utilizing and Optimizing Spring Flood

Scouring" has been conducted again in 2007. To bring the scouring effects of spring floods into full play, thoroughly utilize the steering capacity of Wanjiazhai Reservoir and considers the comprehensive benefits of the Reservoir, shorten the test time, further optimize spring flood processes, adjust the steering indices according to the current boundary conditions and determines that the scouring conditions for Tongguan station is the peak discharge of 2,800 m^3/s and maximum five - day flood volume of 9×10^8 m^3.

2 Test processes

2.1 Breakup conditions and water discharge processes

In the middle ten days of February, 2007, Inner Mongolia reach of Yellow River starts to break up. In the first ten and middle ten days of March, the freezing air moves actively in Inner Mongolia, and the breakup speed of Inner Mongolia reach slows due to the weather conditions. On March 23, 2007, the Inner Mongolia reach has been fully broken up.

Toudaoguai station breaks up on March 16 with a daily discharge of 890 m^3/s, which increases to 1,110 ~ 1,220 m^3/s from March 17 to March 20, rises at 16:00 p.m. on March 20 and reaches the maximum value 1,700 m^3/s at 16:00 p.m. of March 22. Then it gradually falls after rise and falls back to 1,030 m^3/s at 06:30 a.m. on March 25, see fig. 1 for the flood process. The maximum mean daily discharge in spring flood seasons at Toudaoguai station is 1,680 m^3/s, mean daily sediment concentration is 11.6 kg/m^3, while the maximum 10 - day water and sediment content are 10.96×10^8 m^3 and 7.35×10^6 t.

2.2 Optimize spring flood processes in Wanjiazhai Reservoir

Control and apply Wanjiazhai Reservoir at a value no less than 966 m since March 4. In order to meet the demands for water replenishing and to ensure ice prevention safety, Wanjiazhai Reservoir adopts "First Impounding, then Replenishing" method to gradually control the discharge flow from 16:00 p.m. on March 17 and impound, and the highest water level 972.72 m reaches at 8:00 a.m. on March 19 and the water volume impounded is 1.27×10^8 m^3 and the inflowing discharge is 1,230 m^3/s. Control discharge and water replenishing by different discharge standards as 1,500 m^3/s, 2,000 m^3/s, 2,500 m^3/s and 3,000 m^3/s from 13:00 p.m. on March 19 until the completion of replenishing at 15:00 p.m. on March 23, when the water level in Wanjiazhai Reservoir is 952.18 m and the water replenished is 2.82×10^8 m^3. The maximum instant discharge flowing into Wanjiazhai Reservoir is 1,700 m^3/s and the maximum instant outgoing discharge is 3,020 m^3/s, the maximum daily mean drainage discharge is 2,830 m^3/s. See Fig. 1 for the reservoir inflow and outflow discharges.

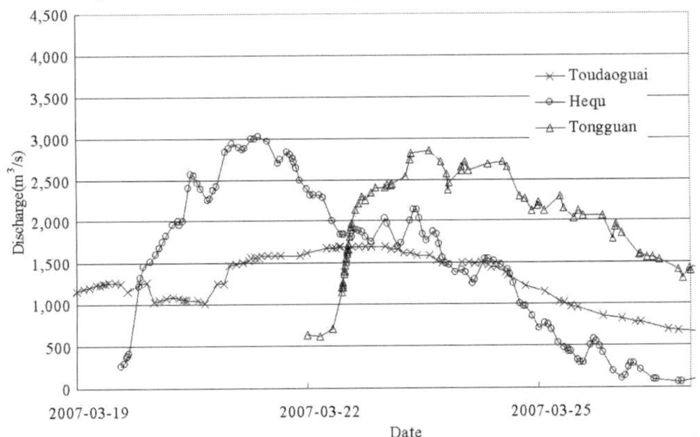

Fig. 1 Flood transmission process in each stations during 2007 spring flood test period

2.3 Fig. 1 Features of spring Flood in Tongguan

The flood peak at Tongguan station rise quickly after adjustment in Wanjiazhai Reservoir, from 620 m^3/s at 4:00 a.m. on March 22 to 2,850 m^3/s at 14:00 p.m. on March 23, then falls back to 2,460 m^3/s at 20:00 p.m. on March 23, but rises again to 2,720 m^3/s at 12:36 a.m. on March 24, and then gradually falls back to about 300 m^3/s on March 31. The maximum daily mean discharge during spring flood season at Tongguan station is 2,640 m^3/s, the maximum daily average sediment concentration is 24.7 kg/m^3, while the maximum 10 - day water and sediment volumes are 12.96×10^8 m^3 and 20.1×10^6 t respectively.

3 Test result

3.1 Scouring against Tongguan cross section

After test, Tongguan reach shows obvious scouring and its main channel is scoured deeply and widely (see Fig. 2). The area of main channel after Tongguan (6) cross section test compared with that before test increases by 525 m^2, the mean bed level of the main channel reduces by 1.58 m, the complete sectional area increases by 458 m^2, the mean bed level increases by 0.79 m; while the water level at the discharge of 1,000 m^3/s reduces by 0.05 m. The complete sectional area of Tongguan (8) increases by 152 m^2, its mean bed value reduces by 0.20 m, and the water level at the discharge of 1,000 m^3/s reduces by 0.08 m.

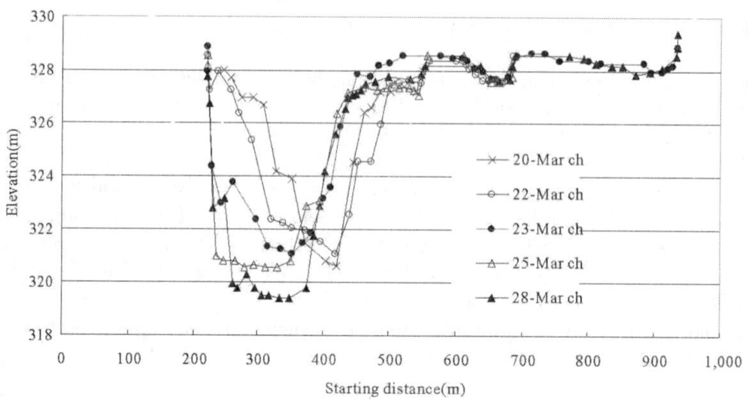

Fig. 2 Scouring against Tongguan (6) cross section

3.2 Improve the silting situations in Wanjiazhai Reservoir

For normal operation of Wanjiazhai Reservoir in the spring flood season, gradually reduce the reservoir stage to 960 ~ 965 m before the spring flood season for the safety of ice prevention, impound water and cut the flood peak during spring flood season until it completes; with an annual impounding volume of 3.00×10^8 t, Wanjiazhai Reservoir generally does not desilt in the spring flood season and the reservoir display siltation situation. According to the test against floods in 2007 spring flood season, the minimum water level of Wanjiazhai Reservoir reduces to 952.18 m and the reservoir has been scoured; the maximum sediment concentration discharged reaches 88.2 kg/m^3, the volume of sediments discharged in the maximum 10 - day flood volumes accordingly is 21.8×10^6 t, while the sediment deposited is 7.35×10^6 t, and 14.45×10^6 t sediments have been scoured in spring flood test against Wanjiazhai Reservoir. In addition, the reservoir is applied at low water level during the late stage of test, the silted delta is scoured, the deposit peak approaches towards the dyke by 5.5 km, and the place about 6.5 km before the dyke has been scoured (see Fig. 3). The belt deposit in front of the dyke forms scouring funnel to effectively improve the silting situations in the reservoir.

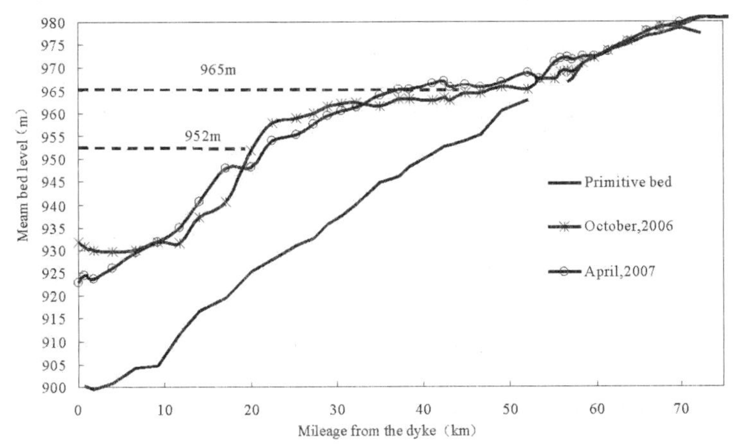

Fig. 3 Longitudinal plan of Wanjiazhai Reservoir

3.3 Scouring against Xiaobeiganliu reach

The filed data analysis suggests that the Xiaobeiganliu reach has been scoured during the spring flood test period. It is calculated that the maximum 10 - day scouring volumes in the Xiaobeiganliu reach reaches 3.47×10^6 t by sediment discharge rate. According to the remote sensing data analysis and field inspection before test and that after test, the Xiaobeiganliu reach hasn't had any floodplain problem, so the scouring against Xiaobeiganliu reach during the test period is mainly acted in the river channels, benefiting from which the flowing capacity of the main channel increases accordingly.

4 Comparative analysis about the scouring effects of Tongguan elevation

2007 spring flood test is the second test, whose test effects and the scouring effects against Tongguan elevation draw much attention. There are 3 methods to calculate the siltation changes of the section: ① identical discharge water head method; ② sectional mean bed level method; ③standard stage below sectional area difference method. Make comparative analysis to the scouring effects of 2006 and 2007 Tongguan elevation in terms of the three aspects.

4.1 Tongguan elevation

The water level at the discharge of 1,000 m^3/s in Tongguan during 2007 spring flood test lowers by 0.05 m and 0.15 m less than that lowers in 2006. From the water level change in the discharge of 1,000 m^3/s in Tongguan we can see that 2007 test produces better scouring effects than 2006. however, the level - discharge relation in Fig. 4 shows that the water level at discharge of 1,000 m^3/s in Tongguan (6) in the early test period in 2006 and that in 2007 are similar, 327.99 m and 327.98 respectively, and their water level are both 328.64 m at the maximum discharge of 2,450 m^3/s. The water levels in Tongguan (8) are 327.45 m and 327.37 m respectively, and are both 327.95 m at the maximum discharge of 2,450 m^3/s. It means that the flowing capacity at large discharge in 2007 is good.

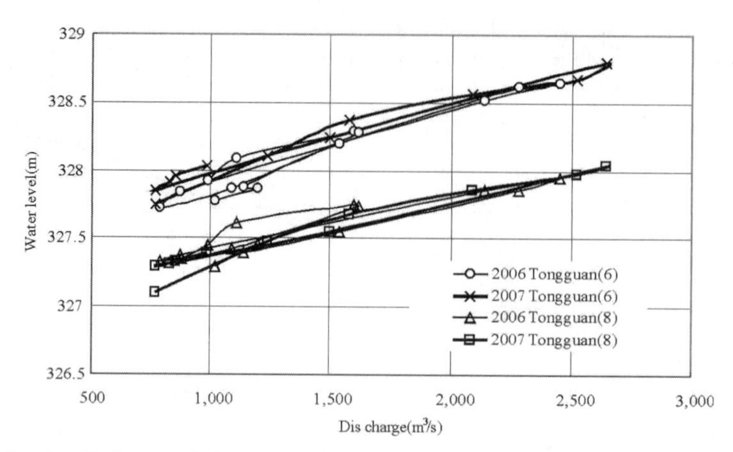

Fig. 4 Level – discharge relation comparison between 2006 and 2007 Tongguan section tests

4.2 Mean bed level of Tongguan section

The complete sectional mean bed level in Tongguan (6) in 2007 lowers by 0.79 m, while that in 2006 lowers by 0.60 m, 0.19 m greater than that in 2006. The complete sectional mean bed level in Tongguan (8) in 2007 reduces by 0.20 m but that in 2006 doesn't change. The variation in mean bed level of Tongguan section suggests that the scouring effects in 2007 spring flood test is better than that in 2006.

4.3 Tongguan sectional area

Tongguan (6) and Tongguan (8) have been fully scoured during 2007 spring flood test, and their complete sectional areas after test are 458 m^2 and 152 m^2 larger than that before test respectively; while the complete sectional areas in 2006 test have been respectively increased by 340 m^2 and 126 m^2 after test, and 185 m^2 and 26 m^2 have been additionally scoured in Tongguan (6) and Tongguan (8) in the complete sectional areas in 2007 compared with that in 2006. According to the scouring variations in the complete sectional areas in Tongguan section, the scouring effects in 2007 spring flood test is better than that in 2006.

5 Causal analysis for the small reduction in Tongguan elevation

According to the unitary water level changes at the discharge of 1,000 m^3/s in Tongguan (6), the scouring effects of Tongguan in 2006 is better; but according to the variation in silting and scouring and that in the mean bed levels of Tongguan (6) and Tongguan (8), the scouring effects in 2007 is better. Which has caused the small reduction of water level at the discharge of 1,000 m^3/s in scouring water channel? The evolution of Tongguan elevation is complex and the factors affecting scouring of Tongguan elevation are multiple. According to analysis, it is found out that the major causes are the channel shape, velocity of flow in Tongguan reach, the roughness factor and the evolution of the upper and lower reach channels.

5.1 Channel shape

Fig. 5 is the sectional drawings for Tongguan (6) before and after test in 2006 and 2007. It can be seen that it is single channel before and after 2007, the shape is good before test and the talweg point is deep, the channel area is largely scoured during the test, but the talweg point only lowers by 1.23 m; while in 2006 test, the channel is 712 m wide and shallow, and the water level displays a high value and the channel is scoured to single channel after test; the width of the main channel reduces to half of its original width and its talweg point reduces by 2.83 m. In view of these facts, it can be educed that the lowering of water level at the discharge of 1,000 m^3/s in 2007 is less than that in 2006.

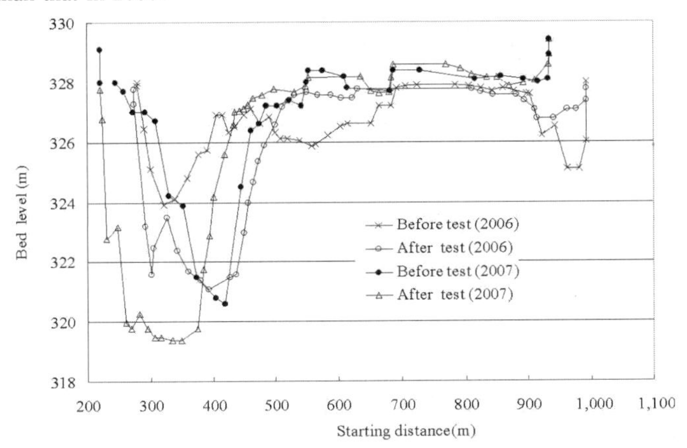

Fig. 5 Graphics of Tongguan (6) section before and after test

5.2 Velocity and roughness factor of Tongguan reach

Analyze and calculate the discharge velocity and roughness factor for the water level at the discharge of 1,000 m^3/s for Tongguan (6) before and after test in 2007, see Tab. 1 for the water level. The gradient listed in the table is worked out by the mean water level of that current day in Tongguan (6) —Guduo, and it can be seen that the channel has been widened and deepened and its area increases by 88% but velocity reduces greatly, from 1.33 m/s before test to 0.71 m/s and the reduction percentage is 47%; the roughness factor of channel is increased from 0.020 to 0.051, increasing by 164%. The large increasing of roughness factor and reduction of velocity produce backwater and result in slow lowering of water level. In addition, the large increasing of roughness factor might be caused by the bed coarsening and the sand wave produced by bed load, etc.

The factors affecting the lowering of water level not only includes channel pattern, enlargement of roughness factor and reduction of velocity, but also includes the scouring in the upper riverway adjacent to the section. In addition, lessening of gradient can also cause high water level. However, we shall not make further analysis in the paper owing to limited literatures.

Accordingly, the variation in siltation of Tongguan reach, with the water level at the discharge of 1,000 m^3/s in Tongguan (6) as one of the representations for Tongguan elevation, sometimes cannot thoroughly reflect the variation in bed siltation.

Tab. 1 Feature value of channel below water level at the discharge of 1,000 m^3/s in Tongguan (6) before and after 2007 test

Item	Rise	Fall	Fall - rise
Date	March 21	March 28	
Water level (m)	327.98	327.93	-0.05
Area (m^2)	751	1413	661
Width (m)	304	333	29
Depth (m)	2.47	4.25	1.77
Velocity (m/s)	1.33	0.71	-0.62
Gradient (‱)	2.02	1.93	-0.09
Roughness factor	0.020	0.051	0.031

6 Conclusions

During the prototype test in 2007, the Xiaobeiganliu reach and Wanjiazhai Reservior were scoured and the Tonggguan elevation fell for 0.05 m. Because the increase in channel roughness, the decrease in the flow velocity in the late of the flood, and also because the good initial channel morphology, the test in 2007 gets lager increase in Tongguan(6) cross section but lower decrease in Tonggguan elevation than those in 2006. The variation of Tongguan Elevation, which is one of the symbols of the riverbed erosion and sedimentation of the Tongguan reach, sometimes can not reflect the riverbed evolution roundly. The evolution of Tonggguan elevation is complicated needing further study.

Acknowledgements

This study was sponsored by the National Special Funds for the Programs for the Public Interests (No. 200701007) and by the "Eleventh Five - Year Plan" of China (Grant No. 2006BAB06B04)

References

Hou Suzhen, Wang Ping. The Impacts of the Spring Flood on Tongguan Elevation[J]. Journal of Sediment Research, 2005(1) :53 - 56.

Hou Suzhen, Lin Xiuzhi, et al. Research of Experiment of Lowering Tongguan Elevation by Utilizing Spring Flood[J]. Science and Technology Report, 2006.

Lin Xiuzhi , Hou Suzhen, et al. Preliminary Analysis for the Effectiveness of the Experiment of Lowering Tongguan Elevation by Utilizing and Optimizing Spring Flood[J]. Yellow River, 2007(3) :14 - 15.

Preliminary Discussion on Implementation of New "Law of Water and Soil Conservation" in Supervision of Soil and Water Conservation in the Yellow River Basin

Zhang Laizhang[1], *Gao Jinghui*[1] and *Dang Weiqin*[2]

1. Upper and Middle Yellow River Bureau, Xi'an, 710000, China
2. Suide Control Bureau for Yellow River Water and Soil Conservation, Suide, 718000, China

Abstract: The second issuance of the *Law of Water and Soil Conservation* (revised edition) provides a forceful legal weapon in prevention and supervision for soil and water conservation. Its new breakthroughs lie in clear exposition of legal responsibility for the watershed management, consolidation of prevention and protection, standardization of supervision and inspection, and reinforcement of legal liability. Its new highlights lay stress on protection of original land features, collision avoiding system, limits of authority in project approval, production and construction project management, ecological compensation mechanism, and measures for supervision and inspection, etc. The watershed management should play an active role in revising relevant laws and regulations based on study and publicity of the new Law, to fully promote functioning of supervision and inspection.

Key words: *Law of Water and Soil Conservation*, the Yellow River Basin, prevention, supervision, production and construction projects

The new *Law of Water and Soil Conservation* was adopted by the 18th conference of the standing committee of the 11th National People's Congress (NPC) on December 25, 2010, and President Hu Jintao formally issued No. 39 President Order on the exact day, that is, the new *Law* would come into effect on March 1, 2011. On comprehensively summing up years of practical experience for water and soil conservation, the newly revised *Law of Water and Soil Conservation* defined the legal duty of the watershed management, put emphasis on prevention and protection, normalized supervision and inspection, and developed legal responsibility. In terms of new requirements on implementing the scientific outlook on development, promoting a conservation culture and administration according to the law, the new *Law* surely will promote water and soil conservation in all aspects. The issuance of the new *Law* is a great event in development of water and soil conservation career in the country. It has realistic significance and will bring profound historical impact on further harnessing of soil erosion, protecting water and soil resources, improving ecological environment, and safeguarding sustainable development of economic society. For protection and supervision on soil erosion, the issuance of the new *Law* will undoubtedly give us more administrative abilities. It is a symbol of sharper legal weapon in soil erosion control. Implementing the new *Law of Water and Soil Conservation*, so as to make legal protection of water and soil resources and legal prevention of soil erosion, is our obligatory duty as well as our sacred mission endowed by law.

1 Issuance and implementation of the *Law of Water and Soil Conservation* has accelerated prevention and supervision on water and soil conservation in the Yellow River Basin

The *Law of Water and Soil Conservation* was issued in 1991 and has been put into practice from then on. The watershed management of the Yellow River Basin and water executive branch on each level has taken a serious attitude toward their law - endowed responsibility and put the law into real effect conscientiously. In such a way, the work of water and soil conservation in the basin has been brought into legal system, and prevention and supervision on soil erosion has been reinforced. In recent years, consciousness for water and soil conservation of the whole society has been intensified; great achievements have been made in preventing & controlling of soil erosion, protecting & rational utilizing of water and soil resources, eco - environment improving, and safeguarding sustainable development of economic society. We could say that our working in water and soil conservation has been stepping forward onto a new step. Firstly, a whole set of legal system for execution and supervision on water and soil conservation has come into being. Prevention and supervision on soil erosion in the Yellow River Basin and inland river area in northwest (including ten provinces or autonomous regions, i. e. Qinhai, Sichuan, Gansu, Ningxia, Inner Mongolia, Shanxi, Shaanxi, Henan, Shandong, Xinjiang) started after the implementation of the *Law of Water and Soil Conservation*. Under the leadership of YRCC, each province or autonomous region in the Basin area has unfolded standardized construction for prevention and supervision on soil erosion, and legal systems for execution and supervision on water and soil conservation of the watershed management as well as in each province (autonomous region), district (city), or county (banner) have been established on the whole. Up to the year 2005, 375 counties (cities, districts, banners) in the severely soil – eroded area, that is, the upper and middle reaches, have set up their own organizations for legal execution and supervision in water and soil conservation, provided with 2,999 persons of full – time staff, and 8,778 for part – time. From 2001 to 2007, 26,094 schemes for water and soil conservation in total have been examined and approved, 18,666 illegal cases have been investigated and prosecuted, and 2,592 cases have been treated with forcible execution. Secondly, a comparatively perfect statutory system for water and soil conservation has been brought into existence. *Law of Water and Soil Conservation* currently available was issued on June 29, 1991, and has been put into practice since then. Then implementary measures for water and soil conservation came on one after another in each province (autonomous region) of the Yellow River Basin. Supporting regulations set by each district on county level or above in the basin came to over 700 pieces. In addition, we worked together with the Ministry of Water Resources (MWR) in revising the *Law of Water and Soil Conservation* and four pieces of regulations issued by the Ministry, also in formulating the *Protecting Measures for Water and Soil Conservation in Developing Large – scale Construction Projects in the Yellow River Basin and Inland River Area in Northwest*, thus further perfecting the supporting statutory system and administrative supervision on water and soil conservation in development of construction projects. Thirdly, a technical service system for water and soil conservation has been upbuilt. Now the technical service framework for scheme drafting, monitoring and supervising in water and soil conservation is growing fast. Nearly 1,000 qualifications in NEC plan making, monitoring and supervising are available by far, and totally 1,200 persons have got professional certificates in scheme drafting and monitoring for water and soil conservation. All these are quite important in standardizing utilization of water and soil resources in production and construction activities, so as to raise our control level in soil erosion through human forces. Fourthly, soil erosion in course of production and construction activities has been greatly reduced. Taking some key construction project, such as west – east natural gas transmission and west – east electricity transmission projects, for example, schemes for water and soil conservation have been compiled, reported and implemented according to the law without exception. Investment on water and soil conservation by production and construction units has amounted to 170×10^9 yuan since 2000, and more than 80,000 m^2 soil – eroded areas has been under control. In other words, the achievements made in the past 20 years since implementation of the *Law of Water and Soil Conservation*, are of three "est", that is, best circumstance, best effect, and fastest developing speed. We could say that such a law is quite necessary and advantageous. Yet along with the speedy development of the economic society, some articles and contents in the original *Law of Water and Soil Conservation* may not adapt to the new circumstances. Therefore, the Ministry of Water Resources has gathered relevant units to begin revisions on the original *Law of Water and Soil Conservation* from 2005.

2 New breakthroughs and highlights of the new *Law of Water and Soil Conservation* may greatly push forward supervision on water and soil conservation in the Yellow River Basin

2.1 New breakthroughs of the new *Law of Water and Soil Conservation*

2.1.1 Defined the legal duty of the watershed management

Firstly, managerial authorities set up in main river and lake basins, which are defined by the state, should take charge of supervision and administration of water and soil conservation within their jurisdiction. Secondly, watershed management should trace and investigate execution of water and soil conservation scheme in project production and construction. Thirdly, watershed managerial authorities have the rights as those granted to the department of water administration under the State Council within their jurisdiction. These newly added stipulations defined the legal duty of the watershed management, which has been inexplicit throughout the years, provided visible legal basis for the watershed management in administration of water and soil conservation, and created better conditions for the Yellow River Conservancy Commission (YRCC) in implementation of the most severe control and monitoring system in water and soil conservation.

2.1.2 Put emphasis on prevention and protection

Firstly, the guidelines for water and soil conservation are *prevention first and proTection preferential* now. Secondly, some stipulations are added, for forbiddance or restriction on activities that may easily induce soil erosion and environment deterioration. Thirdly, systems for scheming, monitoring and verification in water and soil conservation are perfected, and thereby raising the level of prevention and control of artificial soil erosion.

2.1.3 Normalized supervision and inspection

Firstly, duties of supervision and inspection, for water administrative department and watershed management on various levels as well as for staff in charge of water administrative supervision and inspection, are tightened up. Secondly, ways and measures taken in supervision and administration of water and soil conservation are reinforced. Thirdly, investigation and public announcement systems in water and soil conservation are made explicit.

2.1.4 Developed legal responsibility

Stipulations related to legal responsibilities in the original *Law* are excessively principles – based, incomplete and insufficient in punishment, and the new *Law* makes more severe punishment, expands categories of legal responsibility, and enables law enforcement more feasible.

2.2 New highlights in prevention and supervision on soil erosion

2.2.1 Brought about protection on original land features, and further reduced artificial soil erosion

The new *Law* made further explication on protection of original land features, that is, to strictly protect plants, sand shell, surface crusting, and lichen in severely soil – eroded and ecologically fragile regions. This is of vital significance in these regions.

2.2.2 Brought forward that production and construction projects be away from key prevention and key harnessing areas, and further raised control standard of soil erosion

Site and route selections for production and construction projects should be away from key prevention and key harnessing areas; for projects that must pass through these areas, measures should be taken to raise control standard, optimize construction process, and minimize the area of surface disturbance and vegetation damage, to make effective control of possible soil loss.

2.2.3 Made clear of approval authority in water and soil conservation, and further perfected scheme alteration system

It is explicitly stipulated in the new *Law of Water and Soil Conservation* that water administration department now has the authority to "examine and approve" water and soil conservation scheme instead of the original "examine" only, and major alterations in scheme implementation should be reported to the original examination and approval authority for re – approval.

2.2.4 Forbade kickoff of projects without water and soil conservation scheme, and further reinforced management regime for production and construction projects

The new *Law* has perfected the management regime for production and construction projects in aspect of water and soil conservation, and on – stream is forbidden for projects either without water and soil conservation scheme or the attached scheme has not got approval yet.

2.2.5 Restricted the commissioning and operation of unaccepted projects, and further emphasized on the system of "three simultaneities"

The new Law reinforced the system of "three simultaneities" in production and construction projects, that is, water and soil conservation facilities should pass through final acceptance; and for facilities that have not gone through acceptance or have been checked as unqualified, relevant production and construction projects can not be put into use.

2.2.6 Normalized production and construction activities, and further detailed technical route of production and construction projects

Topsoil of occupied land for production and construction activities should go through stratified peeling off, saving and utilizing. Balanced earthworks and minimum disturbance are necessary, and storage locations for disused sand, stone, soil, rock, gangue, and residue should be enclosed, protected on highwall, and prevented from flood. Upon completion of production and construction activities, exposed soil in borrow pit, cut surface and storage locations should be afforested, and closed tailing pond must be reclaimed.

2.2.7 Established ecological benefit compensation system for water and soil conservation, and further speeded up unveiling of eco – compensation system

The revised *Law of Water and Soil Conservation* has established the ecological benefit compensation system for water and soil conservation, emphasizing the state responsibility on prevention and harnessing of soil erosion in river source area, drinking water source protection area and water conservation area. Multiple channels for financing are encouraged to bring the ecological benefit compensation system for water and soil conservation into the nationally established eco – compensation system.

2.2.8 Detailed measures taken for supervision and inspection, and further strengthened ways of supervision

Enforcement personnel should make lawful supervision and inspection; and may access to the site for investigation and evidence collection. The examinee entities or individuals should cooperate well, such as providing truthful reports, relevant documents, materials, certificates, etc. Any examinee entities or individuals who refuse to cease their illegal activities and cause severe soil loss will be reported to corresponding water administration department. And tools used in illegal activities, construction machinery and equipment would be sealed up and detained upon the approval of water administration department concerned.

These new breakthroughs and highlights embody the difficulties and weaknesses of supervision on water and soil conservation in the past 19 years. After revision, the new *Law of Water and Soil Conservation* may forcefully support supervision on water and soil conservation, and may greatly contribute to the public understanding of the man – made soil erosion to a new and higher level.

3 We will fulfill our obligations and fully implement the new *Law of Water and Soil Conservation*

The Upper and Middle Yellow River Bureau, YRCC has completed large quantities of essential work in recent years for the sake of the new *Law* revision. Complete survey on law enforcement for water and soil conservation in the basin has been made and suggestions on revision have been reported to the higher department in charge many a time. We have also given our hands in submission of relevant revised bill by NPC deputies, helped NPC to make law enforcement investigations in succession, and appointed special person to make full participation in drafting of the revised edition. All these are positive in smooth revision of the *Law of Water and Soil Conservation*, and make solid foundation for implementation of the new *Law of Water and Soil Conservation.* At present, what we need to do is to well develop propaganda, training, implementation, and inspection, based on the actual situation of the Yellow River Basin.

3.1 Paying adequate attention to the learning and propaganda of the new *Law*

Adequate attention should be paid to the learning and propaganda of the new *Law of Water and Soil Conservation.* According to the uniform arrangement of the Ministry of Water Resources, all personnel involved in water and soil conservation of the Yellow River Basin should conscientiously learn the new *Law of Water and Soil Conservation*, catch the deepgoing spirit of articles, and improve individual consciousness and initiative in the new Law implementation. In the meantime, water administration departments on various levels should be active in propaganda and education activities oriented to the society, the masses and construction units. Massive and overspread campaigns could be carried through multi ways, such as broadcasting, television, newspaper, magazine, network, short message, banners hanging, and brochures, to publicize knowledge about water and soil conservation, to consolidate legal sense of water and soil conservation, to enhance legal consciousness on water and soil conservation, as well as to create a strong atmosphere in the whole society to support and be concerned with water and soil conservation.

3.2 Making efforts to do well on training

The first, we should further build up our law enforcement and supervision team. Professional capacity building could be realized by carrying on supervision and management capacity building activities for water and soil conservation. On the one hand, we need to provide enough training for our law enforcement and supervision staff, to improve their overall qualities and enforcement abilities as a whole. On the other hand, each system needs be established and perfected to assure our personnel's loyalty and impartiality in law enforcement, and to guarantee that the laws are strictly observed and strictly enforced, and any violations of the law must be investigated and dealt with. Only in this way can we maintain the authority and dignity of the law. The second, according to the arrangement of the Ministry of Water Resources, we are in charge of training relevant personnel of production and construction units for water and soil conservation scheme approved by the MWR, in the effect that the legal responsibility awareness in mind of relevant personnel of those production and construction units could be improved, thus deepening development on production and construction schemes for water and soil conservation.

3.3 Making all - sided revisions of systems of laws and regulations

Constructions on corresponding systems of laws and regulations for water and soil conservation in the Yellow River Basin should be speeded up. Based on the contents listed in the new *Law*, we would make timely revisions and perfections of corresponding regulations and try to form a complete set of institutional system to guarantee lawful enforcement. We are going to revise *Protecting Measures for Water and Soil Conservation in Developing Large - Scale Construction Projects in the Yellow River Basin and Inland Areas in Northwest* to put the new *Law of Water and Soil Conservation* into real practice and bring about a new situation in our water and soil conservation career.

3.4 Greatly pushing supervision and inspection forward

Forceful measures must be taken in supervision and management of water and soil conservation, combining with the publicity and implementation of the new *Law of Water and Soil Conservation.* A through screening should be carried out for those projects, which have been approved and built up yet have got no acceptance check for water and soil conservation facilities, within the basin. Illegal and delinquent production and construction projects should make corrections within the fixed period of time, while those refuse to do so will get legal punishment. Meanwhile, good models discovered in the complete inspection should be popularized. On the contrary, illegal cases will be exposed for more effects. In such a way of spread publicity, the new Law will stand on solid ground and become more influential. Additionally, production and construction units will be more conscious of soil erosion control, the rates for reporting, implementing and approving of water and soil conservation scheme will become higher, and artificial soil erosion will get reduced as a result.

The Research of Our Country's Water Resources Management—Perspective of Rational Bureaucratic System

Zhao Hejing[1], ***Tang Yan***[1], ***Xun Meng***[1] and ***Bi Xia***[2]

1. College of Public Administration/Hohai University, Nanjing, 210098, China
2. College of Marxism/Hohai University, Nanjing, 210098, China

Abstract: In view of Mark Webber's rational bureaucracy, through the analysis of the management system of water resource in China, according to the principles of the theory, division of labor, hierarchy, the compliance of jurisprudence and impersonalization, this paper identifies the existing problems of our management system of water resources, the management institutions purview division is not clear, so as to the specialization division is not distinct; the setting of grades is not standard, and the support system is not consummate; the basis of jurisprudence is not complete. Afterwards, with the guidance of rational bureaucracy, the paper puts forward the specific idea of how to improve our management system of water resources. Firstly, it is to define the division of the authority of the organizations of water resources management. Secondly, it is to explicate the organization structuring of our management of water resources. Thirdly, it is to complete the support system of water resources management system. In addition, the government accountability system of water resources management should be consummated, so as to strengthen the functions and disentangle the responsibilities and increase efficiency.
Key words: water resources, management system, rational bureaucracy

Water resources management system is the organization system of water resources management of the country and the basic system of the division of purview. It's necessary to grasp the elements of water resources system and the relationship between them before the study of water resources management system. The elements of water resources management are the institutions' establishment of water resources management organization and the division of purview of them. Rational bureaucracy system is the organization model widely used in modern society which was put forward by Max Webber. Rational bureaucracy theory contributes to solve the problems of our water resources management system, such as water resources controlled by many services, unclear department labor division and unknown government power and responsibility. Analysis from two levels that internal construction of organizational system and supporting system of water resources management,, the obstacles to the development of water resources management system was found out and the suggestions for improvement was made.

1 The applicability of rational bureaucracy in studying water resources management system

Rational bureaucracy is the industry product in process of western modernization, the system has strict pyramidal hierarchical structure, top – down authority system and a clear division of responsibility mechanism, in which governments and large enterprises and institutions of the administrative management has been widely adopted. The rational bureaucratic system as a modern form of organization represents the rationality and the efficiency, is the most effective, the most popular and the most successful organization form by far.

1.1 China's water resources management system

In August, 2002, the government issued a new *Water Law*, making a significant adjustment on China's water resources management system, and gradually establishing two levels (central and local), four grades (central government – basin – provincial – city and county) of the water resources management system. The new *Water Law* makes clear the central level, namely the national water resources department – the Ministry of Water Resources. The Ministry of Water Resources is responsible for the national water resources management, as the state council department of water administration management, it manage water resources throughout the country.

According to article 12 of the new *Water Law*, "The state applies the system of water resources management that river basins management combined with management of regional administration. The administrative department for water resources under the State Council is responsible for unified management of and supervision over water resources throughout the country. The institutions for river basin management, set up by the administrative department for water resources under the State Council for the main rivers and lakes, shall perform the responsibilities of water resources management, within the limits of their jurisdiction, specified by laws and administrative regulations and assigned to them by the said department. Ministry of Water Resources established institutions to take charge of the main rivers and lakes, and the institutions have the responsibility of supervision specified by laws and administrative regulations which are assigned to them by the Ministry of Water Resources. China has established 7 institutions such as Changjiang Water Resources Commission, Yellow River Conservancy Commission, Huai River Water Resources Commission, Hai River Water Resources Commission, Pearl River Water Resources Commission, Songliao River Water Resources Commission and Taihu Basin Authority, which are directly subordinate to the Ministry of Water Resources. These institutions exercise the *Water Law*, *Flood Control Law*, *Water Pollution Control Law*, *Regulation of River Administration*, and other laws and regulations.

Water resources management system at the local level is divided into two levels; they are provincial and city (county) level respectively. Provincial administrative department of water is the administrative department of water of the people's government of the province, which is responsible for the work of unified management and supervision of water resources of the whole province. Other wading department (such as the department of environmental protection, et al) shall be cooperative for the management of water resources. The administrative department of water of city and country belonging to province is the department of water of the people's government, which is responsible for unified management of water resources that belongs to the city and country.

The relationship between basin management and regional management.

1.1.1 The relationship of guidance and subordination

The overall basin management and local management of administrative areas decided the characteristics of basin management. Administrative area management must obey basin management, basin management guide to administrative region management. Basin management pay attention to the macro management and interest of the whole, but administrative area management focus mainly on the administrative divisions of the local area in the relative benefits, according to the principle of interests of part subordinate to the overall interests, immediate interests subordinate to long – term interests, the principle of administrative area management should be subject to basin management. In particular, it is to point to in the relationship of the whole basin macro management matters, management of administrative area must decide the objects and methods of management of administrative areas according to the requirement and target of basin management, when the regional interests and watershed interest conflict, must be subordinated to the whole interests of the basin, regional planning must obey watershed planning.

1.1.2 The relationship of cooperation and division

Basin management and management of administrative areas have common in the scope of management, which determinate that they must cooperate. Basin management is not the castles in the air, and its management object belongs to the administrative region management within the scope of the local work providing service and support. When the basin management involves upstream and downstream, shore of left and right, inter – area matters, and basin management must be supported by basic materials of water resources, basin management must cooperate with the administrative area management, and the two jointly completed.

Basin work can get support and extension from local work. When they cooperate, river basin management and administrative area management have definite division of labor. Basin management and regional management basin management lay particular stress on the macro, unified management and the treatment of the relation of interested parties; the management of administrative areas focuses on the specific affairs' management in administrative areas. Both of the sides do labor division in the scope of the object, geographic location, influence scope, the degree of importance, the range of benefits, function separately.

1.1.3 The relationship of mutual supervision

The conflicts of interest between different administrative areas in the internal basin decide that basin management must supervise management of administrative areas. For affairs involved or influenced local interest in management of administrative areas, basin institutions must undertake supervisory in order to avoid disputes; to the parties disputed already, basin institutions must also supervise the implementation of the solution both of the sides; at last, administrative areas must be illegal supervision to prevent some illegal affairs happen in the management.

To the fairness of basin management, local agency also needs to supervise whether the river basin management is excessive, and whether neglect their duty.

1.1.4 The relationship of mutual coordination

The object of watershed management and Administrative region management is water resources. The overall goal is consistent, namely improve water conservancy, remove water hazards, bring benefit to people, it is also the foundation and premise of harmonious relationship. Water resources have both the basin and natural unity, needing unified management based on basin. Meanwhile, water resources in a administrative area has a close relationship between local natural resources, ecological environment, economic development and social progress ,which are the standard divide areas. Thus, the way manage water, water mode, the development level of economy and society are influenced by tradition with regional differences. They should manage according to region. The basin characteristics and the regional differences in river lead to harmony and unity of basin management and regional management.

1.2 The theory of rational bureaucracy

The theory of rational bureaucracy was put forward by Max Webber, it's also called the system of administrative centralization, which is a form of organization plays a dominant role in various social organizations in modern society and is featured by division – layering, centralization – unity, command – compliance, equals to administrative organization rule legally in modern society. It is an ideal type of organization highly rational. In Webber's opinion, any organization were eager to exist and operate, must be based in some power. He summarized three legal powers: Firstly, the traditional power . Traditional power came from hereditary or tradition, People had to obey the position the leader occupies and the authority carried by the position, not a leader itself. Secondly, it is the unusual power. The power came from worship and following of others, People obeyed the charm of the leader, no matter how position the leader occupies. Thirdly, it is the legal power. This power was prescribed by law, what people submitted to was authority of rules of organization established under the law system. This organization was called "reasonable – legalization organization". In Webber's opinion, only the power law regulated could be as the foundation of the system of the administrative organization. Thus, he pointed out that, Bureaucracy was a system of organization and a form of management unique to modern society, has specialized functions and fixed rules and regulations which established based on legal rule. It is a reasonably hierarchy organization designed rationally to coordinate activities of individuals, so as to effectively do large – scale management work to achieve organizational goals.

1.3 The applicability of rational bureaucracy in studying water resources management system

The bureaucratic organization described by Webber is in various summaries, while the most important content is division of labor, hierarchy, the compliance of jurisprudence and impersonalization. These four factors make organization stable in large degree. Our water resources management system is the embodiment of such bureaucratic organization, therefore, Webber's rational bureaucracy is in great applicability in studying water resource management system of China.

1.3.1 Division of labor

Labor division according to professional skill is emphasized in Webber's bureaucratic organization. The practical significance of Labor division lies in not only improving efficiency, but also eliminating privilege of grade society. In Webber's division system, the difference of people is of technological capability, instead of social status. Professional ability takes place of faith to one person, power and responsibility belong to position rather than the leaders. Power and responsibility are fixed in the organization in the form of legal system. Organization formed in this way can improve efficiency via labor division and create wealth. Moreover, new organizational relationship following the rule of rationality can be formed.

Our water resource management system is designed right according to this rule. From the evolution of the system, we may get that our water resource management is always in classification and department. At the national level, the Ministry of Water Resources as water conservancy specialty department is mainly responsible for the management of water resources. The organ of the ministry is divided into Department of Water Resources (National Water Conservation Office), Department of Soil and Water Conservation, Bureau of Rural Hydropower and Electrification Development, Office of State Flood Control and Drought Relief Headquarters, and so on, to be responsible for the related business. Additionally, there are 7 basin water resources commissions directly under the Ministry, 13 agencies and institutions in Beijing, and 4 agencies and institutions outside Beijing take charge of related water business. At the local level, the provincial water conservancy bureaus and subordinate water conservancy bureaus as professional water management institutions take charge of local water resources management. There are specific offices in charge of relevant water business. In addition, in China's water resources management system, besides professional water conservancy institutions, other institutions such as departments of environmental protection, agriculture, forestry and fishing are also involving in the use, development, management and protection of water resources. It's good to manage water resources by different departments, for it can give full play to the professional knowledge of each department. It's an effective way to manage water resources.

1.3.2 Hierarchy

In Webber's bureaucratic organizations, bureaucracy means a relationship of command and obedience, which is in short of changes and executed strictly. As an organized system of inequality, bureaucracy is relying on subordinates to obey authority, trust and obey the command develops. Implementing the institutional hierarchy and post grades system principles, powers arranged in strict class hierarchy system from top to bottom, and according to the level given the powers and responsibilities, individual layers are subject to close supervision of officials and institutions.

Current management system of water resources in China is that central government - basin - provincial - city (county), four - level management system. River basin authorities are subordinate to water administrative department and suffer a lower legal status. The national water resources authority, the Ministry of Water Resources, sets basin authorities in the main rivers and lakes defined by the state. Furthermore, it stipulates their responsibilities given by laws, regulations and the ministry of water resources management and supervision. Provincial water administrative department is that of provincial government, which is responsible for the integrated management of water resources and monitoring of the province.

1.3.3 The compliance of jurisprudence

In the ideal state, the formation, division of labor, position setting, selection, as well as operation, power and responsibility of each member, of bureaucratic organization, are all clearly stipulated by the legal system (including not only the written system, but also the unwritten system). These legal rules are arrived by the negotiation of members in the organization or put forward by the upper – class of the organization. However, they are accepted by the members after rational consideration. Therefore, all the rules of the organization are rational. Any member of the organization acts according to the rules, from top leaders to ordinary staff, no one is exceptional. What they obey is the common rules, rather than the measure taken by different people, nor personal preference.

The legislation system of China's water resources management mainly includes relevant provisions of the constitution, laws passed by the National People's Congress and It's Standing Committee, regulations and documents promulgated by the State Council, administrative rules published by departments of the State Council and local regulations made by local people congress and government. These laws and regulations not only design the institutions of our water resources management system, but also define the power and function of each institution. All the institutions act rightly according to these laws and regulations, which ensure the normal operation of them.

1.3.4 Impersonalization

In Webber's eyes, bureaucratic organization is a system of regulations, rather than a system of individual. It should regulate the organization and its members by laws, regulations, documents, etc. As a result, the bureaucracy excludes individual charm. The operation of an organization is not according to personal will, as well as personal emotions. China's water resources management system is in a system of rules and regulations, so as to ensure the normal operation of the organization, rather than influenced by leaders' personal will.

2 Problems of China's water resources management system

In allusion to the internal part of the organization, rational bureaucracy theory points out that the strict division of labor and reasonable institution settings are internal elements of a scientific organization system. However, in China's water resources management system, jumbled and confusing setup, unreasonable structure of power and responsibility go against the request of constructing a good organization, as well as the objective law of water resources. In allusion to the support system of the system of organization, the law system of China's water resources management is not perfect. Departments of water resources management are not conscious of their responsibility, and the absence of government accountability system is the root cause of all the problems of China's water resources management system.

2.1 The management institutions purview division is not clear, and the specialization division is not distinct

In Webber's bureaucracy theory, the aim of specialization division of labor is to improve efficiency and eliminating personal privilege. To achieve this goal, the reasonable division of labor appears particularly important. China's water resources management system has done the division of labor, which can be seen in both national level and local level. However, because of the various grades, overlapping functions and crossing management, the efficiency of this system is restricted. According to the new *Water Law*, water administrative departments take charge of nationwide water resources, but it must be noted that, other institutions such as departments of environmental protection, agriculture, forestry and transportation are also involving in the use, development, management and protection of water resources. It makes the management of water resources in overlapping functions and malposition, which has negative influence on the unity and efficiency of the work of water resources management. The management of water resources is in the possession of water conservancy departments and under the leadership of the State Council. Meanwhile, each administrative area has its own water administrative department, which is directly under the leadership of local government. Similarly, other water resources management departments are also like this, therefore, each administrative area has water resource management department, in charge of water resources development, management, which are directly leaded by local government. Because the administrative decision – making power of each regional is relatively independent, the mutual correlation of basin water resources is cut off. As a result, the unity and integrity of basin management are also cut off. The partition management of water resources leads to every region along the basin only considers its own benefit, making the maximized utilization of water resources in the region. The whole benefit of water resources are difficult to play in such a situation.

2.2 The setting of grades is not standard, and the support system is not consummate

From the viewpoint of rational bureaucracy, strict division of labor and reasonable organization setting are internal elements of a scientific organization system, while in China's water resources management system, complicated and confusing setup is clearly contrary to the requirement of constructing a good system. Although the new *Water Law* defines the comprehensive system of combining basin management and regional management together, due to the lack of specific provisions for how to combine, the partition management of China's water resources is not greatly improved. Every department is involved in water resources management in a certain area, so the scale of water resources management departments is much too great. In the aspect of cross administrative region water resources management, besides the repeating and conflicting management cause by the partition management, the contradiction of basin management and regional management are also increased. Among the 7 basin water resources commissions directly subordinate to the Ministry of Water Resources, except for Changjiang Water Resources Commission and Yellow River Conservancy Commission are of vice – ministerial class (approved by the National Personnel Ministry in 1990), the other 5 are all of department bureau class, say, they are not in the same administrative grade. Since bureaucratic organization acts strictly according to the regulation of power goes in accordance with grade, basin water resources commissions and regional water administrative management are difficult to communicate with each other.

2.3 The basis of jurisprudence is not complete, and the supporting of legal principle is not perfect

According to Webber's rational bureaucracy, the necessary external conditions of organization system's normal operation are the complete laws, regulations and rules to set rule of the operation of the organization, while the he internal condition is that administrative departments carry out their responsibilities actively. The root causes of all the problems of China's water resources management system are the absence of complete law system of water resources management and the absence of government accountability system. That is to say, the basis system to ensure the good operation of water resources management system is not perfect. There are a great deal of laws talk about water resources, but they still exist the problem of definition of water resources is not clear, as well as the division of functions. For example, the function division of each department is not clear, the power of basin water resources management institution are far from enough, there is no law about basin management till now, etc. The problem of the imperfect law about water resources management releases not only in the central level, but also the regional level. In the management of water resources, water resources management departments are unconscious of their responsibility, in the process of making water resources management policies, performing their functions, protecting and managing water resources, the absence of responsibility are in common. Only when power is in accordance with responsibility, can the administrative organization operate well, according to rational bureaucracy. Water resources management departments are the professional departments of providing and managing public products in China. They are responsible for the protection and management of water resources. From the viewpoint of nationwide, the endless affairs of the lack of living water, the pollution of drinking water, rivers and lakes show that the ability of supply and management of water is still a weak spot of our water resources management apartments.

3 Advice on China's water resources management system

According to the rational bureaucratic system, specific advices on China's water resources management system are as follows.

3.1 Define the division of the organizations of water resources management

According to the specialized division of labor of rational bureaucratic system, the division of work must be based on the type and purpose. The organization must divide the work to every unit, reject the part of repetition and pay attention to the part of overlapping. Therefore, authority definition must be reclassified to improve labor efficiency and to eliminate personal privilege.

The Ministry of Water Resources is responsible for the management of national water resources and environment. It has already integrated the water - related functions of departments of construction, agricultural, transportation, tourist, forestry, and environmental protection. Therefore, the function of the Ministry of Water Resources is also including: guiding the city water supply and saving, guiding the fishery water areas, developing the tidal flats and wetlands for farming, organizing the protection of national wetlands and fulfilling international convention.

Basin management organizations are responsible for developing and protecting the basin water resources, making planning of the development of river basin, executing administrative examination and approval and insuring supervision and law enforcement.

The local department of water administration has not only the authority to planning, approving, scheduling, servicing, insuring administrative law enforcement and supervising, but also the functions integrated from the department of environmental protection, construction, agricultural, and forestry.

3.2 Explicate the organization structuring of our management of water resource

According to the rational bureaucracy theory of Max Webber, the clear hierarchy system and organization establishment is to the benefit of establishing a clear management chain. Authority transfers through it and is convenient for internal management and supervision. China's current water resources management system is the typical four level hierarchy system, that is, central government - basin - provincial - city and county. But there is still some problems in the typical four level hierarchy system, strict hierarchy system should be established to set up the authority, clear organization establishment and functional distribution, form the orderly top - down hierarchy.

At the national level, the function of the Ministry of Water Resources should be extended to integrate the other department's authority. Therefore, the ministry can exercise authority in the development, utilization, management and protection of water resources to solve the problem about departmental partition and authority overlapping of water resources management. The process of power integration is also the process of the authority establishment of the Ministry of Water Resources. Responsibilities can transfer through the top - down hierarchy only if the central water management department sets up the real authority. We must know that the function integrity of water sector realizing the unified management of water resources and supervision. At the same time, function expansion also brings responsibility expansion. When the Ministry of Water Resources is performing administrative power, it must strengthen its own construction of responsibility, be responsible to the State Council and the National People's Congress and prepared to accept supervision of superior and the masses.

At the basin level, first, we should complete the setups of organs with unified administrative rank. Say, it is in the leadership of the Ministry of Water Resources. It's a kind of administrative organ and ranks over the water administration department in the region. Second, the legal status of river basin management institutions should be clarified and the authority should be established. Meanwhile, joint conference system capital of river basin to river basin management institutions should be established in order to strengthen the cooperation of river basins and regions. River basin authorities are responsible for conducting the routine work. The joint conference is composed with the office workers of river basin administrative agencies, expatriate personnel of the government and the water management departments, social experts and representatives of water users in proportion. The river basin joint conference system' s major functions are including participating and overseeing the implementation of the comprehensive planning about watershed and regional, protecting and developing resources of the cross – border river basin, solving the main issues of basin management and development, and coordinating the interests between the provinces when developing the basin. Basin commission meeting will be held on regular or irregularly to discuss important technical and administrative affairs of the basin management, coordinate the conflicts between the provinces, and implement the decisions of basin management.

At the local level, provincial water administration department should extend its functions and integrate the functions with the development, utilization, management and protection of water resources in order to solve the problems of vaguer in the responsibilities and duties, crossing functions, incomplete management and management disorder of many government branches policies. Meanwhile, provincial water administrative departments is not only leaded by the local provincial government, but also leaded by the Ministry of Water Resources. The provincial water administrative departments are responsible for the Ministry of Water Resources and must accept the guidance and supervision of the ministry. In addition, the provincial water administrative departments must participate in the basin joint conference and make an overall plan of water resources management with the unify arrangements. The provincial water administrative departments are leaded by the provincial river basin management institutions. The city and county water affairs bureau should promote the reform of water affairs integration and establishment unify water affairs bureau to manage water resources. Functions and personnel related to the water resources management of the city and county water sector, the bureau of land resource, agricultural, forestry, transportations and the city building department should be integrated. Functions of water conservancy construction, water supply, water saving, drainage, flood control, wastewater treatment and recycling should be hand over to the water affairs bureau to ensure the unify management of the city and country water resources.

3.3 Complete the support system of water resources management system

Abiding by the legal rules is one of the important content of the rational bureaucratic and regulations ensuring can insure rational bureaucratic effective operations. The administration has its extent of competence subject to the laws and administrative regulations.

China' s water resources management system must be based on comprehensive procedures, rules, laws and regulations. China' s water resources law system should be based on *Water Law*, Water Pollution Control Law, Flood Control Law, Water, Soil Conservation Law, and so on. The *Water Law* must be the basic law and the conflicts between the laws must be revised to clear the responsibilities and authority of water resources management and to ensure its efficient operation. To basin management, we must have specific basin management laws to coordinate the conflicts between basin and region management.

The accountability system of government must be improved to restrain the behavior of the government in law and to consistent power with responsibility. Their own responsibilities should be strengthened, administrative efficiency should be improved, division of labor should be defined, functions should be strengthened, responsibilities should be clarified, work efficiency should be improved among the water resources department to improve water use efficiency and benefit, establish national examination and evaluation check – up system of water use efficiency and benefit and implement the strict accountability system of government.

References

Max Webber. Economy, Society and Religion – Max · Webber' s Analects [M]. Shanghai: Shanghai Academy of Social Sciences Press, 1997.

Functions of the Ministry of Water Resources. Website of the Ministry of Water Resources[EB/OL]. http://www. mwr. gov. cn/zwzc/jgjs/zyzn/.

Peng Xuejun. Water Resources Management System That Combines Basin and Local Management Together[D]. Ji'nan: Shandong University, 2006.

Sun Qi, Chen Linlin. The Bureaucracy Theory and It's Significance of China's Administrative Reform[J]. Business Culture, 2007(4):55.

Liu Yang. Research on Establishing System of Water Resource Management with Priority of River-basin Management[D]. Shanghai: East China University of Politics and Law, 2008.

Application of Permeable Spur Dike in Middle and Small River Training

Zhou Yinjun, *Jin Zhongwu*, *Wang Jun* and *Zhang Yuqin*

Changjiang River Scientific Research Institute, Wuhan, 430010, China

Abstract: Based on the investigation and indoor experiment, the flow structure and properties of erosion or deposition associated with permeable spur dike were researched. According to the new demands about engineering structure in middle and small rivers regulation, the potential action of permeable spur dike in middle and small river training was analyzed. The results show that, as the permeable structure, the influence of permeable spur dike to main flow and flood level is less than the solid one, and the local scour depth is less too, so the permeable spur dike is safer. In the back of permeable spur dike, it can form a slow flow field, instead of a circumfluence field, and the deposition volume is less than solid one. Therefore, the steady slow flow field can benefit to the small aquatic animal lives. The pervious rate can be adjusted to complete training goals when the project faces different training demands. And the material of spur dike is not limited, it can use the local materials to build, which can reduce the construction cost. The permeable spur dike have a good applicability in the control of middle and small river, it can meet the requirements of flood control, ecosystem, environment, cost saving etc. It has a greatly application prospect.

Key words: river, permeable spur dike, flow, erosion and deposition, integrated training

1 Introduction

At present, with the gradual improvement of large river flood prevention system, the flood prevention and control of middle and small rivers begins to receive more attentions. Moreover, with the enhancement of people's protection awareness, the control in middle and small river, not only achieve the goal of flood control and disaster reduction, but also consider the effect of landscape environment, ecosystem etc. (Zhang, 2005). Of course, in different areas or different types of middle and small rivers, the control demands are different (Chen, 2011). But the mutual basic requirement is not only flood controlling and disaster reducing, but also trying to maintain the natural morphology of river, promote the harmonious coexistence among river, human and aquatic organism, in order to achieve the sustainable development of river ecological environment.

Expect for more attentions on effects of ecological environment, because most middle and small rivers are mountainous river, The flood is fierce and has a quick velocity. The requirement of regulating structure safety itself is high. We should prevent building the structure with a high resistance, to avoid to enhance the flood level and to aggravate disaster(Chen, 2011).

Therefore, the comprehensive control demands of middle and small river are different from the plain large rivers(Miao, 2011). The regulating structure is demanded not only by powerful prevention ability, but also by good ecological environment effect. This demands make the engineering measures cannot be copied. We need renew or optimize the building structure type.

Spur dike is common river or channel training structure, which is applied widely in the control of large rivers and also will be the main control type in the medium and small rivers. In order to adjust to different river regimes and control demands, the structure and layout mode are very different. Permeable spur dike, especially pile permeable spur dike is a sort of new spur dike which is gradually generalized in recent years. Its greatly impact resilience, significant adaption and regulation to flow and river regime, make it has good application prospects in the integrated training in middle and small river (Miao, 2011; Ikeda; 1991; Raudkivi., 1996; Feng, 2002; Zhou, 2007; and Zhang, 2007). However, because the understanding of flow and scour characteristics of permeable dike is lack, it is only on probation in part river regulation, and in most of middle and small rivers regulation, it has not yet been applied.

Base on the indoor experiment, this paper researches the flow structure and properties of erosion or deposition associated with permeable spur dike. According to the new demands about engineering structure, this paper analyzes the potential action and advantages of permeable spur dike in middle and small river training, and discusses the application prospect and adaptability. The study work can offer technical references for taking the groins class engineering measures on the comprehensive management of middle and small rivers.

2 Experiments

In order to study the flow structure and scour properties of permeable spur dike, the model test had 2 steps: fixed bed test and mobile bed test.

The spur dike types of the two models were all pile permeable spur dike. The flume channel of fixed bed test was 20 m long, 0.5 m wide, 0.44 m deep, and the slope can be adjustable. The pervious rate scope of the actual project was for reference (Ikeda; 1991; Zhou, 2007). The pervious rate of experiment was 0, 20%, 25%, 30%, 40%, 50%. The length of spur model was 10 cm, 12 cm and 14 cm, and the angels were all 90°. Flow was designed as uniform flow. In the fixed test, we studied the flow structure, as distribution of velocity and flow pattern etc., of permeable spur dike with each condition.

The flume channel of mobile bed experiment was 10 m long, 1 m wide, 0.5 m deep, and the channel slope was 1/5,000. The cross section was shown in Fig. 2. The channel was filled with sediments to allow for unimpeded development of the scour hole. The bed material is real sediment from the Yangtze River, and we test its properties in lab. The bed sediments had a median size of 0.5 mm and a geometric standard deviation of 1.3. For reflecting the influence of permeable spur dikes to flow and bed more explicitly, inflow was set for clear water, and $v/v_c < 1$, where v was flow velocity, v_c was initial scour velocity of the sediments. The flow is steady and uniform, and Fr was 0.23 ~ 0.37. The angels of flow and spur dike were 35°, 45°, 60°, 75°, 90°, 120°, 135°. Pile permeable spur dike model was made up of circular - shape rebar. The spur dike model was vertical to the bed, their height was equal to the depth of the flume. The purpose of the experiment was to study local scouring around the pile permeable spur dike with different conditions.

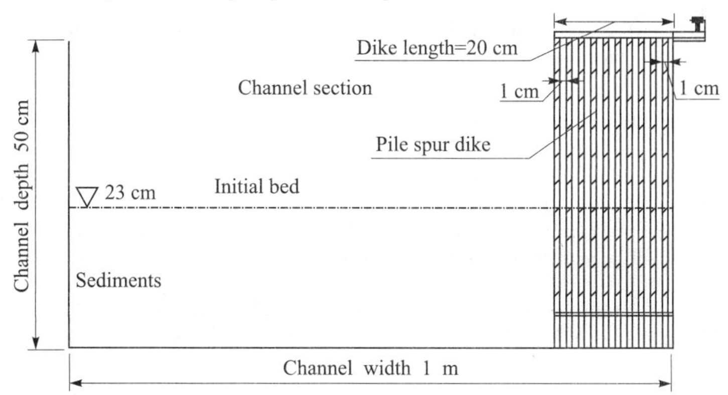

Fig. 1 Schematic plan of permeable pile groin model

Before each group's experiment runs, the bed slope would be adjusted to 1/5,000, and then the model was set. Based on Feng and Zhou' research about local scour of solid spur dike, the scour experiment was continued for 30 h, and the scour hole was stabilized. So after 30 h, the experiment was stopped (Feng, 2002; Zhou, 2007).

3 Flow structure and scour properties of permeable spur dike

3.1 Flow structure of permeable spur dike

Because of permeable body, the flow structure of permeable spur dike was different from the solid spur dike. Besides the effect of flow condition and spur dike size, the flow structure of permeable spur dike was influenced by pervious coefficient.

Being different from the forcedly circumfluence behind solid spur dike (A. Sukhodolov, 2004), the circumfluence of permeable spur dike disappeared away gradually in company with aggrandizement of spur pervious rate. In the fixed bed test, when the pervious rate was at 20% ~ 30% compartment, the circumfluence phenomena disappeared, and a slow flow field appeared, as showed in Fig. 2.

As the permeable structure, the influence of permeable spur dike to main flow, bank flow and flood level is less than the solid one. In the back of permeable spur dike, it can form a slow flow field, instead of a circumfluence field. The flow velocity was not large and flow pattern was steady and the steady slow flow field can benefit to the small aquatic animal lives (J. Geist. ,2011).

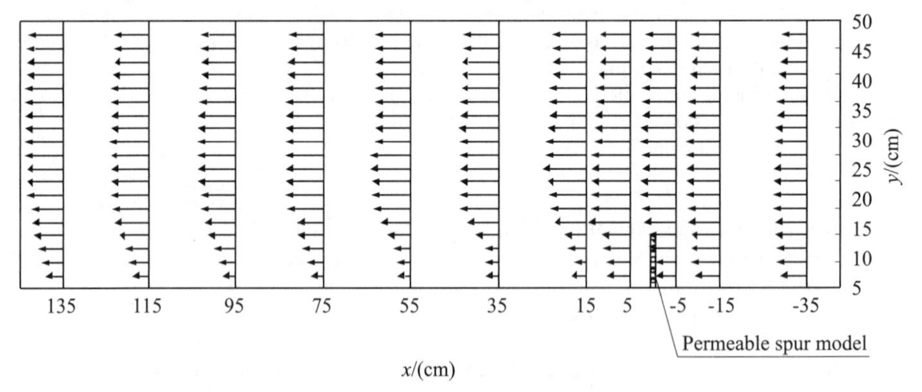

Fig. 2 Measured velocity field near the permeable spur dike with 30% permeable percentage (10 cm/s)

3.2 Scour properties of permeable spur dike

By the mobile bed test, Fig. 3 and Fig. 4 showed respectively actual measured contour map (initial height was 23 cm, and the map was draw by surfer) of district nearby solid spur dike and the permeable one (pervious rate was 20%) which had been scoured in the same flow and sediments condition. The projective length of spur dikes along the width direction were 20 cm, the angle was 60°.

Compared Fig. 3 with Fig. 4, there were several difference about local scour geometry between pile permeable spur dike and solid one. After scoured, the pile permeable spur dike didn't form a scour holes like horse – shoe as solid spur dike, and instead of scour slot as "V" along the dike body. There were local scour existing both before and behind the pile permeable spur dikes. The scouring depth increased from dike root to head with little variation. The depth and range of local bed deformation showed that the permeable spur dike affected the primary channel more slightly and the primary channel could be avoided excessive scour. When permeable spur dikes were used to regulate one region of river, other parts of river were not influenced seriously.

Fig. 5 showed longitudinal section back of permeable spur dike with different pervious rates (all angels were 90°, length was 20cm), when the pervious rate was 0, it was solid spur dike.

From the Fig. 5, the scour channel of permeable spur dike was like "V", which was symmet-

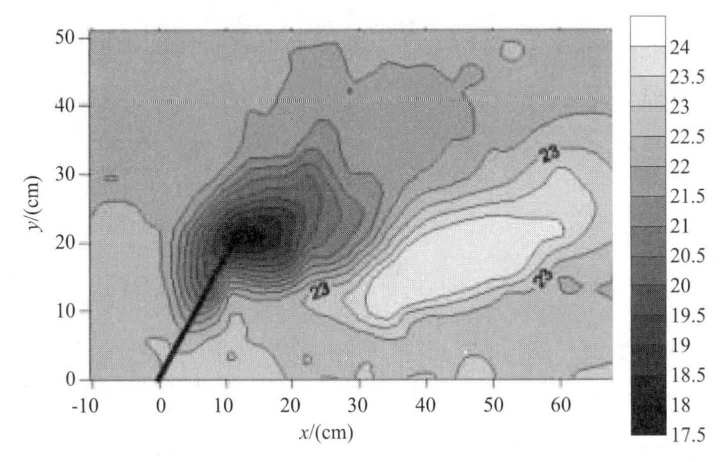

Fig. 3　Actual measured isoline map of district nearby scoured groin

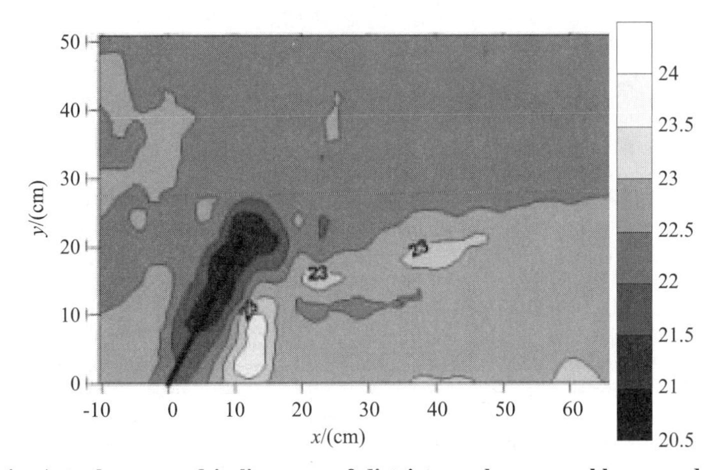

Fig. 4　Actual measured isoline map of district nearby permeable scoured groin

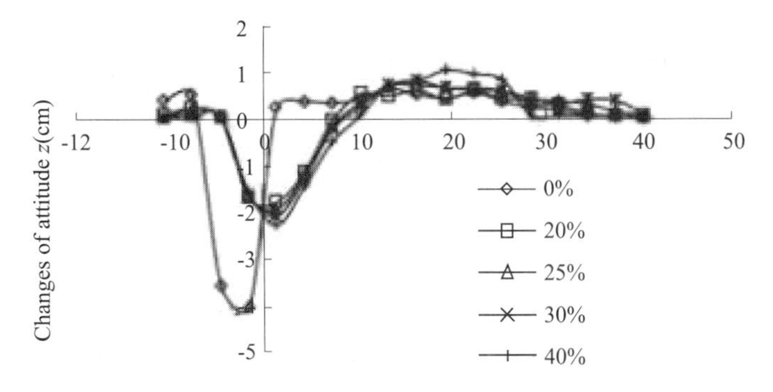

Fig. 5　Altitude versions of lengthwise direction section in district nearby different permeable pile groin

rical with the dike, and the greater part of scour channel of solid one was in upstream of the dike. The scour depth of permeable spur dike was less than solid one obviously, even the pervious rate was not large.

Meanwhile, the location the deposition field back of permeable spur dike was related with the precious rate. Width of scour channel increased with the precious rate. So the larger pervious rate was, the longer distance between deposition field and dike was. Therefore, with the pervious rate increased, length of deposition field first increased and then decreasedI and it was largest when pervious rate was 30%.

4 Discussion

Since river flood system of our country has been completed basically, regulation work of medium and small river will be a focus water conservancy construction in a future period. Most medium and small rivers were mountainous river, and the urban reaches were often urgent management objects. Because the nature of mountainous river and many humanities function of urban river, the regulation project of medium and small rivers are often faced with many requirements, such as, flood and flash flood defense ability, reinforced self - safety, small effect to river mainstream, energy saving, environmental protection, and meeting requirements of ecological and the other aspects.

As a common river regulation structure, the disaster prevention and mitigation role of spur dike is almost indispensable, but when it faces many requirements, traditional solid spur dike cannot meet current development obviously. Because there are many problems, such as (C. Gisonni, 2008), backwater of solid structure that increases the pressure of flood, the adverse effects of circumfluence behind the dike to aquatic organisms, influence of deposition field on riparian environment, increasing of the mainstream velocity, as well as the larger local scour depth (Zhang, 2002). Novel permeable spur is then put forward, and constantly tried.

By the experiment study, the flow structure and scour properties of permeable spur dike was analyzed. For the permeable body, it influenced mainstream slightly. When the pervious rate was larger than 20%, it formed a slow flow field instead of the circumfluence field (solid spur dike) behind the dike. Meanwhile, the local scour depth decreased with increase of pervious rate, and the deposition field could be controlled by adjustment of pervious rate.

Permeable spur is a whole body, and it can satisfy requirements of regulation line to control the dike length. At the same time, it has the permeability, which can avoid excessive interference to the mainstream by choosing a suitable permeation rate, and can create suitable gentle flow for aquatic organisms survival. Permeable spur will not cause excessive deposition field, in order to avoid the influence riparian zone of natural river, and to form wide and narrow river geography. Local scouring depth of permeable groin is relatively small, and it is pile structure, with reinforced self - protection. It can be built with local raw material to reduce the cost, and also can increase the pile space to fill stone, wood, straw bale and other local materials in the space. Obviously, the permeable body brings many advantages, such as wide applicability, reinforced protection, which can meet the different regulation requirements, so that it has a good applicability and development foreground in the medium and small river regulation.

5 Conclusions

(1) With the development of idea about river training, the training engineerings in middle and small river face many demands, such as flood control, ecology, environment, energy conservation and aesthetic. Styles of river training structure will be advanced to adapt these new demands, and to play the integrated function.

(2) For the permeable body, the flow structure and scour properties are different from solid one, and the properties are related with pervious rate. Permeable spur dike influenced mainstream slightly, when the pervious rate was larger than 20%, it formed a slow flow field instead of the circumfluence field (solid spur dike) behind the dike. Meanwhile, the local scour depth decreased with increase of pervious rate, and the deposition field can be controlled by adjustment of pervious

rate.

(3) Permeable spur dike can satisfy requirements of regulation line to control the dike length. At the same time, it can avoid excessive interference to the mainstream by choosing a suitable permeation rate, and can create suitable gentle flow for aquatic organisms survival. Permeable spur will not cause excessive deposition field, in order to avoid the influence riparian zone of natural river. Local scouring depth of permeable groin is relatively small, and it is pile structure with reinforced self - protection. It can be built with local raw material to reduce the cost, and also can increase the pile space to fill stone, wood, straw bale and other local materials in the space. It can satisfy many different regulation requirements to choose suitable pervious rate, so that it has a good applicability and development foreground in the medium and small river regulation.

Progress of river management concept promotes to develop the river regulation technology constantly. Regulation of small and medium river faced multiple demands. Permeable spur dike has demonstrated wide applicability, which can satisfy many different regulation requirements to choose suitable pervious rate in the future. We should reinforce the study on hydraulic characteristics and scouring mechanism of dikes system, in order to learn its suitable condition accurately and estimate its effect rationally and to better play comprehensive benefits including its advantages of ecological environment and the economic cost.

Acknowledgements

Partial financial support of the work presented in this paper from China's National Natural Science Foundation of China (No. 51109011) is gratefully acknowledged.

References

Zhang XL. The Problem and Countermeasure of Medium and Small River Manages in Our Country [J]. Journal of Water Resources Development Research, 2005, 1(s): 68 - 70.

Chen W. S., Chen Z. C. Discussion about Problems of Medium and Small River Manages in Guangdong Province. Guangdong Water Resources and Hydropower, 2011, 8(s): 13 - 16.

Miao J. L., Zhang Y. X., Zhou J. Y. Review of New Structures of Bank Protection and W ater diversion in river Regulation Engineering[J]. Journal of Yangtze River Scientific Research Institute, 2011, 28(3): 1 - 4.

S. Ikeda, N. Izumi, R. Ito.. Effects of Pile Dikes on Flow Retardation and Sediment Transport [J]. Journal of Hydraulic Engineering, ASCE, 1991, 117(11): 1459 - 1478.

A. J. Raudkivi. Permeable Pile Groins[J]. Waterway, Port, Coastal, and Ocean Engineering, ASCE, 1996, 122(6): 267 - 272.

Feng, H. Ch., Shi Z. T. Preliminary Study on Calculate Formula about Scour Depth of Non - flooding Permeable Spur Dikes[J]. China Rural Nater and Hydropower, 2002, 1(5): 46 - 49.

Zhou Y. J., Liu H. F. Discussion on Permeable Spur Design[J]. Yellow River, 2001, 29(1): 65 - 66.

Zhang C. P, Zhang S. C., Yang D. L. Discussion of Permeable spur Dike[C] // Proceeding of 3th International Yellow River Forum. Zhengzhou: Yellow River Conservancy Press, 2007: 68 - 75.

A. Sukhodolov, C. Engelhardt, A. Kruger, etc. Case Study: Turbulent Flow and Sediment Distributions in a groyne Field[J]. Journal of Hydraulic Engineering, ASCE, 2004, 130(1): 1 - 9.

J. Geist. Integrative Freshwater Ecology and Biodiversity Conservation[J]. Procedia Environmental Sciences: Ecological Indicators, 2011, 11(6): 1507 - 1516.

C. Gisonni, W. H. Hager. Spur ailure in River Engineering[J]. Journal of Hydraulic Engineering, ASCE, 2008, 134(2): 135 - 145.

Zhang, B. Sh., Ma J, Y. Local Scour Depth around Spurs in the Yellow River[J]. International Journal of Sediments Research, 2002, 117(3): 243 - 254.

Impact of Check Dams on the Sharp Decrease of Runoff in the Middle Yellow River: Case Study in the Huangfuchuan River Basin

Shi Haiyun*, *Li Tiejian*, *Zhu Jinfeng* and *Zhang Ang

State Key Laboratory of Hydroscience and Engineering, Tsinghua University, Beijing 100084, China

Abstract: Soil erosion in the middle Yellow River is the most serious in China. Recently, due to the increasing human activities for several decades, the condition of runoff and sediment yields has changed obviously, which causes the sharp decrease of runoff in this region. Therefore, deep investigation on the impact of human activities on runoff process appears particularly important. In this study, the Huangfuchuan (noted as HFC hereafter) River basin, one typical tributary in the middle Yellow River, is taken as the study area; and the building of extensive check dams (i. e., sediment trapping dams) are taken as the typical type of human activities in this basin. Firstly, digital drainage network with high spatial resolution was extracted from the DEM (i. e., Digital Elevation Model) by using the TOPAZ module. Then the amount and distribution of check dams were determined through data collection from statistical reports and artificial identification from remote sensing images. Thereafter, the topological relation between the dams and river reaches was identified, which helped to find the check dams, 448 in all, control 45% of the total basin area. To estimate the effect of those dams, periods with weak or significant impact of human activities were defined, respectively, by the analysis of runoff change. Then, the Digital Yellow River Integrated Model (noted as DYRIM hereafter) is used to simulate the response of runoff process on the building of check dams, using the measured precipitation data over the basin from 1976 to 2006. By comparing simulated runoffs with and without check dams, this study shows that the interception effect of check dams is one of the most important reasons for the sharp decrease of runoff in the middle Yellow River, under the condition that the rainfall decrease is obscure.

Key words: the middle Yellow River, the Huangfuchuan River basin, check dam, runoff decrease, DYRIM

1 Introduction

Soil erosion in the middle Yellow River is the most serious in China. A large number of soil and water conservation measures, such as building reservoirs and check dams (i. e., sediment trapping dams), terracing and vegetation restoration, were taken since the late 1950s. Of all the soil and water conservation measures mentioned above, check dams are the most widespread structures in the middle Yellow River. Generally, a complete check dam is composed of the embankment, the spillway and the outlet, which is able to completely or partly intercept the water and sediment from upstream; while a small one is usually simplified without spillways or outlets, which is only able to completely intercept the water and sediment from upstream. Moreover, a group of such check dams built on the same reach with a backbone one downstream can constitute a check dam system, which is effective for preventing floodwater, retaining sediments, forming farmland, storing water for irrigation and so on (Xu et al., 2004).

As a result, the condition of runoff and sediment yields has changed a lot because of those human activities, which significantly impact on the runoff in most tributaries in the middle Yellow River (Ni et al., 1997; Chen et al., 2002; Ran et al., 2004, 2008; Xu and Wang, 2011). In recent decades, runoff of the middle Yellow River decreased sharply as a whole, which has affected the normal life of the people in some regions. However, it seems difficult to be explained only by climate change, and the important effect of human activities should be considered (Xu, 2007; Wu et al., 2010).

This study took the HFC River basin, one typical tributary in the middle Yellow River, as the study area, and aimed to do quantitative research on the impact of check dams on the sharp decrease of runoff. This was done by using the DYRIM based on the measured precipitation data over the basin and hydrologic data of Huangfu station near the basin outlet from 1954 to 2006.

2 Study area

The HFC River basin (110°18′ - 111°12′E, 39°12′ - 39°54′N) is located in the northern edge of the Loess Plateau, with a drainage area of 3,246 km^2 and altitude between 1,000 m and 1,400 m. It is a tributary of the Yellow River and joins into the main stem near Hequ County. The annual rainfall in this region is 406 mm, mainly occurring as short duration high intensity torrential rains in the flood period from June to September. The annual potential evaporation is nearly 1,500 mm. Materials on the earth surface are highly erodible quaternary loess in the flat upper land and Pisha stone (i. e., unconsolidated coarse sand stone) in the incised slopes and valleys. Grassland, farmland and forestland are main types of land use in this region (Ran et al., 2003; Xu et al., 2006).

3 Data and method

3.1 Hydrologic data

Huangfu, Shagedu and Changtan are the three hydrologic stations in the HFC River basin, among which Huangfu is the controlling station with a drainage area of 3,199 km^2 and measured hydrologic data from 1954 to present. There are 14 rainfall stations in the HFC River basin with measured rainfall data from the years when stations were built to present. All the available meteorological and hydrologic data (from the years when stations were built to 2006) are provided by the Hydrological Bureau of the YRCC (i. e., Yellow River Conservancy Commission). Fig. 1 shows the hydrologic and rainfall stations in the HFC River basin.

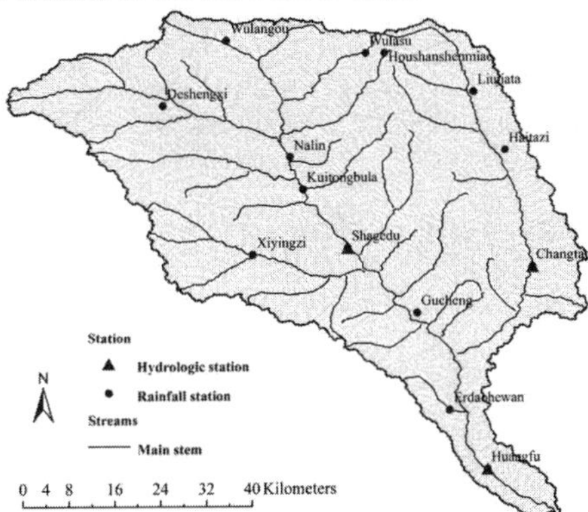

Fig. 1 Hydrologic and rainfall stations in the HFC River basin

3.2 Check dams and their distribution

Data of reservoirs and backbone check dams are provided by the YRCC, including name, longitude and latitude, construction year, control area, total storage capacity, and etc. However, a part of them may have lost effectiveness after years or no longer exist because of extreme floods. Moreover, projects built in the last few years are not recorded in this data set. For this reason, artificial identification of check dams was done from Google Earth (mainly GeoEye images) and other

(e. g. , the CBERS, China – Brazil Earth Resources Satellite) remote sensing images. Fig. 2 shows the distribution of check dams and reservoirs in the HFC River basin and Tab. 1 shows the amount of them.

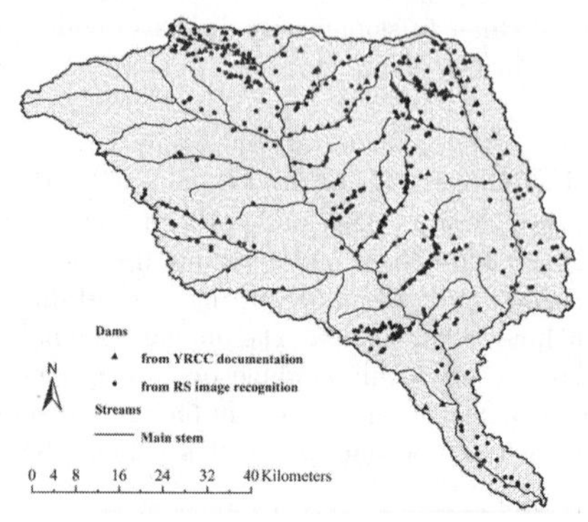

Fig. 2 Distribution of check dams and reservoirs in the HFC River basin

Tab. 1 Amount of check dams and reservoirs in the HFC River basin

①Newly recognized	②Confirmed from the data set	③Excluded from the data set	④To be determined from the data set	Data from the YRCC ②+③+④	Determined ①+②
337	111	70	17	198	448

Images from GeoEye and other remote sensing sources covering the HFC River basin are clear enough as a whole so that there are a lot of check dams newly recognized and confirmed from the data set. Two types of check dams in the data set are excluded: one is that the coordinates of check dams in the data set are out of this river basin, and the other is that no check dam can be found in this river basin where the coordinates point to. Unclear local images lead to the number of check dams to be determined from the data set. As a result, the determined amount of check dams to be used in the simulation of this study is 448, which reflects the current situation basically.

3.3 Brief introduction of the DYRIM

The DYRIM is a distributed model platform developed by Tsinghua University for hydrologic and sediment simulation in large – scale river basins (Wang et al. , 2007). The model is based on digital drainage network, which is extracted from DEM by using the TOPAZ module (Garbrecht and Martz, 1999) and then coded by the modified binary tree method (Li et al. , 2010). Generally, the drainage network is extracted in a high resolution to properly reflect the finely broken topography in the rolling Loess Plateau in the middle Yellow River. Moreover, dynamic parallel computing technology is used in order to speed up the simulation (Li et al. , 2011; Wang et al. , 2009, 2011, 2012).

The hydrologic model in the DYRIM is a distributed physically – based hillslope model running on fine time steps (e. g. , 6 minutes), which basically reflect the infiltration – excess runoff yield mechanism in the study region. The DYRIM takes hillslope – channel as a basic hydrological unit in consideration of essentially different mechanisms of hydrological response of hillslope and channel. It is the basic unit extracted from DEM and is the basis of watershed discretization. There are three slopes when the Horton – Strahler order of a unit is 1, called left, right and source hillslopes, respectively. And there are only two slopes when the Horton – Strahler order of a unit is greater than 1, called left and right hillslopes, respectively. Runoff – yield model is established on the hillslope unit. Runoff of each hillslope flows into the associated channel and then all the channels make up the drainage network.

Hillslope soil mass is divided into top soil layer and deep soil layer in this runoff – yield model. Hydrological processes of a hillslope unit to be simulated include vegetation interception, evapotranspiration, infiltration – excess runoff on the ground surface, subsurface flow in top soil layer and deep soil layer, infiltration of top soil layer and water exchange between the two layers. Fig. 3 shows the relationships between hydrological processes in the runoff – yield model.

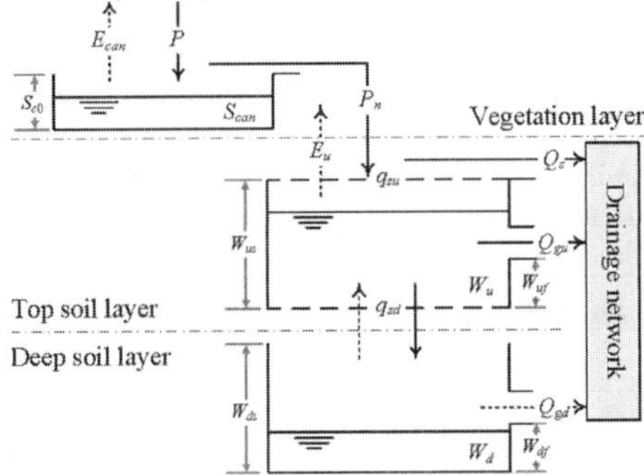

Fig. 3 Sketch map of runoff – yield model (Wang and Li, 2009)

3.4 Extracted drainage network

The extracted digital drainage network with high spatial resolution is necessary to analyze the corresponding relationships between check dams and river reaches. Furthermore, a high – resolution drainage network provides a suitable spatial discretization scale for running hydrologic model where the land surface is incised into numerous hillslope – valley units. The digital drainage network with terrain factors such as slope, aspect, elevation, length, and so on, is extracted by the TOPAZ module from the ASTER DEM with cell size of about 30 m. The two extracting parameters were as follows: the CSA (i. e., critical source area) was 1 ha and the MSCL (i. e., minimum source channel length) was 100 m. There are in total 61,524 river reaches and nearly 154,000 hillslopes in the digital drainage network of the HFC River basin with an average hillslope area of 2.1 ha. Through analyzing the spatial relationship of check dams and the extracted digital drainage network, it is easy to obtain the corresponding relationship of them and calculate the control area of each check dam and to analyze the topological relation between the check dams. Fig. 4 shows the corresponding relationships between check dams and reaches in a sub – basin of the HFC River basin. It is found that the check dams in the HFC River basin have controlled 45% of the total area, coming up to 1,461 km^2.

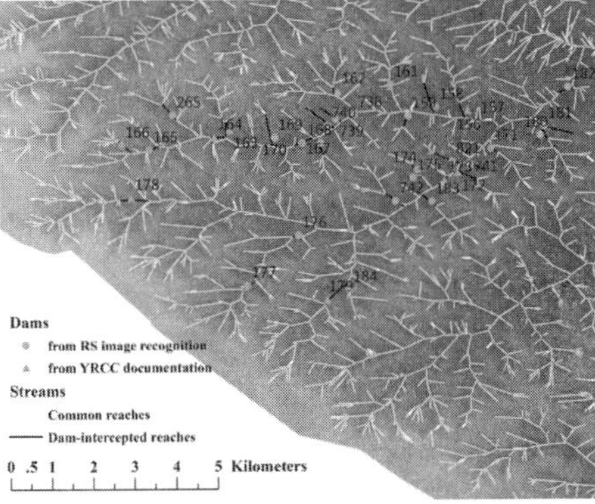

Fig. 4 The corresponding relationships between check dams and river reaches in a sub – basin of the HFC River basin

3.5 Flowchart of the analysis

The DYRIM is used to simulate the rainfall – runoff process of the HFC River basin, as shown in Fig. 5. The measured data of the period with weak impact of human activities are taken for model parameters calibration and verification. And then, the runoff process of the period with significant impact of human activities is calculated with the same parameters. Keeping other inputs unchanged, natural runoff process without the impact of check dams will be obtained if check dams are not considered while disturbed runoff process impacted by check dams will be obtained if check dams are considered. The impact of check dams on runoff process can be quantified by the comparative analysis of the two calculations. It is worth noting that this paper focuses on the extreme situation that water from upstream is completely intercepted by check dams so that the result reflects the maximum impact of check dams on runoff process.

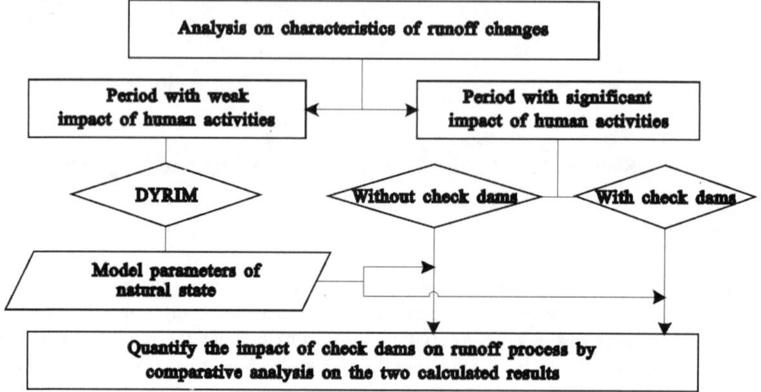

Fig. 5 Flowchart of the analysis in this study

4 Results and discussion

4.1 Analysis on characteristics of runoff changes

Result of trend test for the rainfall and runoff series is shown in Tab. 2, using the linear regression and Mann – Kendall trend test method (Mann, 1945; Kendall, 1975), respectively. The two methods provide the same conclusion that rainfall presents a decreasing trend but not significant ($p > 0.1$), while runoff presents a quite significant decreasing trend ($p < 0.01$). Moreover, slope of Mann – Kendall trend test method is smaller because of ignoring the impact of external unusual values.

Tab. 2 Result of trend test for rainfall and runoff series

Method	Variable	Mean	Trend	Percentage	Significance level
Linear regression	Rainfall	406 mm	-1.50 mm/yr	-0.37 %/yr	$p > 0.1$
	Runoff	43.2 mm	-1.19 mm/yr	-2.75 %/yr	$p < 0.01$
Mann – Kendall	Rainfall	406 mm	-1.12 mm/yr	-0.28 %/yr	$p > 0.1$
	Runoff	43.2 mm	-0.91 mm/yr	-2.11 %/yr	$p < 0.01$

Significant change – points in the rainfall and runoff series are shown in Tab. 3, using the Pettitt change – point detection method (Pettitt, 1979). For the rainfall series, no significant change – point is found. For the runoff series, the first significant change – point is found in 1984 ($p < 0.01$). Moreover, the second significant change – point is found in 1998 ($p < 0.05$) in the sub – series of runoff series from 1985 to 2006.

Tab. 3 Result of change – point detection for rainfall and runoff series

Variable	Change – points	Year	Significance level
Rainfall	None	/	/
Runoff	1	1984	$p < 0.01$
	2	1998	$p < 0.05$

It can be inferred that 1984 is the cut – off point of natural state and disturbed state in the HFC River basin according to the first change – point in the runoff series. Disturbed state can be divided into two periods according to the second change – point, which are from 1985 to 1998 and from 1999 to present, respectively. Annual discharge of the three periods shows a stepped decreasing trend without a significant changing trend of rainfall so that the impact of human activities may be the main reason. Consequently, period before 1984 is considered to be with weak impact of human activities and it can be used for model parameters calibration and verification. Period from 1999 to present is considered to be with significant impact of human activities and it can be used to analyze the impact of check dams.

4.2 Period with weak impact of human activities

Period before 1984 has been selected to be the representative years with weak impact of human activities. As most of the rainfall stations were built after 1976, the period from 1976 to 1984 is taken for model parameters calibration and verification. The measured annual discharges of the nine years were sorted in ascending order, as shown in Tab. 4. The five years in odd orders are taken for calibration, and the other four for verification.

Tab. 4 Ascending sort of the measured annual discharges from 1976 to 1984

Year	1983	1980	1977	1982	1984	1976	1981	1978	1979
Measured annual discharge (10^8 m^3)	0.79	0.92	1.11	1.33	1.63	1.70	2.03	2.65	4.37

Parameters needed to be calibrated can be divided into two types. One type is invariant parameters determined by the climate, geography and other basic features of the watershed. Relevant parameters are from literature and previous studies on the middle Yellow River with the DYRIM (see Tab. 5). The other type is adjustable parameters including infiltration rate of ground surface, vertical infiltration rate from top soil layer to deep soil layer and horizontal infiltration rate of the two soil layers. The underlying surface of the study area, such as soil type and land use, is assumed to be homogeneous in this study.

Tab. 5 Invariant parameters determined by the basic features of the watershed

Parameter	Field capacity of top soil layer	Free water content of top soil layer	Field capacity of deep soil layer	Free water content of deep soil layer	Depth of top soil layer (m)	Water capacity of unit LAI (m)
Value	0.205	0.296	0.22	0.325	0.3	0.003,6

The time step is set to be six minutes when simulating the runoff process of Huangfu station with hour – scale rainfall data. Compare the daily average simulated values against the daily measured data and the relative error of water quantity (noted as RE hereafter) and the Nash – Sutcliffe coefficient of efficiency (noted as NSE hereafter) are taken as the criterion (Nash and Sutcliffe, 1970):

$$NSE = 1 - \frac{\sum_{i=1}^{N}(Q_{i,mea} - Q_{i,sim})^2}{\sum_{i=1}^{N}(Q_{i,mea} - \overline{Q}_{mea})^2} \tag{1}$$

where, $Q_{i,mea}$ is the i – th measured data (m^3/s), $Q_{i,sim}$ is the i – th simulated value (m^3/s), N is the sample size, $\overline{Q}_{mea}$ is the mean value of the measured data.

Fig. 6 shows the comparison of the simulated values against the measured data in the period for model parameters calibration and verification. Tab. 6 gives the values of RE and NSE. It is shown that the simulated values are close to the measured data as a whole. For the five years for calibration, the values of NSE are greater than zero with a maximum 0.9 in 1979 except for a minus −0.2 in 1977; the values of RE are within ±11% except for a large one 42% in 1981. For the four years for verification, good and bad results are half and half, among which 1978 and 1982 are the good years with the values of NSE greater than 0.6 and the values of RE within ±5%. As a result, model parameters are considered to be applicable for the HFC River basin after calibration and verification and can be used to simulate the runoff process in the period with significant impact of human activities (see Tab. 7).

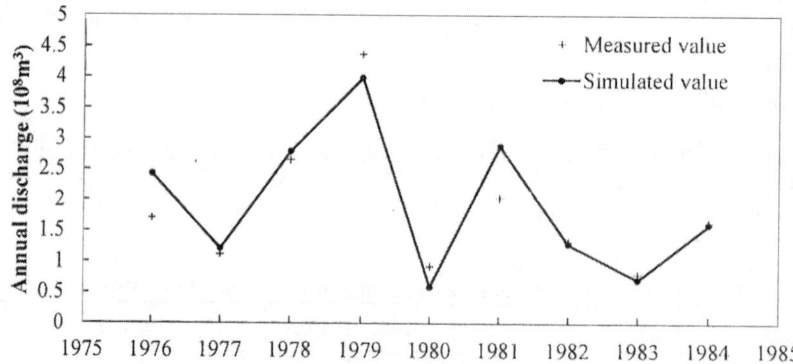

Fig. 6 The comparison of the simulated values against the measured data in the period for model calibration and verification

Tab. 6 Values of NSE and RE in the period for model calibration and verification

Year	Calibration					Verification			
	1982	1977	1979	1981	1983	1984	1976	1978	1980
NSE	−0.2	0.9	0.25	0.012	0.4	0.076	0.61	−5.56	0.76
RE	8.7%	−8.9%	42%	−11%	−2%	42%	5%	−36%	−3%

Tab. 7 Values of adjustable parameters after model calibration and verification

Parameter	Infiltration rate of ground surface (mm/hr)	Vertical infiltration rate from top soil layer to deep soil layer (mm/hr)	Horizontal infiltration rate of top soil layer (mm/hr)	Horizontal infiltration rate of deep soil layer (mm/hr)
Value	0.9	1.0	5.76	4.68

4.3 Period with significant impact of human activities

Period from 1999 to present has been selected to be the representative years with significant impact of human activities. Due to lack of the measured rainfall and hydrologic data after 2007, the period from 1999 to 2006 is used to analyze the impact of check dams by using the DYRIM. Natural runoff process without the impact of check dams is to be calculated with the parameters in Tab. 7. If the extreme situation that water from upstream being completely intercepted by check dams is considered, the maximum impact of check dams on runoff process can be obtained. The comparison of the simulated values against the measured data in the two cases is shown in Fig. 7. Tab. 8 gives the values of RE and NSE, as well as the impact ratio of check dams (i. e., 1 - annual discharge with check dams/ annual discharge without check dams, noted as IR hereafter).

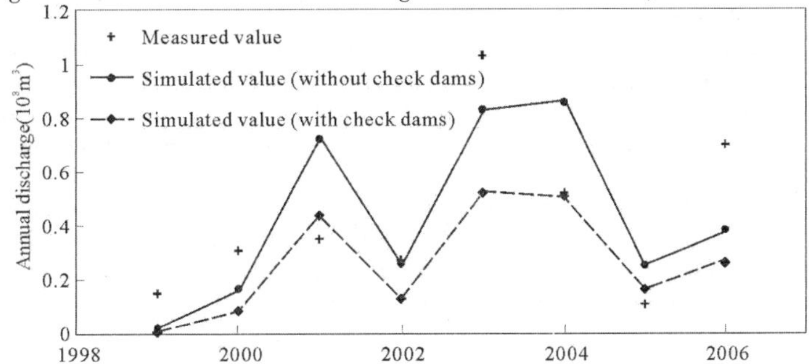

Fig. 7 The comparison of the simulated values against the measured data in the period for model application

Tab. 8 Values of NSE, RE and IR in the period for model application

Year		1999	2000	2001	2002	2003	2004	2005	2006
Measured annual discharge (10^8 m^3)		0.15	0.31	0.35	0.27	1.03	0.52	0.11	0.70
NSE	Without check dams	-0.16	0.63	-4.5	-3.07	0.85	-1.46	-1.0	0.394
	With check dams	-0.06	0.40	-1.12	-0.75	0.79	-0.29	-1.0	0.49
RE	Without check dams	-88%	-46%	106%	-5%	-20%	65%	135%	-46%
	With check dams	-99%	-73%	25%	-53%	-49%	-2%	51%	-63%
IR		92%	50%	39%	51%	37%	41%	36%	32%

Of all the calculated natural annual discharges without the impact of check dams, three of them are greater than the measured data but five are smaller. The values of RE of the three overestimated years are all great than 65% with a maximum of 135% in 2005. Actually, the excess water has not been observed by Huangfu station, which reflects that not all of the runoff can reach the outlet of the watershed impacted by the interception of water conservation projects such as check dams. As a result, check dams are considered as an important factor in this study. The result shows that annual discharge of each year decreases to a certain extent after considering check dams in simulation. It leads to a great change in the values of RE of the three overestimated years 2001, 2004 and 2005, which become 25%, -2% and 51% from original values of 106%, 65% and 135%, respectively, with an average decrease of 77 percentages. Meanwhile, it also has negative effects on the values of RE of the five underestimated years with an average decrease of 26 percentages, which is much smaller than that of the three overestimated years. For the values of NSE, the minus ones whose absolute values were large have turned into values within -1.0 after considering check dams in simulation, except for 2001. It shows that runoff may decrease sharply due to the interception of check dams and considering check dams in simulation can make the result much better. The values of IR fluctuate between 32% and 92%, which may be caused by the change of relative position between the rainfall centre and the sub – basin highly controlled by check dams. The mean values of annual discharge during this period are 0.261×10^8 m^3 and 0.434×10^8 m^3, corresponding to the cases of considering check dams or not, respectively. Thus, the average IR during this period is about 40%, which is a little smaller than the control area ratio of check dams (45%). It may be due to the difference of runoff generation and convergence characteristics in different parts of this basin (e. g., the main stem and the common reaches); while check dams mainly control the common reaches. Consequently, under the condition that water from upstream is completely intercepted by check dams, the control area ratio is the main factor to determine the IR but the difference of runoff generation and convergence characteristics in different parts of this basin also have a certain effect on it.

However, there are still differences between the simulated values and the measured data based on available data. On the one hand, it is caused by the temporal and spatial resolution of rainfall data. There are only 14 rainfall stations in the HFC River basin and each of them controls about 229 km^2 on average, which seems to be insufficient. Most of the time intervals between two rainfall records are 2 hours, which may lead to the bad impact that high intensity rain during short time is homogenized. On the other hand, the underlying surface, such as soil type and land use, is assumed to be homogeneous in this study, which is different from actual condition. Moreover, other water conservation projects, such as terraces, have not been considered yet, and water consumption for mines and towns are difficult to be quantified. All these factors mentioned above may influence the accuracy of simulation.

5 Conclusions

This study quantified the impact of check dams on the sharp decrease of runoff based on the DYRIM, taking the HFC River basin as the study area.

(1) According to the analysis on characteristics of runoff change, runoff in the HFC River basin shows a significant decreasing trend from 1954 to present and it can be divided into three periods, which are before 1984, from 1985 to 1998 and from 1999 to present, respectively. Annual average discharges of the three periods show a stepped decreasing trend under the condition that the rainfall decrease is obscure. Therefore, the impact of human activities may be the main reason and should be considered.

(2) The amount of check dams and reservoirs used in this study is 448 in all, controlling 45% of the total area of the HFC River basin, equal to 1,461 km^2. The two calculations with and without check dams during the period with significant impact of human activities show that runoff decreases sharply due to the interception of check dams, coming up to 40% at most; and better result can be obtained if check dams are considered in simulation.

(3) The accuracy of simulation in this study may be influenced by the resolution of input data (e. g., rainfall data). And it can be improved through data collection and field investigation in further work.

Acknowledgements

The financial support from the Non – profit Fund Program of the Ministry of Water Resources of China (Grant No. 200901016 – 03, 200901019) and the National Natural Science Foundation of China (Grant No. 51109114) are gratefully acknowledged.

References

Chen H, Zhou J X, Lu Z C, et al. 2002. Impact of environmental factors on runoff and sediment variations in middle reaches of the Yellow River. Geographical Research, 21(2): 179 –

187.

Garbrecht J, Martz L W. 1999. TOPAZ overview. Oklahoma: USDA ARS, Grazinglands Research Laboratory.

Kendall M G. 1975. Rank Correlation Measures. London: Charles Griffin.

Li T J, Wang G Q, Chen J. 2010. A modified binary tree codification of drainage networks to support complex hydrological models. Computers & Geosciences, 36(11): 1427 - 1435.

Li T J, Wang G Q, Chen J, et al. 2011. Dynamic parallelization of hydrological model simulations. Environmental Modelling & Software, 26: 1736 - 1746.

Mann H B. 1945. Non - parametric tests against trend. Econometrica, 13: 245 - 259.

Nash J E, Sutcliffe J V. 1970. River flow forecasting through conceptual models, 1. A discussion of principles. Journal of Hydrology, 10: 282 - 290.

Ni J R, Han P, Wang G Q, et al. 1997. Variations of sediment characteristics in the middlereach of Yellow River with respect to regional water and soil conservations: Ⅱ. benefit analysis of reducing the sediment. Journal of Natural Resources, 12(2): 126 - 132.

Pettitt A N. 1979. A non - parametric approach to the change - point problem. Applied Statistics, 28(2): 126 - 135.

Ran D C, Gao J L, Zhao A C, et al. 2003. Analysis on runoff and sediment characteristics and harnessing measure in Huangfuchuan river watershed. Journal of Hydraulic Engineering, 2: 122 - 128.

Ran D C, Luo Q H, Liu B, et al. 2004. Effect of soil - retaining dams on flood and sediment reduction in middle reaches of Yellow River. Journal of Hydraulic Engineering, 5: 7 - 13.

Ran D C, Luo Q H, Zhou Z H, et al. 2008. Sediment retention by check dams in the Hekou - Longmen Section of the Yellow River. International Journal of Sediment Research, 23: 159 - 166.

Wang G Q, Li T J. 2009. River basin sediment dynamics model. Beijing: China WaterPower Press.

Wang G Q, Wu B S, Li T J. 2007. Digital Yellow River model. Journal of Hydro - Environment Research, 1: 1 - 11.

Wang H, Fu X D, Sun Q C, et al. 2009. Method improvement of the parallel computing for the large - scale basin hydrology. Journal of Basic Science and Engineering, 17: 1 - 10.

Wang H, Fu X D, Wang G Q, et al. 2011. A common parallel computing framework for modeling hydrological processes of river basins. Parallel Computing, 37: 302 - 315.

Wang H, Zhou Y, Fu X D, et al. 2012. Maximum speedup ratio curve (MSC) in parallel computing of the binary - tree - based drainage network. Computers & Geosciences, 38: 127 - 135.

Wu R, Chen G F, Zhang J X. 2010. Changes of runoff and sediments in Hekou - Longmen section of the middle Yellow River. Journal of Desert Research, 30(1): 210 - 216.

Xu J H, Lin Y P, Wu C J, et al. 2006. Definition on source area of centralized coarse sediment in middle Yellow River. Zhengzhou: Yellow River Press.

Xu J X. 2007. Impact of human activities on the stream flow of Yellow River. Advances in Water Science, 18(5): 648 - 655.

Xu J X, Wang H B. 2011. Influence of soil and water conservation measures on reducing in frequency of hyperconcentrated flows in the Wudinghe River basin. Environmental Earth Sciences, 62: 1513 - 1520.

Xu X Z, Zhang H W, Zhang O Y. 2004. Development of check - dam systems in gullies on the Loess Plateau, China. Environmental Science & Policy, 7: 79 - 86.

Standardizing Criterion and Regulating Operation to Improve Technical Level of the Yellow River Engineering Management on Full-scale

Cui Yanfeng

Construction Group of Yellow River, Zhengzhou, 450003, China

Abstract: The characteristics of Yellow River engineering is long distance lines of management and large scale and difficult to safeguard. Since 2004, YRCC implemented standard dike construction and management contents are more abundant than before. In Jun. 2006, YRCC finished reformation of 76 water management units and demanded highly level for engineering management due to support in finance. We studied and stipulated management methods and inspecting criterion in light of dike, river training project and sluices as well as file sorting, thus formed criterion system of 4 + 1 which possessing characteristics of Yellow River. This further lays technical foundation for reinforcing engineering management and lead to transformation from Yellow River engineering management to criterion and standardization. To implement overall criterion, standardization and modernization construction and promote the development of engineering management orderly and effectively, YRCC analyzed and studied on work procedure of management and maintenance in main chains, contents and criterion. We raised management and criterion index by sorting various kinds of engineering parts keeping to policy of unified standard, regulated management, fabricate operation. And we put forward specific requirement for safety of engineering, modernization construction and files indoors sorting. During works we strictly kept in management procedure and reinforce supervision management in whole course by catching hold of engineering management plan, overall inspections, fabrication of implementation scheme, subscription of contraction, quality supervision, monitoring, project acceptance, files sorting etc. At same time persisting in construction of nation level unit as momentum, and what we have done is by means of model engineering construction to standardize management criterion and regulating management operation and enhancing the elaborate works and strictly inspecting in order to improve the technical level of Yellow River engineering management and get ideal effect.
Key words: regulation, management, construction, Yellow River

There are many kinds of engineering that managed by YRCC with over 2,400 km length of dike line, 12,800 sites of river training works, 188 sluices and two large-scale reservoirs. The management line is long and quantity is considerable with large scale, so it is difficult to manage and demands highly for flood control safety be set very importance by our nation historically. Since 2004, YRCC implemented standard dike construction and management contents are more abundant than before. In Jun. 2006, YRCC finished reformation of 76 water management units and demanded highly level for engineering management due to support in finance. This needs new and higher demands for engineering criterion, standardization and modernization. What we have done is by means of model engineering construction and consulting experience of other river basin in our country to standardize management criterion and regulating management operation and enhancing the elaborate works and strictly inspecting in order to improve the technical level of Yellow River engineering management and get ideal effect.

1 To establish criterion system and regulate management conduct

After reformation of water management system it regulated management system and got finance support and ascertained relationship of contract between water management unit and corporation of maintenance. It needs to establish management mechanism fitting to state council guidance on water management reformation about profession, standardization and modernization of engineering maintenance to ensure engineering management developing orderly and effectively. Therefore, YRCC studied and analyzed all key parts of work procedure, contents and criterion and put forward many managing methods and established new system running frame. The key work is to determine maintenance items and sign maintenance contract and supervise operation of maintenance and strictly inspect and assess them. According to work thought of determining duty, concreting management criterion, leading to fabricate operation and inspecting strictly, we ensure to finish rating task, at same time we keeping to concrete examination, and encourage and punish thereby promoting overall regulated management.

1.1 To establish criterion system and concrete inspection demands

We studied and stipulated management methods and inspecting criterion in light of dike, river training project and sluices as well as file sorting, thus formed criterion system of 4 + 1 which possessing characteristics of Yellow River. This further lays technical foundation for reinforcing engineering management and lead to transformation from Yellow River engineering management to criterion and standardization. To implement overall criterion, standardization and modernization construction and promote the development of engineering management orderly and effectively, YRCC analyzed and studied on work procedure of management and maintenance in main chains, contents and criterion. For example, in inspection criterion of Yellow River dike engineering it firstly divides dike into 8 parts such as dike top, dike slope, alluvial area, flood controlling trees, dike safeguard strip, engineering safeguard, engineering observation and probing, modernization as well as management yard etc. The inspection criterion almost includes 29 items which are roads of dike top, blocking stone and safeguard stake, drainage at dike top, dike shoulders, dike shoulder trees, abandoning facility, marking plates, dike slope, drainage canals, front and back platform, roads to dike, enclosure levees of alluvial area, trees planting in alluvial area, slope of alluvial area, management of alluvial area, flood controlling trees, dike safeguard strip, border canals, land right determination, safeguard, vehicles abandon, historical carry-over engineering, engineering observation, hidden troubles probing of dike, informatics construction, professional level, architecture facility, greening and beautification and health status and 143 examination points. All inspection points are measured and observed and the total scores are 1,000.

1.2 To grasp key ache and regulate management procedure

Tightly grasp key procedures of engineering management planning, overall inspection, fabricate implementation scheme, signing contract, supervision of quality, supervision, engineering acceptance and handover of indoor files to strictly fulfill managing procedure and reinforce supervision management overall.

Firstly, it lay out uniformly in light of engineering management criterion. Each unit makes their three-year planning of engineering maintenance on base of overall inspection to their engineering and thereby stipulates next year's implementing scheme in Aug. or Sep. of every year. In term of yearly scheme approved, all water management units sign maintenance contract with maintaining corporation.

Secondly, in course of implementation of daily maintenance contract it needs water management unit put forward sorting tables of maintenance items and general table of monthly maintenance items as well as arrange explanation of monthly maintenance, that is two lists and one explanation to specify maintenance task of next moth by observation and inspection. Meanwhile we fetched in constructional item management model and fulfilling procedure of design, checking and approval and carrying out supervision system to improve inspection and acceptance procedure.

Thirdly, it is to improve inspection and examination system. By inspecting and examining periodically and un-periodically for superintendence department it promotes water management unit and maintaining corporation to carry out their work.

1.3 To specify post duty and realize the encourage and punishment

There four key elements about post duty system that is to specify work contents, workload, quality criterion and examining effect. The core is to divide each piece of work into its own post according to its content, characteristics and trait meanwhile checking content, quantity, standard and right to form post duty as well as system related to staff's benefit.

To establish post duty system may enhance staff's responsibility and reduce optional decision-making and sightless operation, therefore, YRCC studied and established double post duty system for Yellow River flood controlling engineering management observation and maintaining staff. Its main content to derive specific assessment from front-line staff of maintenance and observation on their work content, workload, time limitation, quality criterion, work responsibility, inspection method and their vested interest. We demand that each group's post duty, work content and inspection criterion be pasted on wall whereas ensure stall of front-line observation and maintenance to be in their own posts. Meanwhile, we demand all water management units and maintaining corporation to reinforce inspection in front-line and combine there progress with their earnings, thus effectively promoted daily work of maintenance to be carried out.

1.4 To reinforce staff training and raise their technical competence

With the improvement of observation instruments and maintaining tools gradually, considering staff in front-line be aging and many of servicemen and college students waiting for arrange it urgently needs them to be trained and improve their operation competence. By indoors course teaching combined with fieldwork we started training work for staff of water unit and maintaining corporation about relevant criterion, regulations etc. we organize carefully front-line staff to be trained with professional knowledge and post drill. This makes front-line staff grasping operative competence of management criterion, machines and equipments masterly and offered talent support for engineering management into standardization.

2 To implement model engineering strategy and enhance fabricate operation

2.1 To implement model engineering strategy

Considering characteristics of much engineering points, large scale, long dike line, wide scope and lack of repairs as well as much arrear it needs to implement model engineering leading strategy in order to operate fabricate and improve managing level. Arranged by YRCC, each water management unit lain out its model engineering construction plan for five years. Combined with yearly maintenance special item we carried out engineering mending task according to its importance on schedule step by step. In light of demands of constructing one item and accepting it meanwhile keeping it as good condition, by constructing model engineering at high level it promoted engineering appearance to improve continually with large scale stretched from point. Since 2004, water management units of YRCC have established 162 sections of model engineering including dike, sluices and river training works.

2.2 Actively to create water management units of nation level

Water management units of national level are a rule through which can estimate and display synthetic competence, ability and status of managing water project of a unit. To promote management work to get much progress for Yellow River engineering and display level of Yellow River engineering management profession and modernization, YRCC actively carried out creation of water management units of national level thereby dragon head to promote 76 units to heighten synthetic competence. At present, YRCC has got 12 water management units of national level and leading to this work in whole country.

2.3 Guiding and promoting fabricate operation

Fabricate operation is established on base of traditional management which leading traditional management to deep managing conception and managing model. Fabricate operation transfers traditional and extensive management to concrete, standardization and scientific style. There needs a set of strict criterion and regulated inspection procedure in work to put forward quantitative standard. We specified concrete post and duty of maintenance in our work. According to prescriptive procedure and concrete criterion and strict regulations to manage and reinforce fulfillment and supervision as well as performance examination. By professional maintenance it attains to regulated management and fabricate operation, expectant goals and work requirement.

3 To strictly inspect and examine and enhance technology progress

3.1 To inspect at different level and carefully organize to examine

The inspection and estimation was carried out in light of managing methods of Yellow River engineering management and its inspecting criterion. The inspecting contents include dike, river, sluices, reservoir management and mechanism as well as indoors file. YRCC established database of inspection experts of engineering management and organize them to be trained. Before inspection we take name list from database at random to form inspecting group and gather them to study according to uniform arrange. Provincial and city bureau carry out periodical or un-periodical inspection and half year inspection and yearly inspection, at same time they issue file to show the inspection result. And they take the yearly inspection result into yearly examination system of being inspected management units. Water management units and maintaining corporation respectively make out inspection operative methods for their own underling staff in front-line and establish special group to implement monthly. The management effect is relevant to staff' s earnings and prizes and strictly inspecting and fulfill encourage and punishment.

3.2 To set importance on inspection on spot and strictly measuring and examining

It adopts inspection method which overall examination combined with key parts check style of main forms of seeing, hearing, asking and checking. When inspecting on spot we will adopt way of overall inspection combined with section observation measuring at random consulting on engineering management criterion to judge whether the inspecting section indexes meet needs of demands. When listening to introduction of unit we grasp overall managing situation and asking inspection problems or questions. It needs to consult indoors files whether it integrated or not. The record is whether real, standard or not. At same time it needs to check observation data, maintaining dairy, generalized report and probing report of all kinds of engineering technical file in archive. The inspection of engineering management is implemented system of 1,000 points which include main body that is dike, river and sluices accounting for 70% and mechanism accounting for 12% and indoors file accounting for 18%. To encourage and lead to creation in managing work and promote construction of ecology the creating item and creating effect of national water management unit will be prized and add scores when inspecting managing work of units.

3.3 To grasp inspection and rectification and to promote to progress overall

Inspection group kept to inspecting every day and discussed, cleaned and generalized the inspection situation of management units in time. In course of inspection the group would discuss what they had seen with water management unit and finally finished inspecting report after the unit admitting or accepting. After inspection the administering department would issue inspecting report to all units on their inspecting situation in order to encourage advanced unit and promoted lagging unit to go ahead. At same time we would put forward engineering management work developing directions, regulating measurement and suggestions in order to attain to goal of overall progress for whole management work.

4 Conclusion

For recent years, YRCC have set very importance on management work of Yellow River engineering. By strict management and careful maintenance Yellow River dike and river training works have attained to aims of flat top and smooth slope, cleaning and beauty. The dike shoulders trees arrangement is beautiful, sward and trees of flood control grow flourishingly, drainage canals have been improved and marking plates is outstanding, this guarantees engineering integrated and safety use and primarily achieves to managing demands put forward by YRCC that is flood controlling safeguard line, traffic line under emergency and ecological line and gradually transfers to modern managing style of safety, ecology, scenery and culture.

Discussion on Some Technical Issues in Preparation of Water and Soil Conservation Programs for Development and Construction Projects in the Arid and Semiarid Eolian Sand Areas of Western China

An Runlian and ***Chen Haotan***

Xifeng Supervision Bureau for Yellow River Water and Soil Conservation, Xifeng, 745000, China

Abstract: The arid and semiarid eolian sand areas in Western China are suffering from severe natural conditions, where large – scale development and construction are ongoing in recent years. Preparation of water and soil conservation program for the project construction in such areas should follow the principle of avoiding disturbance of the original land surface to the maximum extent. Scientific and feasible technical measures should be taken for protection and rehabilitation, and cost – saving and reasonable programs should be adopted with the purpose of ensuring the normal operation of soil and water conservation facilities as well as preventing and reducing water loss and soil erosion caused by the development and construction activities, by such means to realize harmony of man and nature.
Key words: western, eolian sand area, water and soil conservation, program

Since the reform and opening up, various types of development and construction projects were built around the arid and semiarid eolian sand areas of Western China with the implementation of western development strategic plan. Due to the limitations of natural conditions, the water and soil conservation programs for development and construction projects have significant features in terms of technical content. The author explored some of the present issues in the paper.

1 Profile of western arid and semiarid eolian sand areas

The arid and semiarid eolian sand areas of Western China have a typical continental monsoon climate that is dry, windy, and characterized by a large annual temperature range. The annual precipitation in the arid Gobi desert regions of Hosi Corridor and? Tsaidam Basin is only 40 mm or so, covered by sand and gravel without even a blade of grass. The annual precipitation in the semiarid northern Shaanxi – Gansu – Ningxia is usually less than 300 mm, covered by drift sand or residual sandy soil in a few areas, loose in texture and poor in fertility. The natural vegetation of extremely low coverage mainly consists of drought – tolerant deciduous shrubs (sea buckthorn, lespedeza, caragana microphylla, etc) and herbaceous plants (stipa, festuca, etc). In these areas, the main conditions against wind erosion are surface crusts. Once they are damaged, water loss and soil erosion (caused by wind erosion) will occur accordingly.

These areas, rich in coal, oil, gas and other mineral resources, are the key areas of the Western Development. Large – scale development and construction brings a great pressure to local ecological environment.

2 Technical issues to be noted in preparation of Water and Soil Conservation Programs

(1) As a guiding ideology, minimize disturbance to original earth surface. Limited by hydrothermal conditions and soil conditions, artificial vegetation is generally difficult to survive in these areas, and a high restoration cost is needed when the surface is disturbed. Therefore, one of the important contents of the Water and Soil Conservation Programs is to minimize the disturbed area while meeting the layout of the main engineering facilities, constraint construction operation, and control the disturbance (e. g. walk, trample, etc) outside the project construction areas.

(2) Attention should be paid to the suitability and operability of the measures. Suitability

means that the measures should be suitable for the climate and soil conditions of the areas where projects are located. To achieve this point, it is required to do necessary investigations prior to the design so as to select the types of measures that are suitable for project areas and often used locally. Operability means that the measures should be able to be easily implemented. Raw materials and roads, water, etc. should be taken into full consideration so that the designed measures can be really implemented and play the expected role in water and soil conservation.

Enhancing suitability and operability is the minimum requirement for the preparation of water and soil conservation programs. If permissible, tests shall be performed in a small area for introduction of new measures to provide test data and information for updating and development of regional soil and water conservation measures.

(3) Temporary land use shall be strictly controlled in the design of measures. This can reduce surface disturbance actively and effectively. In general, temporary land use includes borrow sites, spoil grounds, construction camps and roads and so forth. For a given quantity of material to be needed and a given quantity of wastes to be abandoned, the sites/grounds can be big or small, near or far. A smaller site/ground means a smaller disturbed area. So does a shorter distance due to a shorter construction road to be needed.

Prior to the preparation of water and soil conservation programs, field survey shall focus on site selection for? temporary land use. Firstly, investigate the status of land use, and then select the places which are uncultivated and less covered by vegetartion as temporary sites with no or less arable land occupied and no or less vegetation destroyed. Secondly, further investigation shall be conducted for suitability when the said conditions are available. For instance, borrow sites shall be measured and estimated according to required soil conditions and quantities so as to meet the engineering requirements in quality and quantity without an excessive number of borrow sites provided. A small quantity of soil can be borrowed sporadically. Spoil grounds shall be located in lowland free from current rush. Their capacity shall be estimated. Concentrated stacking and protection is preferred. Temporary construction camp shall be arranged in a flat place with no or less cut/fill work to reduce deep disturbance and earth movement. Through these investigations and comparison, select the places close to the works as temporary sites. Temporary land use shall be controlled from such two aspects as temporary sites themselves and the roads to the construction sites.

(4) The following matters shall be noted in the arrangement of revetments and drainage ditches on crossing portions. The masonry revetment of a crossing portion must be beyond the connection between disturbed surface and natural slope (see the sketch below). If not, the foundation of the revetment will be eroded as the disturbed surface isn't fully protected. The revetment may even be destroyed accordingly. The cut-off drain, on the top of the revetment, shall be beyond the both sides of the revetment to a certain length as appropriate. In principle, the drainage ditch connected to the cut-off drain shall be set at a certain distance from the side of the slope to ensure that excavation and masonry activities of the drainage ditch dont affect or damage the foundation of the revetment at the side. The drainage ditch on the slope shall dip outward instead of extending straight down to prevent that the stability of revetment foundation is affected in the event of drainage ditch failure.

Moreover, free overfall shall be designed at the end of steeper drainage ditch to directly deliver water to river/gully channel. River bottom shall be protected by ripraps or other means. Don't arrange the end of drainage ditch on river/gully bank to let water flow into river along the bank, resulting in bank erosion. see Fig. 1.

(5) Natural conditions of the areas where projects are located shall be met in construction design. For the application of temporary measures in these areas, try not to select the measures that are connected with excavation disturbance, e. g. temporary drainage ditch and temporary detritus pit. Since a great slope ratio is needed for loose sandy soil, a temporary drainage ditch causes a considerable disturbed area. Moreover, human trampling in excavation operation also results in unnecessary damage. Therefore, blocking measures shall be appropriately provided in light of actual situation to lead water to flow out in desired direction.

For lack of water sources in the arid Gobi desert regions of Hosi Corridor and Tsaidam Basin, both engineering and vegetation measures are difficult to implement. Gravel coverage is often used

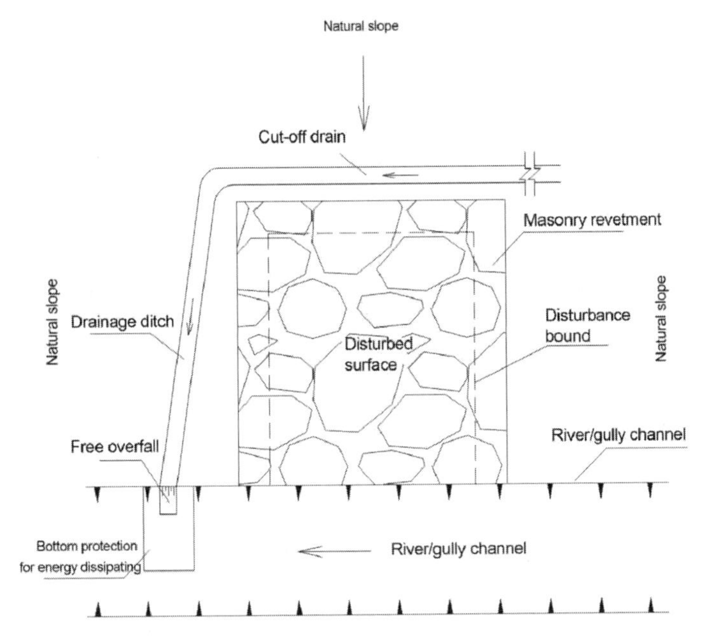

Fig. 1　Schematic diagram for arrangement of revetment, drainage ditch and cut - off drain on a crossing portion

to "control" water and soil loss from disturbed surface. However, the soil and water conservation program shall focus on "prevention" as water/soil conservation and control is difficult to achieve in view of regional fragile ecology. More endeavors shall be placed on engineering arrangement. As mentioned above, investigations and scheme comparison shall be carried out for main works layout and temporary land use to reduce disturbed area. Temporary sites (such as borrow site) shall be placed leeward so that wind erosion is obstructed by mountain. Soil shall be covered during transport. Spoil grounds shall be located under the lee with simple blocking measures provided at slope toe as appropriate to ensure the security of spoils.

Building artificial scenes is also a good measure to improve the environment of arid desert regions. Such regions generally have a rocky landform. The wastes from linear works (e. g. road, railway and pipeline) are largely block stones and rock ballasts. If they are used to build artificial rockwork, livestock, animal or other geometric objects, not only is the spoil disposal issue solved, but the environment is beautified, improving the environmental effect along the works.

(6) Discussion on the issues related to existing Chinese Standards (GB). GB 50433—2008, Technical Code on Soil and Water Conservation of Development and Construction Projects provides, "Borrow sites or spoil grounds for road, railway or other linear works shall be located beyond line of sight from the works." According to years of work practices, this provision shall be optimized in light of actual conditions. In fact, different projects (road or railway) or different sections of a same project are located in different areas. They need different quantities of soil and generate different quantities of spoil. Full enforcement of the provision is bound to cause irrationality in technical economy, and not necessarily conducive to the protection of ecological environment. For road and railway projects, the borrow sites or spoil grounds that are suitable shall be selected no matter whether they are in sight or not. This can reduce the disturbance from road construction and reduce construction cost. Treatment shall be carried out, regardless of in or out of sight, after soil borrow or spoil. However, the reality is that construction units often place much more emphasis on the treatment in sight than that out of sight, or even completely leave it as it is. For this, it is advised to supplement and improve this provision.

3 Conclusions

The arid and semiarid eolian sand areas of western china are harsh in natural conditions with large – scale development and construction and many ethnic communities. All these come up with a stringent requirement for water and soil conservation work of the development and construction projects. It is required to develop a water and soil conservation program in line with the ecological environment protection requirements in light of local actual situation. It is very important to scientifically grasp local actual situations, closely combine them with the engineering characteristics of the development and construction projects, and take the control measures in conformance with technical requirements. In general, reduce disturbance to original earth surface; in technology, take scientific and practical protection and restoration measures; in design, adopt economic and reasonable implementation program to ensure the normal operation of soil and water conservation facilities and achieve the harmony between man and nature.

G. Water Security, Water Transfer and Advanced Water Saving Technology and Monitoring Equipment

Discussion on Ideas and Orientation of Water and Soil Conservation Monitoring

He Xingzhao

Upper and Middle Yellow River Bureau, YRCC, Xi'an, 710021, China

Abstract: Water and soil conservation monitoring is very important because it provides data for assessment of ecological environment improvement and regional economic development. It is also fundamental to promoting the sustainable development of national economy. Through analyzing the present situation of water and soil conservation monitoring, and drawing on the experience of related trades serving for human, social and economic activities, this paper puts forward the orientation and ideas for the development of water and soil conservation monitoring in quite a long time in future, and determines the key tasks of monitoring as well as the importance of setting up a long - term, continuous, stable, and high - efficient monitoring system for water and soil conservation.

Key words: the Yellow River Basin, water and soil conservation, monitoring, ideas

Being fundamental and as one of the "four tasks" in ecological construction for water and soil conservation, water and soil conservation monitoring and forecasting have aroused wider and wider concern of the whole society at all levels. It is required in Law of Water and Soil Conservation of the People's Republic of China issued in 2010 that the water administration department under the State Council should make perfections of the nation - wide water and soil conservation monitoring network for dynamic monitoring of soil erosion in the whole country. Water administration department under the State Council and departments of water administration under the people's government of each province, autonomous region and municipality should make periodical announcement of the following items according to actual water and soil conservation monitoring: types, area, intensity, distribution, and change of water and soil loss; hazards caused by soil erosion; prevention and harnessing for soil erosion. Water and soil conservation monitoring has become the fundamental basis for national macro decision - making in ecological construction. As a key task in water and soil conservation, it has played an active role in ecological environment construction, water resource protection, flood control and disaster mitigation, and new rural reconstruction, providing crucial scientific references for problems solving in soil erosion and ecological environment accompanied by economic and social development. Along with in - depth development of water and soil conservation monitoring, ideas and orientation for monitoring should be made necessary adjustments for more healthy and orderly development of water and soil conservation monitoring.

1 Development on water and soil conservation monitoring of the Yellow River

In recent ten years, centered on ecological construction for water and soil conservation, and under the guidance of down - to - earth reforms and innovations, quite great and all - around progress have been made in water and soil conservation monitoring of the Yellow River, while giving prominence to some key points. And it has played an important role in eco - construction and economic development. In 1999, based on the first and second investigations on soil erosion and accumulated experiences, the YRCC carried out the third investigation by utilizing remote - sensing technique and found out the actual condition of soil erosion on Loess Plateau to a large extent. These goings on have provided scientific foundation for the state in formulating plans for ecological construction and protection, for classification and regionalization of soil erosion in the middle Yellow River, as well as for implementation of a series of key ecological construction projects.

1.1 Foundation works for water and soil conservation monitoring have been reinforced

1.1.1 Significant progress has been made in network construction

First - stage construction of network and information system construction for water and soil conservation monitoring of the Yellow River has been put across, realizing network interconnection and data share among the monitoring center of YRCC, Xi' an monitoring center, and Tianshui, Xifeng, Yulin monitoring sub - centers. Local database, monitoring database and prototype observation database for the Yellow River Basin have been built up.

1.1.2 Managerial system and systematized techniques have been progressively perfected

The Ministry of Water Resources has printed and distributed a series of rules, regulations and office procedures, putting forward relevant specifications and requirements in aspects of monitoring network construction and management, water and soil conservation facilities examination and acceptance for production and construction projects, monitoring qualification administration, communiqué preparation, and annual work reports, to ensure healthy development of monitoring. Totally 22 technical norms have been issued by the Ministry of Water Resources, 5 of which are national standards and 17 of which are industry standards. All these have ensured the criteria for monitoring techniques as well as the quality of monitoring results.

1.1.3 Scientific study has gone deeper step by step

The project of environmental - friendly agriculture dynamic monitoring system technologies in seriously eroded area on Loess Plateau, introduced by the Upper and Middle Yellow River Bureau, has yielded substantial results and won first prize of science and technology progress award of Shaanxi province. In Loess Plateau area, test observations of soil erosion on slopes, small watershed and regional soil erosion and their control efficiency were developed, and forecast models were set up for soil erosion of different scales. The scientific study has deepened our understanding of soil erosion and its control methods, pushed forward forecast of soil erosion. Research on key monitoring technologies by remote sensing for water and soil conservation has accelerated the application and popularization of remote sensing technique in water and soil conservation monitoring.

1.2 Dynamic monitoring for water loss and soil erosion has been gradually developed

1.2.1 Dynamic monitoring and monitoring communiqué has taken the first step

A remote - sensing general survey for water and soil conservation has been made successively in the Yellow River Basin. Observations for small watershed soil erosion and slope surface runoff plots were carried out in Tianshui, Xifeng and Yulin. All these have provided reliable and rich monitoring data for the national annual communiqué of water and soil conservation. In the mean time, we actively promote monitoring communiqué of water and soil conservation in the Yellow River Basin. The announcement of communiqué has opened a window for the masses in acquaintance of the hazardous effect of soil erosion and its harnessing achievements, thus expanding the social influence of water and soil conservation.

1.2.2 A good number of monitoring projects in key regions has come into effect

In recent years, dynamic monitoring for water and soil conservation has been in progress in key regions, for example, the coarse sediment area, the soft rock area, and key tributaries in the middle reaches of the Yellow River, etc. , and large amount of monitoring results have been acquired. They are important in aspects of mastering the dynamic condition of regional soil erosion and conservation, announcing communiqué of soil erosion, and assessing control efficiency of soil erosion in key regions.

1.2.3 Ongoing monitoring continuately for key ecological projects

Water and soil conservation monitoring for key projects, such as ecological engineering for water and soil conservation of the Yellow River, warping dam project on Loess Plateau, dam system engineering in small watershed on Loess Plateau, the first and second phase construction of the WB loan project on Loess Plateau, and pilot project for ecological rehabilitation, etc., has ensured the quality control and made us in timely command of the projects' implementation effect, which lay a solid foundation for project assessment and strongly promote the healthy and stable development in ecological construction for water and soil conservation.

1.2.4 Monitoring for production and construction projects is in orderly running

Water and soil conservation monitoring for production and construction projects in the Yellow River Basin begins from nothing, experiences a rapid development and now strides forward on the way of normalization. Reinforced water and soil conservation monitoring especially for the state large - scale production and construction projects in the watershed, involving pipelines, railways, highways, irrigation works, electric power, petroleum, mineral resources, etc., has effectively controlled soil erosion caused by project construction process, speeded up the implementation of water and soil conservation scheme, and offered detailed data of water and soil conservation for the whole society, thus providing scientific foundation for overall policy - making by government institutions on various levels, as well as for regional economic development.

1.3 Team construction for monitoring has been under sound development

All classes of technical staff serving in the ecological environment monitoring network for water and soil conservation of the Yellow River total 118 people, 48 of which working in the monitoring centers and sub - centers, 12 with senior titles, 19 with medium - grade titles, and 15 with junior titles. Meanwhile, we have a large number of technical advisers in disciplines of water and soil conservation, remote - sensing, geo information and disciplines in related fields.

2 Inspirations from related fields in monitoring service

Environmental monitoring, meteorological monitoring and water and soil conservation monitoring belong to social services without exception, and they own the same service functions and characteristics. By analyzing environmental and meteorological technical system and service system, we have got a clearer idea for the orientation and objective in development of water and soil conservation monitoring. In other words, we should utilize our established and gradually integrated water and soil conservation monitoring network to further perfect the monitoring systematized techniques and managerial system, to promote data share and produce appropriate monitoring service products to cater to social and professional demands on different levels, to enlarge our service scope and field and raise the public identity to water and soil conservation monitoring, to improve our service level, and facilitate deeper and integrated development of water and soil conservation monitoring career.

2.1 Upgrading decision service capability for the government and departments concerned

Water loss and soil erosion is the outcome of natural environment transformation and destroyed ecological environment due to excessive economic activities by human beings. It is quite complex involving nature, society and economy, and its law of development could be made clear only by overall and systematic monitoring while taking account of both ecological environment protection and economic development. At the same time, establishment of a scientific indicator system is the premise of: satisfying requirements on different - scale water and soil conservation service and communiqué, soil erosion assessment in different time and at different places; providing scientific foundation for government at different levels in formulating regional economic plan and watershed management; offering high - quality water and soil conservation monitoring service in aspects of ecological environment construction, water resource protection, flood control and disaster mitigation, and new rural reconstruction, etc.; and dealing well with the relationship between economic and social development and eco - environment protection and soil erosion prevention.

2.2 Carrying the public recognition of the monitoring to a higher level

The system of water and soil conservation monitoring service should reflect human - oriented idea and the concept of harmonious co - existence between human and nature. As the receiver of monitoring service, the public mostly focus on activities that could help them earn more, live more comfortably and of higher quality; and put the influence of soil erosion on their work, life and other social activities, and threat brought on their living environment even caused loss and disaster in the next place. Therefore, the contents for the public service should cover the items mentioned above as much as possible, for example, providing forecast service concerning water and soil conservation and loss, guiding their production activities, and encouraging the public to act consciously to protect and improve ecological environment and supervise those illegal activities in destroying it. At the same time, the channel for information release should be widened in order to let more people acquire relevant data easily.

2.3 Raising level of service for production and construction projects

Water and soil conservation monitoring service for production and construction projects should fully consider environmental requirements in angles of enterprises, integrating with their construction and production and devoting ourselves to dissimilar trades and regions. Meanwhile, we should develop our range of services and seek new opportunities for monitoring service. We could make investigations and give birth to monitoring service products, and try to provide diversified and appropriate service products for all trades and professions.

3 Ideas and orientation of water and soil conservation monitoring

Implementation of the strictest water and soil conservation monitoring system should follow the principle of "demand initiated and application the highest". We should accomplish the following tasks: establishing and gradually perfecting relevant theories and technologies in combination with actual practice, i. e. the monitoring system, the evaluation index system, the mathematic prediction models for soil erosion, the standard specifications for water and soil conservation, and the monitoring data sharing platform; further expanding data elaboration, mining and application. In a word, let water and soil conservation be close to life, let alarm bell of soil erosion keep ringing, let the whole society concern about ecological environment.

3.1 Establishing a long standing, continuous and stable monitoring system

A long standing, continuous, stable and perfect water and soil conservation monitoring system is the base for elevating our technologies, service quality and decision - making capabilities. The first is to center on institution building and personnel training. The second is to emphasize on monitoring network construction to magnify the scientificalness and instructiveness of water and soil conservation communiqué. The third is to reinforce technical system construction of water and soil conservation monitoring to boost its standardization, modernization and informatization. The fourth is to strengthen evaluation system construction of water and soil conservation to ensure scientific decision - making in soil erosion harnessing. The fifth is to set up a long - term investment mechanism to keep the sustainable development of water and soil conservation monitoring, so as to raise its status in national economy.

3.2 Upbuilding a scientific monitoring evaluation index system

The target of upbuilding a scientific water and soil conservation monitoring evaluation index system is for management decision of soil erosion control and regional economic development. Based on spatial clustering and following the principle of scientificalness, systematicness, selfsufficiency, practicability and comparability, the system combines macro with micro, harnessing with developing, the watershed with the regional, tradition techniques with modern techniques, and professional needs with social demands, to make comprehensive analysis of peculiarities of soil erosion, water and soil conservation projects as well as functions of measures taken for the conservation. It deals with water and soil conservation monitoring evaluation systematic structure, filters monitoring evaluation indexes, and set up model framework.

3.3 Speeding up research and development of mathematic prediction model for soil erosion

Mathematic prediction model for soil erosion is an important technical means for determination of overall arrangement of strategic measures, for prediction of soil erosion, for evaluation of water and soil conservation efficacy and improvement of ecological environment. On Loess Plateau, different – scale predictions models used in different time, at different places, and for different erosion types should be developed in accordance with the general layout of the Digital Yellow River Project and relevant technical specifications, and model library and method library should be set up accordingly. Breakthroughs are looked forward to in the period of the "Twelfth Five – year Plan" in fields of quantitative research for soil erosion factors and application research of new technologies.

3.4 Working out a technical standard system for water and soil conservation monitoring

As the principal part of ecological construction in soil eroded area, the meaning of water and soil conservation has got changed along with science progress and rapid development of social economy. Therefore, the contents and methods for conservation monitoring should also keep pace with the times. In other words, we should work out a prospective and sound technical standard system for water and soil conservation, define its goals, and formulate and perfect monitoring evaluation technical standard step by step, for the purpose of realizing scientific and standardized management in monitoring, and facilitating the stable development of water and soil conservation.

3.5 Setting up a monitoring data share and information service platform for water and soil conservation

Taking advantage of modern database technology and information technology, we could develop a data share and information service platform with functions of data management and information distribution, such as data collection, data storage, data updating, data query, data download, etc. The platform should be under a state – level water and soil conservation monitoring data management portal, composed with several professional data platforms integrated in logic while loosened in structure. Each professional platform could set up numbers of distributed sub – platforms of its own. The element considered first in sub – platform setting is the differentiation and uniqueness of data resource, then data security and exchangeability the second, and the consistency of information offered from different sub – platform the last.

3.6 Promoting popularization & application of new technologies and technological innovations

As a comprehensive multi – disciplinary specialty, water and soil conservation monitoring involves technologies applied in agriculture, forestry, mapping, remote – sensing and geo information, etc. The environment for monitoring is always changing, and means and methods for monitoring are with no given pattern. For this reason, we should promote technological innovations, speed up applications of new technologies and transformation rate of scientific achievements, and make more efforts on research and development of new instruments and facilities especially, to seek for fast monitoring techniques and methods in regional soil erosion control and add more technology content into our water and soil conservation monitoring. The first thing we should do is to solidify experimental research and production base. Then we should cooperate more with related industries and departments concerned to enrich the contents of water and soil conservation monitoring. The third is to try utmost in research, development and introduction of new instruments and facilities to increase monitoring efficiency and accuracy of monitoring results, so that the water and soil conservation monitoring in the Yellow River Basin could be steadily pushed forward, realizing a great – leap – forward development.

4 Conclusions

Water and soil conservation monitoring is very important because it provides data for assessment of ecological environment improvement and regional economic development. It is also fundamental to promoting the sustainable development of national economy. By analyzing the practical situation of water and soil conservation monitoring, and using experiences of related industries, accumulated in serving social economic activities, as reference, the ideas and orientations for water and soil conservation monitoring in a rather long period in the future are put forward, giving clear instructions of key tasks in monitoring and emphasizing the importance of establishing a long standing, continuous, stable and highly effective water and soil conservation monitoring system.

References

Liu Zhen. Summing up Experiences, Strengthening Confidence and Carrying Forward National Monitoring on Soil and Water Conservation [J]. Soil and Water Conservation in China, 2006 (1):4 – 5,12.

Guo Suoyan. Fully Implementing the Spirit of the Seventeenth National Congress of CPC and Promoting the Continuous and Healthy Development of Water and Soil Conservation Monitoring[C]// The Third Session of the Special Interest Committee for Water and Soil Conservation Monitoring. Chinese Society of Water and Soil Conservation, 2007.

Zeng Dalin. Pondering on Building of Monitoring System of Soil and Water Conservation[J]. Soil and Water Conservation in China, 2008(2):1 – 2.

Zou Xukai, Zhang Qiang, Wang Youmin, et al. Development on Drought Index Research and Sino – American State – Level Drought Monitoring [J]. Metrological Monthly, 2005,31(7):6 – 9.

He Xingzhao, Yu Quangang, Liang Jianhui. Capacity building of soil and water conservation monitoring in Yellow River Basin [J]. Science of Soil and Water Conservation, 2008, 6 (3):28 – 32.

Liang Jianhui, Zhao Bangyuan, Zheng Hua. Study on Monitoring Assessment Data Share Mechanism in Small Watershed on Loess Plateau [J]. Research of Soil and Water Conservation, 2008(5):210 – 216.

Fan Ping. Development Trend of Environmental Monitoring Data Service [J]. Journal of Social Science of Jiamusi University, 2006(5):26 – 27.

Investigation and Evaluation on Monitoring Data Resources in Small Watershed. Control and Management Projects of Small Watershed by Grants from GB[R]. 2008.

Tian Jing. Survey on Overseas Commercialization of Metrological Services [J]. Gansu Metrology, 2001(3):7 – 11.

Wei Li. Ecological Environment Monitoring and Construction of Information Service System in Jiangxi Province [J]. Jiangxi Metrological Science & Technology, 2003(3):5 – 12.

Zhan Jinyan. Identification and Assessment on Service Functions of Ecological System [M]. Beijing: China Environmental Science, 2011.

Finance Strategie for Drinking Water Supply

Hovhannes V. Togmajyan, Mher M. MKRTUMYAN, Tigran S. MARTIROSYAN and *Gayane G. MADATYAN*

Yerevan State University of Architecture and Construction, 105/1a Teryan Street, Yerevan, 0009, Republic of Armenia

Abstract: Further development of drinking water supply and drainage is of critical importance for the Republic of Armenia. This developoment of drinking water sphere requires economic, technical, institutional, and financial reforms. In this regard, much importance is attached to clarification of the strategic directions of the state's policy and further development is considered. Reliable, safe and continuous water supply through financially healthy, sustainable, and viable water supply services, ensuring the system's safety, improvement of the environmental situation, financial recovery, and making it completely managable, efficiently usable, self-reproductive, and attractive for investors and private administrators require organizational and financial investments.

Key words: water supply, drinking water, subcidy, private administration, strategy

1 Analysis and results of financing the drinking water supply and drainage spheres

One of the critical problems in water resources and water system management is ensuring the thrift and efficient use, recovery, and maintenance thereof. In addition to the said, the investment policies focused on the rehabilitation of the water system is of special importance. In this regard, it should be mentioned that in 2004 ~ 2010 and as of the first half of 2011, according to data provided by the State Committee for Water Management of the Republic of Armenia Ministry of Territorial Administration, with the help of budget financing including through loan and grant projects, 9.22×10^9 AMD worth capital works have been accomplished in the sphere of drinking and irrigation water (other than MCA - funded programs) including 5.22×10^{10} AMD in drinking water. The main founders in the water management sphere are: RA State Budget, World Bank, Asian Development Bank, European Bank for Reconstruction and Development, Abu Dabi Foundation, German KfW Bank, USAID.

Tab. 1 is Financing portions in the sphere of water management system, without MCA project.

Tab. 1 Financing portions in the sphere of water management system, without MCA project ($\times 10^9$ AMD)

Expenses/year	2004	2005	2006	2007	2008	2009	2010	2011/2
Total capital expenses, including	9,06	9,1	8,6	6,3	9,7	18,5	24,6	6,3
Drinking water	4,26	4,3	2,9	3,2	7,0	12,5	12,4	3,6
Total, including operational expenses	19,9	20,0	20,5	20,8	19,7	28,0	35,4	35,0

Capital construction works have been carried out in drinking water supply and drainage sphere in 41 urban and 160 rural settlements of Armenia including the works carried out under the water supply program in Yerevan. Fig. 1 is Capital investments made in the water management sphere in 2004 ~ 2010 and as of first half of 2011.

In 2008 ~ 2010 and as of the first half of 2011, through budget funding, including loan and grant project, 2,200 km - long building, reconstruction and renovation works have been accomplished in drinking water and drainage network and canals, including 1,870 km in drinking water system and 29 km in sewerage system (see Fig. 2):

At the same time, large variety of works have been carried out in construction, reconstruction and renovation of reservoirs (including daily regulation ones), head structures, pump stations,

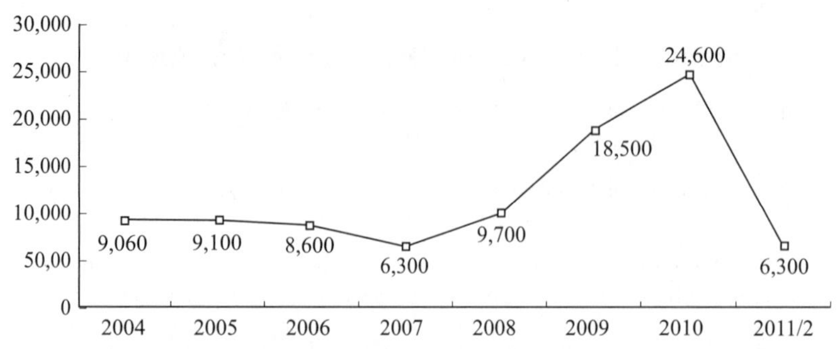

Fig. 1 Capital investments made in the water management sphere in 2004 ~ 2010 and as of first half of 2011 ($\times 10^6$ AMD)[1]

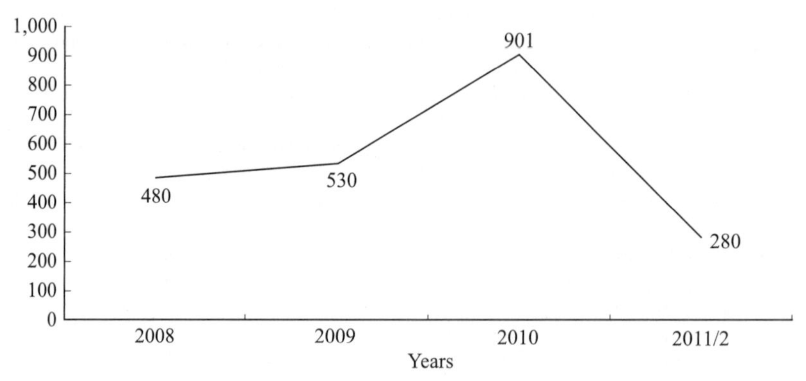

Fig. 2 Length of water pipelines in 2004 ~ 2010 and as of the first half of 2011 (**km**) [1]

deep water wells, treatment stations as well as other hydroengineering structures.

Reforms implemented over the recent years are obvious in the spheres of both water supply and drainage and waste treatment; new systems of sewerage system treatment and sewerage systems themselves have been constructed for the first time following Armenia's independence. In particular:

(1) Within the cooperation with European Bank for Reconstruction and Development (EBRD), loan and grant agreements were signed for "Lake Sevan Environmental Program" in 2007. The goal of the program is construction of mechanical sewerage treatment station in Gavar, Vardenis and Martuni as well as improvement of drainage systems in Gavar, Vardenis, Martuni, Jermuk and Sevan. The investments made under the Lake Sevan Environmental Program will contribute to reduction of waste water emissions and their treatment. Construction works under the program from the very beginning consisted of 2 lots: "Renovation of sewerage systems in the towns of Gavar, Martuni, Vardenis, Sevan and Jermuk" and "Construction of sewerage treatment stations in the towns of Gavar, Martuni, and Vardenis". In Gavar, Martuni, Vardenis, Sevan and Jermuk, 31.4 km – long sewerage systems were built; submission – acceptance works are now over. Under the 2nd lot – "Construction of sewerage treatment stations in the towns of Gavar, Martuni and Vardenis" – construction works are underway in all construction sites of the program. At this point foundations for administrative buildings and treatment station have been finished. July 2012 is specified for the end of works.

(2) Under the water supply and drainage in Yerevan program supported by the Government of France, it is envisaged to implement rehabilitation works in the Aeratsia treatment station. The works will be finished in 2012 ~ 2014.

(3) Within the cooperation with EBRD, it is envisaged to build sewerage treatment stations in Jermuk and Dilijan, with works to be finished during 2014.

(4) Water supplying organizations operating in the system eliminate different current accidents on their own, thus ensuring the waste water disposal for subscribers having connected to the system.

In future the investment policy will continue to focus on rehabilitation and modernization of water management system. For this purpose, it is necessary to develop relevant investment programs, and, involving target loans from international financial organizations, implement reconstruction and other capital renovation of water supply and drainage systems (especially in communities located beyond the service areas of water supply organizations) and drinking water and waste water treatment stations.

The nessecity for attracting investments in (2,3) 2002 ~ 2016 in the sphere of water supply and drainage has been assessed. The mentioned reports are devided into 2 parts: financial strategies for water supply and drainage spheres in large cities and in rural settlements. According to (3), two possible scenarios for funding the water supply and drainage spheres in the Republic of Armenia were prepared: Base Scenario and Development Scenario. The Base Scenario supposes implementation of the program of rehabilitation of water supply system in Yerevan and simple assistance in the state of infrastructures and the quality and volume of the services in the rest of the settlements during the base year. The financing demand will annually make 30.3 bln AMD (60.6 mln EUR) on average and 423.8 AMD (847.6 EUR) during the entire period of implementation of the Scenario. The monetary deficit of the Scenario's financing will exceed 154.7 bln AMD (3,094 mln EUR) in 2015, however implementation of complex measures of optimization of profits and costs will allow to revise the annual funding deficit. In addition to the measures accounted for in the Base Scenario, the Development Scenario provides increased involvement of population in water supply services and ensuring efficient mechanical treatment in all settlements involved in the financial strategy. The funding requirement of this scenarios will make 38.76 bln AMD (77.5 mln EUR) annually and 5,427 AMD (1,085.4 mln EUR) totally - up to 2015 inclusive, of which 21% is the cost of reconstruction and new construction of facilities in the drinking water and drainage system. The monetary deficit of funding in this scenario will exceed 273 bln AMD (546 mln EUR) in 2015, however implementation of measures for optimization of profit and cost will allow, as it is in the case of the Base Scenarios, to revise the annual funding deficit. Optimization of profits and costs of organizations rendering water supply and drainage services will allow the current need for budget subsidies to cover operation costs starting from 2011. Starting from 2011, payments by consumers for water supply and drainage services will completely cover the operation expenses and partially cover expenses of capital renovation and rehabilitation (depreciation deductions) of capital funds. However, the demand for budget financing of modernization and renewal expences of water systems will remain necessary up until 2015; during this time financing of water system modernazation and renewal works should be done mainly by budget and credit funds. According to (2), the financial strategy of the water supply and drainage spheres in the Republic of Armenia addresses three scenarios: those of introduction, of minimum level of water supply, development, and maximum. In the introduction of minimum level water supply scenario, the total costs are estimated 42.0 bln AMD (about 93.0 mln EUR), while in case of possible funding of 33.0 bln AMD (about 73.0 mln EUR), the financial deficit will make 9.0 bln AMD (about 20.0 mln EUR). In the development scenario, the total expences are estimated 45.0 bln AMD (about 100.0 mln EUR), while the volume of possible funding makes 33.5 bln AMD (about 74.0 mln EUR), with financial deficit making 11.5 bln AMD (about 25.0 mln EUR). In the maximum scenario, the toatl expenses are estimated 60.0 bln AMD (about 135.0 mln EUR), while the amount of possible funding makes 35.0 bln AMD (about 79.0 mln EUR), with financial deficit making 25.0 bln AMD (about 56.0 mln EUR). Thus, the analysis performed states once again that the industry greatly needs investments.

In (4), the level of state investments in the drinking water system against GDP has made 0.25% in 2007; 0.4% is defined as a target indicator for 2012 ~ 2018, and 0.3% for 2018 ~ 2021.

The financial strategy for development of drinking water supply system will allow a considerable improvement in the national indicator of continuous average water supply, essential reduction of the national level of water losses, improvement of accounting system (essential increase in the level of setting up watermeters) and increase in the level of state investments against GDP.

While observing the financing strategy in water management system, along with provision of sufficient investments, the role of state subsidies in the sphere of drinking water supply and drainage will be no less important for the years to come.

Tab. 2 Subsidies provided from the RA state budget in 2001 ~ 2010 (mln AMD)

Names of companies	2001	2002	2003	2004	2005	2006	2007	2008	2009	2010
"Yerevan water swerage" CJSC	277	270	1,935	756	1,202.3	252.3	—	—	—	—
"Arm-watersewerage" CJSC	560	1,229	810	952	1,366.5	1,381.26	1,364.5	1,213.5	863	811
"Bor Akunq" CJSC	—	—	—	211.7	63.53	116.55	79.486	73.486	34	26.5
"Lori-watersewerage" CJSC	—	—	—	30.6	25.5	19.89	—	—	—	—
"Shirak- watersewerage" CJSC	—	—	—	35.7	25.5	18.87	—	—	—	—

Subsidies form the RA state budget to organizations rendering drinking water supply and drainage services (see Tab. 2) were provided to cover the financial fissure planned under the current year's financial flows. In this case, the subsidy is provided to the water supplying organizations to compensate the costs of services delivered at lower rates than the cost price.

Over the past years, organizations delivering water supply and drainage services in the Republic of Armenia have been financed from the RA state budget both in the form of direct subsidies and in the form of repayment (or clearing) of accounts payable and cession (or reconstruction) of accounts payable.

As seen in Tab. 2 and Fig. 3, the portions of subsidies provided by years are irregularly allocated, which is a result of wrong planning. As seen from Fig. 4, subsidies have gradually decreased or stopped at all.

Low subsidization from the RA state budget because of wrong planning has led to accumulation of accounts payable in relevant organizations (which have been written off reconstructed later on).

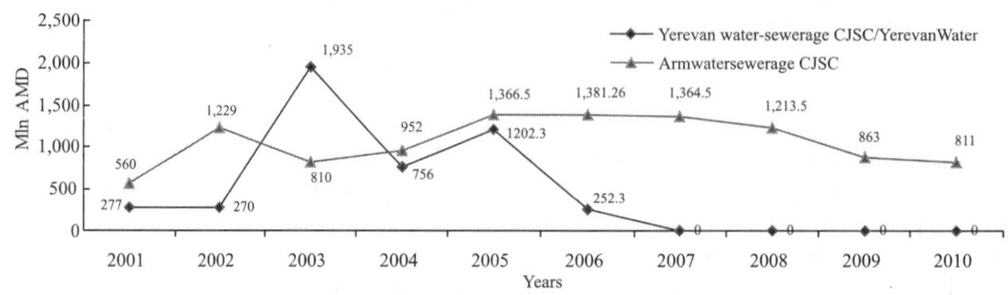

Fig. 3 Subsidies allocated by years to "Yerevan water – sewerage" and "Armwatersewerage" closed joint stock companies

As a result of measures taken over the recent years (including realistic planning of indicators) the portions of subsidies allocated from the RA state budget to organizations have decreased, while on some organizations they have stopped at all.

Fig. 4 provides the amounts of money transferred to the RA state budget by organizations supplying drinking and irrigation water in 2006 ~ 2010, which increase over years.

Overall, along with improvement of financial viability of organizations and solvency of population, the state should gradually refuse to provide subsidies or implement the subsidization mechanism at the expense of improvement of addressability through different social programs.

In providing financial assistance, the goal of the assistance, ensuring equality and excluding discrimination, transparency, the financial status of the assistance receiver, and necessity of protection of water systems are taken into consideration.

Over the past years, organizations delivering water supply services in the Republic of Armenia have been financed from the RA state budget both in the form of direct subsidies and in the form of repayment (or clearing) of accounts payable and cession (or reconstruction) of accounts payable.

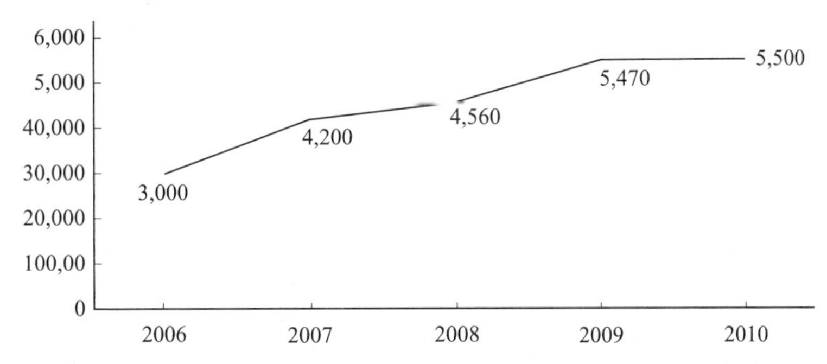

Fig. 4 Amounts of money transferred to the RA state budget by organizations supplying drinking and irrigation water in 2006 ~ 2010 (mln AMD)

2 Further steps to be taken in the strategy of financing drinking water supply and drainage spheres

In addition to ensuring sufficient investment and state financial assistance in the water management sphere, the following are considered important for the years to come:

· Efforts in enhancing the efficiency of the activity of ArmWater Sewerage, Lori-WaterSewerage, Shirak- watersewerage, and Nor Akunq CSJ companies;

· Ensuring the accomplishment of the indicators prescribed by agreements signed with private managers of the organization operatimg in the system;

· Continuity of reforms in management of companies and water systems in the water managment sphere, as well as further development private management;

· Ensuring implementation of loan programs and efforts in attracting new loan programs;

· Completion of the construction of waste water treatment stations in Gavar, Martuni, and Vardenis, as well as Aeratsia treatment station in Yerevan, and launching the construction of waste water treatment stations in Jermuk and Dilijan;

· Improvement of water supply in towns and rural settlements, ensuring the quality of the drinking water supplied, sustainable and reliable water supply, reduction of water losses, gradual increase in the duration of water supply at least by 1 hour compared with the previous year (and in case if it is prescribed by the management agreement – in amount prescribed by agreements), implementation of measures focusing resolution of environmental problems of Lake Sevan, conducting negotiations with donor organizations for new investment programs, as well as implementation of rehabilitation and construction of at least 500 km water pipelines and canals each year;

· Studies focused on improvement of management of water supply in rural areas located beyond the service area of water supplying companies and efforts for further reform thereof.

References

Annual Reports of 2002 ~ 1012 of the State Committee for Water Management of the RA Ministry of Territorial Administration and Current Information. (In Armenian)

National Dialogue on Funding the Water Supply and Drainage in Villages of Armenia – Yerevan, Final Report, OESD, p. 173. (in Russian)

Implementation of a National Finance Strategy for the Water Supply and Sanitation Sector in Armenia, Integrating the Finance Strategy into the Budgetary Process (Task 1) OECD – 2007, p. 107.

Program of Sustainable Development of the Republic of Armenia, Yerevan 2008, p. 316. (in Armenian)

Hydrological Monitoring Techniques and Equipment Applied on the Yellow River

Ma Yonglai and *Zhang Yongping*

Hydrology Bureau of YRCC, Zhengzhou, 450004, China

Abstract: The Yellow River Basin is characterized by the tremendous spatial and temporal changes in runoff, long flood season, large amount of floods, and lack of balance between water and sediment. The paper describes the process of the development of the Yellow River hydro-technical equipment, as well as the new instruments and methods introduced or developed to adapt to flood control and water resource management, such as the self-developed hydrological telemetry terminal (RTU), cableway monitoring and automatic forecasting system, online sediment monitoring equipment, and introduced ADCP, laser particle size analysis and on-line particle size analysis - OPS. And also describes the Key Targets of Hydrological Monitoring Technology and Equipment Development on the Yellow River Basin.

Key words: The Yellow River Basin, hydrological characteristics, monitoring network, technical equipment

As the basis of the Yellow River Basin Management, hydrology has been playing an important technical supporting role in flood prevention and drought control, water resources development and utilization, water ecological protection, as well as governance of water and soil loss. Hydrological monitoring techniques, methods and equipment are also developed in the practice of river management. But there is still a long way to go in the modernization of hydrological monitoring due to the particularity and complexity of the river.

1 Profiles and hydrological characteristics of the Yellow River Basin

1.1 Study area

The Yellow River Basin located in the area of latitude between 32 and 42 degrees, and longitude between 96- and 119 degrees, with catchment area of 752,443 km^2, and full length of 5,464 km. The elevation drop is 4,448 m between the river source and the estuary. Stone Mountain accounts for 29% of total area in the basin, while loess and hilly area accounts for 46%, sandstorm area accounts for 11% and the plain area accounts for 14%. The natural conditions differ greatly within the basin. Especially in the world's largest area of the Loess Plateau, the impact of soil erosion is significant.

The basin belongs to the continental climate with humid southeast, arid northwest and semi-arid central. The average temperature drops gradually from south to north and from east to west. The maximum temperature is 44.2 ℃ (in the Luoyang Basin), while the minimum temperature is −53 ℃ (in the river source area). Precipitation reduces gradually from southeast to northwest, with low intensity and less storms in the upper reach, high intensity, short duration and more storms in the middle and lower reach. The Maximum precipitation is 1,108 mm (at Taishan Summit), while the minimum is below 150 mm (at Hangjin county in the Inner Mongolia Hangjinqi). The annual average natural runoff of the river is 56.3×10^9 m^3, of which 57% comes from the upstream, 31% from the midstream, and 12% from the downstream. The annual average sediment runoff is 1.6×10^9, of which 90.9% comes from the midstream and only 9.1% from the other region.

High sediment load has resulted in the unique morphology in the midstream and downstream region — the secondary perched river. Flushing and siltation changed the riverbed greatly, generally erosion occurred in the canyon regions, while siltation occurred in the broad plain regions. There are 4 siltation regions in the river, respectively located in the Inner Mongolia, Xiaobeiganliu, the lower reaches and estuary.

1.2 The hydrological characteristics

Affected integratively by climate, terrain and topography, the hydrological regime of the river is very complex with the following 4 characteristics:

(1) Tremendous spatial and temporal changes in runoff. The runoff ratio of wet years to dry years is 2.5 to 3.5 times for the main stream, and 2.5 to 40 times for the tributaries. The annual runoff coefficient varies between 0.11 and 0.53. The runoff depth of wet areas is 140 times greater than that of the dry areas. The maximum annual runoff coefficient is 0.7, while the minimum is only 0.01.

(2) Long flood season with many floods. There are 4 flood periods in a year for the Yellow River, namely summer flood season, autumn flood season, ice flood season and spring flood season, which last up to 10 months in total. The summer and autumn are the main flood seasons, in which floods are caused by rainfall from the upper stream, the region between Shanxi and Shaanxi, the region between Longmen to Sanmenxia, the region between Sanmenxia and Huayuankou, and Dawen River catchment. The ice flood occurs mainly in two reaches, Ningxia to Inner Mongolia, and the lower reach. The spring floods are caused by ice melted from Ningxia and Inner Mongolia.

(3) Lack of balance between water and sediment. The average annual runoff of the river is 56.3×10^{9} m^{3}, in which the upper stream contributes 57% of water and only 11% of the sediment, while the middle reaches contributes only 32% of water and 89% of sediment. The maximum sediment load reaches 1,700 kg/m^{3} in the middle reach, which causes the No. 1 sediment laden river in the world. The riverbed changes violently, the river-course wanders and water level changes disorderly.

(4) Global impact of climate change. Affected by global climate change, heavy precipitation increases in local regions. Flood disasters increased intensively.

1.3 The monitoring network

There are 5,314 monitoring stations in the basin, including 416 hydrological stations, 76 water level stations, 2,281 precipitation stations, 252 sediment stations, 161 evaporation stations, 2,128 groundwater monitoring wells, 812 cross sections operated by YRCC for siltation monitoring, and 50 cross sections for water quality.

2 The progress of hydrological monitoring technology and equipment

In the course of 4,000 years of development, hydrological monitoring technology and equipment show obvious stage characteristics. In the early stage, hydrological data were collected in some stationary gauging stations empirically for flood defence. In the interim stage, automatic monitoring technologies were applied, and tour gauging stations and intermittent gauging stations were set up for flood control, disaster relief, water resources management, water environment protection, as well as water & soil loss governance. In the new period, modern hydrological monitoring system was to be established to meet the needs of comprehensive management of water quantity, water quality and aquatic ecosystem.

Due to the complexity and volatility of hydrological regime, foreign advanced instruments and experienced monitoring methods can't be applied directly in the Yellow River. After decades of effort, a series of equipment, methods, technical specifications and industry standards were developed to adapt to the unique characteristics of the river. As to water level monitoring, mobile water gauges were employed combined with fixed gauges, and the different structures of stage recorder were adopted, including traditional island style and Island-shore style, as well as self-developed siphon, sloping, lever and floating style. As to rainfall monitoring, in addition to rainfall barrel and siphon auto-recording gauges, remote transmitted auto-recording gauges and long-term recording gauges were also introduced. So as to water depth measurement, wood was replaced by steel pipe and glass fiber reinforced plastics, traditional sounding weight was replaced by 1,000 kg elliptic-type weigh, in addition, HS-1 hyper-concentrated flow sounding instrument was developed, and dual muddy water depth sounder was introduced. As to discharge measurement, traditional buoys, sheepskin raft and wooden boats were replaced by newly developed buoys dispenser, hydrometric cableways, large motor boats, electric lift trucks and half (full) automatic flow meter cableway. As to sediment measurement, the technique and equipment has developed from utensils, water tube to mode sampler Bladder sampler, online vibrational sediment meter, and isotope sediment meter. As to sediment particle Analysis, the technique and equipment has developed from manual sieve analysis, hydrometer, bottom drain pipe to automatically measured Hutchison photoelectric particle analysis.

3 New progress of the hydrological monitoring technology and equipment

In the last ten years, with the development of science and technology, the hydrological technology and equipment has developed rapidly in the areas such as telemetry, network automation and digitization. In particular, the self-developed hydrological telemetry terminal (RTU), cableway monitoring and automatic forecasting system, online sediment monitoring equipment, and introduced ADCP, laser particle size analysis and on-line particle size analysis-OPS, greatly enhanced the monitoring capacity and the level of automation at complicated water and sediment conditions.

3.1 Telemetry widely popularized and applied

3.1.1 Remote rainfall measuring and forecasting

Based on double dump and a variety of rain sensor, hydrological telemetry terminal (RTU) data processing technology, the rainfall monitoring and forecasting system was established through SMS and satellite communications platform in the entire river basin.

3.1.2 Remote water level measuring and forecasting

According to the local conditions of the river, advanced ultrasonic water level sensor, radar water level sensor, bubble water level sensor, pressure water level sensor and electronic water gauge were integrated with RTU data processing technology, to automatically monitor and forecast water level through SMS and satellite communications platform.

3.1.3 Remote flow monitoring and forecasting

For the gauging stations with stable stage-discharge relation by natural or engineering measures, mathematical model of water level and discharge was established to monitor flow remotely and automatically.

3.2 Telemetry and automatic technology to enhance the capacity of hydrological measurement and prediction

3.2.1 Cableway automatic measurement and forecast system

The Yellow River is a famous sediment laden river. Due to sediment movement, the rivers wandering and apt to change in the middle, lower reach and Inner Mongolia region. The cableway automatic flow measurement system is suitable for the river which has been proven by several decades of practice. The system was self-developed by YRCC, composed of console, elliptic-type weigh, underwater wireless signal transmitter, image monitoring, control and lighting, and remote control devices. The system can be controlled via the touch screen of the console, remote control systems and internet. The system was developed and applied successful, which greatly improved the monitoring and forecasting capacity at complicated water and sediment conditions.

3.2.2 ADCP technology applications

After years of experimental research, Vessel-mounted ADCP was introduced and applied in the gauging stations located in the river-source region or under cascade reservoir dams.

3.2.3 Online sediment monitoring

Sediment measurement has been the problems that plague the Yellow River Hydrology, so it is the focus of research and exploration. In the last ten years, we have developed a new density sensor-vibrational suspended sediment meter, introduced laser particle size analysis instrument based on light diffraction and scattering theory, and introduced OPUS online sediment size and shape sensor based on the principle of ultrasonic attenuation, significantly improved the capacity of sediment monitoring and analysis.

3.3 Advanced instruments to support emergency monitoring system

In order to monitor and respond rapidly to the floods in the middle reach, and the ice disaster in the Ningxia-Inner Mongolia region, state-of-the-art monitoring tools are necessary. In recent years, we developed Independently microwave flow meter, portable depth sounder, ice-drilling flow measurement system, and introduced radar speed meter, dual frequency echo sounder, GPS, laser measuring instrument and other advanced equipment, to implement the emergency and normal monitoring. The rapid hydrological response system of emergency was preliminarily constructed.

4 The key targets of hydrological monitoring technology and equipment development

Although the Yellow River hydrological monitoring equipment has been rapidly developed, It is still arduous to realize automatic and remote because of the diversity of the watershed differences in climate, topography, and the special relationship between water and sediment. To meet this demand, breakthroughs must be obtained in the following aspect.

4.1 Sediment measurement

Through miniaturization of radioactive sources, sensor accuracy can be improved. The isotope sediment sensor should be developed and applied in various environments. Strong field studies of the application of an electric field theory - pole measured sand, the theory has been a breakthrough, the experimental parameters experiments, there has been progress in assembling the experimental design prototype experiment as soon as possible to speed up the research.

4.2 Flow monitoring

The breakthrough has been obtained in the theory and technology of omnidirectional electromagnetic flow sensor and prototype development is nearing completion. Studies of calibration and mathematical modeling are to be continued.

The phased array ADCP can be employed in the sections with low sediment load. Research should be performed to explore the range of sediment load and put it in practice as soon as possible to achieve productivity transformation.

4.3 Information transmission direction

The control and underwater command execution of cableway measurement and boat measurement, as well as data transmission has been the key technologies which bedevils the researches and engineers. Through years of exploration and research, experimental results have been gained in muddy water underwater sonar communication technology. The next step should be to accelerate on-site verification and program appraisal in different aqueous phase.

Macro – invertebrates Composition Study for Water Quality Bio – monitoring in the Brantas River, Brantas River Basin, Indonesia

Harianto[1], ***Alfan Rianto***[2], ***Erwando Rachmadi***[3] and ***Astria Nugrahany***[4]

1. Director of Technical Planning and Development Affair
2. Chief of Business, Management and Technology Development Bureau
3. Chief of Management and Technology Development Unit
4. Staff of Business, Management and Technology Development Bureau

Jasa Tirta I Public Corporation, The Brantas and Bengawan Solo River Basins Management Agency, Jl. Surabaya 2A, Malang, 65115, Indonesia

Abstract: The Brantas River basin with its catchment area of 11,800 km^2 is one of the largest river systems in Indonesia. The basin is located in eastern part of Java Island, functions as the most important source of water supply in East Java Province. In the Brantas River basin, water quality problem has caused a significant impact to the river ecosystem. In the lower reach of Brantas River located a Metropolitan City of Surabaya. The raw water supply for Surabaya City is provided mainly from Brantas River which the Dissolved Oxygen (DO) has dropped to very low level. This causes the standard of water quality for drinking water that produced by local municipal drinking water company often deteriorated.

Macro-invertebrates are small animals that exist in streams, rivers, wetlands and lakes. The term macro-invertebrate describes those animals that have no backbone and can be seen with the naked eye. These animals live in the water for all or part of their lives, so their survival is related to the water quality. If there is a change in the water quality, perhaps cause by a pollutant entering the water, people activities along the river (for example sand mining) or a change in the flow downstream of a dam, then the macro-invertebrate community may also change. Therefore, the richness of macro-invertebrate community composition in a water body can be used to provide the information of water body health.

The sampling was taken on December, 19 to 22, 2008, in five locations in the upper stream of Brantas River i. e., Brantas Origin, Coban Talun, Punten Bridge, Gedhang Klutuk and Pendem Bridge. The activities followed in June, 8 to 11, 2009, in four locations in the upper of the Brantas River watershed i. e., Coban Rondo (Konto River), Mount Kawi (Lekso River), Mount Kelud (Badak River) and Mount Arjuna—Welirang (Krecek River).

According to the water quality classifications for the Hilsenhoff's Biotic Index/BI (Hilsenhoff 1987), the Coban Rondo (Konto River) and Mount Kawi (Lekso River) was categorized as slight organic pollution; the Mount Kelud (Bladak River) and Mount Arjuna-Welirang (Krecek River) was categorized as slight organic pollution and some organic pollution; the Brantas Origin was categorized as some organic pollution and fairly substansial pollution. Coban Talun and Punten Bridge were categorized as fairly substansial pollution, and Gedang Klutuk was categorized as fairly substansial pollution and substansial pollution. Pendem Bridge was categorized as fairly substansial pollution and substansial pollution.

Key words: macro-invertebrates, bio-monitoring, river water quality

1 Introduction

The Brantas River basin, the second largest river basin on the Island of Java, is located in the eastern part of the Java Island, Indonesia, between 110°30′ and 112°55′ East Longitude and 7°01′ and 8°15′ South Latitude. It covers catchment area of about 11,800 km^2 in total and its main stream, the Brantas River, runs about 320 km long. The geological formation of the Brantas River basin is characterized mostly by Pleistocene and Neogene Tertiary with numerous volcanic materials and partly with coral limestone. The plains and delta consist of alluvial soils well suited to paddy cultivation. The basin contains two active volcanoes: Mt. Semeru (3,676 m) to the East, and Mt. Kelud (1,724) near the basin center. Volcanic ash is both a major source of soil fertility and a primary cause of reservoir sedimentation within the basin.

The Brantas River itself originates in Mt. Anjasmoro, located northwest of Malang City and takes its way around the alluvial cone of extinct volcanoes such as Mt. Kawi, Mt. Butak etc. Gathering together many tributaries above the delta include the Lesti (Southeast), Ngrowo (Southwest), Konto (Central) and Widas (Northwest) Rivers along its traveling. At the confluence with the Ngrowo River in the South-western portion of the basin, the Brantas turns north through the agriculturally productive plains region and finally east through the delta, also an important paddy growing area. Then, it finally bifurcates at Mojokerto City to Porong and Surabaya Rivers, both of which pour into the Madura Strait. The annual mean rainfall over the basin is around 2,000 mm, of which more than 80% occurs in the rainy season. Variation of annual rainfall is large: 2,960 mm in a water rich year and 1,370 mm in a drought year. The annual mean rainfall in the high elevation areas is generally high; it reaches 3,000 mm through 4,000 mm especially in southern and western slopes of Mt. Kelud. In term of its water resources potency, the Brantas River basin has a big potency of water resources which is estimated of about $15,821 \times 10^6$ m^3, consist of surface water of about $11,783 \times 10^6$ m^3 dan groundwater of about $4,038 \times 10^6$ m^3.

Population in the basin is quite dense, closing to 15.80×10^6 people in the year 2007. This counts for 43% of East Java population. Population is concentrated in the lower basin, in the major metropolis of Surabaya and surrounding communities; and in the upstream communities of Malang, Kediri and Blitar(Fig. 1).

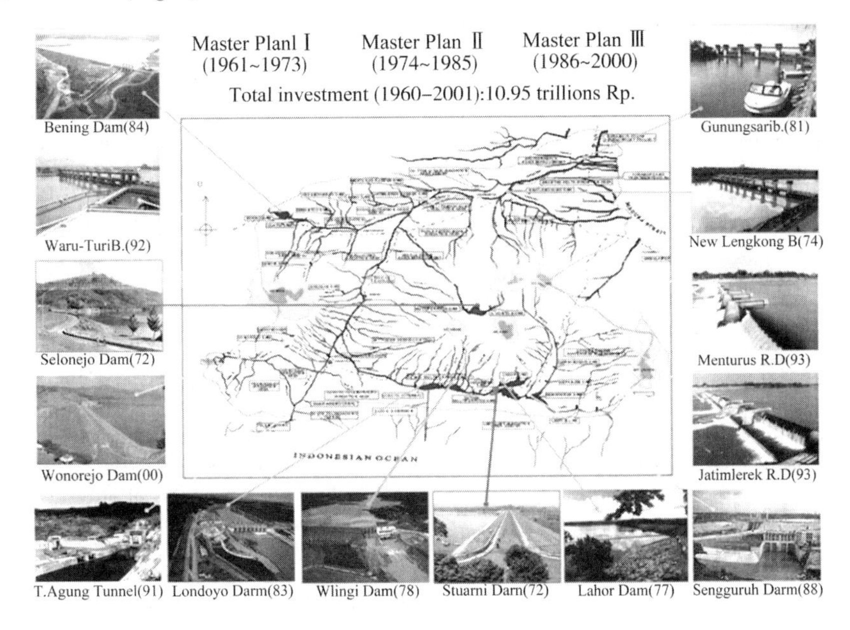

Fig. 1 Brantas River Basin

Water resources in the Brantas River basin have been developed since almost 1,200 years ago. However modern development in water resources took place just as early as the end of the 19th century introduced by the Dutch for agricultural purposes. Comprehensive water resources development in the Brantas River basin was introduced after the Indonesian Government requested the preparation of a water resources development scheme as part of war reparation assistance undertaken by Japanese Government after the Second World War.

The plan was carried out based on One River, One Plan, One Integrated Management principle, and initially prepared for the Brantas River basin in a so called Brantas Plan (1958) that was fortified into the ever first master plan of the Brantas River basin in 1961. This initial master plan until the fourth was then periodically reviewed in order to update the plans in accordance to the national development requirements.

The development of the Brantas River basin is a comprehensive multipurpose project, which uses dams, and reservoir for development of water resources, resulting benefits in flood control, irrigation, and power generation, domestic and industrial water supply. Total development in the basin has resulted into eight reservoirs (Sengguruh, Sutami, Lahor, Wlingi, Lodoyo, Selorejo, Bening and Wonorejo), four river- improvement-schemes, four barrages, and three rubber dams etc.

Multipurpose dams were constructed in the upper reaches, flood control of which could be used to decrease the flood run off in the lower reaches. Around 1958, when the Brantas River Basin Development was commenced, debris produced by the eruption of Mt. Kelud caused the aggradation of riverbed of the Brantas River and its main tributaries. This sediment deposition decreased the river discharge capacity for carrying flood. This incurred flooding almost every year, resulting in personal injury, crop damage and loss of asset.

At present, the major water resources infrastructes in the Brantas River basin are operated and maintained by Jasa Tirta I Public Corporation (PJT I). PJT I was established in 1990, based on the Regulation of the Government of the Republic of Indonesia No. 5/1990 regarding Jasa Tirta Public Corporation. In 1999, this regulation was replaced by the Government Regulation No. 93/1999 regarding Jasa Tirta I Public Corporation to strengthen the organization and permit its jurisdiction to extend to other basins. The Government Regulation No. 93/1999 was replaced by the Government Regulation No. 46/2010 to suit the implementation of Law No 7 of 2004 on Water Resources and the Government Regulation No. 42/2008 on Water Resources Management and to support development of businesses on the development of drinking water supply system and electric power generation.

The benefits from water resources development and management in the Brantas River basin provides significant support to the economic development in the basin, in the region (East Java Province) and National wide which raise economic welfare of the area as measured in GDP. Data from Statistics Central Bureau (BPS) shows that Gross Regional Domestic Product (GRDP) of the Brantas River basin in 2007 amounted to IDR 277,979 $\times 10^9$, which accounted for 59% of GRDP in East Java (IDR 470,600 $\times 10^9$) and 8% of the National GDP (IDR 3,536,797 $\times 10^9$).

2 Macro-invertebrates as Water Quality Bio-monitoring

Streams, rivers, wetlands and lakes are home for many small animals generally include insects, crustaceans, molluscs, arachnids and annelids. The term macro-invertebrate describes those animals that have no backbone and can be seen with the naked eye. Some aquatic macro-invertebrates can be quite large, such as freshwater crayfish, however, most are very small. Invertebrates that are retained on a 0.25 mm mesh net are generally termed macro-invertebrates.

These animals live in the water for all or part of their lives, so their survival is related to the water quality. They are significant within the food chain as larger animals such as fish and birds rely on them as a food source. Macro-invertebrates are sensitive to different chemical and physical conditions. If there is a change in the water quality, perhaps because of a pollutant entering the water quality, perhaps because of a pollutant entering the water, or a change in the flow downstream of a dam, then the macro-invertebrate community may also change.

Various types of biota groups have been used for bio-monitoring purposes and macro-invertebrates are the most widely used type. The several reasons that macro-invertebrates commonly use in bio-monitoring:

(1) can be found in several zones of aquatic habitats, with different water quality conditions and consists of several species that may respond differently to water quality;

(2) relatively easier to recognize than the species of microorganisms;

(3) it is easy to gather or collect it, since it takes only a simple tool to make it;

(4) Macro invertebrates are usually settled, with a ratio of long life spans, so it can be used for assessing the quality of water at a place on a longer period, because it can determine the impact of domestic waste, industrial, petroleum, agriculture, and the impact of land use change.

With many advantages above, the macro invertebrates can be used for long-term water quality analysis, both routine and incidental, with various concentrations of pollutants, from a single pollutant or complex, which has a synergies or antagonistic effect on macro invertebrates itself.

In this study, to determine the richness of macro-invertebrates is using the taxa richness, *EPT* (Ephemeroptera, Plecoptera, Trichoptera) taxa richness, *EPT* percentage (*EPT%*) and the modified Family Biotic Index (*FBI*).

(1) Taxa richness (the diversity of taxa)

The number of taxa (Family / Species) is the number of types of organisms in the family or species in taxon level. Generally the higher of the species diversity of taxa, the healthier of the river ecosystem, on the contrary the smaller the number of taxa indicated the declining of water quality.

(2) *EPT* taxa richness (the diversity of *EPT* taxa)

This number is the total number of species/families of Ephemeroptera, Plecoptera, and Trichoptera (*EPT*) because most families / species of this group are sensitive to pollution. The high diversity of *EPT* taxa can be use as the indication of the good water quality and the healthier of the river ecosystems.

(3) *EPT* percentage (*EPT%*)

Calculate the ratio of the percentage of types of Ephemeroptera, Plecoptera, and Trichoptera (*EPT*) in the total number of individuals that was taken in the sample.

(4) The Modified Family Biotic Index (*FBI*)

The Modified Family Biotic Index (*FBI*) has been widely used to indicate the level of organic pollution in the waters, where every macro-invertebrates family has a specific value that indicates the level of tolerance to organic pollution. The calculation of biotic index values in this study site using the following formula.

$$FBI = \sum_{i=1}^{i=n} \frac{x_i \cdot t_i}{N}$$

where:

FBI = biotic index values;

i = order of the families which determine of the macro-invertebrates community;

x_i = the number of individuals, groups, families to – i;

t_i = tolerance level to – i family;

N = total number of individuals which determine of the macro-invertebrates community.

Tab. 1 shows the macro-invertebrates response shown againts environmental disturbance in the river ecosystem.

Tab. 1 The macro-invertebrates response shown againts environmental disturbance in the river ecosystem Hilsenhoff, 1988

Environmental stress	Standing crop (number of individuals or biomass)	Number of taxon
Toxic materials	Decrease	Decrease
Drastic changes in temperature	Decrease / Increase	Decrease
Sedimentation	Decrease	Decrease
Low level of pH	Decrease	Decrease
Increase in inorganic nutrients	Increase	Increase / Decrease
Increase in organic content (low level of DO/Dissolved Oxygen)	Increase	Decrease
Level of silt (non toxic)	Increase	Decrease

Water quality assessment conducted by a unit or a particular metric as the macro-invertebrate response to pollution and changes in aquatic environments, as shown in Tab. 2 below.

Tab. 2 The unit or metric use to predict the macro-invertebrates response to water pollution (Mandaville, 2002)

Metric	Prediction of macro-invertebrates response
Number of taxa	Increase
Number of EPT taxa	Increase
Number of 'clinger' taxa	Increase
'scrapper' percentage	Increase
Dominant taxa percentage	Increase
EPT percentage	Decrease
Diptera percentage	Increase
Oligochaeta percentage	Increase
Hilsenhoff's Biotic Index (HBI)	Increase

Interpretation of the biotic index to determine the level of organic pollution carried out by following the provisions in Tab. 3.

Tab. 3 Interpretation of the FBI biotic index to assess water quality (Hilsenhoff, 1988)

Family Biotic Index	Water Quality	Organic Pollution Level
0 ~ 3,75	Excellent	Organic pollution unlikely
3,76 ~ 4,25	Very Good	Possible slight organic pollution
4,26 ~ 5,00	Good	Some organic pollution probable
5,01 ~ 5,75	Fair	Fairly substantial pollution likely
5,76 ~ 6,50	Fairly Poor	Substantial pollution likely
6,51 ~ 7,25	Poor	Very substantial pollution likely
7,26 ~ 10,00	Very Poor	Severe organic pollution likely

3 Macro-invertebrates as water quality bio-monitoring in the brantas river basin

3.1 Location

The location of macro-invertebrates as Water Quality Bio-monitoring in the Brantas River Basin is in the Upper of Brantas River (Fig. 2 ~ Fig. 3). The sampling was taken on December, 19 to 22, 2008, in five locations in the upper stream of Brantas River i. e., Brantas Origin, Coban Talun, Punten Bridge, Gedhang Klutuk and Pendem Bridge. The activities followed in June, 8 to 11, 2009, in four locations in the upper of the Brantas River watershed i. e., Coban Rondo (Konto River), Mount Kawi (Lekso River), Mount Kelud (Badak River) and Mount Arjuna-Welirang (Krecek River).

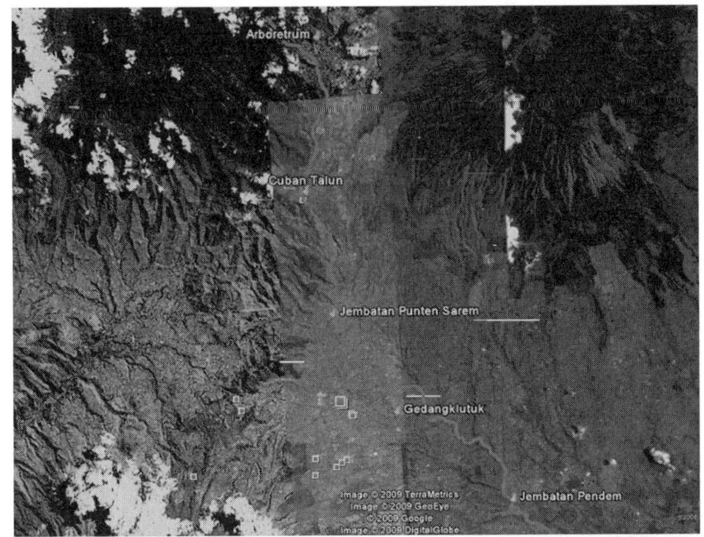

Fig. 2 The sampling location in the Upper of Brantas River Basin in 2008

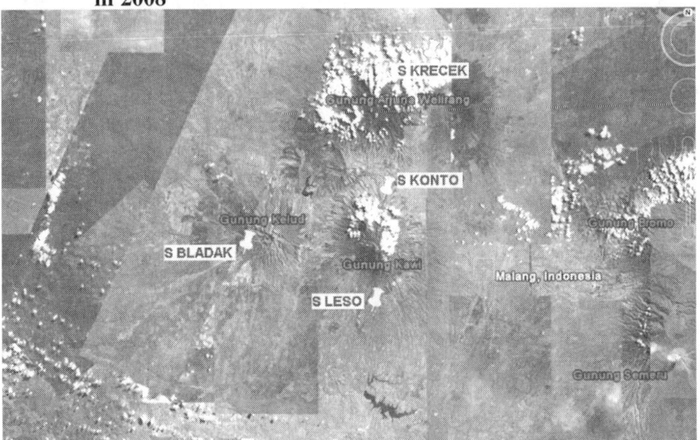

Fig. 3 The sampling location in the Upper of Brantas River in 2009

3.2 Result

3.2.1 Physical-chemical parameter in the sampling location

In the table below are the summarized of the water quality monitoring using the physical-chemical parameters. The parameters are temperature (℃), Dissolved Oxygen/DO (mg/L), Total Dissolved Solids/TDS (mg/L), pH and Turbidity (NTU).

In overall, it can be concluded for parameters Temperature, TDS and pH already meet the standard, except in some locations due to its location near to the city or centre of the tourism area. For TDS parameter majority in location study, the upper limit was exceeding the natural condition for TDS value in the Asian rivers naturally are 123.5 mg/L (Allan, 1995). For turbidity parameters in most of the location, the value has meet the standard, except for Pendem Bridge location has the highest level due to its location is quite near with the residential area m Tab. 4.

Tab. 4

No	Sampling Station	Temperature (℃)	DO (mg/L)	TDS (mg/L)	pH	Turbidity (NTU)
1.	Brantas Origin	17.9 ~ 19.2	5.91 ~ 6.49	69.1 ~ 134	7	5 ~ 6.9
2.	Cuban Talun	18.5 ~ 19.4	7.87 ~ 8.90	92.9 ~ 122	6	66.8 ~ 79.7
3.	Punten Bridge	20.9 ~ 22.6	7.36 ~ 9.65	143 ~ 149	7	16.1 ~ 17.6
4.	Gedhang Klutuk	21.2 ~ 25.0	7.19 ~ 9.80	183 ~ 191	7	18.64 ~ 19.10
5.	Pendem Bridge	22.5 ~ 22.8	6.78 ~ 8.73	187	7	160.80
6.	Coban Rondo (Konto River)	18.2 ~ 19.2	8.10 ~ 9.10	62 ~ 63	7.8 ~ 8.1	1 ~ 4
7.	Mt. Kawi (Lekso River)	22.3 ~ 23.1	8.90 ~ 14.1	11 ~ 92	7.4 ~ 7.9	1 ~ 10
8.	Mt. Kelud (Badak River)	25.3 ~ 27.6	7.0 ~ 10.10	105 ~ 293	7.4 ~ 8.1	< 0,1 ~ 35
9.	Mt. Arjuna-Welirang (Krecek River)	20.7 ~ 23.9	7.90 ~ 10.9	110 ~ 115	7.8 ~ 8.2	< 0.1 ~ 5
	Standard	20 ~ 30	3 < DO < 5	Natural: 123.5	7 ~ 8.5	Normal: 50

3.2.2 Taxa richness, EPT taxa richness, %EPT dan FBI

The result from the macro-invertebrates bio-monitoring to determine of the water quality in the study location can be explained as follows:

(1) Most of the sampling location in the study especially for mountainous area, the taxa richness can be classified that the river ecosystem is still in good condition. The fewest number of taxa richness in Cuban Talun area due to its location was use as tourism area with many human interference activities.

(2) For the EPT taxa richness, the condition of the sampling location such as Punten Bridge and Pendem Bridge has the fewest value of the EPT taxa richness which indicate that the condition of the water quality is deteriorate. This condition also applied in the %EPT.

(3) For the FBI value according to Hillsenhoff, 1987, it can be described that for the location Coban Rondo (Konto River) and Mount Kawi (Lekso River) was categorized as slight organic pollution; the Mount Kelud (Bladak River) and Mount Arjuna-Welirang (Krecek River) was categorized as slight organic pollution and some organic pollution; the Brantas Origin was categorized as some organic pollution and fairly substansial pollution. Coban Talun and Punten Bridge were categorized as fairly substansial pollution, and Gedang Klutuk was categorized as fairly substansial pollution and substansial pollution. Pendem Bridge was categorized as fairly substansial pollution and substansial pollution.

Tab. 5

No.	Location	Taxa richness	EPT taxa richness	%EPT	FBI
1	Brantas Origin	24	8	48.95	4.04 ~ 5.41
2	Coban Talun	11	6	21.01	4.31 ~ 4.94
3	Punten Bridge	14	4	44.29	4.20 ~ 4.89
4	Gedhang Klutuk	17	5	41.41	4.53 ~ 5.89
5	Pendem Bridge	12	2	7.22	5.67 ~ 5.97
6	Coban Rondo (Konto River)	22	10	79.42	3.58 ~ 4.24
7	Mount Kawi (Lekso River)	40	12	81.21	3.57 ~ 4.17
8	Mount Kelud (Badak River)	30	14	51.97	4.07 ~ 5.04
9.	Mount Arjuna-Welirang (Krecek River)	37	10	63.68	3.96 ~ 4.66

4 Conclusion

(1) Macro invertebrates use as bio-indicators of stream health because they are easier to locate and study than the more mobile organisms like fish. Many of them have specific requirements in terms of water quality. When only pollution tolerant species are present, they provide an indication of how fit the water is for other organisms that are harder to study (like fish).

(2) In this study, the sampling location is in upper stream of Brantas river basin, some of results indicate that the condition of the river health is good but in some location shown that the pollution has occurs. This may cause by the use of the river as tourism activities and human interference to the ecosystem.

(3) To maintain the sustainability of ecological life of the river, we should consider the ecologic function of rivers as natural habitat of the river biota. Also consider the friendly model for water infrastructures such as barrages, check dam and river revetment.

(4) Need to raise the public awareness on the health of the river ecosystem and actively participated in this activity using macro-invertebrates as the indicator of the water quality because it is easy and inexpensive technique.

References

Anonim. Rapid Bioassessment Protocols for Use in Streams and Wadeable Rivers: Periphyton, Benthic Macroinvertebrates, and Fish. Second Edition.

J. Boone Kauffman, Robert L. Beschta, Nick Otting, et al.. An Ecological Perspective of Riparian and Stream Restoration in the Western united States[J]. Fisheries Special Issue on Watershed Restoration Vol. 22, No. 5.

Jasa Tirta I Public Corporation and Ecological Observations and Wetlands Conservation (ECOTON), The Macro-invertebrates Composition Study in the Upper stream of the Brantas River (Studi Komposisi Makroinvertebrata di Kali Brantas Bagian Hulu). 2008.

Jasa Tirta I Public Corporation and Ecological Observations and Wetlands Conservation (ECOTON), The Macro-invertebrates Composition Study in the Upper Stream of the Brantas River Basin (Lahor River, Badak River, Brangkal River, Konto River) (Studi Komposisi Makroinvertebrata Daerah Hulu DAS Kali Brantas (Kali Lahor, Kali Badak, Kali Brangkal, Kali Konto)). 2009.

Kathleen Suozzo. The Use of Aquatic Insects and Benthic Macroinvertebrate Communities to Assess Water Quality Upstream and Downstream of the Village of Stamford Wastewater Treatment Facility, SUNY Oneonta biology candidate enrolled in Biol. 644 - Topics in Entomology. SUNY Oneonta.

Nyman, D. C.. Restoration Design for an Urban River: Some Lessons Learned, Proceedings, Wetlands Engineering & River Restoration Conference [J]. ASCE, March 1998, Denver, Colorado.

S. M. Mandaville. Benthic Macroinvertebrates in Freshwaters-Taxa Tolerance Values, Metrics, and Protocols, (Project H-1)[J]. Soil & Water Conservation Society of Metro Halifax, 2002 (6).

Preliminary Study on Supply Target of Xiaojiang Water Transfer Scenario

Wang Yu*, *Zhang Mei*, *Cao Tingli* and *Yang Lifeng

Yellow River Engineering Consulting co. , Ltd. , Zhengzhou, 450003, China

Abstract: Xiaojiang Water Transfer Scenario which is one of the water transfer projects studied in the recent years is planning to transfer water from the Three Gorges Reservoir to the Weihe River, a tributary of the Yellow River, with the dimension of $4 \times 10^9 \sim 10.3 \times 10^9$ m^3 annually. Starting with the analysis on the water demand – supply situation of the Yellow River Basin and the water demand due to the economic and social development, this paper studies the water shortage problems of different regions in the Yellow River Basin and the possible solutions and measurement as well. Based on the characteristics of water resources allocation and management and the project routine, the Supply Regions of Xiaojiang Water Transfer Scenario is determined. Through studying the water use of sediment flushing for Lower Reaches of the Weihe River and the Yellow River, analyzing the water demand – supply situation, the requirements of the Supply Regions is studied. The Supply Target of the Xiaojiang Water Transfer Scenario put forward is to supplement the eco – environmental water use of Lower Reaches of Weihe River and Lower Reaches of Yellow River, and to increase the base – flow for sediment flushing and eco – environment of river channel.

Key words: Xiaojiang, Water transfer, Yellow River Basin, Supply target

1 Introduction

In 2001, some experts and scholars put forward the transfer scenario, which transferred water from the three gorges reservoir area in Yangtze River, which is called Xiaojiang Water Transfer Scenario for short. In 2005 the Ministry of Water Resources arranged Yellow River Conservancy Commission and Changjiang Water Resources Commission to jointly develop a preliminary study on the Xiaojiang Water Transfer Scenario which transferred water from the mainstream and some tributaries of Yangtze River.

Based on preliminary study results, the Xiaojiang Water Transfer Scenario pumps water from Xiaojiang in the Three Gorges reservoir with the pumping head of 380 m, and it linked two reservoirs on the tributaries, transmitted water mainly by tunnels, crossed the Renhe River, a tributary of Hanjiang River by aqueduct, passed through Qinling mountains by several tunnels , and then entered Fenghe River a tributary of the Weihe River, regulating by Fenghe plain diversion and storage reservoir, at last entered the mainstream of the Yellow River in Tongguan, with the total length of 453 km, of which is 307 km tunnels. Analyzed according to the runoff process of the Three Gorges reservoir and the operation requirements of the reservoir and power station, under the premise that it doesn't affect shipping in dry season in the Yangtze River and decrease the generating capacity of the Three Gorges power station, three water transfer schemes was studied out, the scale of which are 4×10^9 m^3, 5.5×10^9 m^3 and 10.3×10^9 m^3. Considering the demand of the water diversion areas, engineering and economic index and water transfer effect etc. , the water transfer scheme with the scale of 5.5×10^9 m^3 is considered as the representative one. In this paper, the water supply target of the Xiaojiang Water Transfer Scenario was studied according to situation of water shortage in the Yellow River Basin and characteristics of Xiaojiang water diversion scheme.

2 Analysis on the Water Supply Regions of the Xiaojiang Water Transfer Scenario

In general, to identify the Water Supply Regions of the water transfer project is not only related to its line arrangement, water transfer scale, and water transfer process and storage capacity and so on, but also related to the natural geography and the development and utilization and management of water resources. To determine the Water Supply Regions of the Xiaojiang Water Transfer Scenario, not only the water diversion scheme itself should be considered, but also the unified distribution, unified management principles of the water resources in the Yellow River should be abided, and the characteristics of water resources utilization and water shortage should be considered and analyzed.

2.1 The situation of water shortage of Yellow River Basin

According to the relevant outcomes about the Yellow River Basin Integrated Water Resources Planning, taking into account the ecological environment and the national demand for water outside the river, analyzing the water supply and demand situation, the Yellow River water shortage in 2000(base year) and 2030 are 81.9×10^9 m^3 and 14.45×10^9 m^3.

According to projections, under the conditions that the sediment is up to 9.00×10^8 t and the lower Yellow River each year siltation is maintained to 1.50×10^8 t, the average ecological environment water demand for years of the lower Yellow River will be 22×10^9 m^3. By supply and demand analysis and calculation, the average water into the sea for many years in 2030 will be 18.5×10^9 m^3, and the water shortage inside river will be nearly 3.5×10^9 m^3.

The total volume of water shortage or the river outside in the Yellow River Basin will be 11.03×10^9 m^3, water shortages are mainly distributed in the upper and middle reaches above Sanmenxia, especially the two reaches, one of which is from Lanzhou to Hekou, the other of which is from Longmen to Sanmenxia. The water shortage in the reach from Lanzhou to Hekou is 44% of the total volume of water shortage, and the water shortage in the reach from Longmen to Sanmenxia is 26% of the total one. The water shortage in the reach from Longmen to Sanmenxia is mainly concentrated in the Weihe River Basin, and the water shortage of Weihe River Basin will be 22.2×10^9 m^3 in level year 2030, 80% of which is distributed in the Guanzhong area.

2.2 The approaches to solve the water shortage of Yellow River Basin

The Yellow River Basin is Resource dry – land, taking full account of the conditions for implementing the water conservation measures, the water supply and demand gap is still great in every level year, and the fundamental way to solve the problem of water shortage is water transfer. Because of different water transfer scenario has different engineering layout, water diversion scale and process, the focus and extent of the settlement of the Yellow River problem are different. Therefore, it is necessary to start from the whole Yellow River Basin, consider the economic and social development and environmental water need at different sections on the middle and lower reaches, analyze the situation of water shortage in different regions and different sectors, study different ways to solve water shortage in different regions, and then determine the supply region and supply objects of the water transfer scenario.

2.2.1 The upper reaches

The upper reaches of the Yellow River is a vast territory with abundant resources. With the implementation of the strategy of the western development, the region's economy will have a sustained and rapid development. Given full consideration to water saving measures, the volume of water shortage of the upper reaches of the Yellow River region will still amount to 6.19×10^9 m^3 in 2030 level year, 59% of the total water shortage outside river, and it is one of the major water shortage regions of the Yellow River Basin. Upstream water shortage mainly manifests in two aspects. On one hand, the contradiction between supply and demand is obvious, and the economic and social development of water is inadequate, on the other hand, lacking of water within the river results in serious siltation atrophy in Ningxia and Inner Mongolia reaches, then to form a "ground hanging river".

The upper reaches of the Yellow River, particularly the area that above Lanzhou reaches is the main source area of the Yellow River runoff. In accordance with the requirements of the uniform configuration of the Yellow River Water Resources, the upper reaches of the Yellow River runoff is used to support the economic and social development in the upstream river, at the same time, it is also necessary to ensure a certain amount of water discharged in the river, in order to maintain the water channel shape and sediment, as well as to meet the water requirements in the small North River and Yellow River downstream. According to the analysis, the ecological environment of water at the section of Hekou town is at least 20×10^9 m^3, of which 12×10^9 m^3 is used to maintain the water shape and sediment in water in Ningxia and Inner Mongolia reaches in the food season, and other 8×10^9 m^3 of which is used to consider the ice needs and ecological base flow requires in non – flood season. The mean annual natural runoff at the section of Hekou town is 33.17×10^9 m^3, and the amount of water in the river outside of the status upstream region is close to 13×10^9 m^3, which has reached the upper limit of available water in the river outside, as a result, water resources have been unable to support sustainable economic and social development of water demand, therefore, Inter – basin water transfer must be used to resolve the development of water in the upstream river and river eco – environmental water.

If increased by adjusting the upstream water indicators of the national economy amount of water in the upper reaches of river, it will further squeeze the ecological and environmental water at the section of Hekou town, and then exacerbate the siltation of the reaches from Ningxia and Inner Mongolia, the small North River and the Yellow River. According to preliminary analysis, If upstream water increased by 4×10^9 m^3, the average annual deposit volume of the Ningxia and Inner Mongolia reach would be from 0.07×10^9 t to 0.097×10^9 t, and the situation of the river siltation in Inner Mongolia would be more severe, and also bring adversely affect on the use of Guxian Reservior and the Tongguan control. As a result, in order to solve the water problem of the upper reaches of the Yellow River, it is necessary to increase the total water resources in the upper reaches and shape co – ordinary relationship of water and sediment combing with the regulation and storage of the large and medium reservoirs. When water is transferred into the Yellow River, the position is so low that water transferred can't increase water resources in the upper reaches of the Yellow River, and it is difficult to solve the problem of water shortage in the upper reaches of the Yellow River and reduce the silting in Ningxia—Inner Mongolia Reach and Xiaobeigan Main Stream of the Yellow River.

2.2.2 The middle reaches

The middle reaches of the Yellow River is considered as the main source area of sediment in the Yellow River Basin, also is one of the economical developed and mineral concentration areas. In the middle reaches of the Yellow River Basin, the volume of water shortage outside river will be 3.57×10^9 m^3 of the 2030 level year, 34.3% of the total water shortage outside river. The water shortage of the middle reaches is mainly distributed in the Weihe River. The water shortage of the 2030 level year is 2.23×10^9 m^3. The Weihe River water shortage is mainly concentrated in the Guanzhong area. Due to the shortage of water resources and the obvious contradiction between supply and demand, ecological and environmental water in the Weihe Rive has been diverted, resulting in serious silting of the Lower Weihe River, drainage capacity decline and frequent floods, at the same time, water pollution is becoming more intensified, threatening the human condition.

To solve the problem of water shortage in Weihe River Basin, the water diversion from the Hanjiang River to the Weihe River and the Xiaojiang Water Transfer Scenario can be taken into account, which water is transferred directly from Hanjiang River or the main streams and tributaries of The Yangtze River to Weihe River, also transferred via the west line of the South – to – North Water Transfer Project into the Yellow River, and then supply Weihe River by the Taohe River. The directly water transferred into Baoji—Xianyang reaches of Weihe River can not only help to alleviate the most serious problem that water shortage outside river in Guanzhong Region of Weihe River.

2.2.3 The downstream region

It has a wide range of water supply region in the downstream of Yellow River where the economic developed rapidly, and it is also known to all as a "Ground hanging water". The volume of water shortage will be 5.45×10^9 m^3 in the downstream region in level year 2030, 70% of total water shortage, and downstream water shortage mainly manifests in ecological environmental water shortage inside river.

The East Line and Middle Line of the South - to - North Water Transfer Project cross from the Yellow River downstream, which may supply water to the Yellow River, however, there was no mission to supply water to the Yellow River in the project planning stage, and if the conditions and scale of the engineering left unchanged, while under the conditions that there is no storage project, they may be very limited to supply water to the Yellow River.

Both of the Xiaojiang Water Transfer Scenario and the west line of South - to - North Water Transfer Project can supplement water to the lower reaches of the Yellow River, combing with possible water and sediment control measures, which can create conditions for the sedimentation reduction in the downstream region of the Yellow River. The Xiaojiang water transferred mainly concentrated in the flood period, the water diversion flow is large, which can increase sediment water inside river and slow the deposition of the Weihe River and the Downstream of Yellow River. The west line of South - to - North Water Transfer Project can supplement ecological environment water to the lower reaches of the Yellow River, combing with water and sediment control measures in the main stream of the Yellow River to make appropriate flow and sediment process, which can improve the efficiency of the sediment transport and shape channel and slow down silting of the Lower Yellow River.

2.3 The determination of the Water Supply Regions of Xiaojiang Water transfer scenario

The Xiaojiang water transferred into the Weihe River in Baoji—Xianyang reach of the main stream of Weihe River, enter into the main stream of the Yellow River. Affected by geographical conditions and the water transferred location into the Yellow River Basin, it is difficult to solve the problem of water shortage in the upper reaches above Tongguan of the Yellow River and reduce the silting in Ningxia—Inner Mongolia Reach and Xiaobeigan Main Stream of the Yellow River. According to the pressing issues and situations of the water shortages in different regions, water diversion project layout and its development and utilization conditions, analysis of the ways to solve water shortage problems, the supply region of the Xiaojiang Water Transfer Scenario is determined, which is Guanzhong area of the Weihe Basin and the following Tongguan reaches of the main stream of the Yellow River.

3 Analysis on water use demand in the Water Supply Regions

In the paper, the author focus on the ecological environment water demand and process requirements in the Lower Weihe River, receiving the volume of water shortage of the Water Supply Regions.

3.1 The ecological environment water demand in the Lower Weihe River

3.1.1 The sediment water consumption in the Lower Weihe River

The incoming water and sediment mainly concentrated in the flood season, flood water accounts for more than 60% of the annual amount of water, sand accounts for more than 90%, therefore, it is necessary to study the sediment water consumption in the flood season and non - flood season separately.

According to the motion regulation of the water and sediment and sediment transport characteristics, using the measured data in flood season between 1974 and 2003 in the Lower Weihe River, through the analysis the relationship between sediment water consumption in the Lower Weihe River and the main influencing factors (incoming sediment, sedimentation amount, bank - full discharge), the formula of the sediment water consumption in the lower River (the Huaxian section, the same below) is put forward, and the calculated results are shown in Tab. 1.

$$W = (17W_S - 50.6\Delta W_S) \times (1 - e^{(-1.24\times10^{-3}\times Q)})$$

In the formula: W is the sediment water consumptation in flood season, $\times 10^{11}$ m^3; W_S is the incoming sediment at Xianyang and Zhangjiashan Points in flood season, $\times 10^{11}$ t; ΔW_S is the sediment amount in flood season in the Lowever Weihe River, $\times 10^{11}$ m^3; Q is the bank – full discharge at Huaxian section before the flood season, m^3/s.

Tab. 1 The sediment water consumption in flood season (the situation with the bank – full discharge 3,000 m^3/s at Huaxian section)

the sedimentamount in flood season ($\times 10^{11}$ m^3)	The incoming sediment in flood season($\times 10^{11}$ t)				
	4.0	3.5	3.0	2.5	2.0
0.4	46.6	38.3	30.0	21.7	13.4
0.1	61.4	53.1	44.8	36.5	28.2
0	66.4	58.1	49.8	41.5	33.2
-0.1	71.3	63.0	54.7	46.4	38.1
-0.4	86.1	77.8	69.5	61.2	52.9

Through analysis of the changes of the sediment water consumption in flood season after the implementation of water conservation measures, it is put forward that the incoming sediment in flood season is 0.318×10^9 t in level year 2000 and 0.255×10^9 t in level year 2030. From multiple aspects which include no floodplain in the medium water river bed, convergence to the river training works standard and coordination to the river flood reconstruction project, it is proved that the bank – full discharge of Huaxian section corresponding to the medium water river bed is 3,000 m^3/s (actually 1,500 m^3/s in the level year 2000) in level year 2030. Calculated using the above formula, the sediment water consumption in flood season that meet the median water river bed in the Lower Weihe River is 5.23×10^9 m^3 in level year 2000, and 4.24×10^9 m^3 in level year 2030.

Small discharge floods of hyper – concentration sediment from Weihe River occurred frequently, which have a greater impact on the river silting. Certain sediment water consumption is needed to ensure there isn't sedimentation in the river in June. The incoming sediment in June is 0.034×10^9 t in level year 2000 and 0.025×10^9 t in level year 2030, and the sediment water consumption is 0.58×10^9 m^3 in level year 2000 and 4.24×10^9 m^3 in level year 2030. The incoming sediment is relatively low in non – flood season November to next May, and the sediment water consumption isn't to be considered.

Calculated through the above analysis, the sediment water consumption in flood season in the Lower Weihe River is 5.81×10^9 m^3 in level year 2000, and 4.75×10^9 m^3 in level year 2030.

3.1.2 The sediment water consumption in the Lower Weihe River

The sediment water consumption is not only need a certain amount of water, but also need a certain discharge, therefore, this paper studied the critical conditions between and sediment of non – floodplain flooding and the sediment water consumption. According to the measured data of the 100 non – floodplain in the lower Weihe River, this paper studied the characteristics and distribution (vertical distribution along the process and horizontal beach slot distribution) of erosion and deposition in different periods of the Lower Weihe River, the erosion and deposition characteristics of the water period in flood season and floodplain flood period, calendar year bank – full discharge and its affecting factors, and the erosion and deposition characteristics non – floodplain flood period under the conditions of different water – sand combination, combing with analysis of the relationship among river sediment ratio, average flow and maximum daily average flow in the flood period of the Lower Weihe River, the critical condition between erosion and deposition in non – floodplain period of the Lower Weihe River are put forward, which is shown in Tab. 2. As shown in Tab. 2, for the higher sediment concentration flood, the average flow must be more than 1,000 m^3/s to reach the balance of erosion and deposition.

Tab. 2 Non – floodplain flood critical flow and sediment condition in the Lower Weihe River

Flood type	Low Sediment concentration	Medium Sediment concentration	Higher Sediment concentration	Highest Sediment concentration
Sediment concentration (kg/m^3)	<20	20 ~ 100	100 ~ 300	>300
Average flow(m^3/s)	400	600	1,000	800
Peak flow(m^3/s)	650	1,000	1,700	1,400

In order to improve the conveyance efficiency, this paper studied the relationship among the sediment water consumption in the flood period, the flood event siltation ratio, incoming sediment and siltation. Contrast to the sediment water consumption of the flood season and runoff season, when the incoming is 0.3×10^9 t, the Huaxian medium discharge is 3,000 m^3/s and the balance between erosion and deposition is maintained, the sediment water consumption is 4.98×10^9 m^3 in the natural season. If regulated peak flow to 1,500 m^3/s, the sediment water consumption is 3.23×10^9 m^3; if regulated peak flow to 3,000 m^3/s, the sediment water consumption is 1.8×10^9 m^3, which show that water diversion or construction of reservoirs, increasing flood flow as far as possible helps to reduce river sediment water consumption, of course. To make peak not only need engineering conditions, but also need to consider flood control, water resources use and other factors.

3.1.3 Non – flood season river low eco – environmental water demand

When the river lower limit eco – environmental water demand is determined, several aspects are mainly considered ,which include maintaining the basic form of the Weihe River, to guarantee a certain base flow, covering the amount of groundwater, maintaining certain dilution and self – purification capacity and the basic ecological environment of the Weihe River , and to meet the entertainment landscape water, etc. According to the basic needs of the ecological environment in the Weihe River, this paper initially set the lower limit eco – environmental water demand in the non – flood season, the minimum discharge flow is 10 m^3/s at Linjiacun section, 15 m^3/s at Xianyang section and 20 m^3/s at Huaxian section, due to the sediment water consumption can meet the lower limit ecological environment water demand inside river in June and flood season, the minimum discharged volume of three sections are separately controlled at 0.175×10^9 m^3, 0.394×10^9 m^3 and 0.613×10^9 m^3 from November to next May.

3.1.4 Eco – environment water demand in the Lower Weihe River

The eco – environmental water demand in the Lower Weihe River include three ones, which are the low limit eco – environmental water demand in flood season, in June (non – flood season) and in non – flood season (November to next May). Based on the forgoing analysis, the eco – environmental water demand in the Lower Weihe River (the Huaxian section) is 6.42×10^9 m^3 in level year 2000 and 5.36×10^9 m^3 in level year 2030(shown in Tab. 3).

Tab. 3 River eco – environmental water demand in the Lower Weihe River (the Huaxian section) (Unit: $\times 10^8$ m^3)

Level year	Flood season	June	November to next May	total
2000	52.3	5.8	6.1	64.2
2030	42.4	5.1	6.1	53.6

3.2 Eco – environment water demand in the Lower Yellow River

According to the research results over the years, at the Xiaolangdi normal run – time, when the incoming sediment is 0.9×10^9 m^3 and the siltation is 0.15×10^9 m^3, the years of average eco – environment is 22×10^9 m^3 in the Lower Yellow river (the Lijin section), 17×10^9 m^3 in flood season and 5×10^9 m^3 in non – flood season.

In order to reduce deposition of the Lower Yellow River, in addition to meet the water requirements, a certain flow and sediment concentration are prerequisite, and it is necessary to avoid adverse flow level of 800 ~ 2,600 m^3/s. The water and sediment regulation test and the application over the years show that water and sediment regulation is not only one of the important means to solve siltation problem, but also the key measure to shape, restore and maintain the medium water river bed. Since 2002, water and sediment regulation mainly use the large water and sediment regulation capacity of Xiaolangdi Reservoir in sediment retention period to shape flow and sediment process conducive to the sediment transport of the Lower Yellow River and increase river flooding capacity. However, with the siltation in the Xiaolangdi Reservoir gradually increased, the capacity that can regulate water and sediment is getting smaller and the difficulty of water and sediment is growing larger. To further the implementation of water and sediment regulation, it is necessary to use water and sediment regulation system in the main stream of the Yellow River and the outer basin water transfer as supports.

3.3 Water demand projections in the receiving area

The water demand outside river is divided into three types, which are production water demand, living water demand and eco – environmental water demand. In this paper, development indicators and quota act are used to forecast water demand in level year 2030. Water demand outside river forecast need to take full account of the transformation of industrial structure and water – saving, strictly control the water demand growth. The forecasting results of water demand of the Weihe River are shown in Tab. 4. Because the contradiction between supply and demand of the Yellow River Basin has been very prominent, in addition to considering a certain growth of the living water demand, other water demand are basically in accordance with the status water supply of the Lower Yellow River in the future. The forecasting results are shown in Tab. 5, which are consistent with the integrated water resources planning of the Yellow River Basin.

Tab. 4 The Weihe River basin water demand projections

(Unit: $\times 10^8$ m^3)

Level year	Life	Production	Ecology (River inside and outside)	Total
2000	7.2	69.7	64.6	141.5
2030	14.7	77.4	54.8	146.9

Tab. 5 The lower reaches of the Yellow River water demand projections

(Unit: $\times 10^8$ m^3)

Level year	Basin			Basin outside	Total
	Life	Production	Ecology (River inside and outside)		
2000	3.5	59.2	220.1	85.0	367.8
2030	5.1	46.3	220.3	85.0	356.7

4 Analysis of the water shortages and water transfer requirement in receiving areas

4.1 Analysis of the water shortages in receiving area

Using the water demand prediction results, the balance between water supply and demand in different level years are calculated by "the Yellow River Basin Water Resources Allocation Decision

Support System". In this paper, the basin - wide water supply and demand balance calculation that considered 240 compute nodes of the Yellow River Basin, using a total of 45 years runoff data from 1956 to 2000, selected month as the calculation period, considered the water demand requirements which include anti - ice, anti - drying and sediment transport, etc. The priority order of water supply of departments is that taking precedence to meet the urban life, rural drinking and industrial water requirements, the following one is eco - environmental water demand in the river and the last ones are agricultural irrigation and eco - environmental water demand outside river. The river water supply order is that upstream first and downstream after, the tributary first and main river after.

The Water Supply Regions of the Xiaojiang Water Transfer Scenario are Guanzhong area in the Weihe River Basin and the Lower reaches of the Yellow River. According to the analysis and calculation results of water supply and demand, water shortages in the receiving area of the Xiaojiang Water Transfer Scenario are shown in Tab. 6.

Tab. 6 Supply and demand situation on the Water Supply Regions of the Xiaojiang Water Transfer Scenario (Unit: $\times 10^8$ m^3)

Level year	Partition	water shortage outside river	water shortage inside river	Total volume of water shortage
	Guanzhong in Weihe River Basin	12.0	13.3	25.3
2000	The lower Yellow River	7.9	13.3	21.2
	subtotal	19.9	13.3	33.2
	Guanzhong in Weihe River Basin	18.6	7.6	26.2
2030	The lower Yellow River	7.3	34.2	41.5
	subtotal	25.9	34.2	60.1

The water shortage in the Water Supply Regions is 3.42×10^9 m^3, and it will be 6.01×10^9 m^3 according to the prediction, therefore, the water shortage situation is more serious. Both of Guanzhong region and the lower reaches of the Yellow River are serious water shortage regions, the lower reaches of the Yellow River is a main water shortage area, the volume of water shortage of which account for 69% of the total volume of water shortage in the Water Supply Regions. Water shortage outside river and water shortage inside river coexist. The water shortage inside river is the main one, the volume of which is 56.9% of the total volume of water shortage inside river.

4.2 Water transfer requirement in the Water Supply Regions

The water shortage in Guanzhong area of Weihe River is 2.62×10^9 m^3 in level year 2030, the main of which are national economic water scarcity, the water shortage is 1.86×10^9 m^3, and the water shortage rate has reached 27.5%, as a result, to solve the water shortage problem has become urgent. Alleviating the water shortage of the Weihe River need to transfer water 2.5×10^9 m^3, about 1.5×10^9 m^3 water is used for increasing the national economic water demand, and about 0.7×10^9 m^3 water is used for increasing the eco - environment water demand. The national economic water demand, especially the domestic and industrial water demand, the processes of which are so evenly that the regulation requirements of the transferred water is relatively simple compared to the sediment water, easy to reduce the scale of the water diversion project. However, realization of the sedimentation reduction of the lower Weihe River requires not only certain transferred water, but also a large flow process.

The water shortage in the lower Yellow River is 4.15×10^9 m^3 in level year 2030, the main of which is ecological and environment water shortage, the water shortage is 3.42×10^9 m^3, and the gap is large. The above – mentioned water shortage is drawn under the condition that consider the 22×10^{9} m^{3} water, the lower limit eco – environmental water demand of the lower Yellow River. If further reduced the siltation of the downstream, the sediment water consumption will be increased and the water shortage will be greater. Therefore, to solve the water shortage problem of the lower Yellow River in 2030 would need at least 4.2×10^{9} m^{3} transferred water, which is mainly used to supplement water demand for transferring sediment and reducing siltation in the river. Reducing siltation not only need a certain amount of water, but also requires a large amount of flow process.

Therefore, the requirements of the water transferred for the Water Supply Regions is not only include the total volume of the water transferred, but also the flow process. It is relatively easy and has a small requirement of diversion to solve the national economy water shortage problems, and the capable of transporting water flow is uniform. However, it is relatively difficult to solve the sediment deposition in the Lower Reaches of Yellow River and Weihe River, not only need a large quantity of water, but also need to create a certain flow process, involving many factors.

5 The supply target of the Xiaojiang water transfer Scenario

It has been already very urgent to solve the water shortage problem in the Weihe River, the national economic water shortage problem in Guanzhong area of the Weihe River should be solved at first. At present, Shanxi province is to carry out the preparatory work of the water diversion from Hanjiang River to Weihe River, which plans to transfer 1.5×10^{9} m^{3} water from Hanjiang River and its tributary Meridian River. The length of the route of water transfer is 79.2 km. The scale of the project is relatively small. The water shortage problem in the Weihe River can be solved via the water diversion Hanjiang River to Weihe River.

In addition to the water diversion project from Hanjiang River to Weihe River, considering the requirements of the sedimentation reduction in the lower reaches of the Weihe River and the Yellow River, as well as to alleviate the national economic water shortage in the lower Yellow River, in the case of the lower Yellow River is still silting 0.15×10^{9} m^{3}, at least 4.2×10^{9} m^{3} water is needed which can be solved by Xiaojiang water transfer. The water transfer should primarily concentrate in the flood season and form certain flow process through regulating, to increase sediment transport flow in the lower Weihe River, maintain the medium water river bed and also supplement the sediment water consumption in the lower Yellow River, to meet the requirement about using water at a low limit.

By comprehensive consideration the water shortage problems and pressing issues of management and development of the Lower reaches of Weihe River and the Lower Yellow River water, the Supply Target of the Xiaojiang Water Transfer Scenario put forward focus on to supplement the eco – environmental water use of Lower Reaches of Weihe River and Lower Reaches of Yellow River, and to increase the base – flow for sediment flushing and eco – environment of river channel, to reduce siltation in the lower Weihe River, to curb the trend of ecological deterioration, to recover and ultimately maintain the basic functions of the middle and lower reaches of the Weihe River; to reduce siltation in the lower Yellow River , in order to improve and maintain the mecium water river bed, to supplement the lower Yellow River water for edological use in non – flood season, contribute to the maintenance of healthy life of the Yellow River.

6 Conclusions

Starting from water shortage situation of the Yellow River, from a macro and holistic perspective, this paper studied the measures to solve the water shortage problems in various regions of the Yellow River Basin, put forward the Water Supply Regions of Xiaojiang Water Transfer Scenario, which are Guanzhong region of the Weihe River Basin and reaches downstream of Tongguan of the main Yellow River, studied the sediment water consumption in flood season in the lower Weihe River, combining with the requirements for Water Requirement for Sediment Transport in flood season and in June and the ecological base flow in non – flood season, proposed that the ecological environment water demand of the Lower Weihe River is 6.42×10^{9} m^{3} in level year 2000 and $5.36\times$

10^9 m^3 in level year 2030. Analyzed the supply and demand situation in receiving regions, the water shortage in receiving regions of the Xiaojiang water transfer scenario is 3.32 × 10^9 m^3 in level year 2000 and 6.01 × 10^9 m^3 in level year 2030, studied the requirements of the receiving area on the water diversion, Finally the Supply Target of the Xiaojiang Water Transfer Scenario put forward is to supplement the eco – environmental water use of Lower Reaches of Weihe River and Lower Reaches of Yellow River, and to increase the base – flow for sediment flushing and eco – environment of river channel.

References

Yellow River Engineering Consulting Co., Ltd. Research on the Schemes of Yangtze River – to – Wei River – to – Yellow River Water Diversion Project[R]. 2011.

Yellow River Conservancy Commission of the Ministry of Water Resources. Comprehensive Water Resources Planning of the Yellow River Basin[R]. 2010.

Yang Lifeng, Wang Yu, Chen Xiongbo, et al.. Study on Sediment – carrying Water Volume in the Lower Weihe River[J]. Journal of Sediment Research, 2007(3): 24 – 29.

Experience of Development of Scientific and Methodical Bases of Design of Selective Water Intakes of Hydroelectric Power Stations

V. N. Zhilenkov and ***S. Y. Ladenko***

RusHydro, Vedeneev VNIIG, 21, Gzhatskaya Str., St. Petersburg, 195220, Russia

Abstract: Changes of hydrothermal regime of the river in the tail water are one of the negative results of hydrosystem construction in climate zones with contrast seasonal cycles: in summer water is becoming considerably colder, in winter - warmer. The main reason of such phenomenon depends on location of water intakes of HPP turbine disposition at low elevations, where the water temperature is almost constant during a year: not higher than 4/5 ℃ and not lower than 3 ℃. As a result, during the winter in tail water of all high-head and medium-head hydroelectric power stations there are multikilometer ice-holes over which in frosty days the dense fog keeps observed, and in the summer low water temperature in the river negatively influences river flora and fauna. It conducts to deterioration of a microclimate of the coastal territory and conditions of the population dwelling.

Ecological situation in tail water may be improved by selective intake of water from the surface layer of the reservoir. But in many cases the selective water intake at run-of-river HPP arrangement is practically unrealizable, because at limited width of water intake front and large discharges of high-head HPP it is impossible to provide necessary unit discharge of water inflow from surface layers of the reservoir, water from surface and deep layers mixes during water intake.

To exclude water intake straight in front of the turbine channel, Zhilenkov V. N. proposed principally new approach to solve this task: the selective water intake unit is located on the bank slope, it provides necessary width of the water intake front.

To determine the parameters of selective water intake and its arrangement (direct-flow or bank) we have to know the conditions of formation of hydrothermal regime in the particular reservoir, first of all the position of the thermal wedge (layer of temperature rise) in which water temperature changes abruptly and the conditions to maintain stability of water layers with different density approaching the water intake unit.

The authors suggested and validated the application of density Froude number as the criterion of non-mixing of layers.

To study the conditions of hydrothermal regime formation in reservoirs, the authors analyzed the information of in-situ survey by 11 deepwater reservoirs of HPP in the Siberia, Far East, North of RF, Eastern Kazakhstan. At that an interesting regularity was found out: for every reservoir, during the whole period of observations, the position of the upper boundary of temperature changes by depth (thermal wedge) was constant for every month. Basing on it, one can conclude that if the depth of the reservoir does not change considerable, its temperature regime remains constant. Averaging the observation data by years there may be obtained the typical diagrams of water temperature distribution by depth for every month for the particular reservoir. These data may also serve as analogue data to design reservoirs for new HPP with selective water intake.

Key words: Reservoir, hydrothermal regime, ecological situation in tail water, selective water intakes

It is known that one of the negative consequences of hydroelectric power station construction in climatic zones with contrast seasonal cycles is change of a hydrothermal regime of the river in the tail water of hydraulic structure where in the summer water becomes much colder, and in the winter is warmer. The main reason for such change is that water intakes of hydroelectric power station turbine conduits, taking into account low-water evacuation of a reservoir, are situated on low marks where water temperature remains almost constant during the whole year not exceeding 4/5 ℃ and not falling below 3 ℃. As a result, during the winter in tail water of all high-head and medium-head hydroelectric power stations there are multikilometer ice-holes over which in frosty days the dense fog keeps observed, and in the summer low water temperature in the river negatively influences river flora and fauna . It conducts to deterioration of a microclimate of the coastal territory and conditions of the population dwelling.

The most effective remedy of providing ecologically favorable conditions in the tail-water of hydroelectric power station is selective water intake from the reservoir surface layer. Nowadays there is a rather wide range of calculated and constructive and technological solutions of this problem but, unfortunately, they didn't manage to be realized at design and building of hydro energy projects in Siberia and in the Far East.

Application of a water intake of selective type will allow to regulate quality of water (a hydrothermal regime) in the tail-water. However, it is necessary to remember that in many cases (roughly, at pressures about 100 m and the established capacity of the station more than 500 MW) the selective water intake of direct-flow type is almost impracticable, as at limited width of the water intake front it is impossible to provide necessary specific rate of water inflow from the reservoir surface layer.

It is considerable that if the established capacity of medium-head ($H < 50$ m) hydroelectric power station doesn't exceed 500 MW, possible and more acceptable on technical and economic indicators there can be an option of selective water intake of not derivational, but pot-head (direct-flow) type.

But more often, at consumption of medium-head hydroelectric power stations of 1,000 m^3/s, overall performance of a water intake construction can be provided only at the corresponding lengthening of the front of water inflow and preservation thus in a reservoir of stability of layers with various temperature.

To exclude water intake arrangement right before turbine conduits at which it is impossible to provide selective water abstraction at high level of consumption, an essentially new approach to the solution of this task which feature is in an arrangement of the selective water intake device on a beach approach that allows to provide necessary width of the water intake front was offered.

So, two schemes of water intakes of selective type essentially different in configuration are possible: direct-flow and derivational, each of them is necessary to choose depending on hydrological characteristics and topography in an hydroelectric site, hydroelectric power station capacity, pressure and other less significant parameters that are usually considered at its design and operation.

Practical implementation of decisions on selection of hydroelectric complex configuration and construction should be based in the course of design on basic data about conditions of formation of a hydrothermal regimes in a particular reservoir, first of all, about the position of a thermocline (a layer of temperature jump) within which there is a sharp change of water temperature, and about conditions of preservation of its stability on the way to the water intake construction of this or that type.

Quantitative regularities of a hydrothermal state for rivers and reservoirs, as a rule, can be received on the basis of the analysis and generalization of materials of long-term natural observations and calculative and theoretical studies. At the heart of processing of initial information there is, mainly, the statistical analysis of long-term sets of observations. The revealed regularities of ice and thermal processes can be used as analogs for the understudied water streams and reservoirs taking into account their features when forecasting a thermal regime of a reservoir in similar climatic conditions.

For this purpose we studied materials of natural observations on 11 deep-water reservoirs of hydroelectric power stations located in Siberia, in the Far East, the North of the Russian Federation, East Kazakhstan. The longest series of observations were conducted on four reservoirs: Krasnoyarsky (1968/1973), Bratsky (1964/1972), Vilyuysky (1968/1971; 2001/2003) and Sayano-Shushensky (1980/1993; 2003). Based on these observations temperature profiles were done for every month according to the depth of these reservoirs. As an example on Fig. 1 temperature profiles are shown on the depth of Vilyuysky reservoir for 1968/1971; 2001/2003.

Therefore, it was possible to reveal some interesting regularity: for each reservoir during all years of observations, the position of the top border of change of temperature on depth (thermocline) remained constant for every month. And for a reservoir of Vilyuysky hydroelectric power stations Ⅰ, Ⅱ (Fig. 1) during the periods the position of this border of temperature change remained constant, especially for the winter period. Serious changes in distribution of temperature on depth occurred only in the Sayano-Shushenskaya reservoir. It is connected with reservoir filling, and hydroelectric power station service conditions (the provision of water intakes). So, in a temperature regime of the Sayano-Shushenskaya reservoir two periods were defined: 1980/1984 and 1985/2003.

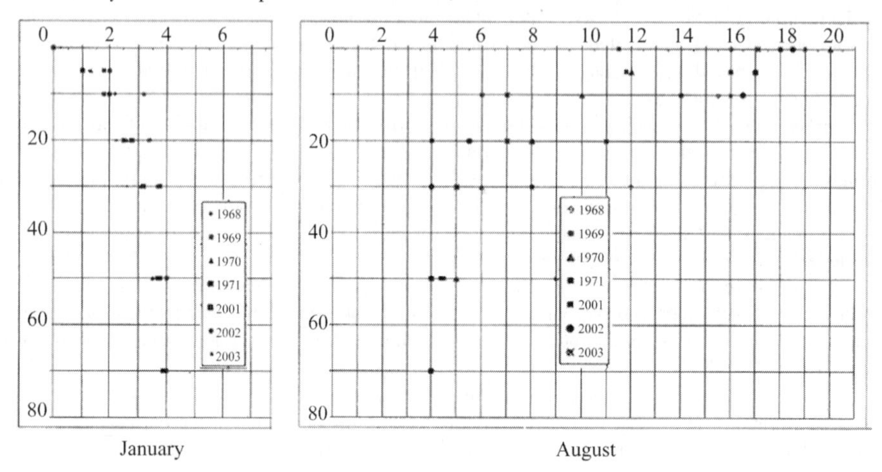

Fig. 1 An example of temperature profiles on depth of the Vilyuysky reservoir for January and August 1968 ~ 1971; 2001 ~ 2003

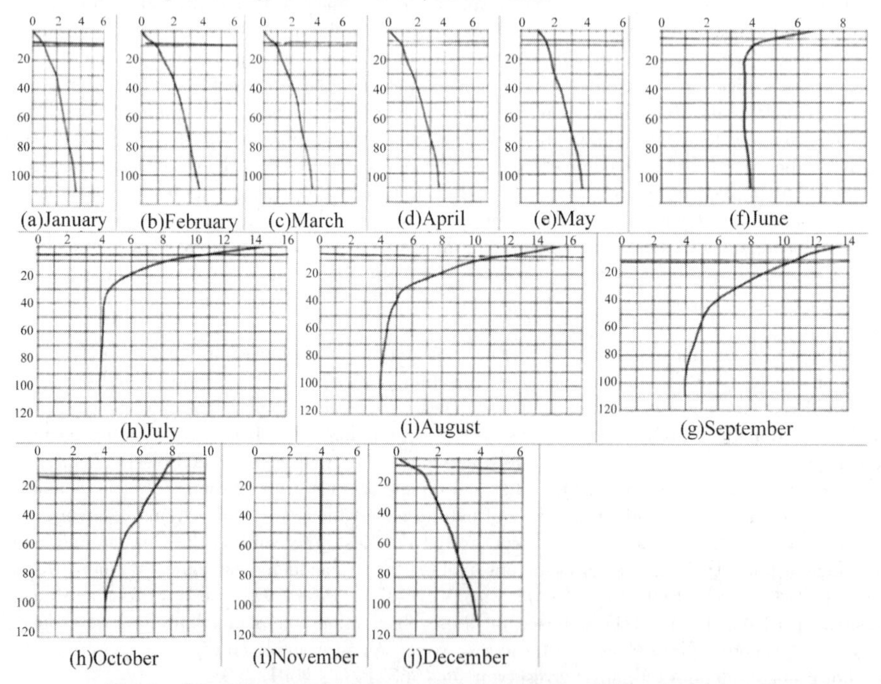

Fig. 2 Monthly temperature profiles on depth of Bratsky reservoir

On the basis of this it is possible to assume that if the reservoir depth didn't change essentially, its temperature regime remains constant. Averaging the available data of observations for these years it is possible to receive standard water temperature profiles on depth shown for months for the particular reservoir. These data can also be used as analogue when designing any new hydroelectric power stations. As an example shown in Fig. 2 are provided temperature profiles on depth of the Bratsky reservoir on months, averaged for the period of 1964 ~ 1972 where the top border of the thermocline is defined.

It should be noted that for establishment of position of the top border of thermocline layer it is necessary to formalize an approach to its definition. According to the researches carried out by us, the top border of thermocline can be considered as value of water temperature in reservoir 0.2 · | t bottom-t surface |. So, during regressive temperature stratification, when bottom water temperature t = 4 ℃ and surface t = 0 ℃, the top border of the thermocline will be situated at temperature level of 0,8 ℃.

We have executed temperature profiles on all available monthly observations for 11 reservoirs which can be used at selection of any particular hydrothermal regime for the analog reservoir.

For selection of an analog reservoir tables are provided, using which it is possible to define the closest analog for the hydroelectric power station designed (Tab. 1, Tab. 2).

Tab. 1 Analog reservoirs to determine stratification periods

Reservoir	Years of observations	Depth (in meters)	HPS capacity	Volume of reservoir (in square km)	Stratification period			Air temperature		
					direct	return	Total (months)	Cold period 0.95	Warm period 0.95	Annual average
Krasnoyarsky	1968 – 1973	68 ~ 101	1,500 ~ 6,000	35 ~ 73.3	January ~ April	June ~ October	9	~ 22	22	0.5
Buhktarminsky	1964 ~ 1967	49 ~ 55	675	49	January ~ April	May ~ October	10	~ 23	26.8	2.8
Viluysky Ⅰ—Ⅱ	1968 ~ 1971, 2001 ~ 2003	48 ~ 68	340 ~ 648	7.4 ~ 26.0 35.9	November ~ May	July ~ September	10	~ 41	21.7	~ 9.0
Serebryansky	1972 ~ 1974	98 ~ 110	204.9	4.17	November ~ May	July ~ September	10	~ 13	11.1	0.6
Kolymsky	2001 ~ 2003	97 ~ 115	900	14.5	December ~ May	July ~ September	9	~ 42	19.8	~ 11.4
Zeysky	1977 ~ 1979	61 ~ 75	680 ~ 900	68.4	November ~ May	July ~ September	10	~ 35	23.1	~ 4.1
Bratsky	1964 ~ 1972	98 ~ 108	4.515	114 ~ 169.4	December ~ May	June ~ October	10	~ 26	21.1	~ 1.6
Ust ~ Ilimsky	1977 ~ 1979	89 ~ 92	3.600	52 ~ 60	November ~ April	July ~ October	9	~ 30	23.1	~ 3.6
Mamakansky	1963 ~ 1964	34 ~ 45	100	0.1 ~ 0.2	November ~ May	July ~ September	10	~ 36	21.9	~ 5.6
Sayano ~ Shushensky	1980 ~ 1984	120 ~ 135	2560 ~ 4480	23.0	January ~ April	June ~ October	10	~ 25	23.8	0.3
	1985 ~ 1993, 2003	140 ~ 220	6.400	31.3	February ~ April	June ~ November	10			
Ust ~ Kamenogorsky	1954 ~ 1955,	39 ~ 45	248.4	0.65	January ~ March	June ~ August	6	~ 23	26.8	2.8
	1964 ~ 1967	39 ~ 45	331.2	0.65						
Moksky (under design) *	~	100	1200	51.4	November ~ May	July ~ September	10	~ 34	22.2	~ 5.6

Note: * Supposed stratification period was received by the analog – Zeysky reservoir.

Tab. 2 Deepening of the top layer of thermocline in reservoirs

Reservoir	Years of observations	Deepening of the top layer of thermocline, in meters towards the surface											
		1	2	3	4	5	6	7	8	9	10	11	12
Krasnoyarsky	1968 – 1973	9	9	10	10	-	4	10	20	36	42	-	-
Buhktarminsky	1964 – 1967	5	6	4	—	10	9	9	12	18	21	—	—
Viluysky Ⅰ—Ⅱ	1968 – 1971, 2001 – 2003	4	4	4	4	4	—	4	9	20	—	4	5
Serebryansky	1972 – 1974	4	5	4	4	4	—	11	17	24	—	4	8
Zeysky	1977 – 1979	9	8	9	8	7	—	12	15	30	—	4	9
Kolymsky	2001 – 2003	10	10	10	10	10	—	10	16	50	—	4	10
Bratsky	1964 – 1972	9	10	9	9	8	5	6	7	11	12	—	4
Ust – Ilimsky	1977 – 1979	24	33	42	48	50	—	7	14	12	—	10	10
Mamakansky	1963 – 1964	28	28	26	26	24	—	7	10	10	—	20	28
Sayano – Shushensky	1980 – 1984	28	32	30	21	—	42	40	30	40	62	80	—
	1985 – 1993, 2003	—	9	9	10	—	10	18	30	42	56	60	90
Ust – Kamenogorsky	1955 – 1956,	38	35	34									—
	1964 – 1967				—	—	10	10	10	—	—	—	—

Besides, that at carrying out researches on identification of temperature regimes of reservoirs, attempts to formalize an analog reservoir choice were made, considering its morphometric characteristics, climatic features of the territory, hydroelectric power station capacity, etc. So, it is noticed that air temperature (the minimum, maximum provided and mid-annual values) rather precisely define the approach of the periods of direct and return temperature stratification of water in a reservoir.

But to reveal accurate regularities and to establish sequence in an analog choice at this stage it is not obviously possible. Insufficiency of data on available reservoirs, and also multiple-factor dependences of formation of a thermal regime can be the reason of it. Thus, selection of an analog reservoir is carried out by the method of ? the greatest compliance? with the available tabular materials.

The Moksky hydroelectric power station reservoir was chosen as an illustration in order to show the analog of forecasting of temperature regime. Zeysky, Ust-Ilimsky and Mamakansky reservoirs are the closest reservoirs according to their air temperature indicators common for this region. Ust-Ilimsky, Zeysky, Bukhtarminsky and Krasnoyarsky reservoirs are the closest on volume indicators. On hydroelectric power station capacity the closest are Krasnoyarsky, Kolymsky and Zeysky reservoirs . On the reservoir depth the closest are Krasnoyarsky, Serebryansky, Kolymsky, Zeysky, Bratsky, Ust-Ilimsky. Thus, the closest values on all indicators has the Zeysky reservoir which has been accepted as an analog one.

In order that water arrived only from the upper layer, should be met a condition of preservation of stability (immiscibility) of layers with various temperature which at a selective water intake is carried out at value of Froude density number less than the critical:

$$Fr'' = \frac{q}{h^2}\sqrt{\frac{h\rho_0}{g\Delta\rho}}$$

where q is a specific consumption of water; h is the overall depth; ρ0 — bottom water density ; $\Delta\rho$ is difference in density between the surface and bottom; g is acceleration of free fall.

According to the careful studying of materials of natural observations and the corresponding laboratory experiments, we have found out that violation of stability of layers with different density occurs at $Fr'' > 0.28$.

The selective water intake device should be structurally executed so that to provide:

(1) water intake from a reservoir surface layer;

(2) work of a water intake construction in the conditions of winter operation;

(3) termination of access of water at inspections, repair works, in case of failure;

(4) water intake regulation at changes of hydroelectric station load;

(5) position of water intake channel and an outline of its approach should provide a smooth input of water in a water intake channel with the minimum loss of pressure;

(6) execution of general requirements for all hydraulic engineering constructions: stability, durability, convenience in operation and availability for supervision.

The usual types of gates falling from top to bottom, can be flat panel boards on wheels, segment, for which lengthening of piers is required, but lower position of the auxiliary bridge, flap gates, roof weirs, etc.

1— concrete fragment of the water intake; 2— pier dividing the bay; 3— vertical grooves in piers for placement a sheeting dam boards and trash-rack structures; 4—vertical grooves in piers where the ends of rods fastening among shutters of the gate; 5—the maneuvering mechanism for lifting and dropping of the flap gate; 6—joint fastening plates; 7— a rack holding the folding screen of the gate in the stream direction; 8—a traverse holding the screen; 9—resilient seals (brush or rubber) at the flat end of the shutters; 10—water intake openings of turbine conduits.

Aspiration to achieve the greatest profitability compels to resort to constructive solutions when the width of space occupied by the gates will be probably smaller and which give the chance for lifting the gate from the down position at the most insignificant difference between the head water and tail water. For this purpose V. N. Zhilenkov developed a compact design of a maneuverable overflow joint and flap gate (Fig. 3), allowing carrying out selective water intake.

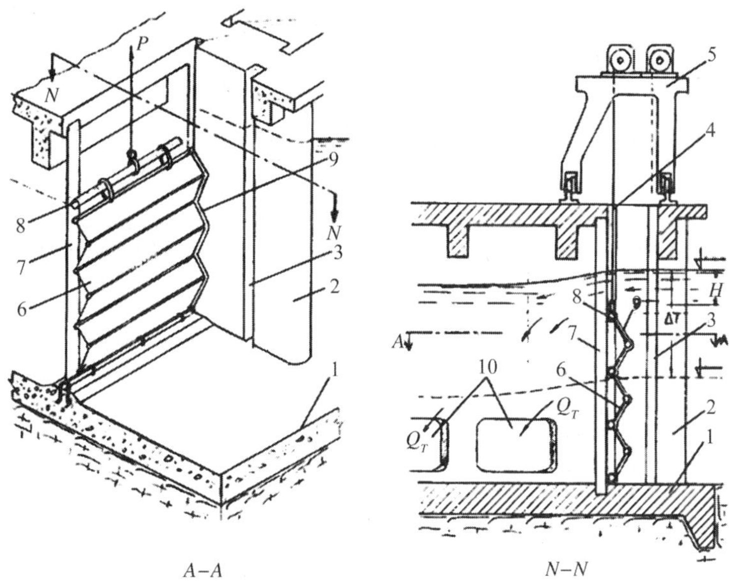

Fig. 3 Overflow joint and flap gate designed by. V. N. Zhilenkov

As a result of the carried-out works we offered a technique of definition of key parameters of the selective water intake taking into account hydrothermal conditions determined by the analog reservoirs which main positions are given below.

Technique of definition of key parameters of the selective water intake

Considering characteristics of formation of hydrothermal regimes in reservoirs of the hydroelectric power stations constructed on the rivers of Siberia, the Far East, and also located in similar climatic conditions, design of selective water intake constructions for future power plants is expedient to be carried out according to the following methodical recommendations:

1. Definition of a critical specific discharge.

1.1. According to the previously established power parameters of hydroelectric power station - capacities of N (kW) and pressure of N (m) total consumption of water passing through all turbines is defined:

$$Q \approx N/(8 \div 10) \cdot H \quad (\mathrm{m}^3/\mathrm{s}) \tag{1}$$

1.2. According to the known characteristics of a reservoir (depth before the dam h, draft and temperature regime) critical specific discharge of q is found, at which Froude density number

$$F\gamma'' = \frac{q}{h^2}\sqrt{\frac{h\rho_0}{g\Delta\rho}}$$

is less than its critical meaning that equals 0.28.

$$q_{xp} = F\gamma'' h^2 \sqrt{\frac{g\Delta\rho}{h\rho_n}} \tag{2}$$

The design temperature of the surface and bottom layers should be set (for example, for winter conditions density $\rho_w = 999.87\ \mathrm{kg/m^3}$ corresponds to the surface water layer temperature of 0 ℃, for summer conditions such density water has at temperature of 8 ℃, temperature of the bottom layer 4 ℃, in this case its density equals to $\rho_H = 1{,}000\ \mathrm{kg/m^3}$);

$$\Delta\rho = \rho_H - \rho_B = 0.13 = 0.13\ \mathrm{kg/m^3}$$

2. Definition of a shutter crest (spillway weir) position of selective water intake.

To define the position of a shutter crest (spillway weir) of the selective water intake we use the recommended depth of water abstraction from the surface layer determined by the analog reservoirs. Selection of the analog is carried out based on the maximum correspondence with parameters of a projected reservoir to the analog (Tab. 1, Tab. 2).

3. Choosing an option of hydroelectric complex configuration. Calculation of total width of the water intake front.

3.1. By the known critical specific discharge of qKr we determine in a first approximation the total width of the water intake front B:

$$B = Q/q_{kr}$$

3.2. If the received value of B allows to arrange a water intake construction in a direct – flow configuration of hydroelectric power station, we design hydroelectric power station with the channel configuration (the direct – flow scheme of a water intake).

3.3. If the width of the river bed and depth of a reservoir draft does not let to execute direct – flow configuration of the selective water intake device, headrace canal of hydroelectric power station at the coastal scheme of configuration (the derivational scheme of a water intake) has to be designed.

4. Choosing the scheme of a water intake with gates or without gates.

Definition of number of bays of a water intake construction.

Proceeding from the optimum width of each bay blocked by the gate (in case of need), we find total number of bays in a water intake construction.

Conclusions

(1) The selective water intake from the surface layer of reservoirs of hydroelectric power station is the most perspective instrument for ensuring of conditions of ecologically acceptable thermal regime in the tail water.

(2) Considering negative experience of violation of natural hydrothermal regime of a waterway after hydroelectric power station construction in areas with severe climatic conditions, it is inaccept-

able to ignore ecological aspects at design of new hydroelectric power stations.

(3) On the basis of natural and experimental studies, authors developed a method of calculating of superficial selective water intakes of hydroelectric power stations taking into account hydrothermal regime of a reservoir which can be determined by the analog reservoir.

(4) The design of the overflow joint and flap gate, known by its high maneuverability was offered.

(5) The procedure of parameters calculation of superficial selective water intakes of hydroelectric power station can be used at design of hydroelectric power stations which are under construction in common for Siberia and the Far East climatic conditions.

Analysis on Operation Mode of Inter-basin Water Transfer Project Under Various Constraints

Jia Xinping[1] , *Liu Gang*[2] , *Fang Jing*[3] and *Zhang Yongyong*[2]

1. Department of Planning and Programming, Yellow River Conservancy Commission, Zhengzhou, 450003, China
2. Yellow River Engineering Consulting Co. , Ltd. , Zhengzhou ,450003, China
3. China University of Mining and Technology, Xuzhou, 221116, China

Abstract: Inter-basin water transfer project is a new water resources allocation system, which is jointly composed of water exporting region and in-take area, the operation mode of water conveyance is key link for the new water resources allocation. The change of region water resources allocation pattern is faced by various constraints, such as social、environment、technology and economy, the appropriate operation mode can balance all kinds of constraints, reduce the influence of water exporting region as far as possible, satisfy the water demand of in-take area to the great extent, and the scale of engineering construction is not oversize, the final objectives is the water resources system with minimum cost for maximal benefit. Take West-Route Project of South-to-North Water Transfer as case study, the analysis method for operation mode of inter-basin water transfer project under several kinds of constraints is discussed.

Key words: operation mode, inter-basin water transfer, water reservoir, water conveyance route, constraint condition

The Inter-basin water transfer project research involves in-take area, water exporting region and engineering region, and the facing problem and content are different. The in-take area and water exporting region research are based on analysis of supply and demand of water resources, the water demand and water transfer' s function, and the transferable water volumes and water transfer' s effect are researched. The engineering region research is based on arrangement of water conveyance line, the construction condition and scale of engineering are researched. Through the water conveying volumes, the three regions are linked into an organic whole, which combine the water resources problems and the engine problems, then a new water resources allocation system is constituted, the core of which is the operation mode of the water conveyance engineering. Through studying the operation mode of the West-Route Project of South-to-North Water Transfer, the research method for operation mode of inter-basin water transfer project among the water resources system is discussed.

1 The operation rules and main constraints

The basic rule of the operation mode of the Inter-basin water transfer project is as possible as to reduce the impact on water exporting region, and as far as possible to meet the water demand of the in-take area, and the scale of engineering construction is not oversize, the final objectives is the water resources system with minimum cost to get maximal benefit.

1.1 Operation rules and ideas

The first is the water demand of the in-take area. By analysis of the economic and social development, the water supply and demand situation, and the water transfer quantity allocation and function, the water transfer total quantity demand and process of the in-take area is proposed. Considered the limited water transferable quantity and fully and efficiently use, the water transfer total quantity shouldn' t be more than the water demand of the in-take area. Meanwhile, in order to meet the water demand of different water objects, the transfer water process should be adapted with the water demand process of water objects.

The second is water transferable quantity of water exporting region. By analysis of the economic and social development and water demand of ecological environment in different level year, the water transferable quantity and process of water exporting region is proposed. In order to reduce the impact on economic and social development and ecological environment, the water transferable total quantity should be less than the water transferable quantity of water transfer river. Meanwhile water transferable quantity of each water reservoir must be less than the water quantity of the river, the reservoir sluice quantity should meet the water demand and process of each industry.

The third are technical and economic conditions of water transfer project. Considered all factor, such as the runoff characteristics of water reservoir, the construction condition, arrangement the water transfer tunnel, and so on, it is as far as possible to reduce the construction scale of the each water reservoir and the water transfer tunnel, meanwhile, the west route project is located in the Qinghai-Tibet plateau of minority inhabit region, it is complex to solve the problem of submerged immigration, it should be as far as possible to reduce submerged damage.

1.2 Model establishment

The first stage of West-Route Project of South-to-North Water Transfer is from Dadu River in the upstream of Yangtze River and a tributary of Yalong River to the upstream of Yellow River, the water transfer project is composed of several water reservoirs and several section of water transfer tunnel, and among which the hydraulic connection between the in-take area and water exporting region is established.

The movement equation of a water reservoir in the system

$$I_i^t + V_i^t + D_{i-1}^t = D_i^t + \sum Y_i^t + Q_i^t + E_i^t$$

where:

I_i^t — The natural runoff of the i-th reservoir in the period t;

V_i^t — The reservoir storage capacity of the i-th reservoir in the period t;

D_{i-1}^t — The water transfer quantity of the(i-1)-th reservoir in the period t;

Y_i^t — The ecological water demand of off stream of the i-th reservoir in the period t;

Q_i^t — The ecological water demand of inner river of the i-th reservoirs in the period t;

E_i^t — Water loss of the i-th reservoir in the period t.

1.3 Main constraints

The operation mode should in addition to meet the main construction scale of water transfer project, the main constraints is come from in-take area and water exporting region.

(1) Reservoir capacity.

$$V_{i\,\min} \leqslant V_i^{\,t} \leqslant V_{i\,\max}$$

where: $V_{i\,\min}$, $V_{i\,\max}$ is the maximum and minimum water storage of water reservoir.

(2) Diversion capacity of water transfer tunnel.

$$D_i^{\,t} \leqslant D_{i\,\max}^t$$

where: $D_{i\,\max}^t$ is the design diversion capacity of water transfer from the i-th reservoir to the next reservoir.

(3) Water supply of outer river.

$$\sum Y_i^{\,t} \geqslant Y_{i\,\text{demand}}^t$$

where: $Y_{i\,\text{demand}}^t$ is the water supply of outer river of the reservoir among the transfer river.

(4) Water demand of inner river.

$$Q_i^{\,t} \geqslant Q_{i\,\text{demand}}^t$$

where: $Q_{i\,\text{demand}}^t$ is the water demand of inner river of the reservoir among the transfer river.

(5) Water demand of the in-take area.

$$\sum D_i^{\,t} \leqslant W_s$$

where: W_s is the water transfer quantity of the in-take area.

(6) Water transferable quantity of water exporting region.

$$D_i^t \leqslant W_{di}$$

where: W_{di} is the water transferable quantity of each transfer river.

2 Water demand of the in-take area

The West Line Project of the South-to-North Water Transfer could transfer water to the source region of the Yellow River, but it is located in high position and its coverage scale is wide, could ease the water resources shortage situation of the Yellow River Basin, and the in-take area is the whole Yellow River Basin. However considered the limited water transferable quantity and fully and efficiently use, the main in-take area of outer river of the West Route Project of the South-to-North Water Transfer is the up and middle stream of Yellow River and the neighboring inland river areas, the water objects are key cities, energy bases, Heishanxia's ecological construction region and Shiyanghe River, etc., and the inner river water object is mainstream of Yellow River.

2.1 Water transfer quantity demand

According to analysis of the water supply and demand situation of water resources, the new water demand of the 14 key cities of the in-take area in 2030 is 2.43×10^9 m^3, the new water demand of the 4 large energy and chemical industry base in 2030 is 1.8×10^9 m^3, and the Heishanxia's ecological construction region is the important national reserve cultivated land, it has increased 3.64×10^6 acres, the water demand is 1.86×10^9 m^3, the Shiyanghe River of the important northwest inland river, the water shortage in 2030 is 4.25×10^{10} m^3. So the water demand of outer river of the first stage of west route is 6.51×10^9 m^3; and in addition to considering the sediment transport of inner-river and ecological water demand, the water shortage is 3.82×10^9 m^3, the whole water demand of the first stage of West Route Project of the South-to-North Water Transfer is 10.33×10^{10} m^3.

2.2 Water transfer process demand

Among the water objects, the water guarantee rate of urban domestic and energy bases is high, and the annual water consumption quantity and water utilization process is relatively stable, sediment transport and silt reduction of inner river requires flood process with concentrated flood seasons and high requirements for water quantity regulation degree, the Heishanxia's ecological construction region mainly water consumption is irrigation water, and it is of certain seasonality.

Seven large mainstay water control engineering have been built and planed, which are Longyangxia, Liujiaxia, Daliushu, Qikou, Guxian, Sanmen Gorge and Xiaolangdi, and the reservoir storage capacity is up to 9.32×10^{10} m^3, the regulation storage capacity is 4.72×10^{10} m^3, among them the reservoir storage capacity of leading reservoir called Longyangxia is 1.935×10^{10} m^3, it is a multi-year regulating reservoir, the storage coefficient is above 0.9, it will remain the multi-year regulation capacity even after considering the first stage of West Route Project. By analysis of the water regulation engineering after transferring water to Yellow River, the regulation capacity of the reservoirs in the mainstream of Yellow River can fully meet the water demand of the water objects, and the water transfer process only needs to consider the total water demand, not to consider the constraint of the water transfer process.

3 Water transferable quantity of transferred river

The transferred rivers of the first stage of the West Route Project are Yalong River and Dadu River in the upstream of Yangtze River, and the water reservoirs are located in the mainstream of Yalong River and its tributaries named Daqu and Niqu, and Dadu River's tributaries named Sequ, Duke River, Make River and Ake River, the sevse diversion dams are respectively located at Reba,

Aan, Renda, Luoren, Zhu'anda, Huona and Keke.

3.1 Current water consumption situation and water demand forecasting of transferred river

Yalong River is the largest tributary above the Yibin of Yangtze River, its river basin area is 1.284×10^8 km^2, and the water quantity is 6.004×10^{10} m^3. Dadu River is the largest tributary along the left bank of Jinsha River, its river basin area is 7.71×10^7 km^2, and the water quantity is 4.755×10^{10} m^3. Because of the natural conditions discrepancy, the social and economic situation in the up-middle-down steams of Yalong River and Dadu River is largely difference. The up-and-middle stream is mainly of altiplano and upland, it belongs to semi-agricultural and semi-pasturing area, mainly in animal husbandry, and is sparsely inhabited, the middle-and-down stream is a transition terrain from high mountain to low mountain and hilly-gully, it has a concentrated population, and is developed in industries and agriculture.

The current total population in Yalong River and Dadu River basin is respectively 26,720,00 and 22,080,00, GDP is respectively 1.535×10^{10} Yuan and 1.253×10^{10} Yuan, and the cultivated area is respectively ,316,066.667 hm^2 and 19,393.33 hm; above 80% population, GDP and cultivated area is concentrated in middle-and-down stream.

The current total water consumption quantity in Yalong River Basin is 2.45×10^9 m^3, among which the water consumption quantity in the down stream accounts for 92.0%. The current total water consumption quantity in Dadu River Basin is 0.9×10^9 m^3, among which the water consumption quantity in the middle and down stream respectively accounts for 49.5% and 92.0%.

It's forecasted that the total water demand in Yalong River Basin in 2030 would be 3.05×10^9 m^3, among which the water demand in the middle and down stream respectively accounts for 15.7% and 80.7%, and the total water demand in Dadu River would be 1.17×10^9 m^3, among which the water demand amount in middle and down stream respectively accounts for 43.4% and 28.2%. The current and future water consumption in Yalong River and Dadu River is mainly distributed in the middle and down stream.

In terms of natural and economic characteristics in Yalong River and Dadu River, water resources sectionalization of each river basin is provide and the water resource exploitation and utilization and water demand in different sections in the up-and-down stream of dams is forecasted.

3.2 Ecological water demand of inner river

Being less effected by human being, most parts in the upstream of Yalong River and Dadu River is still in a nature state, and the river biodiversity remains relatively sound. The middle and down stream of Yalong River and Dadu River has plentiful water and numerous tributaries, while with formation of reservoir's type channels by the hydropower cascade construction, the ecologic environment background has been changed, and the ecological water usage is subject to hydropower cascade operation. Therefore, the key research on ecological water usage of the transferred river is basically keep natural conditions in the upstream.

According to the ecological features of the transferred river, the main ecological protected objects are aquatic organisms of inner river, aquatic plants and aquatic environment. The reauirments of ecological water demand and its process are different. The ecological water demand of water transferred river would be divided into two parts: one is the basic ecological water demand, and the ecological water demand amount under general discipline conditions is to be analyzed; another is the water demand of key protected objects, namely represented by fishes, and the water demand in critical period during the growth process of aquatic lives is to be analyzed. Both take the line of ectocyst as the ecological water demand in the water transfer sections of West Route Project.

3.2.1 Basic ecological water demand

Basic ecological water demand is the fundamental condition for the water flow continuity, and is the most fundamental water demand for various protected objects. We make multiple methods are take to analyze, and the basic ecological flow is: 35 m^3/s in Reba, 5 m^3/s in A' an, Renda, Zhu' anda and Huona, and 2 m^3/s in Luoruo and Keke.

3.2.2 Water demand of key protected objects

Among the protected objects in river system, fishes are the most sensitive to water changes. According to water demand in the periods like spawning, foraging, wintering and migration for typical fishes in transferred river, the ecological flow is 2 ~ 40 m^3/s in the normal period and 4 ~ 55 m^3/s in the breeding season.

In order to meet the ecological water demand by different protected objects, the ecological water demand in the water transfer sections takes the line of ecotocyst in each month, namely 4 ~ 55 m^3/s form March to June, and 2 ~ 40 m^3/s from July to February. The average annual ecological flow of inner river is: 40 m^3/s in Reba, 6 m^3/s in A' an, Renda, Zhu' anda and Huona, and 2.7 m^3/s in Luoruo and Keke.

3.3 Water transferable quantity

After fully considering the social and economical development and ecologic environment protection for water resource demand in the transferred river, the largest possible transferable water quantity from river transverse profile is water transferable quantity.

The total water demand of outer river in 2030 in Yalong river and Dadu river respectively accounts for 5.7% and 2.6% of the runoff, among which the water demand of each river transverse profile of inner-and-outer river accounts for 14.1% ~ 22.1% of water inflow; during 90% dry years, the water demand of each dam transverse profile accounts for 17.2% ~ 25.9% of water inflow. It shows that after meeting transverse profile water demand, the diversion profile of each transferred river has enough water transferable quantity. See Tab. 1.

Tab. 1 The water transferable quantity of each diversion dam in planning level year

(Unit: $\times 10^9$ m^3)

Water transferred river	Diversion dam	Nature runoff	Water demand of outer river	Water demand of inner river	Surplus water quantity	Water demand account for water inflow
Mainstream of Yalong River	Reba	6.07	0.08	1.26	4.73	22.1
Daqu	A' an	1.0	0.014	0.19	0.8	20.3
Niqu	Renda	1.15	0.017	0.19	9.4	17.9
Sequ	Luoruo	0.412	0.005	0.48	0.32	21.7
Duke River	Zhu' anda	1.15	0.014	0.19	1.24	14.1
Make River	Huona	1.11	0.011	0.19	0.91	18.0
Ake River	Keke	0.66	0.005	0.08	0.52	14.6
Total		1.179	0.146	2.19	9.46	19.8

The water demand by each section in the downstream of dam also accounts for little percentage of interval runoff, except that the water demand of Tongzijie in the downstream of Dadu River accounts for 32.2%, and the water demand of Wali in the down stream of Yalong accounts for no more than 10.7%, and other sections just accounts for no more than 3%, it shows that each section in the downstream of the dam, just interval confluence can meet the water demand of outer river. After meeting the water demand of each section of outer river, the surplus water quantity in interval rivers of each section of outer river in the down stream of Yalong River is up to 89.3% ~ 99.2% before water transfer, and the surplus water quantity in interval rivers of each section of outer river in the down stream is up to 67.8% ~ 99.5% before water transfer, so the surplus water quantity in interval rivers of each section of the two rivers can fully meet the ecological water demand of inner river.

It can be seen that the water demand of outer river in the downstream of dams, just internal runoff can meet the requirement, and no more water supply of outer river in the downstream of river by source water reservoir. The multi-year average transferable water quantity of Reba, A'an, Renda, Luoren, Zhu'anda, Huona and Keke is 9.46×10^9 m^3.

4 Technical conditions of engineering

The first stage of West Route Project is composed of seven water reservoirs and multiple water transfer tunnels, and it adopts the method of open-channel flow conveyance, through which reservoirs and water transfer tunnels are interrelated as a whole. The engineering technical conditions should be proper, indexes such as reservoir storage capacity, dam height and water transfer tunnel diameter of water reservoirs should be appropriate and feasible. Combined with the arrangement of West Route Project, the main factors that affect the engineering technical conditions are the regulation mode of water reservoirs and the design water flow guarantee rate.

4.1 Regulation mode of water reservoirs

The regulation modes of water reservoirs have 3 manners: multi-year regulation, annual regulation and no regulation. During the water transferable quantity scope, take a water reservoir with water transfer quantity of 8×10^9 m^3 for instance, analyze the different operation modes and tunnel dimensions of all water reservoirs, and see Tab. 2. From the table, it can be seen that, multi-year regulation mode has the characteristics of good reservoir storage capacity, sufficient water regulation, balanced water regulation process and small tunnel dimensions; while no-regulation mode has the characteristics of no the design reservoir storage capacity, large the design water flow, and unbalanced water regulation process. Comparing with the multi-year regulation mode, annual regulation mode has the characteristics of lessened reservoir storage capacity and increased the design water flow. The West Route Project has a long line of water transfer tunnels, and the tunnel investment accounts for 70% of the total investment, so in the view of lessening engineering scale, cutting project investment and enhancing water regulation assurance rate, source water reservoirs appropriately adopt multi-year regulation mode.

Tab. 2 Calculation results of runoff regulation by different reservoir regulation modes

Water resevoir	Dam height (m)	Dead level (m)	Multiple average water transfer quantity ($\times 10^9$ m^3)	Regulation storage($\times 10^9$ m^3)			Normal level(m)			Net water transfer tunnel diamete(m)		
				Multi-year regulation	Annual regulation	No regulation	Multi-year regulation	Annual regulation	No regulation	Multi-year regulation	Annual regulation	Ao regulation
Reba	3,527	3,660	4.2	2.25	0.178		3,707.0	3,700.1	3,660	8.4	8.7	11.8
A'an	3,604	3,640	0.7	0.49	0.32		3,718.9	3,705.7	3,640	9.0	9.3	12.5
Renda	3,598	3,635	0.75	0.415	0.34		3,702.7	3,695.8	3,635	9.4	9.8	13.2
Luoruo	3,747	3,758	0.25	0.127	0.12		3,791.1	3,790.0	3,758	9.6	9.9	13.4
Zhu'anda	3,539	3,575	1	0.459	0.42		3,634.2	3,631.1	3,575	10.2	10.5	14.1
Huona	3,544	3,570	0.75	0.398	0.32		3,632.5	3,625.1	3,570	10.6	10.9	14.7
Keke	3,474	3,510	0.35	0.139	0.14		3,559.2	3,559.4	3,510	10.8	11.1	14.9
Total			8	4.27	3.44							

4.2 Design water flow guarantee rate

The design water flow guarantee rate means the degree of the diversion water flow reaching the design water flow after reservoir regulation, which reflects the reservoir regulation capability and the utilization ratio of water transfer tunnels.

Tunnel investment occupies a large scale of the West Route Project, and properly enlarging reservoir storage capacity, strengthening conditioning capacity, raising the design water conditioning assurance rate, and decreasing tunnel dimensions is beneficial for the total project investment. But when the design water flow guarantee rate. has been raised to certain degree, the reduced investment caused by lessened water transfer tunnel dimensions will be less than the increased investment caused by dam height increase, then it is uneconomic to further improve reservoir conditioning capacity and cut the design water conditioning inflow. So the choice of the design water flow guarantee rate, both of reservoir and water transfer tunnel dimensions should be taken into consideration to have comparison of multiple plans.

In order to analyze the propionate the design water regulation assurance rate of each water reservoirs, several figures are drawn up: 50% , 60% , 75% , 80% , 90% , 95% , 99% , etc. , the relation between the design water flow guarantee rate and reservoir storage capacity and design water conditioning inflow, the relation between the design water flow guarantee rate and the total water transport project should be analyzed. See Fig. 1 and Fig. 2.

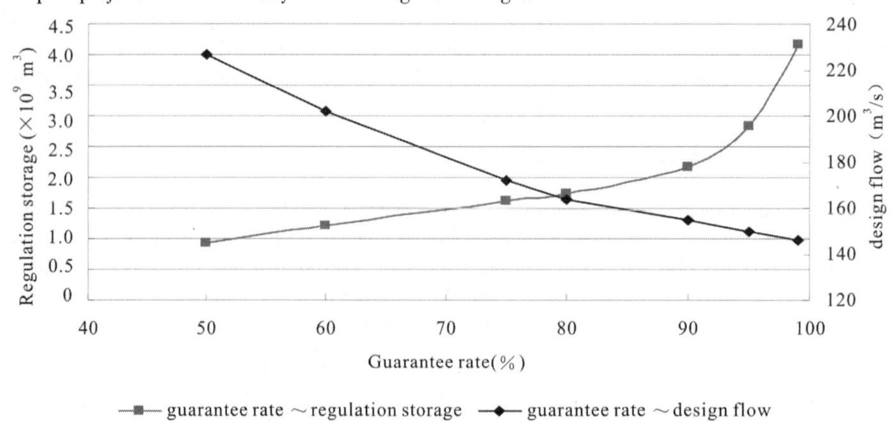

Fig. 1 The relationship between water transfer guarantee rate, regulation storage and design water retransfer tunnel flow of Reba reservoir

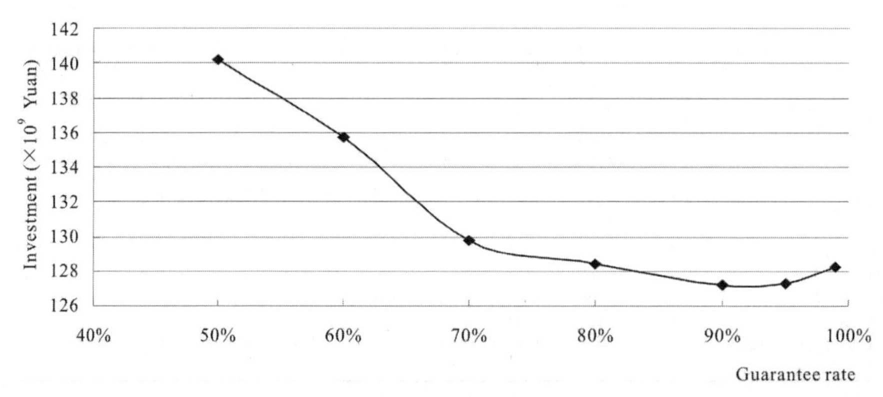

Fig. 2 The relationship between water transfer guarantee rate and engineering investment

Under fixed water conditioning flow, proper the design water flow guarantee rate can make the storage reservoir capacity and the design water flow not too excessive, namely, the reservoir scale and water transfer tunnel is both appropriate and the total water transport project scale is relatively small. According to the comprehensive analysis of the design water flow guarantee rate of seven reservoirs, the reservoirs scale and water transfer tunnels, and the change rule of project investment, appropriate water flow guarantee rate is about 90%.

According to the above analysis, under adoption of multi-year regulation mode and the water flow guarantee rate is 90% , the water reservoir storage capacity is $0.14 \times 10^9 \sim 2.25 \times 10^9$ m^3 , its accounts for 22.8% ~ 49% of multiple-years average runoff; in accordance with dam height and normal level, except Reba dam height reaching 180 m, the heights of the rest reservoir dams are all forecasted as about 100m; and the net water transfer tunnel diameter is 8.4 ~ 10.8 m. All the above engineering indexes are feasible in the current engineering technique.

5 Water reservoir inundation effect

The West Route Project of South-to-North Water Transfer is located in the Qinghai-Tibet plateau area of minority inhabit a region, the main effect objects of reservoir construction are villages and towns and religious facilities in the reservoir area. When the important submerged object within the reservoir area, it need to adjust the mode of operation mode , control the characteristic water level. Therefore, during the water transferable quantity range of each dam, different water diversion scheme should to analyze the dispatching operation mode. Take Reba reservoir as an example, during the water transferable quantity range, the water transferable quantity of $2.5 \times 10^9 \sim 4.6 \times 10^9$ m^3 between various water diversion tunnel and reservoir capacity are analyzed, see Fig. 3.

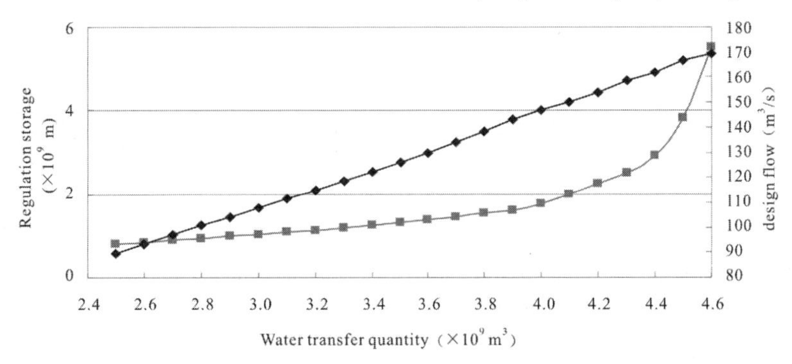

Fig. 3 The relationship between water transfer quantities, regulation storage and design water retransfer tunnel flow

From the relationship between water transfer quantity and regulation storage, with the increase of water transfer quantity, regulation storage, height increases, design water retransfer tunnel flow is increased, regulation storage and the dam height with the water transfer quantity increase is larger, the design water retransfer tunnel flow uniformly increase. When water transfer quantity is less than 4×10^9 m^3, the unit volume of water transfer quantity increases needed to increase regulation storage, dam height rarely, water transfer quantity should not be less than 4×10^9 m^3; when the water transfer quantity is more than 4.5×10^9 m^3, the unit volume of water transfer quantity increases needed to increase the regulation storage capacity, high dam are increased substantially, from the engineering economy, water transfer quantity should not more than 4.5×10^9 m^3.

From the reservoir inundation situation, Reba dam's important submerged location in the upstream is Axu Town, the elevation is about 3,710 m. The water transfer quantity 4.5×10^9 m^3, the normal level is 3,729 m, it effects Axu Town, and it impacts on the King Gesar Temple (3,735 m) traffic. The normal level is 3,710 m, can meet the water transfer quantity of 4.3×10^9 m^3, it has no effect on King Gesar temple . Therefore, considering the economy of the project and reducing the loss of submerged, Reba dam's water transfer quantity should be between $4 \times 10^9 m^3$ and 4.3×10^9 m^3.

Similarly, take the water reservoir comprehensive analysis, initially determine the Raba, A'an, Reda, Luoruo, Zhu'anda, Gongjie, Huona and Keke's water reservoir suitable water transfer quantity range, respectively, should not exceed 4.3×10^9 m^3, 0.7×10^9 m^3, 0.8×10^9 m^3, 0.25×10^9 m^3, 1×10^9 m^3, 0.75×10^9 m^3, 0.4×10^9 m^3 (Tab. 3).

Tab. 3 The water reservoir suitable water transfer quantity in 2030 (Unit: $\times 10^9$ m^3)

Water transfer river	Water reservoir	Nature runoff	Inflow	Min water discharge	Water transferable quantity	Suitable water transfer quantity range
Water transferred tiver	Reba	6.07	5.991	1.261	4.732	4 ~4.3
Mainstrean of Yalong River	A'an	11	0.987	0.189	0.797	0.65 ~0.7
Daqu	Renda	1.15	1.132	0.189	0.943	0.78 ~0.8
Niqu	Luoruo	0.412	0.406	0.084	0.322	0.2 ~0.25
Sequ	Zhu'anda	1.45	1.43	0.189	1.241	0.95 ~1
Duke River	Huona	1.11	1.097	1.89	0.909	0.7 ~0.75
Make River	Keke	0.606	0.601	0.084	0.516	0.35 ~0.4
Total		11.79	11.647	2.186	0.946	7.6 ~8.2

6 Conclusions

Considered the constraints of the water demand of in-take area, the effect on the water exporting region and engineering technology condition, the first stage of West Route Project should be adopted by way of reservoir and the tunnel joint dispatching operation, water reservoir should adopt multiple regulating operation, water transfer tunnel should adopt open channel, and design water flow guarantee rate is 90% , it can fully regulate the water quantity, and improve water transfer tunnel utilization efficiency. In the water reservoir dispatching, it should give priority to ensuring the upstream water demand in the upstream and the ecological water demand of cross-section, and the reservoir operation can not consider the production and living water supply in the downstream of dam.

Through analysis of the water transfer project operation mode under various demand constraint, the recommend operation mode with appropriate water transfer scale is about 8×10^9 m^3, is can meet about 70% of the water shortage in the Yellow River Basin. By the key reservoir regulation, can meet the water demand of the in-take area. The water transfer does not effect on the national economic water consumption, and its effect on ecological environmental is in the downstream of dam near the river section. Reservoir operation should consider maintain the certain ecological water demand of inner river, it can basically meet r ecological water demand near the rive section, and the recommend operation mode and appropriate water transfer scale can minimize the reservoir inundation area as far as possible, and appropriate scale of the water reservoir and water transfer tunnel can be achieved, under the current engineering technical level.

References

Zhang Mei, Jia Xinping, Wei Hongtao. Ecological and Environmental Water Demand in River Channels of the First Stage Water Diversion Areas from the South to North Via the Western Course [J]. Resources Science, 2005, 27(4), 180 - 184.

Zhang Mei, Zhang Wei. Research on the Major Issues in the Analysis for Transferable Water by Means of the Western-Line South-to-North Water Transfer Project [J]. Hydrology, 2002, 22 (4): 32 - 36.

Status and Future Development Direction of Water Right Transfer of Yellow River

He Hongmou, ***Yin Huijuan*** and ***Huang Kui***

Yellow River Institute of Hydraulic Research, Zhengzhou, 450003, China

Abstract: Constrained by water resources conditions, water issue has become 'bottleneck' of restricting socio – economic development of Yellow River basin. So in 2003, water right transfer was carried out, which improved water use efficiency and effectiveness. Water right transfer being implemented in Yellow River is the transfer from agricultural water to industrial water within a city. Based on water resources management in recent years of Yellow River, water resources conditions and water demand for socio – economic development, 4 kinds of water right transfer modes are proposed, including of modern agriculture water right transfer, cross – city water right transfer, water right replacement of cross – basin water diversion, water right transfer for soil and water conservation, which provide support for water resources management and socio – economic development.
Key words: agricultural water conservation, modern agriculture, water right transfer, soil and water conservation, cross – basin water diversion, Yellow River

Water shortages and highlight contradiction between supply and demand are the problem in Yellow River basin. In 1987, 'Yellow River Water Allocation Plan' was approved by State Council, which allocated 37×10^9 m^3 water to 9 provinces (autonomous regions) and Hebei province and Tianjin city, and provided the basis for the unified management and scheduling of water resources of Yellow River basin. Recent years, with the implementation of China's western development strategy, socio – economic of provinces (autonomous regions) along Yellow River basin has developed rapidly, and water use has been also a rapid increase. In accordance with the 'increase in abundance years and decrease in dry years' principle, the water consumption index for most of provinces has been close to or over water allocation index approved by the State Council in 1987, and the red line of total control of Yellow River is facing severe tests and challenges.

Shortage of water resources and enrichment of mineral resources is the significant features of Yellow River basin. Yellow River water resources are facing tremendous pressure in protecting food security, ecological security and energy security. Under the guidance of the red line of total control of '87' water allocation plan, how to coordinate water resources management and water demand for socio – economic development is the problem that urgent need to address. So in 2003, water right transfer began to implement, and to coordinate the contradiction between water shortage and socio – economic development. At present, the agricultural water – saving water right transfer has achieved remarkable success, at the same time the feasibility of other forms of water right transfer has been actively explored. The status and future development trends of various forms of water right transfer are shown in Fig. 1.

1 Status of water right transfer of Yellow River

Since April 2003, approved by Yellow River Conservancy Commission, 5 projects have been as pilots of water right transfer of Yellow River, whose capacity is 7230 MW and total additional water volume is 8.383×10^7 m^3. Corresponds to the 5 pilots, the irrigation districts to transfer water rights involve in South Bank Irrigation District of Yellow River in Inner Mongolia and Qingtongxia Irrigation District in Ningxia, whose volume of water – saving is 9.833×10^7 m^3 and The total investment of Water – saving renovation project is about 3.3×10^8 Yuan.

To the end of 2010, 37 projects of water right transfer have been examined and approved, among of which 28 were implemented in Inner Mongolia and 6 were implemented in Ningxia. The to-

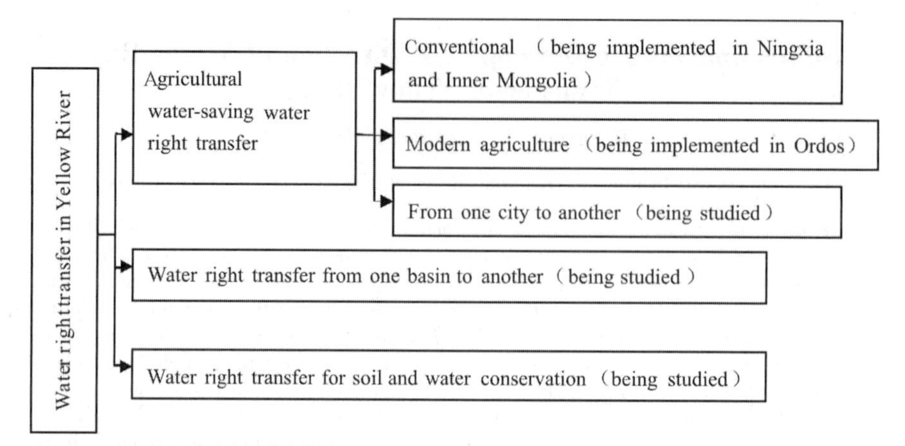

Fig. 1 Status and future development trends of water right transfer of Yellow River

tal volume of water which is converted is 3.2×10^8 m^3. The cumulative volume of water - saving of saving projects is 3.7×10^8 m^3. The total investment of projects is 23.4×10^8 Yuan, and the mean investment is 7.3 Yuan/m^3, which have increased significantly water resources utilization efficiency and effectiveness. For example, in south bank irrigation district in Inner Mongolia, after water - saving renovation project is completed, the canal water utilization coefficient would be improved from 0.348 to 0.636. In Yellow River basin of Inner Mongolia autonomous region, industrial water use efficiency is 83.8 Yuan/m^3, agricultural water use efficiency is 0.78 Yuan/m^3, and industrial water use efficiency is about 110 times higher than agricultural water use efficiency. 14 projects of water right transfer have been implemented in south bank irrigation district in Inner Mongolia, the total amount of water which is converted has been up to 1.2×10^8 m^3, and the added value of the net benefits of water use has been up to 9.3×10^8 m^3.

In April 2003, the pilots of water right transfer of Yellow River were carried out in Ningxia and Inner Mongolia Autonomous Region. In May 18, 2004, in order to guide, regulate and promote the work of water right transfer, the Ministry of Water Resources published "Guidance on Water Right Transfer of Yellow River in Inner Mongolia and Ningxia", based on which "Yellow River Water Right Transfer Management(For Trial Implementation)" and "Water - saving Projects Verification of Yellow River Water right transfer(For Trial Implementation)" is formulated by YRCC. In 2009, "Yellow River Water Right Transfer Management (For Trial Implementation)" is amended and "Yellow River Water right transfer Management" is published, which further standardized behavior of water right transfer in Yellow River.

In process of practice and explore of water right transfer in Ningxia and Inner Mongolia, the concept of the Yellow River water right is defined, and the management system, technical system and monitoring system of water right transfer with the characteristics of Yellow River is formatted initially, all of which further enrich theoretical system of water right and water market in China.

2 Study on the direction of future development

Because of the restricts of technology conditions as well as the complexity of water right transfer, in recent years water right transfer in Yellow River focus on the transfer from agricultural water to industrial water use, the quantity of convertible water is what saved by water - saving facilities in irrigation district, and water right transfer is in progress mainly within a province (area). With the mature of the technology, management and monitoring development, the water right transfer mode would enrich ceaselessly, and the transfer range would expand unceasingly. So based on the characteristic and the problem of water right transfer in Yellow River, the future development direction of Yellow River water right transfer should be centered on the following kinds.

2.1 Agriculture water – saving water right transfer

2.1.1 Modern agriculture water – saving water right transfer

That modern agriculture water – saving water right transfer is to take the water right, the water market theoretical as the guidance, and depend on high – effect water – saving modern agriculture, to promote the adjustment of water utilization structure by the redress of economy structure. In the field of industrial water use, follow the principle of "high starting point , high technology, high beneficial result, height , further, high added – value, altitude energy conservation environmental protection", strengthen industrial water administration, develop circulating economy, adopt new methods for industrialization improve water resource efficacy and economic effect. Apply advanced water – saving irrigation technique to agriculture water use, combining with development high grade high – effect modern agriculture , adopt the adjusting of the agriculture and rural Industrial Structure to improve water resource utilization ratio and beneficial, and realize agriculture facilities modernization. Design the mode that "government leads to gear to the demand of the market, explore points of economic growth + faucet enterprise + the modernized pattern of farming + agricultural infrastructure + that enterprise adapts to each other with the faucet has the pattern definitely cultivating a technology's , having the employee dignity peasant team's in hand", form the production mode of the enterprise base adding peasant household, ultimately realize the purpose of market – oriented , putting to use in production , scale – rization and intension, and achieve effective utilization of water resources.

At present modern agriculture water – saving water right transfer already begins to be put into effect in south bank irrigation district in Ordos City in Inner Mongolia. In 2009 YRCC has shifted an official reply to < the plan of water right transfer and modern agriculture high – effect water – saving project of Yellow River Irrigation Area in Ordos City > , the total investment is 1.422×10^9 Yuan, the water volume transferred is 9.96×10^7 m^3.

Modern agriculture water – saving water right transfer already has begun to be put into effect, bur only at pilot stage at present. and needs to further discuss the problems such as the stability of water flow, time limit, the impact on soil environment of high – effect water – saving and so on.

2.1.2 Cross – city agriculture water – saving water right transfer

According to the complexity of cross – regional water right transfer and the restriction of the clarity degree of water right, in recent years water right transfer in Yellow River Basin is in progress mainly within a province (area), and do not stride over a field (city, alliance), which has brought about certain problems although having got certain achievement, such as the unbalance of regional development. Take the Inner Mongolia Autonomous Region as example, because of the enrich mineral, Ordos City has been a new developing industrial park, many industries projects which around resource development, resource transformation, deep processing demands to start, and economy displays forceful development tendency. But there are certain water – saving amounts to transfer only in south bank irrigation district in Ordos City Inner Mongolia. According measures and calculates, the south bank irrigation district in Ordos City could save water volume of 1.68×10^8 m^3 through project measure, and water volume of 1.24×10^8 m^3 already has been transferred at the end of 2009. But still magnanimous industrial projects have no way to start because of lack of water. In the other hand, the water – saving potential by project measure in Hetao Irrigation Area of Inner Mongolia is 1.036×10^9 m^3, water – saving potentials are enormous, Because of the restriction of districts, there are no industrial enterprise to invest water – saving reform and water resource use efficiency is lower. One aspect is that the Ordos City industrial enterprise can not starts because of water resource restraint, another aspect is Hetao Irrigation Area of Inner Mongolia water resources waste is serious, and water right transfer was restricted seriously by the jurisdiction of administrative region, mainly embody the imbalances between the water saving potential region and the water demanding area. So in order to break administrative divisions restricting, it is very necessary to explore cross – city water right transfer.

2.2 Water right transfer for soil and water conservation

2.2.1 Basic idea

For 50 years of harnessing Yellow River, large – scale harness to water and soil erosion has been carried out in the loess plateau. Many soil and water conservation measures have been made, such as repairing dam, constructing terraces, planting trees and grass and so on. According to these measures, a number of remarkable achievements have been obtained, such as decreasing annual 3×10^9 t soil to Yellow River, slowing down the uplift of the lower Yellow River. But, at present, the speed and scale of harness soil erosion in the Loess Plateau area is much slower than the urgent needs of Yellow River management. The main reason is not engineering problems, but the scale and amount of investment.

Soil and water conservation construction projects are public projects, investment is get only from government, and lack the innovation of investment and financing system, and so it is difficult to effectively mobilize the enthusiasm of community investment in water and soil conservation. Therefore, we should comply with the requirements of the scientific development concept, solve the bottleneck problem of investment in harness soil erosion as soon as possible, and accelerate the speed and scale of harnessing soil erosion.

Yellow River is a sediment – laden river, and the lower Yellow River is suspended river. So in order to harness Yellow River and control flood, it is necessary to leave certain amount of water to put sediment into sea. It is provided in Yellow River water allocation plan approved by State Council in 1987 that "annual average into sea from Lijin station should keep 210×10^9 m^3." In 2006, Yellow River Water Dispatch Regulation was promulgated by the State council, which further clarify that 21×10^9 m^3 water include 15×10^9 m^3 water consumption for sediment transport in flood season, 5×10^9 m^3 ecological water consumption in non – flood season and 10×10^9 m^3 natural losses." One of the Yellow River outstanding problems is that there are too much sediment, so, it is possible in theory to displacement some water consumption for sediment transport of downstream under the prerequisite that the amount of sediment transported to downstream is reduced by taking effective measures.

Drawing lessons from water right transfer being implemented in Yellow River Basin, the general idea of water right transfer for soil and water conservation is proposed. In the government's guidance, industrial enterprise invest to construct water and soil conservation projects in the loess plateau so as to achieve the goal of reducing sediment to Yellow River, and some water consumption for sediment transport is displaced to the industrial enterprise. With water right transfer for soil and water conservation, a multiple financing channel is built, and the purpose is realized that industrial enterprise invest to construct some public welfare projects such as soil and water conservation project. The system of water right transfer for soil and water conservation is the supplement and improvement to the current investment and financing system of harness soil and water conservation.

2.2.2 Possible modes

Through the implementation of water right transfer for soil and water conservation, effective use of water resources should be achieved under the premise of effectively intercept the coarse sediment. There are 4 available modes, including of the mode of scattered collection and unified water supply, the mode of scattered release and getting water from the main channel in different place, the mode of scattered release and replaced water amount for sediment delivery in the downstream of Yellow River and so on.

(1) The mode of scattered collection and unified water supply.

Through constructing large scales of dams, the sediment and water resources are intercept in gully tributaries and don't enter the Yellow River mainstream. Then choose the appropriate place to built storage project, send water resources intercept by dams to storage project, and supply water to industrial enterprises.

(2) The mode of scattered release, getting water from the main channel in different place.

The model, which is invested by industrial enterprise, is applied to the tributaries channel of the loess plateau for the purpose of avoiding soil erosion. All of the sediment in gully tributaries is intercepted by some engineering measures, and the clear water flows out of the ditch and gets into the mainstream of Yellow River without sediment. Then the clear water is extracted from the suitable place and used by the industrial enterprise to realize the water resource utilization.

The replaced object of the model is the water production in tributaries. Especially, the water consumed for soil and water conservation should be considered before the determination of the replaced water. Then the produced water of tributaries will be used to by the industrial enterprise.

(3) The mode of scattered release, replaced water amount for sediment delivery in the downstream of Yellow River.

The model, which is invested by industrial enterprise, is applied to the tributaries channel of the loess plateau for the purpose of avoiding soil erosion. All of the sediment in gully tributaries is intercepted by some engineering measures to reduce the sediment getting into Yellow River. As a result, part of the water amount for sediment delivery, which is replaced, will be divested to industrial enterprise, while the clear water flows out of the ditch and gets into the mainstream of Yellow River.

The replaced object of the model is the water amount for sediment delivery in the downstream of Yellow River. It is suggested that the determination of replaced water should be according to prevent river channel sedimentation through intercepting coarse sand, while reduction of water amount for sediment delivery in the downstream. The relationship between the water amount for sediment delivery in the downstream and the intercepted amount of coarse sand should be quantified, and then the saving water amount for sediment delivery in the downstream is estimated, which will be used as replaced water amount. The practical replaced water amount is the replaced water amount without the consumption in the tributaries.

There are some technological and policy problems, which need to be researched, in the models above. The main technological problem is the relationship between the intercepted amount of coarse sand and the water amount for sediment delivery in the downstream, while the key policy question is a certain operational management system, which should be workable, need to be established.

2.3 Water right replacement of cross - basin water diversion

Cross - basin water diversion is an effective way to solve the shortage of water resources of the Yellow River basin. There is an important practical significance to guide cross - basin water diversion project construction and management of water resources using the water rights and water market theory.

At present, the west - line project of South - to - North Water Diversion and water diversion from the Han to the Weihe River have good previous work in cross - basin transfer project. In which the water diversion from the Han to the Weihe River has being constructed, and plans to diverse water volume of 1×10^8 m^3 into Weihe River in 2020.

Using water right and market theory can solve the problem about investment and financing of main body of the project effectively. Meanwhile, it can realize water resources efficient allocation by market means.

As water diversion from the Han to the Weihe River for example, according to the "Yellow River Basin Integrated Water Resources Plan" and the GDP target of 2×10^{12} RMB of Shaanxi province in 2020, water requirement will be up to 1.1×10^{10} m^3 from Yellow River. Combined with distribution index of Yellow River, the existing and planning water conservancy project , water supply will be 8.5×10^9 m^3/a at Shaanxi province in 2020 , the total water shortage around 3.5×10^9 m^3 , The gap between water supply and demand is serious. The following measures should be applied to relieve water shortages.

Firstly, it must be taken super strong water - saving measures to inhibiting the growth of water demand. For increasing the water - saving strength according to supernormal water - saving model, we can save industrial water volume of 5×10^8 m^3.

Secondly, it must be speed up the cross - basin diversion project to improve bearing capacity of regional water resources. Base on Shaanxi province integrated water resources plan, projects of South – to – North Water Diversion in Shaanxi province would divert water resources volume of 1.15×10^9 m^3 in 2020. Adding the water – saving displacement in southern Shanxi, water resources volume could be reach 1.4×10^9 m^3.

Thirdly, diverting water from the Yellow River mainstream combined with water index to support social and economic development of energy industry base in northern shaanxi. Energy industry base in northern shaanxi is the energy bases of china in the 21 century, where have rich mineral resources, but lack of water resources and fragile ecological environment. Based on the overall plan, it will be a high – speed development before 2020, water demand will be growing rapidly and the local water resources can't meet it. We think FuGu, YuShen, YuHeng industrial zone will have $8 \times 10^8 \sim 1 \times 10^9$ m^3 water shortage.

Fourthly, reduce agricultural water consumption. Extraction the agricultural water around 5×10^8 m^3 by various measures, which could be configured to industry by the water right transfer with appropriate engineering adjustment.

Through the above measures we can meet the shaanxi province water demand volume about 3.3×10^8 m^3, in which there are water volume of 5×10^8 m^3 from water saving, 1.4×10^9 m^3 from water diversion, 9×10^8 m^3 from water divert and 5×10^8 m^3 from water replacement.

From the overall assumption of water resources allocation of shaanxi province, the key project to realize the economic social development and basic coordination of water resources in shaanxi is water diversion from the Han to the Weihe River, the difficulty of the water resources allocation is in energy industry base in northern shaanxi. Before the operation of South – to – North Water Diversion and under the restriction of the Yellow River water supply capacity index, there is one of the effective ways to solve the problem of using water in energy industry base in northern shaanxi that is water rights replacement of water diversion from the Han to the Weihe River. Which means leaving some water into the Wei river to meet the water demand of ecological environment in the lower Weihe River and replacing some water from Yellow River to solve water requirements of energy industry base in northern shaanxi.

How and what to replacement is the key question of water right replacement of cross – basin water diversion. Many problems should be researched, include different of enter water and import water, apply allocation between import water and Yellow River raw water, and adjust the water resources regulation scheme current order for the water river. Building management system of water right replacement system and monitoring and evaluation system based on the water right allocation research.

3 Conclusions and suggestions

Water right transfer is an effective way to improve water use efficiency and effectiveness, and to solve the contradiction between water supply and demand. In process of practice and explore of water right transfer in Ningxia and Inner Mongolia, the management system, technical system and monitoring system of water right transfer is formatted initially, all of which ensure the implementation of Yellow River water right transfer. Based on water resources management in recent years of Yellow River, 4 kinds of water right transfer modes are proposed, including of modern agriculture water right transfer, cross – city water right transfer, water right replacement of cross – basin water diversion, water right transfer for soil and water conservation. To actively explore the future direction of Yellow River water right transfer, it is recommended that the following preliminary studies should be carried out.

(1) Studying on the technical aspects. Focus on studying time limit of water right transfer, the fee structure and cost calculation method, the quantitative relationship between sediment reduction and water which is converted, the compensation mechanism for stakeholders, the impact on ecological environment and so on, all of which would provide basis for management decisions.

(2) Studying on management policies. Learning from the experience of agricultural water – saving water right transfer, to design management policies with strong operability, and ensure the implementation of water right transfer.

(3) Studying on monitoring and evaluation system. Accurate water metering monitoring is the basis to ensure stakeholders' interests and to manage water resources. We should combine the different modes of water right transfer, propose the monitoring program involving the main information of water right transfer, and carry out periodic assessements based on monitoring results.

References

YRCC of MWR. System Construction and Practice of Water Right Transfer of Yellow River[M]. Zhengzhou: Yellow River Conservancy Press, 2008.

Assessment and Improvement to Water Right Transfer of Yellow River[R]. Zhengzhou: Yellow River Institute of Hydraulic Research, 2011.

Li Guoying. Explore and Practice of Water Right Transfer of Yellow River[J]. China Water Resources, 2007(19):30 - 31.

Water Right Transfer and Efficient Water - Saving of Modern Agriculture in Irrigated Areas in Ordos City[R]. Huhehaote: Inner Mongolia Water Resources and Hydropower Survey and Design Institute, 2009.

Water Right Transfer Planning Report of Inner Mongolia Autonomous Region[R]. Huhehaote: Inner Mongolia Water Resources and Hydropower Survey and Design Institute, 2005.

Study on Efficient Long – Term Water Saving Mechanism in Irrigation Region of the Lower Yellow River

Yang Jianshun[1], ***Zhang Ying***[2], ***Zhang Kai***[2] and ***Han Tao***[1]

1. The Administration Office of Yellow River Conservancy Commission, Zhengzhou, 450003, China
2. The Yellow River Basin Water Resources Protection Bureau, Zhengzhou, 450004, China

Abstract: Since early 1950s, people began to irrigate using the Yellow river water in the lower River regions which include 23 cities of Henan and Shandong Provinces. Nowadays, the Yellow River water has been used for irrigation, industry and city life. The contradiction between water supply and demand in the lower Yellow River is becoming sharper and sharper. The reasons include two parts which are the unreasonable proportion and lower efficiency of water consuming in the lower Yellow River Basin. Therefore, it is urgent to define reasonable water demand of irrigation and make series of laws and regulations which are aimed at establishing long – term water saving mechanism for irrigation regions of the lower Yellow River. Aiming at water supply of irrigation region in the lower Yellow River, the article analyses current status and the key issues. Then the article presents some mechanism and guideline to save water permanently in irrigation region in the lower Yellow River.

Key words: the Yellow River, irrigation, mechanism, water saving

Since early 1950s, people began to irrigate using the Yellow River water in the lower river regions which include 23 cities of Henan and Shandong Provinces. Nowadays, the Yellow River water has been used for irrigation, industry and city life. The contradiction between water supply and demand in the lower Yellow River is becoming sharper and sharper.

The reasons include two parts which are the unreasonable proportion and lower efficiency of water consuming in the lower Yellow River Basin. Therefore, it is urgent to define reasonable water demand of irrigation and make series of laws and regulations which are aimed at establishing long – term water saving mechanism for irrigation regions of the lower Yellow River.

The article analyses current status and the key issues, while the article presents some mechanism and guideline to save water permanently in irrigation region of the lower Yellow River.

1 The current status of irrigation in the lower Yellow River

Nowadays, water consuming of the lower Yellow River bases on laws and regulations, such as *the Yellow River Water Supply Distribution Plan*, *the Regulations on Allocation of the Yellow River*, *the Allocation Plan of water supply and main stream of the Yellow River*, *the Management Method on Allocation of the Yellow River*. These laws and regulations defined that nation have the duty to allocate the Yellow River water. The principles include control in water quantity and water discharge control at a section, level – to – level administration and level – to – level responsibility.

In 1987, the State council approved *the Yellow River Water Supply Distribution Plan*, which was to allocate the remaining 3.7×10^{10} m^3 of water to nine provinces or regions along the Yellow River after subtracting water for flushing sediment and ecological purposes from the total 5.8×10^{10} m^3. The quantity allocation water of the lower Yellow River provinces is 1.254×10^{10} m^3, which includes 5.54×10^9 m^3 for Henan Province and 7×10^9 m^3 for Shandong Province.

In December 1998, *the Allocation Plan of water supply and main stream of the Yellow River and the Management Method on Allocation of the Yellow River* were approved by the State Council and were issued by the State Development Planning Commission and the Ministry of Water Resources. The regulations authorized Yellow River Conservancy Commission (YRCC) to control the water resources of the Yellow River. Since 1999, the water resources of the Yellow River had been brought under unified management.

The Yellow River water was main water resources flowing from outside region for Henan and Shandong Provinces which are consumed by 23 cities and 96 counties. There are 84 irrigation regions using the Yellow River water resources and there are 39 irrigation regions with more than 0.3×10^6 mu (1 mu = 0.0667 hm^2). The design irrigation area is 5.8×10^7 mu and the actual area is 4.465×10^7 mu which include 3.373×10^7 mu in Shandong Province occupying 49% irrigation area by surface water. The irrigation regions consuming the Yellow River water have become main food production region in Henan and Shandong Provinces being more and more significant in irrigation, city, oil field, manufacture, locals living and ecological water supply.

Fig. 1 Sketch map of water supply from the lower Yellow River

The lower Yellow River water used for irrigation occupied 80% of total water supply, while the residual part is used for industry and city living. During 2006 to 2010, there was approximately 8.53×10^9 m^3 of water in the lower Yellow River. 7.23×10^9 m^3 of water occupying 84.8% of total water supply was consumed in irrigation. The average number of water supply in Henan Province is 2.21×10^9 m^3. The quantity of water used in irrigation and the others field are 1.91×10^9 m^3 (86.43% of total water supply) and 0.3×10^9 m^3 (13.57% of total water supply). The average number of water supply in Shandong Province is 6.32×10^9 m^3, while the quantity of water used in irrigation and the other fields are 5.32×10^9 m^3 (84.2% of total water supply) and 1×10^9 m^3 (15.8% of total water supply).

2 The challenges of the irrigation in the lower Yellow River

There are several challenges of the Irrigation in the lower Yellow River, such as the low efficiency of water consuming, the low water price, the low level of technical application and the low awareness of water saving etc. All these challenges limit the development of water saving in agriculture.

2.1 The low efficiency of water consuming

The majority of irrigation structures in the lower Yellow River have being worked for more than 20 years. The general standards of these works and matching construction are low. Most irrigation regions are gravity irrigation area, where commonly used the broad irrigation and string irrigation, and is rugged land which makes water waste severely. The average water efficiency is less than 0.45 which leads to the low irrigation ability and low economic benefit.

2.2 The low water price

The present water price of the irrigation in the lower Yellow River is composed of the water price of head works, the water price of the irrigation administration department and the water price of local management organizations.

The water price of head works is 0.012 CNY/m^3 from April to June and is 0.012 CNY/m^3 in other months, which was authorized by the State Development Planning Commission in 2000.

The water price of the irrigation administration department and local management organizations are different for irrigation regions because of different matching conditions, different operation and maintenance cost and different planting structure. The terminal water price of the irrigation is 0.036 ~ 0.085 CNY/m^3 in Henan Province and 0.081 ~ 0.137 CNY/m^3 in Shandong Province (Zhang Xia, Cheng Xianguo and Hu Yawei, 2007). For example, the water price of branch canal was 0.096 CNY/m^3 in 2000 in Weishan irrigation region, while the cost was already 0.012 CNY/m^3 in 1998. According to the investigation in Dongchangfu district and Gaotang County, the water cost accounted for 3% ~ 6% of the income of agriculture and 10% to 20 of the cost of agriculture, which were all on the low side (Hang Jun, Jiang Haibo and Liu Xinbing, 2001).

2.3 The low level of technical application

There are short of resources in the lower Yellow River and there were flooding in most counties frequently in history. The local economy mainly relies on agriculture which is just so so. In these regions agriculture is in a weak position rare of capital, technology and labor. Especially, the technical application level of water saving is low. Local farmers would not pay more for the advanced irrigation technology because of the low water price and the expensive reform and maintenance cost of irrigation technology which is approximately several hundred CNY per mu.

2.4 The low awareness of water saving

The water saving is a social operation which not only requires a series of engineering measures, technical measures and price lever but also requires the initiative of the consumers in irrigation regions. At present, the water consumers are rare of water saving awareness in the lower Yellow River due to the low developing level of irrigation regions, the excellent flow irrigation condition, the low price and the traditional irrigation method. The majority of farmers consider that is the government's duty to save water and has nothing to do with them.

3 The fundamental contents of permanently mechanism of irrigation water saving in the lower Yellow River

Nowadays, the irrigation in the lower Yellow River Basin is extensive form on account of no lining canal system. More than 30% of water was leaked into the soil during the transport through the canal system. It will benefits from establishing the efficient long – term water saving mechanism to increase the efficiency of water usage. It is necessary to formulate reasonable plan, establish the mechanism of multi – channel investment, enhance the administration management of agriculture water saving, perfect technology system and accelerate legislation development.

3.1 To formulate reasonable plan of irrigation water saving

The government of Henan and Shandong Provinces should formulate a medium and long – term plan to save water and bring it into the plan of national economic and social development cooperating with local agriculture development plan, water resource plan and ecological environment development plan.

The local government should identify the proper development pattern for every region basing on the nature condition, the mode of agriculture production and operation, crop species and the level of economic development. Since 1982, the household – contract responsibility system has been implemented for 30 years which means the rural land has been hold by individual families. The reform emphasis of government is to increase the scale of agriculture units and encourage farmer to save water resources integrating the adjustment of planting structure and urbanization.

3.2 To establish the mechanism of multi – channel investment

The government of Henan and Shandong Provinces should establish a special fund for agricultural water saving. The capital of the fund can originate from the structures fund of governments, the financial agriculture fund, small agriculture operating expense, part of water conservancy constructing fund, the water resources fee, the agriculture comprehensive development fund, poverty relief fund and the compensation fund of occupying agriculture water resources. The fund should be used to build agricultural structures, demonstrate overall water saving, develop the water saving technology and service and train farmers.

3.3 To enhance the administration management of agricultural water saving

It is time to establish agricultural water and water saving management system. Agricultural water should be controlled in rules of the total amount and quota. Basing on the overall consideration to life, the production, the ecological water use, the regional total amount of agricultural water and irrigation water quota should be definite as the foundation of agricultural water management.

To reform and perfect the management and assessment system of irrigation regions in the lower Yellow River should be executed according to indexes such as irrigation land area, the irrigation quota of water saving, the maintenance management of works etc. The local government should explore the new method of investment and management for irrigation regions in order to increase income not mainly relying on selling water resources.

The local government should establish declare, examination and approval system for large scale water saving works setting up works management commission for investors and workers. The foundational systems include bidding system, engineering construction supervisor system, financial auditing system and lifelong responsibility system of engineering quality.

3.4 To perfect technology system

The local government of the lower Yellow River should invest more to technology of agriculture water saving, and support to transform more scientific and technological achievement of agriculture water saving. The approaches include encouraging the cooperation between the universities and enterprises, encouraging the cooperation between the research institutes and enterprises, jointly developing new technology and new products, accelerating the transform of the scientific and technological achievements and increasing the technology ability of enterprises.

It also needs to formulate the industrialization development plan of water saving irrigation equipment, define clear objects and emphasis, choose leading enterprises and prompt scale and standard production.

To formulate the technical proposal of water saving in irrigation regions of the lower Yellow River is another method. It needs to establish the index system of agricultural water saving and irrigation quota basing on different water and soil resources, simultaneously to integrate agricultural water saving technologies and to demonstrate irrigation technologies. The modern technologies of water saving, such as sprinkling irrigation and micro – irrigation, need to be popularized in zones which are prosperous regions, commercial crop growing regions and intensive agricultural regions.

It is urgent to establish the technology transfer system of agricultural water saving. That means to reform the current technology transfer system, to encourage farmers and enterprises to take part in the technology transferring, to form a new system with initiative of government and market.

3.5 To accelerate legislation development of water saving in agriculture

Henan and Shandong Provinces need to formulate the Agricultural Water Saving Regulation immediately. It needs to define the principles and demands legally, define the responsibility and duty of government and farmers, define the property and legal status of big and medium size irrigation regions and define the formulating principles of water price.

It needs to establish the technical standard system and assessing index system of agricultural water saving in order to realize the scientific and institutionalization of engineering construction and management of agricultural water saving.

4 Conclusions

The contradiction between water quantity and demand of the lower Yellow River is becoming more and more serious. The consuming agricultural water occupies 84.8% of total water resources of the lower Yellow River. There are several challenges of the irrigation in the lower Yellow River, such as the low efficiency of water consuming, the low water price, the low level of technical application, the low awareness of water saving et al. Therefore, It is necessary to formulate reasonable plan, establish the mechanism of multi - channel investment, enhance the administration management of agriculture water saving, perfect technology system and accelerate legislation development.

References

Zhang Xia, Cheng Xianguo, Hu Yawei. The Analysis of Terminal Water Price of Farmer in the Lower Yellow River[J]. Water Science and Economy,2007(11):803 - 806.

Hang Jun, Jiang Haibo, Liu Xinbing. The Reform Discussion of Water Price of Weishan Irrigation District[J]. Water Economy,2001(6):40 - 46.

On Integrated Flood Control Management of the Yellow River Basin

Zhang Xinghong, *Zheng Shixun*, *Tan Shengbing* and **Guo Guili**

1. Flood Control Office, YRCC, Zhengzhou, 450003, China
2. Yellow River Sanxi Bureau, Yuncheng, 044000, China

Abstract: In history, the Yellow River has arisesed frequent floods and caused severe calamities, so rulers of past dynasties have regarded it as China's mortal malady. Since the foundation of New China in 1949, thanks to the particular importance attached to the Yellow River by the Party and the State, the career of the Yellow River harnessing, development and management has made great progress, and the Yellow River flood control management system has also kept being adjusted and improved. Integrated Yellow River flood control management is a complex and huge systematic engineering, which not only has to solve the problem of flood control and disaster reduction, but also has to coordinate the interests of provinces (autonomous regions), the power sector and reservoir management unites within the basin, and consider multiple objectives of water supply, power generation, silting reduction and ecological water consumption. This paper describes the evolution of Yellow River flood control management system, the present status of, and the tremendous achievements in, the integrated the Yellow River flood control management, points out the main problems existing in the present flood control management system, and proposes countermeasures to further strengthen integrated the Yellow River Basin flood control management.

Key words: the Yellow River, basin, integrated management, flood control

The Yellow River is the second largest river in China, with vast basin area and complex river situation. According to the time of occurrence, the Yellow River Basin flood is divided into four types: summer flood, autumn flood, ice run and spring flood. In terms of cause of formation, the Yellow River Basin flood is divided into rainstorm flood and ice flood. The rainstorm flood that occurs in July and August is called summer flood, and that in September and October is called autumn flood, both collectively known as the large summer – autumn flood. The flood occurring in winter is called ice flood, and that in spring is known as spring flood. Customarily, both are collectively called ice flood. In a nutshell, the Yellow River flood control refers to the corresponding work carried out to defend against the Yellow River rainstorm flood and ice flood, and prevent or alleviate disaster losses.

In history, the Yellow River caused disasters frequently with rainstorm flood, and on average, it "burst twice in three years and changed its course once in one hundred years". Records that "flood spread wide and the dead drifted everywhere" never ended in books. The flood spread to Mengjin in the west, Tianjin in the north and Jianghuai in the south, and the floodplain involved the area of 2.5×10^5 km^2 in five provinces of Hebei, Shandong, Henan, Anhui and Jiangsu on the Yellow River Plain, the Huaihe River Plain and the Haihe River Plain. Each burst or course change would bring about huge losses and severe disasters to people's lives and properties, and cause severe damage to, and long – term adverse effect on, the ecological environment. The ice floods of the Yellow River evolve in a very complex way, and they change rapidly, so they are difficult to forecast and defense. They often flooded the floodplains and villages, and even burst to cause disasters. In history there were saying that "the summer flood is easy to deal with, but the ice flood is difficult to defend against", and that "river officials should not be accused of ice flood burst". The safety of the Yellow River concerns the overall situation. Therefore, the Yellow River is often known as the "Chinese nation's mortal malady".

The Yellow River flows through nine provinces (autonomous regions), across three large stairs of the Qinghai – Tibet Plateau, the Loess Plateau and the North China Plain. The upper and lower reaches, the river itself and its tributaries, and the left and the right banks of the river are closed related. Rulers of past dynasties all attached great importance to the Yellow River safety. Since the founding of New China in 1949, the Party and the State have attached particular importance to the Yellow River flood control, and thus great progress has been made in the cause of Yellow River harnessing, development and management. To be adapted to the requirement of Yellow River flood control management, the Yellow River Basin flood control system has gone through a process of continuous adjustment and improvement.

1 Evolution of the Yellow River Basin flood control management system

Before the founding of New China in 1949, a number of organizations were set up by rulers of past dynasties for engineering defence of the lower Yellow River, but most of them were mainly responsible for overall management of the lower Yellow River, and each was responsible for a part of the river individually. During the Republic of China, though Yellow River Conservancy Commission (YRCC) was nominally established, it was only charged with the lower Yellow River, and was led by provinces Henan, Hebei and Shandong respectively. Due to separatist warlord regimes and turbulent situation, no real coordinated management of the whole river could be achieved. This kind of management system that could not coordinate control of the upper and lower reaches constituted the cause of dyke disrepair, infrastructure weakness and frequent flood occurrence, and it greatly aggravated Yellow River flooding. Since 1946 when the people took over the Yellow River control, the Party and the State have attached great importance to the harnessing, development and management of the Yellow River, and the Yellow River flood control management system has gone through three major reforms, which reflected the development direction of the Yellow River management system, namely expanding gradually from the lower reaches to the whole basin, from flood control management alone to overall management of flood control and drought relief.

1.1 The first major reform—lower Yellow River flood control management system established at the early stage after the founding of New China in 1949

At the early stage after the founding of New China, the task of the Yellow River control was arduous, and all aspects of basic work were very weak. In this situation, establishing the Yellow River management system and improving river control organizations and teams were urgent for the cause of the Yellow River control. Viewing the river as a whole, the Party and the State established the Yellow River management system. On January 25th, 1950, the Ministry of Water Resources (MWR) forwarded the Administration Council's No. 1 Decree, which formally defined the institutional nature and management limits of Yellow River Conservancy Commission (YRCC), and decided to change YRCC from a joint river control organization into a basin administrative organization, which shall coordinate water business of the whole river. All the Yellow River control organizations of the three provinces Shandong, and Henan shall be under the direct leadership of YRCC and shall still be guided by the people's governments of their respective provinces. In order to strengthen organizational leadership of the Yellow River flood control and to improve flood control organizations at all levels, on June 6, 1950, the Central People's Government Administrative Council issued the "Decision on Establishment of Flood Control Organizations at All Levels" (No. [50]709), which gave clear instructions: flood control of the upper Yellow River shall be taken charge of by the provinces the river flows through, and for the three provinces Shandong and Henan on the lower Yellow River Flood Control Headquarters (YRFCH) shall be set up, which shall be under the leadership of the Central Flood Control Headquarters, and whose director and vice-directors shall be taken up by the leaders of the People's Governments of provinces Shandong, and Henan and the director of YRCC. The office of the Yellow River Flood Control Headquarters shall be based in YRCC and shall take charge of routine work. Each of the three provinces shall set up the Yellow River flood control headquarters of its own, whose director shall be taken up by the chairman or vice-chairman of the People's Government of the province, and whose two vice-directors shall be acted as by the representative of the provincial military command and the director of

Yellow River Bureau of the province, and which shall be under the leadership of YRFCH. On June 26, 1950, the establishment ceremony of YRFCH was held in Kaifeng, Henan Province. Before the flood period that year, the provinces Henan and Shandong set up their provincial flood control headquarters (or flood control and drought relief headquarters), the prefectures (cities), counties (districts) along the Yellow River also set up corresponding flood control directing organizations, and Yellow River Flood Control Offices were set up in all Yellow River prefecture agencies and county agencies to take charge of flood control of the river sections under their jurisdiction. The Yellow River flood control leading groups were set up in production teams and villages. At the early stage after the founding of New China in 1949, the four - level Yellow River management system comprised of YRFCH, provincial, prefecture (city), and county (district) agencies was formed, and the basic management structure of the Yellow River was established initially.

1.2 The second major reform—middle and lower Yellow River flood control management system was established

In 1960, the Sanmenxia Reservoir was put into operation, which linked the flood control of the upper and lower reaches of the River. In 1961, the Central Committee called the persons in charge of provinces Shanxi, Shaanxi, Henan, Shandong and Hebei to study the problem of the Yellow River flood control. In 1962, the State Council decided that the director and vice - directors of YRFCH shall be acted as persons in charge of the four provinces Shanxi, Shaanxi, Henan and Shandong and the director of YRCC, with office place based in YRCC. The prefectures and counties concerned in the Sanmenxia Reservoir area also set up corresponding flood control organizations, which shall take charge of flood control of the reservoir area, the lower Weihe River and the Xiaobeiganliu (namely the reaches between Yumenkou and Tongguan) of the Yellow River.

In 1983, in view of the fact that the persons in charge of provinces concerned and YRCC changed now and then, and to prevent flood control work from being interrupted, the Ministry of Water Resources and Electric Power reported a proposal to the State Council for approval that the general director of YRFCH shall be acted as by the governor of Henan Province, and the vice - general - directors shall be taken up by vice - governors in charge of agriculture of provinces Shandong, Shanxi and Shaanxi; and that, from then on, those posts shall be no longer reported to the State Council for appointment every year due to change of leading members, but shall be taken over naturally by the above - mentioned leading comrades. On June 14, 1983, the State Council pointed out in the No. [83]124 Notice that, to ensure continuity of flood control work, leaders of the Yellow River Flood Control Headquarters shall be no longer appointed year after year due to personnel changes, but shall be acted as by the comrades taking over corresponding jobs, and shall be reported to the State Council and the State Flood Control Headquarters for record. In June 1996, the vice - governor of Henan Province in charge of agriculture and the deputy chief of staff of Ji'nan Military Command were added as vice - general - direcotors of the headquarters, and Ji'nan Military Command became a member unit of YRFCH.

In order to further give play to the basin organizations' coordinating role to facilitate implementation of flood control decisions and scheduling schemes, approved of by the leaders of the State Council, on April 21, 1997, the State Flood Control and Drought Relief Headquarters issued "Notice on Adjustment of Leading Members of Yellow River and Changjiang River Flood Control Headquarters" (No. [1997]4), which decided that the director of YRCC and the director of the Changjiang Water Resources Commission shall act as executive vice - general - directors of Yellow River Flood Control General Headquarters and Changjiang Flood Control General Headquarters respectively.

1.3 The third major reform—the Yellow River Basin flood control and drought relief management system was established

Over the past fifty - odd years, the YRFCH had played an important role in flood fighting and emergency handling and flood control management of the Yellow River, and had provided a strong organizational guarantee for the Yellow River safety for 60 successive years. Yet, with economic and social development, the basin management system became unable to meet the requirement of flood control work. In view of the new situation, new problems and new requirements with respect to the Yellow River Basin flood control and drought relief, and the administrative duties given to basin organizations by laws and regulations of the State, in order to spread the work of flood control and drought relief to the upper, middle and lower reaches of the Yellow River and to conduct more effectively unified operation of the reservoirs on the river and to carry out integrated management of the Yellow River flood control and drought relief, on March 14, 2007, the State Flood Control and Drought Relief Headquarters issued to the provinces (autonomous regions) along the Yellow River and YRCC "Approval on Establising Yellow River Flood Control and Drought Relief Headquarters" (No. [2007] 2), deciding to establish Yellow River Flood Control and Drought Relief Headquarters (YRFCDRH). On the basis of the original headquarters, the new headquarters had an added function of drought relief in the basin, and had four provinces (autonomous regions) of Qinghai, Gansu, Ningxia and Inner Mongolia and Lanzhou and Beijing Military Commands as newly added member units. YRFCDRH shall have the governor of Henan Province as general director, have the director of YRCC as executive vice - general - director, and shall have the vice - governors (vice - chairmen) of provinces (autonomous regions) Qinghai, Gansu, Ningxia, Inner Mongolia, Shanxi, Shaanxi, Henan and Shandong and the vice - chiefs of staff of Lanzhou, Beijing and Ji'nan Military Commands as vice - directors. The office of YRFCDRH shall be based in YRCC and shall take on routine work of the headquarters. The establishment of the new directing organization indicated that the Yellow River flood control and drought relief had entered upon a new era of integrated basin - wide management.

2 Present situation of integrated the Yellow River Basin flood control management

2.1 Connotation of integrated the Yellow River Basin flood control management

The Yellow River flood control refers to the work carried out to defend against the Yellow River flood and prevent or mitigate flood disasters. It comprises engineering measures and non - engineering measures, the former including construction and protection of flood control works, the latter including flood control directing and scheduling, danger checking and handling, establishment of rules and regulations, organization of flood control teams, resettlement and rescue of floodplain residents and construction of hydrological information, et al. Faced with new situations and new requirements, the connotation of the Yellow River flood control has been constantly enriched and developed. In recent years, YRFCDRH have been thoroughly implementing scientific concept of development and actively practicing the "two changes" concept with respect to flood control and drought relief. In flood control and flood fighting, we should change from the previous flood control alone in the past to management of flood and sediment, and change from the previous single objective of flood control and calamity reduction to flood control as the main objective with multiple objectives of water supply, power generation, sedimentation reduction and ecological restoration. Therefore, the integrated the Yellow River Basin flood control management means that YRFCDRH organize, coordinate, supervise and direct the provinces (autonomous regions) and departments and units concerned within the basin to carry out flood control and calamity reduction on every side and at multiple levels with both engineering and non - engineering measures. We should exert effective control of large floods and extraordinary floods to ensure that the Yellow River dykes would not burst within design standard. In case of floods beyond standards, we should make our best to reduce calamity losses to the minimum. With medium and small floods, while ensuring flood control safety, we should coordinate the relation between flood control and water supply, power generation, sediment transport and ecological water consumption, to achieve maximum overall benefits. To achieve integrated the Yellow River flood control management, we have to take overall measures such as administrative, engineering, legal, technical and economic measures, properly deal with the relation between flood control and water supply, power generation, sedimentation reduction and ecological water consumption, and turn flood into resources.

2.2 Great achievements in the integrated Yellow River flood control management

The new basin management system has more effectively coordinated the relations among all sides, given consideration to both flood control and drought relief in the Yellow River Basin, strengthened unified directing and scheduling of the Yellow River flood control and drought relief, and strongly promoted normalization and standardization of flood control and drought relief. It has played an important role in guaranteeing Yellow River safety, speeding up the Yellow River harnessing and development, and promoting economic and social development in the basin. YRFCDRH have established a member liaison meeting system. Every year, the Headquarters called relevant provinces (autonomous regions), military commands, electric power departments and reservoir management units to hold a Yellow River flood control meeting and a Yellow River ice flood control meeting to earnestly sum up experience of the Yellow River flood control and ice flood control of the previous year, to analyze and estimate the situation of the Yellow River flood control and ice flood control of the same year, and plan flood control and ice flood control of the same year. In view of the grim situation of ice flood control in the Ningxia – Inner Mongolia reaches, since 2008, the Headquarters have held a consultation on ice flood control during the key ice break period in Ningxia or in Inner Mongolia every year to coordinate and arrange key ice flood control work during ice break period, which has strongly promoted orderly, effective and vigorous development of basin ice flood control.

2.2.1 Ensure the Yellow River flood control safety, and achieve multiple objectives of flood control, drought relief and power generation

Though there have not occurred basin – wide large floods in the Yellow River since 1996, localized rainstorm floods have arisen oftentimes. In 2003 the Yellow River Basin generated the autumn flood rare in history, and in 2005 and 2011, autumn floods again occurred in the Weihe River and the Yiluohe River. During the flood periods of 2001 and 2004, the stage of the Dongpinghu Reservoir exceeded the warning water level many times. Faced with the complex flood control situation, the Office of YRFCDRH made full use of the existing flood control and disaster reduction system, made unified arrangements for flood control of the basin, and scientifically operated reservoirs of Sanmenxia, Xiaolangdi, Luhun, Guxian and Dongpinghu, et al., thus greatly reduced flood control pressure on the lower reaches, and guaranteed safety of people's lives and property.

During the ice flood period from 2007 to 2008, some reaches of the upper Yellow River happened the highest water stage in history, in some parts the ice jam was very severe and dangerous situations occurred frequently, and the Ningxia – Inner Mongolia Reaches encountered the severest ice flood in 40 years. Faced with the severe situation, YRFCDRH made earnestly various schemes for ice flood control, studied and arranged various flood control measures and strengthened field supervision and guidance to ensure real implementation of measures. During the ice break period, a major danger situation occurred in the reaches with Inner Mongolia. YRFCDRH scientifically operated the reservoirs of Liujiaxia and Wanjiazhai, made effective use of the Hetao Irrigation District and the Sanshenggong Water Control Project to carry out ice flood diversion, and properly guide and vigorously support local flood control headquarters in emergency handling, thus timely and effectively dealt with dangerous situations, achieved "zero death" and ensured safety of people's lives.

In the time from 2008 to 2009 when extraordinary drought arose in the Yellow River Basin, YRFCDRH coordinate the upper, middle and lower reaches of the river, gave consideration to water needs of all sides, carried out joint reservoir operation to support drought combating of the basin with water resources of the whole river, thus achieved summer grain increase and harvest in drought years in provinces of Henan and Shandong, and ensured grain security of the country's major wheat producing areas.

During the autumn flood period of 2011, the Weihe River had generated the largest flood since 1981 and the Yiluohe River had encountered the largest flood since 1983. Due to returning residents and the restrictions on farmland flooding in the reservoir area, the Luhun and Guxian Reservoirs had flood control capacity of mere 3.0×10^7 m^3/s, while the discharge was required to be no larger than 1,000 m^3/s to ensure safety of the lower reaches, besides, the forecast ahead time was very short, there was a hard choice between guaranteeing reservoir safety and turning flood into resources. Faced with the severe flood control situation, YRFCDRH made elaborate plans, made arrangements in advance, strengthened forecast, held timely consultations and scheduled carefully. Within a period of near 20 d, YRFCDRH sent more than 80 orders and telegrams, and sent work teams three times to direct flood fighting and danger handling in the field. In this way, they managed to achieve historic breakthrough that the Weihe dyke and the Nanshan tributary dyke did not burst at the same magnitude of flood, and to avoid resettlement of more than 200 thousand people in the area of the Luhun and Guxian Reservoirs and on the Yihe – Luohe – Jiahetan floodplains. Through flood retention and peak staggering, the peak flow in the lower Yellow River was reduced from 7,800 m^3/s under natural condition to 3,120 m^3/s, 2,900 km^2 of floodplains, 3.3×10^6 mu of farmland and 1.15 million people were prevented from being flooded, and economic losses were reduced by 1.54×10^{10} yuan. In the experiment of scouring Tongguan elevation by utilizing and optimizing spring flood, they optimized regulation schemes and reduced the surplus water from the Wanjiazhai Reservoir to the minimum. During the process of water and sediment regulation prior to the flood period, through overall considerations and all – round arrangements, they supported Henan Province in coping with summer power consumption peak, and were cited for this by the Government of Henan Province. In September, seizing the opportunity time when there was abundant incoming water in the reaches between Liujiaxia and Lanzhou, by operating the Liujiaxia Reservoir to mold an advantageous flood hydrograph, they carried out Wanjiazhai Reservoir desilting experiment. During flood regulation of the autumn flood period, they carried out reservoir risk operation in a decisive manner to make full use of the flood water kept in the reservoirs during the flood period. With 5.8×10^9 m^3 of flood storage increment, the reservoirs of Luhun, Guxian and Xiaolangdi stored ample water for drought combating and achieved a win – win objective.

2.2.2 Open a new way to turn flood into resources

In accordance with the requirement of the concept of maintaining Yellow River's healthy life, the Office of YRFCDRH has been actively exploring the new mode of controlling, utilizing and molding the Yellow River flood. The first, to actively carry out experiment of dynamic control of flood limit level of reservoirs of Wanjiazhai and Guxian, et al. to save valuable water resources as far as possible. The second, to make full use of the water above the flood limit level discharged out of reservoirs prior to the flood period and the incoming water and sediment during the flood period to carry out Yellow River water and sediment regulation. Since 2002, water and sediment regulation has been carried out 13 times for 10 years running, the whole main channel of the lower Yellow River has been kept being scoured, 7.5×10^8 t of sediment has been transported to the Bohai Sea, and the main channel of the lower Yellow River has been lowered by 1.5 m on average. The sedimentation locations of the Xiaolangdi Reservoir and the cross – section pattern of the lower reaches have been effectively adjusted. The discharge capacity of the main channel of the lower Yellow River increased from 1,800 m^3/s before water and sediment regulation to 4,000 m^3/s, which improved the situation of "floodplain being submerged even at small flood", and played an important role in life and production of the people and social stability along the Yellow River and on the floodplains. The long – standing altitude of ignoring the fact that Yellow River water for sediment transport was otherwise used in quantity has been changed, and improvement of ecological environment in the river channel and the estuary region has been promoted. The third, to make the best of spring flood water by operating reservoirs of Wanjiazhai, Longkou, Sanmenxia, et al. to carry out experiment of scouring Tongguan elevation. From 2006 to 2011, six times they carried out experiment of reducing Tongguan elevation by scouring by means of utilization and optimization of the spring flood process, so that Tongguan elevation is lowered 0.05 m to 0.30 m respectively, the sedimentation manner of the reservoirs of Wanjiazhai and Sanmenxia is improved, and the understanding of the movement pattern of the Yellow River water and sediment is further deepened. The fourth, since 2004, the YRFCDRH office has carried out warping experiment in the Lianbotan floodplain at Hejin on the Xiaobeiganliu reaches in the middle Yellow River, deposited $5,727.9 \times 10^6$ t of coarse sediment, of which the coarse sediment of grain size larger than 0.05 mm accounted for 26.8%, so that they preliminarily achieved the aim of "silting the coarse sediment while discharging the fine sediment", and accumulated experience of retaining coarse sediment from the middle Yellow River. The fifth, while ensuring flood control safety, they stored water as much as possible for water supply and irrigation along the Yellow River and for diverting water to Tianjin, Hebei, Qingdao, et al.

2.2.3 Strongly promote integrated basin flood control management

2.2.3.1 Strengthen basin flood control management and strongly support other basins in emergency handling and relief

Adapted to the extension of management limit and expansion of management duties of YRFCDRH, the YRFCDRH office has carried out general survey of flood control works in provinces (autonomous regions) within the Yellow River Basin, sorted out basic data of flood control and ice flood control. They have carried out special survey of non-flood-control works, standardized examination and management of flood control effect evaluation, and grasped overall information about flood, works, drought and calamities in the Yellow River Basin. They have established a basin flood control and emergency handling expert database and strengthened guidance, coordination and supervision of major work of flood control and emergency handling. Faced with severe natural disasters such as the "May 12th" devastating Wenchuan earthquake and Zhouqu extraordinary mudslides in Gansu Province, YRFCDRH incorporated the forces of the whole basin, responded quickly and assembled teams rapidly, satisfactorily completed rescue and relief tasks such as reservoir danger elimination, dyke plugging and river scouring. In view of major dangerous situations frequently occurring in the basin, such as the dangerous situation of the Ge'ermu Reservoir in Qinghai, YRFCDRH sent without delay working groups and expert groups to aid and guide local authorities and departments in flood fighting and emergency handling.

2.2.3.2 Strengthen construction of overall basin ice flood control capability, and improve basin ice flood control level

In 2008, YRFCDRH called together flood control offices of Ningxia and Inner Mongolia Autonomous Regions, Operations Departments of Beijing, Lanzhou and Ji'nan Military Commands, and relevant units under YRCC to work out the "Implementation Plan for Construction of Overall Yellow River Ice Flood Control Capability", and energetically promoted its implementation, thus improved overall ice flood control capability of all levels and laid a foundation for success in ice flood control over the years.

2.2.3.3 Put forth effort to establish a integrated, coordinated, standardized and efficient basin flood control management operation mechanism

In accordance with changes in the State's policies on water conservation and the requirement of flood control situation, YRFCDRH have revised and perfected "Rules on Yellow River Flood Control Directing and Scheduling", worked out rules and regulations such as "Regulations on Operation Management of Reservoirs on Yellow River and its Major Tributaries", and issued and enforced a series of work regulations such as "Responsibilities of Yellow River Flood Control and Drought Relief Headquarters", "Responsibilities of Office of Yellow River Flood Control and Drought Relief Headquarters", "Regulations on Flood Control and Drought Relief Propaganda Work of Yellow River Flood Control and Drought Relief Headquarters", "Information Handling Procedure of Office of Yellow River Flood Control and Drought Relief Headquarters", "Standards of Yellow River Ice Flood Control Work" and "Regulations on Directing and Scheduling of Ice Flood Control in Yellow River Ningxia-Inner Mongolia Reaches", et al., all of which have guaranteed efficient operation of the Yellow River Basin flood control management system.

2.2.3.4 Steadily improve the mechanism of the Yellow River Basin information exchange, release and sharing

YRFCDRH have developed a reservoir operation information platform to strengthen unified management of major reservoirs on the Yellow River and its tributaries. Relying on the State flood control and drought relief directing system, and jointly with flood control headquarters of province within the basin and operations departments of the three military commands, the headquarters completed construction of Yellow River Basin flood control and drought relief information platform, thus preliminarily achieved basin flood control and drought relief information sharing.

3 Problems existing in integrated Yellow River Basin management and Countermeasures

The integrated Yellow River Basin flood control management is an organic whole. The Yellow River flood control belongs to the category of productivity, and its main work object is floods; while the integrated basin flood control management belongs to the category of production relations, and its main task is to standardize and coordinate behaviors and interests between relevant provinces (autonomous regions), units and departments. With constantly occurring new situations and new problems concerning the Yellow River, the challenge, the integrated basin management is faced with, has become more serious. We should be fully aware of the problems existing in the integrated basin flood control management, actively plan countermeasures and energetically promote integrated basin management to provide vigorous management support for maintaining the Yellow River' s healthy life.

3.1 Basin management lacks a reasonable power structure, and basin organizations should be given the power of integrated basin administration through legislation

The new "Water Law" stipulates that the State shall, with respect to water resources, adopt a system that organizes the administration by watersheds as well as by administrative areas. Thought the legal status of the basin management organization has been established, the limits of administration by watersheds and that by administrative areas have not been clearly defined, and to which some particular duties should belong has not been made clear. Division of work between YRFCDRH and the headquarters of provinces (autonomous regions) has not been clearly defined. In some areas, it was often the case that the administration by watersheds was stressed, while the administration by watersheds was weakened, and basin flood control scheduling often could not be carried out thoroughly and completely. For example, in reservoir flood control operation and in river obstacle clearing, the administration by the basin often met with resistance from reservoir management units and provinces (autonomous regions). Therefore, we should accelerate the formulation of "Yellow River Law". We should, through legislation, give basin administrative organizations the power of coordinating basin flood control, water supply, power generation and ecology scheduling, and should, by means of law, define the duties of basin organizations, so as to completely solve the problems that the duties of basin administrative organizations do not conform to their power.

3.2 Unified reservoir operation is difficult, and we should establish a major water project management and scheduling mechanism

Major water projects on the main stream and its major tributaries have multiple functions such as flood control, ice flood control, irrigation, power generation and sedimentation reduction, and they have an important status and role in controlling the Yellow River flood, coordinating water - sediment relation, achieving integrated water resources management and scheduling, improving river ecological environment and ensuring no flow cutoff of the Yellow River. However, since reservoirs on the Yellow River are under the jurisdiction of different departments and enterprises, their operation aim is to chase maximum power generation benefit, which often conflicts with flood control, water supply, drought relief and ecological scheduling, and seriously affects normal implementation of flood control scheduling. Therefore, we must establish a mechanism for the basin administrative organization managing and scheduling the major water projects on the main steam and its major tributaries. Major water projects on the main stream and its major tributaries with flood control and sedimentation reduction as the main aim that have an important position in the Yellow River water and sediment regulating system, such as the Xiaolangdi Multipurpose Project, should be brought under the direct administration of the basin administrative organization. With respect to the water projects such as Guxian and Qikou water projects that are being planned, the construction should be organized by the basin administrative organization, so as to promote formation of a rolling development and integrated management mechanism; and after completed, these projects should be directly administered and scheduled by the basin administrative organization. The water projects on the main stream with power generation and other benefits as the main aim and also with tasks of flood (ice flood) control and sedimentation reduction should be under unified scheduling by YRF-CDRH in terms of flood (ice flood) control.

3.3 The Yellow River flood control operation mechanism is yet to be perfect, and the system of consultation for basin management should be further perfected

As yet, a perfect flood control management operation system has not been established in the Yellow River Basin. Construction of flood control standardization should be further promoted. The system of consultation for the basin management should be further perfected in accordance with the principle of organizing the administration by watersheds as well as by administrative areas. The Yellow River Basin administrative organization organizes establishment of a consultation platform, the governments of provinces (autonomous regions) and relevant areas, departments and beneficial owners participate in, and they together negotiate about major issues and important affaires in the basin harnessing and development, coordinate benefit distribution relations among regions and departments, so as to make the major decisions about the basin based on the democratic consultation, and make the requirements of the basin, regions and departments fully reflected in the decision-making process. We should establish and perfect step by step rules of consultation, joint meeting system and information sharing system.

3.4 IT application level of the Yellow River Basin flood control has to be improved urgently, and we should make efforts to build a "Digital Yellow River Basin"

Integrated Yellow River Basin flood control management requires a solid information and technology base for support. However, the existing flood control IT application level is unable to meet the needs of integrated basin flood control management. Therefore, we should, on the basis of construction of the State Flood Control and Drought Relief Directing System, coordinate provinces (autonomous regions) within the basin to establish step by step a platform based on basin flood control information sharing system, so as to strengthen the relation between the basin administrative organization and provinces (autonomous regions) with respect to water regime forecast, directing and scheduling, technical exchange, basic information sharing and supervision and coordination in flood control work, achieve flood control information sharing and provide technical support for the basin flood control and drought relief.

References

Li Guoying. Maintain Yellow River's Health Life[M]. Zhengzhou: Yellow River Conservancy Press, 2005.

Li Guoying. Q & A on Yellow River[M]. Zhengzhou: Yellow River Conservancy Press, 2009.

Yellow River Conservancy Commission. Sixty Years of Yellow River Control by the People[M]. Zhengzhou: Yellow River Conservancy Press, 2006.

Li Guoying. Report to River Work Conference 2010[R]. Zhengzhou: Yellow River Conservancy Commission, 2010.

Chen Xiaojiang. Report to the Conference on Yellow River Flood Control and Drought Relief 2011 [R]. Zhengzhou: Yellow River Conservancy Commission, 2011.

Zhao Yongxin. Promote Integrated River Basin Administration[N]. People's Daily, 2004 - 11.

Liu Hongbin, Li Yuelun. ABC of Yellow River Flood Control[M]. Zhengzhou: Yellow River Conservancy Press, 2001.

Yellow River Conservancy Commission. A Dictionary of Yellow River Control[M]. Zhengzhou: Yellow River Conservancy Press, 1995.

Anticipated Revenue, Opportunity Cost and Transaction of Water Property Rights between Agriculture and Industry
—Take the transaction of water property rights between agriculture and industry in Ningxia Autonomous Regions an example

Ma Xiaoqiang[1] and *Zhang Lu*[2]

School of Economics and Management of North West University, Xi'an, 710069, China

Abstract: The variable anticipated revenue and opportunity cost of water resource from the different industrial agencies are the fundamental reason of the transaction of water property rights between agricultural and industrial. The transaction of water property rights between agricultural and industrial in Ningxia Area has been taken as an example, and the internal mechanism and policy connotations has been argued.

Key words: original water property rights, the transaction of water property rights, the transaction of water property rights between industrial agencies

1 Anticipated revenue, opportunity cost and transaction of water property rights between agricultural and industrial

The transaction of water property rights among different industrial agencies in China now refers mainly to those departments or projects which cannot get access to the allocation of quota of existing water resources, especially industrial sectors and projects, gain new water rights saved by agricultural sector in water saving projects like canal lining, and finish paid transaction of water usage rights while maintaining the original configuration of water property rights. Such transaction is the water right conversion between industries, the background of whose formation is that water utilization structure relies heavily on agriculture.

The Yellow River Conservancy Commission (YRCC) has begun to explore the use of market instruments to optimize the allocation of the Yellow River Water Resources and propose a compensation mode of "Transform from Agriculture to Industry" since 2003. In order to improve water utilization, optimize the water resources using structure, meet the new demands of water usage of industrial projects and make the water resources to be a more important role in the regional economic development, the owners of units of the industrial projects invested the Water-saving transformation projects of Yellow River Irrigation in some provinces which have no surplus water resources. Theoretically, the fundamental reason of the transfer of water rights between different industry sectors is that there are different expected returns of water resources in different industrial sectors. Ningxia, for example, although its water resources consumption accounts for 95%, the agriculture sectors provided less than 36% of GDP. Industry and the tertiary industry consumed only 5% of the water resources, but they provided more than 60% of GDP. if such a disparity in water use effectiveness can be retained under the planned economic system, then under market economy conditions, it is contrary to the basic law of the allocation of resources and the exploitation and development of the region will be subject to significant constraints. Coupled with the water resources, unlike grain and other agricultural products which can be commodity to be allocated in a wider range of space, in addition to the small number of large-scale inter-basin water transfer, mainly were allocated in an optimal way in the river basin because the water circulation and the transaction subject to the constraints of the drainage basin factors. More importantly, as the regional resource development, especially the speed of development of mineral resources, the powerful driving force generated by the acceleration of the development of regional economic gave birth to the conversion of water resources between the different departments.

The essence of water right transition between industry and agriculture is the inevitable choice product of the related interests object in the multiple constraints. These constraints mainly are the principles of keeping the status quo of water right transition, besides the change of surface runoff in natural factors. On one hand, agriculture is the foundation of the national economy, the development and prosperity of agriculture have strategic meaning to our social and economic development. On the other hand, the development of the industrial projects is rigid. Ningxia is undeveloped area; both regional economic development and regional coordination result in increased demand of water resource. In the end, the water resource allocation situation is hard to change. In practical terms, water rights system revolution is restrained by following terms: Firstly, water supply is not stable, some times reduced, Secondly, new demands from industry and other non - agricultural sectors continue to increase, the contradiction of water resource between supply and demand in Ningxia and Neimeng is becoming increasingly acute. Thirdly, the current structure of water usage between industrial and agricultural is irrational . Ningxia hope to increase the industrial water consumption by saving from agriculture. Agricultural water consumption accounts for more than 95% of the total water consumption, while industrial water consumption only accounts for 3% , which is 20% lower than the national average level. Fourthly, the waste of agricultural irrigation has sharpen the contradictions between agricultural and industrial. And non-agricultural industries are increasing the demand for water resources with the upgrading of industrial structure, the development of resources and the increase of the number of water-needed projects.

In this paper, it's anticipated revenue, opportunity cost and initial allocation that determine water right transition among departments. It is not only logical basis but also important factors that should be considered in water rights transition among departments.

Firstly, anticipated revenue is the economic benefit and social benefit that can be foreseen and measured after water rights transition. Direct economic benefit of potential water right can help industrial project realize output and obtain benefit. Direct social benefit of potential water right can help industrial project gain tax benefits. What's more, the higher the expected benefits, the greater the power. According to the budget from the WRD of Water Resources Department of Ningxia, at present each unit of agricultural irrigation water earn 1 yuan while equal unit of industrial water earn 57.9 yuan. The present agriculture - to - industry water transfers on the area of Ningxia could bring 3 billion yuan. The water rights conversion goal has been achieved by2010. It achieved more than 18 billion yuan, accounting for over half of revenues industrial added value of 2006 in Ningxia. Anticipated revenue is primary cause of water rights transition among departments.

Then, traditional agricultural irrigation methods lead to flooding, sediment canal lead to water infiltration. In consequence, the use of water resource is inefficient. Meanwhile, industrial project required water lose the opportunity of the development for water resources total supply remains. Too much water infiltration make industrial project required water lose the chance to gain profit. The chance can be measured by expected return and input - output ratio. As for these industries, in case elasticity of demand of water resources is low and economic output is high - water resources are more likely to be configured in agriculture. Meanwhile, water infiltration and runoff is more likely to happen.

As a matter of experience, obvious transformational motives always come along with high expected returns and opportunity cost. Consequently, water rights conversion among departments means externalize this opportunity cost to a large extent, which can push forward paid circulation of water resources property rights.

Lastly, initial allocation of water rights set pattern and degree for function of water rights trading market. Characteristics of the initial allocation limit the domain of water resources reallocation. As a result, the water rights conversion pilot emerges. It's initial allocation of water rights that determine the degree and extent of water right conversion among the departments. If initial allocation of water rights is open, democratic and dynamic, water right conversion among the departments will be more efficient.

The markets provide basic institutional frameworks for water resource flow among departments. Under the planned economy system, the lack of consideration of expected return and opportunity cost almost make paid circulation impossible. After the establishment of market economy system, the role of the market in resource allocation enhanced, which make water resources paid circulation among departments or regions according to the expected benefits possible. In general, market emerges and develops following commodity exchange, it play essential roles in economic development and social progress. According to the research, the structure of the water rights market in China is summed up as" two levels and three types". Two levels refer to primary market and secondary market. Three types refer to regional water rights trading market, departmental water rights trading market and water rights trading market between farmers. Level is vertical concept, type is horizontal concept. They are form the three-dimensional structure.

According to the survey, the transformation of water rights between departments has following feathers: ①it's highly effective, timely and specific as relying on certain projects. ②The operation and management of projects is more efficient and accounting for the benefits and costs is more detailed and accurate. ③The transaction market of water property rights is affected by the progress of technology significantly, with technology making progress, industry reduces its dependence on water resources or improves the efficiency of water using or uses other resources to replace, then the market may shrink until it disappears. ④On the level of operation, the transaction of water property rights is special, the water allocation scheme should be reviewed and approved by the management agents. The transaction market of water property rights has advantage on that industry invest funds to support agricultural water-saving facilities and water saved by agriculture help the industrial project run more effectively. The powerful push of the market comes from enterprise, especially from the interest of certain projects in enterprise. Accordingly, it's more scientific to account in this market than water transfers on the area of origin. ⑤Active support from the government is necessary. They can not only invest to guide and encourage the transaction of water property rights between different departments, but also produce relative policies.

2 Analysis on the case of water rights transition Ningxia

"Thousands of miles of the Yellow River make Ningxia wealthy" mainly refers that irrigated agriculture is well developed in Ningxia owing to the Yellow River. However, high proportion and low efficiency of agricultural water is an existing problem. The ratio that agricultural water consumption in the total water consumption has reached about 95%, and serious waste happened in the agricultural irrigation process, more than half of the water is wasted during the transportation. The important reason is that low efficiency in water use caused by too much infiltration in sediment quality canal and poor irrigation facilities. Lining rate of the channel at all levels is 4.5% to 20% in irrigation district. Buildings have been damaged to varying degrees. Effective utilization coefficient of the irrigation channels is about 0.46 and the utilization coefficient of the irrigation water is about 0.46, which is away from the national average level. Therefore, irrigation districts still have great water-saving potential. Water consumption per acre irrigated areas of Ningxia and the Yellow River south bank in Inner Mongolia is up to more than 1,000 m^3, which is 2.4 times as the national average.

The water consumption of Ningxia is based on the program "87", at present The indicator of water taking from the Yellow River is no more than 4×10^9 m^3 for the whole region and drop to 3.6×10^9 m^3 for Ningxia. Ningxia's total irrigated area is about 700 million acres, including pumping irrigation. per capita own 1 acre of irrigated cropland, this situation is rare in the northern region. Energy construction and city development are mainly centered in the northern and central five cities, which has accounted for two - thirds of the total population in the region. There are eight state-level poverty - stricken counties in the southern mountain regions, the central arid zone and the southern mountain regions have developed by pumping irrigation, the irrigated area has covered more than one million acres, which can solve the problem of poverty in the central arid zone. Now the development of Yinchuan, including its drinking water, mainly relies on the groundwater resources on the Helan Mountain. The Shizuishan and Wuzhong also mainly rely on groundwater, the development of Ningdong basically relies on the Yellow River, and Sun Mountain Industrial Zone partially rely on the Yellow River.

The strong momentum of development of chemical industries makes an unprecedented water demand. To develop, process and convert resources are the most direct and effective way for the resource-rich region to develop rapidly. The main resource is coal in Ningxia along the Yellow River, whose coal reserves rank sixth in the country, while the discovered reserves of Ningdong account for 88% in the whole region. Such resource endowments make Ningxia choose energy and chemical industry based on coal, while the energy and chemical industry such as generating electricity with coal need much water. Water is the blood of energy development. The national initial water rights allocation pattern limit the development of the economy in Ningxia. At the beginning of new century, Ningxia decide to build Ningdong Energy and Chemical Base to achieve leap forward development of the region. The industrial water at this stage in Ningxia is groundwater and surface water. The existing plan show that, water consumption of Ningdong, Sun Mountain, including the the Shizuishan Industrial Park will be quadrupled. In the first phase Ningdong water supply quantity is 1.50×10^8 m^3, by 2010 it would double to 3.30×10^8 m^3. In accordance with the regional economic development strategy and the distribution of productive forces, Ningxia plan to build industrial base along the Yellow River. Seven large-scale thermal power plants will be built and total installed capacity will increase 1.5×10^7 kW by 2020. According to the water saving reform plan for Qingtongxia Irrigation and Weining Irrigation District enacted by Ningxia, 7.00×10^8 m^3 agricultural water has been saved owing to engineering water-saving measures by 2010. According to preliminary forecasts, if 2.00×10^8 m^3 water are used for water rights transition, there will be enough water for industry in 2010.

The Project of Water Supply In NingDong: In the energy and chemical industry base near Yinchuan, agricultural water is transformed into industrial water by water supply project, which provides security for the sustainable development of the base. The water supply project is made up of two parts. One is water source project, the other is water purify and distribute project. The first stage of water supple in Ningdong started in 2003 December and took 3 years before finish. The total expense was 790 billion yuan. So far, the rate of emission has reached 2.8 m^3/s in the first stage, inside of which contains 1.357×10^8 m^3 for industry and 2.4×10^7 m^3/s for ecology. The total expense of water source project is 550 billion yuan. The water source project has been finished. The reservoir can hold more than 1.5×10^9 m^3 water. The water purify and distribute project which can purify 100 thouthand had been finished by October, which is mainly supply water for producing alcohol and generate electricity. The output water is about 5.0×10^5 m^3 to 6.0×10^5 m^3 and accumulates 2.0×10^6 m^3. The water purify and distribute project which can purify 300 thouthand had been finished by August this year. The user in Ningdong base can apply water supply and sign the protocol of water convert possession. They can get industrial water paid from the total water distributed for the Yellow River in Ningxia Province by our nation. The transforming expense is used for agricultural water. The second stage of water source project and water purify and distribute project in Ningdong's energy and chemical industry base had been started in May, 2012. The scale is about 4.0×10^5 m^3 per day. The finished first and second stage project can supply 8.0×10^5 m^3 water per day, which can meet the need of water for industry, ecology and life before 2020 in the base and provides guarantee for the construction and development of the base. The water supply and source project: this project estimates 389,716,500 yuan. The second project can supply 2.11×10^8 m^3 water, and get 2.39×10^8 m^3 water from the Yellow River. The second water supply project in Ningdong of Ningxia autonomous region which is investment 1.153 billion yuan is started on May 9th, 2012, it indicated the project started building formally. After finished, the water source project gets 2.39×10^8 t water from the Yellow River. The reservoir could hold 3.522×10^7 t water. The water shop could supply 8.0×10^5 t water. The project could guarantee the water need of Ningdong's energy and chemical industry base—the national energy and chemical industry base.

The project of Lingwu's electric power plant in Ningxia. Based on the protocol of water rights transition, the Lingwu's electric power plant invested 4.0 million yuan for irrigation channel reformation in Ningxia. After waste of water resources are effectively reduced, enterprises could obtain 1.44×10^7 m^3 water using index. At the end of the year of 2005, the transforming water rights of Lingwu's electric power plant had been finished. The plant invested more than 2.9 million yuan for the Tanglai irrigation channel reformation project. Annualy 2.0×10^7 m^3 of water can be saved. The Tanglai irrigation channel transfers 1.4×10^7 m^3 water using index every year. 150 million yuan Dam are invested on the three pilot projects of Power Plant Phase Ⅲ expansion project, Ningdong Maliantai power plant and Lingwu power plant. By April 2008, 72.4 million yuan have been spent on the three projects. 20.57 km of trunk sewers and 155.84 km of Branch Canal assemble carewere have been completed. 1,485 buildings are newly constructed, renovated, transformed. 2,200 acres are set as the area of completion of the well and canal combined with the project. As return, these three companies can transform 5.39×10^7 m^3 of water each year. The project of water rights transaction ensures the water use of Ningdong, Maliantai, spiritual state power plant and 2.5×10^5 t of methanol.

Relevant supporting policies published by River basin authorities and local government effectively promote water rights transition among departments. During the conversion work, Ningxia take total control and quota management. Some systems were enacted, such as "Views on the implementation of water rights transition between Ningxia and the Yellow River", "Measures for the Administration on the use of funds for water rights transition between Ningxia and the Yellow River". water standards are formulated, such as "Industrial water quota", "Water quota of urban life". Clarify the allocation of water rights from city further to country. YRCC has initiated a water right transfer pilot project in Ningxia, Promulgated "Management and implementation approach for water rights transition of the Yellow River" and "Water – saving projects verification approach for water rights transition of the Yellow River", and established a water rights transition system with the characteristics of the Yellow River.

In 2010, it is the first time that both water taken from the Yellow River and water consumption are not excessive in Ningxia. Total water consumption decreased by 5.80×10^8 m^3/s than compared to 2005, and agricultural water consumption reduced by 6.10×10^8 m^3/s, utilization coefficient of irrigation water increased from 0.36 to 0.43. Water consumption decreased by 65.8% for per 10,000 yuan GDP, water consumption of industrial added value down 53.6% for per 10,000 yuan GDP, both of which exceeded the decline in 30%. Ningxia has organized a special group to research, they draw the following conclusions: Water-saving projects not only greatly reduce the loss in the water delivery, but also improve canal conveyance conditions, which accelerate conveyance speed. Irrigation time of each round has been shorten by 3 ~ 4 d, irrigation water consumption from farmers has reduced, 2.8 yuan is saved per acre. Since April 2008, there is no rain for more than continuous 80 d in Ningxia, it has not affected agricultural irrigation in the region taken water from the Yellow River. Xu guangru said that the water rights transition has played a very important role in this regard.

Throughout the water rights transition case of Ningxia, water rights transition was still in the pilot and initial implementation. But it was really necessary both on theory and practice and had brought many benefits to all the parties and achieved the goal of win – win. In the aspect of supporting the development of regional economic and in the premise of not taking more water from the Yellow River, it provided enough production water supply for the proposed project, broke through the resource constraints of the development of enterprises and promoted the speed of development of the regional economy. For water – saving projects, it expanded the financing channels of water resources project, raised investment funds effectively and improved the projects quality significantly by implementing Irrigation District Project. In terms of farmers' interests, it protected the farmers' legal water interests, reduced the loss during transportation, dropped the expenditure of water usage and increased income and welfare of the farmer in some degree. In water resource utilization respect, it improved the efficiency and effectiveness notably, achieved the optimal allocation of water resources, provided an important foundation to build water-saving, conservation-minded society. To take the coordination of agricultural and industrial relations into consideration, ensured the 1.30×10^8 m^3 of water which was conversed used in industrial, not only achieved the optimal allocation of water resources, improved water use efficiency, but also supported high – speed operation of the industrial economy, explored the effective model to realize the industry to support agriculture and to promote the coordinated development of industry and agriculture, and it boosted comprehensive regional socio – economic developments at the same time. In Ningxia province, the increase of revenue was expected to reach 1 billion yuan to 20 billion yuan.

3 Strategies and thoughts on improving the transaction market of water property rights between agricultural and industrial agencies

Based on the analysis above, transaction of water property rights between agricultural and industrial depends on both logical theory and reality. In order to speed up the establishment and improvement of transaction market of water property rights between agricultural and industrial, we propose the following five strategies and thoughts.

3.1 Make the legal status of transaction market of water property rights between agricultural and industrial agencies more clear

To make the transaction market of water property rights between agricultural and industrial agencies has the same legal status as other factor markets, law to confirm and clarify its legality and legitimacy is currently needed . This ordinary point is the legal basis of the relevant rules and regulations. The problem has been solved a long time ago in countries and regions that is active in the transaction of water property rights. Therefore, it is important to amend the "Water Law" and accelerate the introduction of a special law, which can provide legal protection for the transaction market of water property rights. The establishment of the transaction system of water property rights has two key issues: one is legal definition on the rights to use and transfer, the other is adjustment on the functions of relevant government departments and the establishment and improvement of independent oversight mechanism.

3.2 The relation of stakeholders need to be balanced in the process of the transaction of water property rights between different agencies

We should not only achieve the interests of industrial investment projects and agricultural sector, but also concern about the ecological benefits. In the above-mentioned factors, the interest of industrial investment projects is affirmative and timely, and the agricultural sector is monopolistic, while the ecological effects come with an apparent delay. Therefore, the fundamental problem is to deal with interests relationship among the three factors, especially when setting a price at the water property rights. These three factors are closely linked, on one hand, on the other hand, the adjustment of economic structure, who provides a stable demand-side and supply-side, is an important driver of water rights trading, on the other hand, water rights trading can give a push when adjusting the economic structure.

3.3 Technical measures to improve the conversion of water rights

To improve the technical measures of the trading of water rights, it should be focused on the construction of the IT platform, the innovation of canal lining and water measurement technology, improving the lining rate, the innovation of the technology of measuring water, providing a more accurate technical support for water rights conversion and protecting the vital interests of the parties to the transaction. It is a priority for the investors to accelerate the project modification of sub – channels, the branch canal and lateral canal to save water resources. At the same time, they should resolve the problems of canal leakage, evaporation during water conveyance. The other aspects of the technical measures are controllable hydraulic engineering. The mobility of water resources determines that the trading of water rights should be monitored and measured in dynamic way. This is the key factor that makes the water rights trading market different from other factor markets. Therefore, this will determine that the water rights trading market should put more emphasis on technical equipment and technical measures. The Yellow River, for example, has 12 large and medium-sized hydropower stations which has been built or are under construction in its main stream. These main stream storage projects will give full play to the role of runoff regulation to optimize the allocation of the Yellow River water resources. They will also provide important material carrier for the Yellow River water rights trading and support the establishment and running of the Yellow River water rights trading market.

3.4 To establish a scientific system of pricing for water property rights

The pricing of water rights is an important part of building the water right conversion market. The existing water right prices exclude ecological economic compensation, renovation of the canal facilities, operation and maintenance costs, and engineering construction investment which is now as the main investment. At present, water resources fee and the opportunity cost of water resources is neither included in the price of the water rights. Therefore, the price of water rights trading in the third class water rights trading market is actually a narrow and incomplete water rights price which can not fully reflect the value of water resources. The key to build third class of water rights market is the establishment of an operable water right transfer price formation mechanism and pricing the water rights trading and the conversion scientifically. In the future, it will be allowed greater flexibility in the trading period and long and short - term water rights trading co - exist. nd encourage the temporary short - term water rights trading. Transferring the water rights from agricultural to industry and other areas is in accordance with principle of efficiency.

The basic principles of pricing in the sale of water rights, agricultural water users must recover to at least the cost of the quantity of water (water charges and other related costs). This can be achieved by a variety of ways such as agreement pricing, auctions and so on.

3.5 To establish a comprehensive regulatory system

In addition to strict vetting to water permit, assisted mechanism for disclosure of information and public participation, mechanism for evaluating by the third-party ,and mechanism of benefit compensation and insurance funds, all of which are basic conditions of water rights trading , should be established to make water rights trading effective. It is the government that provides compensation currently, which is not conducive to internalize the cost of externalities from the transaction. Influence from the third-party in the process of water rights trading is multiple, covert and delayed, therefore water rights transaction management agencies can charge on a certain percentage, while the government provide financial subsidies to establish a dedicated insurance fund for water rights transaction, which can meet the ecological crisis that may occur and weaken the damage from stakeholders.

A Study on Water Rights Transaction

***Duan Zhaochang*[1] and *Wu Yichong*[2]**

1. Xiaolangdi Multipurpose Project Construction & Administration Bureau, Zhengzhou 450000, China
2. Pingdingshan Municipal City Hall Office Affairs Administration Bureau, Pingdingshan 467000, China

Abstract: Water, the source of life, plays a vital role in the survival and development of human beings. However, the situation of water resources in China is not optimistic. On one hand, the total volume of water resources is decreasing substantially; on the other hand, waste of water resources by artificial factors is very common. Also due to the unbalanced water distribution, and deterioration of water ecological situation, China's sustainable development is being hindered. Therefore, facing such a prominent water crisis, in-depth studies have to be made to the water rights system in our country; a way to the protection of water resources as well as a way to the scientific and reasonable utilization, and development of water resources need to be worked out. Water rights transaction system is widely accepted by many countries because of its advantages of improving the utilization of water resources, optimizing the allocation, and alleviating conflicts in water use, and it's also practiced in some places of our country. Through review of water rights transaction and the theory basis of its legal system, this paper makes a study on the legal part of water rights transaction in our country, and provides some positive assistance to the construction of the water rights transaction system.

Key words: water rights, water rights transaction, legal system

1 Definition and classification of water rights transaction

The earliest water rights transaction emerged in western America and the obligee of water right sold the surplus water in the market. After decades of development, the transaction method has been gradually accepted by many countries and regions. Hence, the system of water rights transaction has formed and improved.

1.1 Concept of water rights transaction

There is no uniform understanding of water rights transaction by the academia. According to Huang Xisheng, water rights circulation, in narrow sense, refers to the market players in equal status, for certain economic purposes, transfer with compensation the use rights and the water-drawing rights obtained from the state, and the product water rights gained by adding labor after the water-drawing rights were obtained. While Lin Long holds that the water rights transaction refers to the activity by people who enjoyed "the initial country's right to use water" to transfer (sell, exchange or give) their right, or people who obtained the use rights of state-owned water resources by the aforesaid method re-transfer their rights.

Although their views are stated differently, we can still find some similarities in the definitions of water rights transaction after a close comparison. Based on the initial water rights allocation, the water rights transaction mainly refers to the transfer of use rights, and the other conditions are mostly details to make the two concepts more specific or definite. Therefore, the concept of water rights transaction can be summed up as follows: under the micro-control of the government, the water rights players legitimately transfer their legal use rights under the market mechanism.

1.2 Types of Water Rights Transaction

Water rights transaction can be classified into various types according to different criteria:

there are temporary trading, permanent trading and the water rights leasing if classified by the trading time; the trade within the basin, outside the basin if classified by theregion. The author agrees with the classification of water rights transaction in terms of the basin of water together with the length of time: temporary trade within the basin/region, permanent trade within the basin/region, cross-basin/region temporary trade, cross-basin/region permanent trade.

Specifically, temporary trade within the basin/region is the most widely used one: the water rights are not required to be transferred nor registered, and the price is low. While permanent trade within the basin/region entails the water rights partly or wholly transferred after a series of legal process, which is both time-consuming and costly. Cross-basin/region temporary trade refers to the temporary water rights transaction across different basins. Cross-basin/region permanent trade refers to the permanent water rights transaction across different basins; it's similar to the permanent trade within the basin/region but also slightly different from it. Due to the differences of the water resources management law and water rights transaction procedures and principles in different basins/regions, it is necessary to study in advance the relevant laws, the property rights, cost recovery and pricing, exchange coefficient in water rights transaction, and also including environment issues as anti-salinization etc.

2 Theoretical basis for water rights transaction

2.1 Theory of resources scarcity

The scarcity of resources mainly refers to the imbalanced supply and demand of the resource. The total demand of the resource outstrips the total supply. For one thing, the resources per se fail to meet the demands; for another, although the quantity of resource can meet the total demand, the supply falls short of the demand because of the limits of physical conditions.

As to our country, although the total supply of water resources is abundant, the per capita water resource is poor and the distribution is very uneven in both time and space. Furthermore, serious water pollution and the serious waste of water deteriorate the current situation. Our country faces water shortage at both absolute and relative levels. Therefore, in order to protect the water environment and resources, the scarcity of the resources must be stressed; we should not only prevent the waste and pollution of water resources, but also emphasize the controlled plan of exploration and use of water resources, and as well as balanced supply and demand of water resources. Water rights transaction is the most feasible way in this regard, and it can improve the utilization efficiency of water resources.

2.2 Societal cost theory

With "externalities" as the starting point, social cost theory attains a structure of rights through further limitation of the litigants' right. The allocation of rights is varied and different structures result in different societal costs. The current allocation of resources mainly falls into two categories: market-oriented and plan-oriented. How to diminish the societal cost to the lowest becomes the most significant question.

The plan-oriented allocation is based on administrative measures. There is no market transaction cost, but it does not mean there is no cost at all. The cost of government is varied, which covers costs of information collection, establishment of laws and policies, and guarantee of implementation thereof etc., and the cost is administration cost which replaced market transaction by means of administration orders and decisions. Although this allocation has innate advantages of absolute monopoly, avoiding market challenging, and strong support of coercive power, tremendous cost existed in reality, rather than no cost. At the same time, planned allocation of water resources will make the water price mechanism impossible to be realized, utilization efficiency of water resources unable to be improved, and reduction of cost impossible, and resulting in problems as "tragedy of the commons" and power rent-seeking.

Under market-oriented model, price is fully reflected by the supply-demand relationship. Water users, to a large extent, will adapt their way of using water actively based on the water prices, improve the utilization of water resources, reduce cost and avoid waste, and maximize the value of water resources. To water users, the more water they use, the more cost they have to pay. Au contraire, saving watcr resources is tantamount to obtain an extra income through market.

Just as market allocation laws, there are also failures in market when allocating water resources. There might be public crises in flood control and protection of water ecological environment, etc. caused by the monopoly of water resources and pursuit of productivity, and intergenerational inequity caused by the accompanying negative "external" problems.

From this comparison we can arrive at the conclusion that the market-oriented model is from bottom up and consistent with our self-interests. Whereas the administrative model is from top down and the expectation and reality is inconsistent. These two models have their own advantages and can complement each other. Hence, if we can combine the market-oriented model with the plan-oriented model in a scientific and efficient way and have the "visible hand" of government to protect water resources, control floods, manage residents water prices, maintain the sustainable development of water resources, and the "invisible hand" on the supply-demand mechanism, price mechanism and competition mechanism of market, the allocation of water resources will become more reasonable.

2.3 Theory of environmental and natural resources law

"Agenda 21" defines sustainable development as follows: the development not only meets the needs of contemporary people, but also will not cause damages to the capabilities of future generations to satisfy their demands. The theory views that economic development and environmental protection are equally important. The economic development should not be based upon the destruction of environment or unlimited expropriation of natural resources. We should guarantee that our future generations also have equal opportunities to enjoy the development and social wealth consumption. Water resources are irreplaceable in human beings' survival and development and are the basis of societal economic development. The exploitation of water resources should conform to natural laws within an affordable range, otherwise it will lead to disaster.

The water rights transaction shall be guided by theory of sustainable development and the relationship between the exploitation of water resources and environmental protection shall be well managed, and water rights of contemporaries and future generations be balanced. Only by so doing, we can gradually intensify the idea of "environmental human" in the masses, and have due consideration to benefits of environment, society and economic development, and ensure a benign circulation of water rights transaction, build a society with harmonious relationship between man and man, man and society, man and nature.

3 The necessity of establishing water rights market

Scarcity is the key for the necessity of trade. For quite a long time our country regards the water resources as "ownerless resources", so the constraints on water taking only have the symbolic significance. With growing demand for water resources and the continuous decrease of available water resources, conflicts among many regions and departments for water resources are becoming serious day by day; the scarcity of water resources becomes more and more obvious. How to make the allocation of resources more reasonable becomes the urgent reality that we have to face with water resources.

3.1 Water rights transaction is the inevitable requirement of our country's market economy development

Market-oriented economy is established after the reform and opening-up of our country, which requires respect for the law of value and allocation of resources by market. However, in the field of water resources allocation, there is a big inertia of planned-oriented economy, although it's based on the consideration of the specialty of water resources, and this relatively simple configuration does have conflict with the market-oriented economy.

At present, our country is at a stage at which the economy develops rapidly, demands of industrial water and urban water will further increase and the original water resources allocation system cannot meet the needs in reality, so the productivity and production relations have come into conflict, which cannot be resolved completely by the administrative management and democratic consultation in the traditional system. The severe situation of water shortage calls for a new configuration, and the introduction of market mechanism will definitely become a mainstream of the historical development.

3.2 Water rights transaction remedies the defects of water resources allocation by government

Water resources have double properties—social goods and economic goods and need to be regulated by two means—government regulation and market regulation. China's water resources have been collocated by administrative means for a long time, so the scarcity of water resources cannot be reflected normally. With no earnings by saving water or no increasing cost by wasting water, the water resources allocation is out-of-balance, the alarming waste of water resources in utilization coexist with serious shortage of water.

The reason why many projects with good prospect in market and urgent needed in our country could not be put into action or use is the limitation of water permit; on the other hand, some enterprises saved a large amount of water by means of technical reformations, recycling, and reuse, but these water resources could be in no way put on the stage of market by water rights transaction and be regulated to meet demands by others, it's a waste of resources.

The practices of water rights transaction at home and abroad tell us that it is impossible to reach the goal of protecting, saving and managing water resources comprehensively only depends on the administrative distribution. Water rights transaction can realize the dynamic allocation of water resources among the users, so the "government failure" will be avoided.

3.3 Water rights transaction can improve the utilization of water resources, optimize the allocation of water resources and ease the conflict in water use

Due to the characteristic of river, the upstream has a congenital advantage in using water when compared with the downstream, therefore, there exists the conflict between upstream and downstream in the free and low water distribution mechanism. The conflict also exists within the agricultural departments themselves and between agricultural departments and other departments.

In the current water rights system, both the encouraging mechanism for saving water and improving the efficiency of water use, and the restriction mechanism for destroying water environment and low efficiency of water use are extremely crumbling, even are totally blank. There is no physical makeup for the water resources saved by the construction of water-saving facilities, and the cost for wasting water is quite low, besides, it does not need to undertake the social costs of water shortage at the downstream or to others. Therefore, waste of water resources by the excessive water resources development and blindly use, and the worsening water environment occur frequently in reality.

However, the amount of water resources in a particular region is almost a constant, so the possibility of change is unlikely. In order to solve the problem of scarcity of water resources, increasing the utilization rate becomes the inevitable method. Put it into details: first, "increase" the total volume of water resources indirectly by saving some water resources; second, improving the efficiency of water using and reducing the demand for water. The motivation of aforesaid increase and reduction comes from the pursuit of interests by the "economic man", while water rights transaction system is a platform of increase and reduction.

In this case, user of water resources will invest in the construction of water saving facilities, improvement of water supply equipment, and will get more benefit from the improvement of utilization of water resources; waste of water resources means waste of money and lost the opportunity of selling water rights. In this way, water resources are turned to reasonable flow from low-efficiency use to the high-efficiency use, and the exchange between water users can effectively relieve the conflicts between the supply and demand of water resources.

3.4 Water rights transaction is required by the environment protection

The use of water resources mostly emphasizes on economic value rather than the ecological value, however, ecological value is the foundation of economic value. If water is polluted, the function will decline, and one of the consequences is the undoubted reduction of economic value. Therefore, using traditional administrative measures to protect the environment, and using the negative economic stimulus to prevent and control the pollution will cause a lot of problems such as: the rigid executive order, poor operations, and even the cost of some environmental protection administrative orders are higher than expected. So it is a good idea to coordinate economic development and environmental protection by economic means.

Economic measures have the advantages of higher economic benefits and smaller risks, and it can combine with the market mechanism effectively and provide stable financial resources, etc. The use right of water resources, after the deal, would make the water users internalize the cost as soon as possible. The "economic man" will try his best to improve the utilization rate of water resources and water quality, and pursue the rest of the water resources and the realistic interests, which objectively reduced consumption of water demand and waste of water resources, and give active protection and "externalities" to the water resources. In the end, it will provide the objective space for reasonable allocation of water resources.

4 The attempt of water rights transaction in our country

Although now in China water rights transaction are not mature in theory and legislation, and are still in a "shy and misty" situation, it does exist in real life and the water rights transaction becomes more active and frequent. Especially" Opinions Regarding Water Rights Transfer by the Ministry of Water Resources" that promulgated and implemented in China in January 2005 indicates the start of water rights market construction in our country. In the "Real Right Law" promulgated in 2007, it's regulated that the right of water taking is the usufructuary right enjoyed by the civil subject; even though it is not detailed enough, it still provides a very large space for water rights transaction system, water distribution mode of " market plays the major role and with plan as auxiliary support" is becoming gradually clear. The spring of water rights transaction is coming soon.

4.1 Water rights transaction between Dongyang and Yiwu

On November 24, 2000, the agreement signed between Dongyang and Yiwu for the compensated transfer of partial water rights of Hengjin Reservoir created the first water rights transaction in our country.

Dongyang and Yiwu are located at the upstream and downstream of Jinhua River, and there are big differences between total water resources and per capita amount. In order to achieve a win-win situation, the two city governments signed the water rights transfer agreement. Yiwu one-time buys perpetuity of 49.999×10^6 m^3 cubic meters of water in Dong yang Hengjin Reservoir every year, the water quality shall meet the drinking water standard. The buyer should pay comprehensive management fees ¥0.1 per cubic meter according to the actual amount of water supply and is responsible for water pipeline engineering. Dongyang is responsible for the operation, management and maintenance of the reservoir. The original ownership does not change after the transfer of use right. So far, our country's first water rights transaction gains success in material meaning.

Although the nature of the water rights transaction is on debate, some hold that the deal is essentially water commodities trading, both Dongyang city and the local governments at all levels are not representatives of the state property of water resources, however, the water rights transaction takes a big step forward. It is the qualitative change of water rights transaction from scratch, and undoubtedly is of great significance to the water rights system reform.

Scholars, as AnGang Hu, made the right and precise comments on this water rights transaction. Firstly, it breaks the tradition of water rights distribution by administrative monopoly means; secondly, it marks the formal birth of China's water rights market; thirdly, it proves that the market mechanism is an effective means of water resources allocation.

4.2 Ningxia, Inner Mongolia water rights transaction

Ningxia, Inner Mongolia's water rights transaction is clearly a higher-level project, which realized the compensable transfer of water rights among industries. In short, the enterprises invest in agricultural water-saving facilities and in exchange for the saved water resources to expand production.

Though the industrial water consumption is less in Ningxia and Inner Mongolia regions, the agricultural water consumption here is excessively large. There is a great amount of water waste in agricultural irrigation and water transport. It means that there is a great potential to improve the agricultural water use efficiency, and then the saved water can be convincingly made up for the industrial water shortage.

In December of 2002, the Yellow River Conservancy Commission put forward the method of water rights transaction regarding the serious canal seepage, waste in agricultural water use, and the great potential of water-saving, and proposed to compensable transfer the saved water in the water-saving projects to some energy projects as industrial water. The government and enterprises respectively invested 1/3 and 2/3 into the water saving reform, then the enterprises got, with payment, the use right of the saved water resource, and the agriculture got the financial support from the enterprises. Eventually, this system increased the investment in agriculture, the problem of industrial water use is basically settled, the efficiency of water use has greatly improved, and the actual water consumption greatly reduced. Agriculture and industry can be seen in the coordinated development.

4.3 The south-to-north water diversion project

The distribution of water resources in China is not balanced as the south has much more water than the north. If the project of transferring south water resources to the north succeeds, water shortage in the 13 provinces along the way can be solved; the economic, social and environmental development in the local regions, even the whole nation can be secured effectively.

The south-to-north water diversion project is operated in combination of macro control by the central government, the participation of the local governments along the way, and operation by the enterprises. The enforcement of this project can speed up the construction of the new water resources management system and make some theoretical explorations possibly in practice, which lays a practical foundation for further water rights transaction.

The above water rights transaction is special and representative in China. Lots of water rights transactions have appeared in some regions like Zhejiang Province, Liaoning Province, and so on. The transaction system has been accepted and operated by more and more people and places.

5 Proposals for the legal system of water rights transaction in our country

The legal system of water rights transaction in our country has already achieved a considerable progress both in theory and in practice. However, there are still some problems. In order to ensure a healthy development of water rights transaction, the related systems must be improved.

5.1 Improve the related systems of water rights transaction

Water price occupies a significant position in the water rights transaction. The water price in

China has been too low for a long time, and a series of social and ecological problems were caused thereby. Establishing a reasonable water price system becomes a more and more urgent problem to be solved.

It is unnecessary to debate on whether water resource has value or not. Water resource not only has the economic value which has been recognized traditionally, but also has the ecological value. Because of the unreasonable water price, for a long time, water resource was supplied free of charge or low price, which has been regarded as a kind of "social welfare". This action of ignoring the market attribution of water resource and violating the laws of economics led to enormous waste of water resource, which objectively caused many social and ecological problems.

The traditional standard of water price merely by water consumption is not scientific. The distribution of water resources varies in time and space, and the extent of scarcity is also different on the local scale. As a kind of commodity, water resource should follow the law of value and change its price along with the supply-demand relations. However, it does not mean that the water price should be too high to bear in areas where water resource is extremely insufficient or too low in areas that have rich water resource. It means that water price should have an objective reflection on the condition of supply and demand of water resources and its extent of scarcity.

It is essential to take multiple standards in pricing industrial water. Besides the scarcity of water resources, many other factors should be considered, such as degree of contamination, duration of water supply and the existence of engineering measures to improve utilization efficiency or improve water quality by water conservation. The agricultural water management in our country has been long-term disordered and the proportion of agricultural water use is oversize in the total water consumption. Extracting ground water to irrigate cultivated land free of charge or just by taking some symbolic cost is a very common phenomenon, for which, our government has not come up with any effective measures to control or supervise.

Although it's stated in the provisions of Regulations on Administration of Water Abstraction Licensing and Collection of Water Resources Charges to charge for water resources, its legal hierarchy is apparently not on the same order with the problems to be solved. With the deficiency of operability and government's extensive management, its effect is limited in reality. Low water price not only caused inadequate performance and loss of the water supply company, but also restrained the implementation of water-saving policy and hinder the development of urban water supply.

The low water price not only made people lack of motivation for water saving, for water conservancy department; it is also a stupendous financial burden. Water rate revenue is the main source of funds for water conservancy department, as the water rate can hardly balance the cost of operation and management, the necessary maintenance of many water projects couldn't be carried out, and the efficiency of water related operation is low. Thus agriculture and related industries have been influenced in different levels. The cost recovery after processing of water resources directly concerns the motivation of construction and management of this project, meanwhile, payoff and profit of the investment are considered with no doubt before construction of the project.

The factor of environmental value in price mechanism of water rights transaction should also be taken into account. First of all, the water resource itself is valuable, and the consumption thereof is required to be compensated; secondly, the sewage produced in the water rights transaction will directly lead to environmental contamination and destruction, and objectively, it will also indirectly make negative "external" effects; it is no doubt that the pollution treatment and ecological protection require payment. Therefore, in accordance with the "polluter deals with it" principle of environmental and resources law, environmental water price should be charged.

In order to solve the problems of unscientific water price, many factors should be considered when pricing the water, such as supply-demand relations, the particularity of the space and time, production value of water, resource value of water and environmental value of water. The scientific water price can improve people's attitude toward the consumption of water resources, adjust people's motivation to develop water resource and promote the harmonious and sustainable development of society, economy and environment.

5.2 Establish a perfect registration mechanism for water rights transaction

Water rights transaction needs identifying its ownership by "chattel" public, meanwhile the country needs to arrange the management of water resources rationally. Under the precondition that water rights transaction contract is completed, it would be a necessary process in the water rights transaction to record the water rights transfer at the water rights registration authorities. Water resources as a kind of "chattel" should make real property registration in the water rights transaction, because the water rights transaction must have credibility.

Credibility aims at serving as the transaction security and trust protection basis in case of errors in the publication, and hence maintain the transaction expectation. In the process of registration and reviewing, in order to ensure the safety of the water rights transaction it is necessary to ensure the accuracy of the information and to perform substantial examination of items required to be recorded. A party, which intended to take part in the water rights transaction, can get the related water rights information through the publication of the registered water right transaction. Based on the trust of the registration, the party does not need to spend too much time or energy on investigating conditions of water rights, so that cost of water rights transaction is reduced for the involved parties and efficiency of the water rights transaction improved.

At the same time, the registration mechanism of water rights transaction is open to a third party; the third party can have access to the status of water rights changes and can anticipate the possible effects to him and environment. So any adverse consequences can be avoided effectively.

To the country, as a manager, the registration authority of water rights transaction should be the same as the authority that issues the water taking license; through the registration mechanism, the water rights transaction can be guided scientifically and supervised effectively.

References

Huang Xisheng. A Study of Water Rights Transaction [M]. Beijing: Science Press, 2005: 117. in Chinese.

Lin Long. A Study of China Legal System Construction of Water Right Transaction [D]. Fuzhou: Fuzhou University, 2005. in Chinese.

Zhang Rentian, Chen Shoulun., Tong Lizhong. Water Right Distribution and Water Right Transaction System in Water Right Market [J]. University Journals of North China Institute of Water Conservancy and Electricity, 2002, (6). in Chinese.

WCED. Our Common Future (Chinese Version) [M]. Changchun: Jilin People's Press, 1997: 52. in Chinese.

Hu Angang., Wang Yahua. The Optimal Allocation of Water Resources in The Transition Period [N]. Guangming Daily, 2001-5-25. in Chinese.

Comparative Application of Methods for Water Resources Allocation of Trans – cantonal Rivers in Southwestern China

Gu Shixiang, *Yang Xiao*, *Zhang Yurong*, *Ma Pingsen* and *Mao Changshu*

Yunnan Survey and Design Institute of Water Conservancy and Hydropower, Kunming, 650021, China

Abstract: This paper mainly uses a case study of the Dianxi River in the southwest region of China, to discuss three methods respectively, which are based on water – use quota (WUQ), integrated watershed water resources planning (IWP) and analytic hierarchy process (AHP), for allocating water resources of trans – cantonal rivers on the principle of "fairness and justice first, efficiency moderately considered". The comparative results indicate that WUQ – based approach with advantages of clear conception and less data needed, may be applied to make a broad – brush allocation for reference. IWP – based approach, featuring high complexity of computation and more and detailed requirement of data, takes full account of the balance among the supply, depletion and emission of water, as well as socioeconomic development. However, this approach may result in benefit encroachment of vulnerable groups or developing areas. AHP – based approach, considering the priority ranks of water uses, pays more attention to the comprehensive influence of multi – index in each study sub – region when giving priority to fairness and efficiency. But the complexity of the factors involved in the AHP – based approach, the subjectivity of determining the index system and the arbitrary of obtaining the data may affect the rationality of the result. It is thus suggested that all the three methods should be synthetically applied to ensure the scientific and rationality of water allocation scheme at the next stage of administrative consultation and arbitral confirmation.

Key words: water resources allocation, water – use quota, watershed water resources integrated planning, analytic hierarchy process

1 Introduction

Water scarcity is already causing revolution in economic growth mode and impelling many countries to take actions for insuring sustainable utilization of water resources, such as the European water framework directive, south Africa's national WR strategy, USA Colorado RB Compact, Pecos river basin compact , Pakistan – Indus water accord, et al. Besides controlling extaction or in-stream water users, these directives or compacts propose water quality target and environmental water requirements. Some areas in China have developed water resources allocation methods, such as availability and water allocation scheme for the main stream of the Yellow River, water allocation scheme of Heihe River, water allocation scheme of Yongding River, initially allocating water resources among administrative regions in the Daling River Basin, water allocation scheme of Poyang Lake, water allocation scheme of Dongjiang River, et al. The methodology of water resources allocation includes prediction method of water – use quota, classification weight method, analytic hierarchy process, et al. Wang et al. presented two methods for allocating initially water rights in their case study of the Daling River, which are based on the principle of distribution and mathematical model of water allocation respectively.

Ministry of water resources proposed the main goal, index system, every river basin's decomposable limit and key tasks of "water resources developing and using red line", "water – use efficiency controlling red line", "water function area pollution discharge limitation red line" during 2010 ~ 2015 in "working scheme about carrying out the strictest water resources management system". Every province must determine each city's and county's water abstraction licensing of water quantity, accelerate making water resources allocation scheme, and strive for completing water resources allocation scheme of important rivers before 2015. Yunnan province situated in southwest region of China, total water resources amount is relatively abundant, but the wet and low water of precipitation is very great, and the water resources are mal – distributed in space and time. The development of economic, social and water resources between upstream and downstream, left and right bank is not equilibrium. Resources water shortage, engineering water shortage and water quality – induced water shortage exist in local area. Study on water resources allocation was almost at water deficient areas recently. Systematic study and practice in view of characteristics of southern area's water resources development and utilization are less commonly. Because of the differences of water resources 's natural endowment and development and utilization level between different areas, the involving influencing factors on carrying out water resources allocation are multiple and complex, and the technical difficulties are great. So, this study has great significance for carrying out water resources allocation in the southern region.

2 Data and methods

2.1 Regional general situation

Yunnan province started the water allocation of trans – cantonal rivers with six experimental unit rivers at February, 2011. On the basis of "Yunnan water resources integrated planning", "Yunnan river basin comprehensive planning revision", "The twelfth five – year plan of Yunnan province water conservancy development ", "Key water source project at five provinces in southwest China's recent construction planning" which are finished in recent years or are being carried out, different allocation methods is used. Comparison of suitable scheme is chosen, which should be approved by water managing administrative unit at all levels of province, city and county. Dianxi River is one of the six experimental unit rivers.

Dianxi River, which is the tributary of Nanpanjiang River on the left bank, originates at Shashipo, Xiongbi town, Shizong county, Qujing city, then affluxes Nanpan River at southern Lingluo village, Pengpu town, Mile county, Honghe state. Its basin area is 4,416.1 km^2, and total water resources amount is 1.28×10^9 m^3. Dianxi River basin involves Luliang county and Shizong county of Qujing, Shilin county of Kunming, Luxi county and Mile county of Honghe. The total population of this area is 8.49×10^5, and the Gross Domestic Production (GDP) is 12.8 billion Yuan. The product value ratio of primary industry, secondary industry, and tertiary industry is 14:66:20. The cultivated land area is 13.1×10^4 hm^2, and the effective irrigation area is 4.64×10^4 hm^2, so the degree of effective irrigation land is 35.2%. The total water supply of current hydraulic engineering is 3.196×10^8 m^3, so water supply ratio of water storage projects, diversion projects and lifting project is 71:23:6. In total social water using amount, the domestic water is 4.09×10^7 m^3, industrial water is 1.832×10^6 m^3, and agricultural water is 2.604×10^8 m^3. The water consumption per capita is 376 m^3, and the water consumption level is 250 m^3 per 10,000 Yuan's GDP, but the water resources amount per capita is only 1,507 m^3. The utilization percent of water resources is 25%, which far exceeds Yunnan's and national average utilization percent of water resources, so this region have severe problems of water shortage.

2.2 Principle of water resources allocation

According to actual conditions of southern area, especially Yunnan's river basin, the general principle of water resources allocation is determined as: ①Sustainable utilization of water resources should coordinate with economic and social development; ②Previous and present water – use condition should be fully considered; ③Integrated plan water resources on the principle of fairness and justice first, efficiency moderately considered; ④Domestic water and environmental flows should be guaranteed first; ⑤ Democracy consultation and administration confirmation should be considered simultaneously; ⑥ As the principle of protecting groundwater resources, groundwater exploitation should be limited in the local area or at special emergency period, so this paper didn't consider the groundwater resources.

2.3 Basic data

(1) Hydrological data, meteorological data and water resource amount of Dianxi River basin, which is recorded monthly from 1956 to 2000.

(2) Social and economic data, land using statistical data of Dianxi River basin from 1980 to 2008, "Yunnan province water resources bulletin" from 1998 to 2008.

(3) Design data of hydraulic projects which are finished or planned to construct in Dianxi River basin, such as storage capacity curve, operation rule line, water supply quantity, reservoir runoff and evaporation of reservoirs of which the storage capacity is larger than 1,000,000 m^3.

(4) Yunnan Provincial standard "water - use quota" (DB53/T 168 - 2006), in which inhabitant and municipal, industrial, livestock, agricultural irrigation and eco - environmental water - use quota of every administrative region are contained.

2.4 Calculation method

2.4.1 Water - use quota predicting

This method is based on present water consumption of each region (city) in Dianxi River basin. Water - use quota is the control index. Through predicting the water consumption scale and controlling water - use quota, using this method can calculate the water resources allocation proportion of each region. The key of water - use quota predicting is reasonable for predicting water requirement in the future years. It reflects the principle of fully considering present water - use condition, acknowledging the difference of water - use quota and water - use efficiency in each region.

2.4.2 River basin water resources integrative planning

River basin water resources integrative planning adopts the water resources allocation method proposed by "National water resources integrative planning technology detailed rules". On the basis of investigating and analyzing economic and social development, water supply, water consumption and water discharge balance, this method predicts economic and social development, water saving and protection in future, then makes water resources supply and demand balance analysis three times for present and future years at each region in the river basin combining with the water supply scheme. Through combining and analyzing possible measures of reasonably restraining water demand, effectively increasing water supply and actively protecting ecological environment, reasonable water resources allocation scheme is got, so the water resources allocation proportion is obtained. This study uses MIKE BASIN as calculation tool, which is usually used in planning and managing water resources of river basin or scientific research aboard. Recently, MIKE BASIN is used in water resources integrative planning, water resources assessment for the project construction, analysis on transferable water volume of diversion projects and lifting project and reservoir scheduling in China.

2.4.3 Analytic hierarchy process

Analytic hierarchy process (AHP) is a system engineering method which integrates qualitative analysis and quantitative analysis. It is suitable for complicated fuzzy comprehensive evaluation system, and it is widely used for determining the weight. The main steps are: ①Establish hierarchy structure diagram, choose rule hierarchy and index hierarchy according to principles and requirements of water resources allocation; ②Through comparing the index of every hierarchy, construct scheduling and judgment matrix of every single hierarchy index weight, then make consistency check for the matrix; ③ On the basis of calculating the index weight of single hierarchy, sort the hierarchy from the upper layers to the lower layers according to the hierarchy structure diagram, then make the consistency check; ④ Through calculating evaluation value of each water consumption region's index, construct the scheme decision matrix to calculate water allocation weight. The main calculation formula is :

$$G = W \times R = \begin{bmatrix} W_1 \\ W_2 \\ \vdots \\ W_n \end{bmatrix} \times \begin{bmatrix} r_{11} & r_{21} & \cdots & r_{m1} \\ r_{12} & r_{22} & \cdots & r_{m2} \\ \vdots & \vdots & \vdots & \vdots \\ r_{1n} & r_{2n} & \cdots & r_{mn} \end{bmatrix} = (g_1, g_2, \cdots, n) \tag{1}$$

Where, n is the numbers of index; m is the numbers of the water consumption region; W is the index weight vector; R is the evaluation value matrix of water consumption region; G is water allocation weight matrix of water consumption region.

3 Results and analysis

3.1 Water resources allocation result

In Dianxi River basin, the land area of Qujing accounts for 17.1% of the river basin area, the land area of Kunming accounts for 9.8%, and the land area of Honghe accounts for 73.1%. The GDP of Qujing accounts for 7.1% of the river basin's GDP, the GDP of Kunming accounts for 1.5%, and the GDP of Honghe accounts for 91.4%. Honghe's land area is the largest, the economic and society is the most developed. Dianxi River is one of Honghe's main water sources. What Kunming involves is only part villages, water consumption is small. Present water supply of Qujing, Kunming and Honghe respectively are 5.655×10^7 m^3, 5.800×10^6 m^3, 5.57×10^8 m^3. The water supply of Honghe is the largest, and water resource is the most developed, which is 29.6%, over than that of Dianxi River.

The result of Water - use quota predicting for Dianxi River basin in 2020 is that the water requirement of Qujing is 9.489×10^7 m^3, the water requirement of Kunming is 1.234×10^7 m^3, the water requirement of Honghe is 9.489×10^7 m^3. According to this predicting results, the water resources allocation weight of these regions are 0.185, 0.024, 0.791.

On the basis of the river basin's development situation and anural layout about saving, protecting and developing in future, water resource system general network can be constructed by MIKE BASIN. Through inputting basis data, calibrating the parameters, simulating the configuration scheme and outputting the result, the result of river basin water resources integrative planning for Dianxi River basin in future year of 2020 was obtained (Tab. 1).

Tab. 1 The result of river basin water resources integrative planning for Dianxi River basin in 2020

State (city)	Water supply($\times 10^4$ m^3)					Water requirement($\times 10^4$ m^3)					Water shortage($\times 10^4$ m^3)					Water shortage rate(%)
	urban	industry	agriculture	country	total	urban	industry	agriculture	country	total	urban	industry	agriculture	country	total	
Qujing	312	94	8,324	529	9,259	312	94	8,553	530	9,489	0	0	-229	-1	-230	-3%
Kunming	63	169	804	178	1214	63	173	819	179	1234	0	-4	-15	0	-20	-2%
Honghe	2,537	3,795	31,625	2,015	39,973	2,538	3,797	32,231	2,016	40,581	0	-2	-605	-1	-608	-2%

According to the principle and requirement of water resources allocation, fairness and efficiency is chosen as rule hierarchy index. The 13 influencing factors of index hierarchy can be roughly classified into two kinds, namely fairness index and efficiency index. The former includes the population size, agricultural acreage, effective irrigation acreage, total number of livestock, gross agricultural output value, gross industrial output value, gross construction industrial output value, gross service industrial output value and ecological water requirement. The latter includes gross agricultural output value, total grain yield, agricultural output value per 1 m^3 water consumption, grain yield per 1 m^3 water consumption, gross industrial output value, and industrial output value per 1 m^3 water consumption. Based on analytic hierarchy theory, the judgment values which represent the importance degree of rule hierarchy index is chosen, then scheme set can be generated, and analysis and optimization should be done. In this study, the scheme set is composed of three feasible schemes (Tab. 2). Scheme 1 represents fairness first, efficiency moderately considered. Scheme 3 represents fairness and efficiency is equaled important. Scheme 2 lies between scheme 1 and scheme 3. The water resources allocation weight of these schemes is shown in Tab. 3. In Tab. 3, there are water resources amount proportion of three states (cities) shown to compare with the water resources allocation weight.

Tab. 2 Scheme set of analytic hierarchy process

Schemes	Weight of rule hierarchy index	
	fairness	efficiency
Scheme 1	0.7	0.3
Scheme 2	0.6	0.4
Scheme 3	0.5	0.5

Tab. 3 Water resources allocation weight for three methods

Methods	Water resources allocation weight		
	Qujing	Kunming	Honghe
Water resources amount proportion	0.221	0.099	0.680
Water-use quota predicting	0.185	0.024	0.791
River basin water resources integrative planning	0.184	0.024	0.792
Scheme 1 of analytic hierarchy process	0.161	0.068	0.771
Scheme 2 of analytic hierarchy process	0.167	0.076	0.756
Scheme 3 of analytic hierarchy process	0.173	0.085	0.742

When using river basin water resources integrative planning, water inflow series of the area above junction of Dianxi River after deducting the water consumption of social economic development can be got according to the result of water resources integrative planning in 2020. Using the results of environmental water requirement of instream flow as a constraint, the guarantee degree of minimum environmental water requirement can be judged. There are two method to calculate environmental water requirement: the method of "water resources assessment for construction projects" (SL/Z 322—2005) and the method of Tennant. According to the "water resources assessment for construction projects" (SL/Z 322—2005), instream flow should be guaranteed not less than 10% of mean annual flow in principle. According to the method of Tennant, when river flow status is moderate, instream flow should be guaranteed not less than 30% of mean annual flow from June to November, and 10% of mean annual flow from December to next May. When using the method of "water resources assessment for construction projects", in the flow series of 1956 ~ 2000 per month, the guarantee degree of environmental water requirement at the junction of Dianxi River is 99.6%, and when using the method of Tennant, the guarantee degree is 93.5%. Obviously, when using the method of "water resources assessment for construction projects", the environmental flows can be basically guaranteed. Besides, the data of environmental water-use of Dianxi River is very few, so the research about the environmental water-use has not been developed.

3.2 Rationality analysis

(1) The general trend of the results of three methods is similar, and the correlation of each region's water resources allocation weight is big. The supply and demand balance for water resources integrative planning is based on the water demand prediction, which are constrained and guided by "water - use quota" (DB53/T 168—2006). So the result of water - use quota predicting and river basin water resources integrative planning is more similar.

(2) As the result of three scheme of analytic hierarchy process, when the weight of efficiency is larger, the water resources allocation weight of Honghe is smaller. The reason probably is: the efficiency of Honghe is poorer than Qujing and Kunming, so when the weight is larger, the water resources allocation weight is smaller. The water resources allocation weight of Honghe for analytic hierarchy process is less than the weight for the other two methods. The reason is, in analytic hier-

archy process, most index of index hierarchy focus on agriculture. There are two irrigation districts named Jinmaba and Miyang located on Luxi and Mile in Honghe, of which the total area is 2.359×10^4 hm^2. They are main planting area of Honghe and even Yunnan province at which grain, grape and tobacco are planted. The area of these two irrigation districts will be 2.627×10^4 hm^2 in 2020. The agricultural water use accounts for 79.4% of Dianxi River basin, and the present water efficiency of irrigation is only 0.53, so the potential of agricultural water saving is great. These are why when the weight is larger; the water resources allocation weight is less.

(3) The result of scheme 1 of analytic hierarchy process is the most similar with the other allocation, and the index of duty hierarchy for scheme 1 according with the principle of "fairness and justice first, efficiency moderately considered", so the scheme 1 is optimal in analytic hierarchy process.

(4) The comparative results demonstrates that water resources allocation range of each region can be basically determined, which can provide suggestions for the next stage of administration consultation and arbitral confirmation and impel each water – use region to propose reasonable expectation, and then improve the efficiency of consultation and confirmation. On the basis of the water allocation result of three methods, government of the water-use region and relative experts should evaluation the expectation of each method, then collect all opinion, and determinate the finally reasonable and feasible water resources allocation.

4 Conclusions

Water allocation of rivers is the allocation about resources and benefits, so it should reflect the principle of democracy consultation. In this study, three water resources allocation methods were adopted to provide scientific and reasonable reference for democracy decision – making. Prediction method of water – use quota features clear concept and less data needed, can be simple line reference for water allocation. Method based on river basin water resources integrative planning takes fully consideration of economic and social development, water supply, water consumption and water discharge balance, but it can cause the benefit expropriation with disadvantage groups or underdeveloped areas, and has features of complex computation and more and detailed data needed. Analytic hierarchy process considers priority of water consumption in different categories, and pays more attention to the comprehensive influence of water utilization unit's multi – index, when allocation follows the principle of fairness, efficiency can be moderately considered. But it involves numerous factors, both the index system and the data obtained are imperfect and subjective, that can affect the rationality of the result. At the next stage of administrative consultation and arbitral confirmation, all the three methods should be comprehensively used to ensure the scientific and rationality of water allocation scheme.

References

Porter J W. International Practice and Trends in Water Resources Management & Allocation [C]// International Seminar on Water Resources Allocation. 2010. Beijing.

Liu C M, et al. Water Science & Technology in China: A Roadmap to 2050 [M]. Beijing: Science Press, 2009.

Li R C, Beek E. Equity Principles in Integrated Water Allocation in the Yellow River Basin, Part 1. Philosophical Basis and Guidelines[C] // Proceedings of the 4th International Yellow River Forum on Ecological Civilization and River Ethics. Zhengzhou: Yellow River Conservancy Press, 2010.

Beek E, Li R C. Equity Principles in Integrated Water Allocation in the Yellow River Basin, Part 2. Allocation Algorithm and Implementation [C] // Proceedings of the 4th International Yellow River Forum on Ecological Civilization and River Ethics. Zhengzhou: Yellow River Conservancy Press, 2010.

Cao J T. Review of Domestic Practice [C] // International Seminar on Water Resources Allocation. 2010. Beijing.

Wang H, Dang L W, Qi Y L, et al. Theory and Practice on Allocation of Initial Water Rights in Watershed[M]. Beijing: China WaterPower Press, 2008.

Chang B Y, Xue S G. Water Resources Reasonable Allocation and Optimization Scheduling of Yellow River Basin[M]. Zhengzhou: Yellow River Conservancy Press, 1998.

Gui F L, Hu T S, Xu X F. Inter－regional Water Right Distribution of Fuhe River in China[J]. Engineering Journal of Wuhan University, 2007,40(3):27－30.

Hu S Y, Wang Z Z, Wang Y T, et al. Total Control-based Unified Allocation Model for Allowable Basin Water Withdrawal and Sewage Discharge [J]. Sci. China Tech. Sci: 2010,40(10): 1130－1139.

Gu S X, Wu L Q. Study of Central Yunnan's Water Resources[M]. Kunming: Yunnan Science & Technology Press, 2005.

Gu S X, Li Y H, He D M, et al. Watershed Water Resources Three-allocation Based on MIKE BASIN[J]. Journal of Water Resources & Water Engineering, 2007,18(1):5－10.

On the River Sustainable Development in View of Third – party Effects of Water Trading

Han Jinmian and ***Ma Xiaoqiang***

School of Economics and Management, Shaanxi Xi'an, Xi'an 7710127, China

Abstract: The scarcity of water resource makes the people in the arid areas take water trading to meet the increasing water demand in the region. In the process, the efficiency of water resource allocation and utilization has increased. However, water trading often gives priority to the interests of both the buyers and the sellers, whereas the third parties affected by water trading do not enter the game process of water trading because there is no institutional protection. Thus in the perfectly market – oriented economy, it is difficult for the third – party victims to give pressure to the water buyers and sellers through the market mechanism, and it is difficult to obtain compensation from water trading parties. The presence of third – party effects of water trading gives a serious challenge to water markets and sustainable development of the river. On the basis of studying a large number of domestic and foreign literatures, this paper explores the definition and types of third – party effects of water trading. In particular it studies various third – party positive and negative effects of water trading, and analyzes how the positive effects of water trading promotes the river sustainable development and how the negative effects of water trading hinder the river sustainable development. It finally analyzes the compensation for the positive third – party effects and the governance for the negative third – party effects in order to promote the river sustainable development in water trading areas.

Key words: river sustainable development, water trading, third – party effects

1 The definition and nature of the third – party effects in water trading

The third – party effects in water trading refer to that the water buyers and sellers, who are two primary parties in water trading, bring losses or benefits to the interests of the third parties in their water transaction, the third parties assume these losses or accept these benefits, and do not get deserved compensation for the losses or do not pay the corresponding costs for the benefits. Specifically, the third parties in water trading include those who hold vested water rights but that may be at risk in the process of transaction, and those who are non – trading parties but their relevant economic, social and environmental benefits are involved in the water trading.

To consider the efficiency and fairness in water trading simultaneously, the third – party effects become an inevitable topic. Water resource is a kind of common resources, and water trading is different from the trading of private goods. The two primary parties in water trading, that is water buyers and sellers, negotiate in their own best interests, and may sacrifice the interests of the third parties. For example, many countries including China stipulate the basic principles of water trading, which include water utilization must conform to the public interests, and not to cause losses to the third parties. But in the real economic life we can often see the phenomenon that unlimited water trading causes the large – scale reduction of instream water quantity and the abatement of its diluting and purifying function. Furthermore, this may worsen water quality and it is likely to cause a series of ecological problems, such as the aquatic organism survival security problem, wild animals drinking water problem and desertification problem.

However, the third parties affected by water trading do not enter the game process of water trading because there is no institutional protection. Thus in the perfectly market-oriented economy, it is difficult for the third – party victims to give pressure to the water buyers and sellers through the market mechanism, and it is difficult for the third – party victims to obtain compensation from water trading parties. The effective operation of the market inevitably requires that the third – party effect in water trading be recognized. Only when all the relevant costs in the process of water trading can be reflected, water trading market can be called an effective market, so the existence of the third – party effects is a challenge to the market.

Water trading changes the original way of water utilization, and has an influence on the water using parties and water environment. If water trading leads to the increase of consumptive water, the river flow will get less in the long run. Only when the efficiency of consumptive water is improved, it is possible to create more water supply. If we do not improve water utilization efficiency and allow the water trading, it will inevitably affect the third parties' interests. From the nature of the effects, the third – party effects in water trading can be divided into two types: positive effects and negative effects. The positive third – party effects in water trading refer to that the water trading has a good effect on the third parties, but the third parties don't offer the primary parties corresponding compensation. It may cause the increase of the third parties' utility or the decrease of the primary parties' cost, which means that the primary parties pay for the cost and increase the interests of the third parties. The positive third – party effects also mean that in the water trading, social benefits are greater than private gains. The negative third – party effects in water trading refer to that the water trading brings others losses, but the primary parties do not pay for the costs. This may cause the reduction of the third parties' utility or the increase of their cost, which means that the third parties pay the price and increase the primary parties' benefits. The negative third – party effects also mean that in the water trading, social cost is greater than private cost. In practice, more water trading has had negative third – party effects than positive third – party effects, which reduces the efficiency of water trading, and causes a deviation of "Pareto optimality" state.

2 The positive third – party effects in water trading and its promotion to the river sustainable development

2.1 The positive ecological effects

Water trading can be regarded as an important means of increasing river flow and recovering the ecology. When the initial water rights have been distributed completely by different water using parties, the new added ecological water needs such as to save endangered species, to recover wild species habitat can only be met by water trading. The government or the social organizations which protect water resources can serve as the transaction parities to purchase the water, and then control the water in their hands or to let the water flow back to the river, to achieve the goal of reducing water using amount and maintaining water ecological balance. In the western regions of America, Idaho state water bank redistributed the water in the upper reaches of Snake River to the downstream areas and Columbia River to save the endangered salmon there, this is the reason why the federal government is the main water buyers of Idaho water bank, and it purchased all water rights of several sections of the river for salmon protection.

2.2 The positive local economy effects

When the water can be traded, the enterprises that are not assigned to the initial water recognized they can get the water resources by buying the water of farmers, so that they will invest in the region. This is beneficial to the development of local economy because the investment of enterprises makes local employment opportunities and taxes increase. The revenue of selling water in local the region further promotes the local economic development, and changes local agriculture – based industrial structure. In internal agriculture industry, there is also adjustment and optimization of the industrial structure, because water trading makes the farmer understand the potential economic value of water resources, and induce them to invest in high-efficiency and water – saving agriculture in order to get more water selling returns.

2.3 The positive technological progress effects

To improve the efficiency in the utilization of water resources and gain profits from water trading, agricultural water holders will adjust crop structure, use water – saving facilities, and improve water supply equipments, which will stimulate the technological innovation of water supply equipments and water facilities. After water trading, the urban water suppliers also improve water supply equipments and sewage processing facilities, so as to resell the excess water resources to urban residents, which also induce technological innovation of the urban water supply equipment. Water trading also needs accurate measurement of water resources, timely water – loss control equipments, which all require the innovation of water – measuring facilities. In addition, water trading also requires more standardized canals to decrease the water loss rate, so the lining of original sediment canal becomes essential. Data show that the water leakage of ditches in NingMeng irrigation zone of the Yellow River basin reduced by 50% after the lining.

3 The negative third – party effects in water trading and its hindrance of the river sustainable development

3.1 The negative water supply reliance effects

The reliance of the water supply refers to the possibility of water rights holders to get a certain amount of water resources in certain time. If water rights are not clearly defined among the water users, the water transferred or transacted to consumption use by water holders in the upstream or main stream of a river may affect water quality of water quantity of the downstream water rights holders, and have negative third – party effects on the water users in these regions. The effects water trading brings to other water rights holders include the decrease of water supply quantity, the destruction of the water quality, and loss of water use flexibility. That is, the water trading will affect the reliability of the water supply of downriver water users.

3.2 The negative local farmers and agricultural economy effects

Farmers sell the water that they are assigned to industrial or other departments, and get temporary increase of income, but in the long run, if a large amount of farmers sell their water, it could lead to the reduction of agricultural irrigation area and farming area, and the decline of agricultural production, which may cause the land retirement and the reduction of agricultural economy based on the water conservancy. Decreasing of irrigation agriculture would cause negative multiplier effects, including the decline of local employment, and the negative effect on the local retail, restaurants and service industry. Some economists consider water trading will not create sudden great influence on local economy, because only the low – benefit "marginal land" that is not suitable for farming and irrigation is sold, and the farmers concentrate on working on the high quality and high irrigation benefit land. But it has proved that not only those "marginal land" was bought, the water buyers also bought the water rights of high quality land. In Arizona and Colorado, the high – profit pecan garden was brought, and the land of low-economic value was kept. In Arkansas River valley of the United States, from the 1960s, when agricultural water was sold to cities, the local agriculture has declined. In the western United States, in order to avoid the damage of water trading to the third – party, if famers expect that their water right will be affected, the states usually allow them to protest (or stop) any parameters change of their water rights. However, because of the hydrologic complexity, the affected water users may not realize that what kind of change will affect their water supply.

3.3 The negative return flows effects

Return flows refer to the situation that water resources are not fully used in the consumption, and become the surface water or groundwater in still good quality, and continue to be used by other water users. The maintaining of natural ecological system needs a minimum return flows, but the water trading parties may not pay attention to this, water trading may cause the decrease of river return flows. If the high – return – rate upstream water is sold to the low – return – rate downstream water users, it will reduce the water flow of the river downstream. If the low – return – rate water is sold to the high – return – rate upstream water users, it may reduce the water flow in the river of both parties. The life of many animals and plants relies on the wetland formed by the return flows, and the reduction of return flows threatens their living. For example, in the San Joaquin valley in California, the lacking of return flows of agricultural irrigation water leads to the gathering of toxic selenium and later the pollution and closing of Kesterson national wildlife refuge.

3.4 The negative water quality effects

Although many countries and regions have paid great attention to water quality change and make special laws and provisions on it, water trading agreements are usually based only on the water quantity, and ignore the possible water quality change led by water trading. Where, when and in what manner of getting water has a profound influence on the water quality. If water trading is from the general agricultural water to industrial water or chemical industry water, it may cause the worsening of water quality of the river lower reaches. Water flow affects the dissolved oxygen levels and other water quality parameters. When the water dries up, the water quality will be also damaged. The waste digestive ability of a flow provides economic benefits related to sewage treatment costs, if there is not enough water, these costs will be burdened by the dischargers and the downstream water users. We can use "damage avoided approach" to evaluate water quality of different water source. Through the water trading of different qualities of water, it can also help us to assess the economic value of the high quality water.

3.5 The negative water origin area effects

Because the people of water origin areas are rarely buyers and sellers of water, their interests are less considered by the water trading parties. When the water of a river is less controlled by the residents in water origin, they will lose the decision abilities, which may have influence on the local economy, society and environment. The impacts water trading has on the water origin areas occur in different time, some of which occur during the water trading, some of which occur after the land retirement because the land can not be irrigated, and still some of which occur after water resource is sent to a new area. In many water origin areas, even before water trading, the rural economy appears to decline, but water trading increases the declining process. In water trading region of the Arkansas River valley, the population of six rural counties in southeast Colorado in 1930 was 68,576, and the population of Pueblo was 66,038. By 1987 the population of the six rural counties decreased to 50,000 and the population of Pueblo increased to 132,000. Local fiscal losses often appear after the water resources are transferred. For the water sources, "perceived water supply" is also very important, if water origin areas can't control the water supply and the water there is considered to dry up, this kind of "perception" will prevent its future development, and the spreading weeds on the retired land deepens this view of water origin. If water buying increases the land prices a area, the availability of land that can be developed will be affected. Many people in water origin consider that the water produced in their living areas should belong to them, and transferring the water out of there may cause they suffer from great losses, which can not be compensated through the fiscal transfer payments.

3.6 The negative ecological environment effects

The impact of water trading on the environment is broad. It will not only influences the water flow that fish, wildlife and plant needed, but also affects the integrity of the river channel, the potential of hydraulic power. It can also cause the change of ecological system structure and health state of the river basin. Sometimes the ecological value of the water in the basin is equal to or greater than its value of consumption use, especially in some important areas for water entertainment and wildlife. Water trading should consider such utilization in the basin and the environmental consequences. The Mono lake, the second largest lake of California, showed the negative effect caused by the water drying up in the basin. Water trading makes the ecological environment of the fish, wildlife habitats along the Arkansas River and south Pratt river in the United States and the coast areas deteriorate continuously. Cottonwood trees along the coast of the river died because of lacking of irrigation, which also led to the reduction of wild animals and plants, the wetland dried up because of the lack of water.

4 Regulating the third – party effects in water trading to promote the river sustainable development

The third – party effect of water trading affects the effectiveness and fairness of water market. From the perspective of system theory, human water flow, water quality, and water environment have reached a stable and dynamic balance in the long natural evolution. In this system, the water resources carrying a threshold of human activities, once the threshold is broken by humans, parts of or all the balance in the system will be broken, which endangers humans' development and living. This means there exists problems that the water resources allocation and utilization from the "economic man" perspective, and it needs the government to play the regulatory function in the water trading market.

4.1 The key points to regulate the third – party effects in water trading

4.1.1 To scientifically assess the third – party effects

Water resources and water ecological environment is one of the important human welfare. The impacts of water trading on water quality, water environment and other water users need scientific assessment, which is the basis for the implementation of compensation of the positive effects and the government for the negative effects in water trading. We should strengthen general designing, dynamic tracking and effective improvement of the third – party effects assessment system in water trading. The index we select should not only reflect the important attributes of water resources and water environment, but also identify and consider the benefits and losses that the third parties will face. We should provide the scientific information for decision – makers to understand the third – party effects. We should assess the smallest water flow that the river basin needs and the minimum returns flow in order to maintain water ecological environment and water quality, assess the influence of water trading so as to recover the flow instream and the vegetation and ecology along the coast, because studies show that the benefits of water insteam use are more than outstream use and protecting specific flow for water quality, ecology and entertainment can promote regional economic development.

4.1.2 The government laying down regulations on water trading

Since the third parties haven't participated in the water trading process, and they not even have qualifications to game, for this specifically weak group, it needs the government to make institutions to protect their interests. Besides, for the positive third – party effects, beneficiaries get water trading income freely, this also need the government to coordinate, to compensate the contributors of the positive effects such as the trading parties and the people of water origins. The government should pay attention to the hidden cost when examining and endorsing the water trading agreement. The government can set up relevant agencies to limit the scale of the water trading and prevent possible negative effects, and can also strictly limit the water trading that seriously affect the third parties' interests such as water quality hazards, and can also put forward to coordinate the relationship between the primary party and the third parties.

4.2 The compensation measures for the positive third – party effects in water trading

The compensation for the positive third – party effects in water trading aims to the incentive. If river upstream people saving water improves the downstream water ecological environment, promote the downstream social and economic development, they should be compensated. Especially, sometimes in order to guarantee the water quality that traded to lower reaches, upper river or water origin people take the responsibility to protect the ecological environment, close some of the industries and sacrifice their own developing rights, the loss of the development opportunity should also get compensation. People of water origin hope to get compensation for tax losses and economic losses, and also want to get revenue of water selling to the city, but their most urgent hope is to leave enough water to support local economic development and population growth.

4.2.1 To adhere to the Benefit Party Paying (BPP) compensation principle

Water resource is a type of common resources, which is scarce and valuable. Water saving through water trading brings the third - party benefit, and the beneficiaries should compensate to the contributors of positive effects. The BPP principle not only ensures the beneficiaries' stable and sustainable benefits, but also is a kind of incentive for the provider of the positive effects.

This principle specifically includes: river downstream provides compensation for the investment of the direct costs that the upstream region people contribute; to provide compensation or allowance for the opportunity cost for the lost development opportunities of the river upstream water environment protectors; to compensate to the group or individual that contributes to the ecological construction according to their investment and opportunity cost. In the water trading of Colorado river, the water origin asked the purchaser, NCWCD, to build a ten million dollar "compensatory storage reservoir" to compensate for the outputted water resources, and later it became a twelve million direct economic compensation. In addition, it also includes a series of other compensation and donation to regional future economic development and animal research.

4.2.2 To establish the reasonable compensation paths

At present, the compensation paths mainly include the following: ①Fiscal transfer payment. When the responsible party of the third-party effects is not clear, or it is unable to compensate, the fiscal transfer payment can be used to ensure the positive effects provider is compensated. ②Setting up a special fund. It emphasizes more on the specialness of the funds, such as the special compensation to the water quality protection in water origin and water conservation and other special purposes. ③Loans and even an interest - free loan for ecological protection, and technical transformation of the positive third - party effect providers in water trading all belong to the choices of compensation paths.

4.2.3 To explore multi - channel compensation methods

Compensation methods will usually affect the compensation function and effects achieved. It can be cash, policy, technology and human resources, et al. At present, the methods of compensation mainly include the monetary compensation and the goods compensation. Monetary compensation is the most typical compensation method. It is mainly put into practice by some concrete forms such as water resources fee compensation, water environmental pollution compensation. Its advantages are clear, exact and convenient. Goods compensation is to take a definite object as compensation. It is very practical. In addition to the two basic compensation methods, the compensation methods at groping stage also include technology and intelligence investment compensation, policy compensation. Technology and intelligence investment are more suitable for providing the special knowledge, technology consultation and technical service, conveying staff of specific skills, and providing the technical materials and related equipments. Policy compensation are mainly used by the central government.

4.3 The governing measures for the negative third - party effects in water trading

4.3.1 To enact laws to protect the third parties' interests

The government should enact laws to protect the interests of the third parties, and protect the water quality, water origin, ecology along the river, wetlands and endangered species. The water law in the western United States develop towards two directions: to reduce transaction costs and to reduce the negative third - party effects in water trading, but the two directions have contradictions.

Although China's current legal system begins to tend to encourage water trading, it still does not meet basic requirement of protecting the interests of the third parties. Specifically, we should clear the role orientation for water using parties by legislation, give them corresponding rights and obligations. At the same time, we should pay attention that during this transition from planned economy to socialist market economy, water trading still needs a lot of conditions. It is not pure market behavior, water market must be on the basis of regulating by the laws and regulations, the purpose of which is to balance personal interests and the social benefits, and to control the behavior that will damage the third parties' interests in water trading.

4.3.2 To promote the third Parties' participation in the process of water trading

An important way of governing the third – party effect in water trading is to encourage the third – party to participate in the trading process. The government promoting ways of the third – party participation include: ①To allow the third parties to attend hearings. Water trading should hold hearings to let the third – party representatives state their opinions and views to ensure the fair and comprehensive consideration about their interests. ②To have legal power to affect the water trading conditions and to postpone their consent of the water trading. When the third – party representatives request to delay the approval of the deal, and change the trading conditions, we should make their "bargaining" effective. It will inspire both trading sides and the third parties to negotiate their interests. ③To specify an agency to represent the third parties' interests, so that it can exert their function of interests representation and interests proposition. ④To provide support on funds and legal aspects. The third – party that is economically disadvantaged may need some financial aid to do the research and collect the evidence of the trading impacts, and hire lawyers and experts to take part in. The government can consider providing these kinds of aid.

4.3.3 To set up the protection mechanism for the third parties' rights and interests

Market mechanism has its limitations when dealing with the third – party effects, so it needs the government to set up the protection mechanism. In China, the Ministry of Water Resources in 2005 regulate that "water trading can't harm the interests of the third parties ", which is the base to protect the interests of the third parties. But it is broad and vague. The approaches to establish the protection mechanism includes: ①To identify and evaluate the third – party losses, especially in the choice of methods and tools. Many states in the western United States require public interest review of water trading. The third parties need hydrological scholars, lawyers, engineers and other experts to confirm their potential losses. ②To establish and complete the compensation mechanism and to pay for the third parties. Water sellers can directly pay the compensation fund to the third – party victims, and we can also let the water acquirer build "compensatory reservoir" in water origin. ③To innovate the water trading mechanism to internalize the external cost. The cost of reducing the negative third – party effects should be internalized as the water trading costs.

References

The Detailed Instruction of This Method refers to Gibbons[R] // The Economic Value of Water. Washington, D. C. Resources for the Future, 1986.

McGinnis B G M, Rait K, Wahl R. 1990.

Transferring Water Rights in the Western States: A Comparison of Policies and Procedures[R]. Boulder: University of Colorado, Natural Resources Law Center, 1990.

Liu Hongmei, et al. The Return Flows in Water Trading[J]. Fiscal and Economic Science,2006 (1):64 – 71.

Green G, et al. Water Allocation, Transfers and Conservation: Links Between Policy and Hydrology [J]. Water Resources Development, 2000,16(2).

Tarlock D. The Endangered Species Act and Western water rights[J]. Land and Water Law Review,1985.

Anna Heaney, et al. Third – party Effects of Water Trading and Potential Policy Responses[J]. The Australian Journal of Agricultural and Resource Economics. 2006,50:277 – 293.

Charney A H, Woodard G C. Socioeconomic Impacts of Water Farming on Rural areas of Origin in Arizona[J]. American Journal of Agricultural Economics, 1990, 72(5):1193 – 1199.

Howe, et al. The Economic Impacts of Agriculture – to – urban Water Transfers on the Area of Origin: A Case Study of the Arkansas River Valley in Colorado[J]. American Journal of Agricultural Economics, 1990, 72(5):1200 – 1204.

Whittlesey N, et al. Water Policy Issues for the Twenty – first Century[J]. American Journal of Agriculture Economics, 1995(5).

National Research Council. Irrigation – induced Water Quality Problems: What Can be Learned from the San Joaquin Valley Experience[M]. Wash., D. C.: National Academy Press, 1989.

Getches, et al. Controlling Water Use: The Unfinished. Business of Water Quality Protection[R]. Boulder: University of Colorado, Natural Resources Law Center.

Young R A, Gray S L. Economic Value of Water: Concepts and Empirical Estimates. Springfield, Va. National Technical Information Service, Accession No. PB – 210356, 1972.

Howe C W, Lazo J K, Weber K R. The Economic Impacts of Agriculture – to – urban Water Transfers on the Area of Origin: A Case Study of the Arkansas River Valley in Colorado[J]. American Journal of Agricultural Economics, 1990, 72(5):2300 – 2304.

Nunn S C, Ingram H M. Information, the Decision Forum and Third – party Effects in Water Transfers[J]. Water Resources Research, 1988, 24(4):473 – 480.

Oggins C, Ingram H. Does Anybody Win? The Community Consequences of Rural – to – Urban Water Transfer: An Arizona Perspective[R]. Issue Paper No. 2. Tucson: University of Arizona, Udall Center, 1990.

Colby B G, McGinnis M, Rait K, et al. Transferring Water Rights in the Western States: A Comparison of Policies and Procedures[R]. Boulder: University of Colorado, Natural Resources Law Center, 1990.

Botkin D, et al. The Future of Mono Lake—Report of the Community and Organization Research Institute Blue Ribbon Panel for the Legislature of the State of California[R]. Davis: University of California, California Water Resources Center, Report No. 68, 1988.

National Research Council. The Mono Basin Ecosystem: Effects of Changing Lake Level[M]. Washington, D. C.: National Academy Press, 1987.

Committee on Western Water Management, National Research Council. Water Transfers in the West Efficiency, Equity, and the Environment[M]. Washington, D. C: National Academy Press, 1992.

Colby B. Sources of Water: Agriculture—The Deep Pool? In L. MacDonnell, ed., Moving the West's Water to New Uses: Winners and Losers[R]. Boulder: University of Colorado, Natural Resources Law Center, 1990.

Gregg F. The widening circle: The Groundwater Management Act in the Context of Arizona Water Policy Evolution[C]// In Taking the Arizona Groundwater Act into the Nineties, the Proceeding of a Conference. Tucson: University of Arizona, Udall Center for Studies in Public Policy and Water Resources Research Center, 1990.

Committee on Western Water Management, National Research Council. Water Transfers in the West Efficiency, Equity, and the Environment[M]. Washington, D. C: National Academy Press, 1992.

MacDonnell, Lawrence J. The Water Transfer Process as a Management. Option for Meeting Changing Water Demands[R]. Volume I, USGS Grant Award No. 14 – 08 – 0001 – G1538, University of Colorado, Boulder, CO, 1990.

Colby B G, McGinnis M, Rait K, et al. Transferring Water Rights in the Western States: A Comparison of Policies and Procedures[R]. Boulder: University of Colorado, Natural Resources Law Center, 1990.

Key Technological Issues of Designing Flood Control Schemes of the Yellow River

Liu Hongzhen[1], *Li Baoguo*[1], *Wei Jun*[2], *Fu Jian*[1], *Li Chaoqun*[1] and *Song Weihua*[1]

1. Yellow River Engineering Consulting Co., Ltd., Zhengzhou, 450003, China
2. Yellow River Conservancy Commission of the Ministry of Water Resources, Zhengzhou, 450003, China

Abstract: There are significantly serious flood and sediment disasters in the Yellow River catchment. In recent years, compared with the previous situation, water and sediment situations, flood control projects and flood control ideas have changed. Therefore, in order to solve the main problems of flood control and flood control schemes designing, supporting for decision-making, it is necessary to re-design the flood control scheme of the Yellow River. In this study, analysis on recent changes of flood control situation of the Yellow River was carried out firstly. Analysis results show that the main problems of flood control of the Yellow River are as follows: ①flood disasters occurred in the floodplain areas of the lower Yellow River are rather serious; ②flood control operation modes of the Xiaolangdi reservoir during its later sediment-retaining period is not definitely clear. Considering these issues, the major focus of this study is on characteristics analysis of common and small floods of Huayuankou station under current projects condition, and the joint operation modes of the flood control projects system in the lower Yellow River during the later sediment-retaining period of Xiaolangdi reservoir and so on. Therefore, for the middle and lower Yellow River, issues were Studied as follows: ①characteristics common and small floods of Huayuankou Station was analyzed; ②inundation loss of beaches in the lower Yellow River under different flood magnitudes was calculated; ③control indexes and operation modes of Xiaolangdi reservoir during its later sediment-retaining period was presented for common and small floods; ④for different types of floods, the control conditions and joint operation modes of Sanmenxia reservoir, Xiaolangdi reservoir and Dongpinghu project during the later sediment-retaining period of Xiaolangdi reservoir were studied, and the draining sequences of Luhun reservoir, Guxian reservoir and Hekoucun reservoir were analyzed. For the upper Yellow River, the joint flood control operation modes of Longyangxia reservoir and Liujiaxia reservoir for upstream of the Yellow River was proposed, after analyzing the current flood control situation and existing problems. At last, issues to be studied furthermore and relevant suggestions were put forward in this study.
Key words: the Yellow River, flood control, later sediment-retaining period of Xiaolangdi reservoir, common and small floods, flood control for floodplain areas

1 Introduction

The problems of floods control and sediments management are complicated in the Yellow River catchment, and there are significantly serious flood and sediment disasters in the history. Since the founding of China, the Communist Party and the government have taken the flood control and disaster reduction as rather serious issues. Flood control projects have been built, including embankments, reservoirs, and flood storage and detention projects. The flood control project system, which is based on the ideas of *Upper-blocking*, *Lower-draining*, *both banks divided-blocking*, has been formed. In order to control floods of the Yellow River, the State Council gave a written reply to *Extreme Large Flood Control Schemes in the Yellow River*, where Control Schemes for large floods in the Yellow River were designed, and which has supported for flood control decision-making and played an important role in flood management.

However, since 1986, many large reservoirs, such as the Longyangxia reservoir, the Guxian reservoir, the Xiaolangdi reservoir and so on, have been built successively in the Yellow River watershed. The State Council gave a written reply to *Flood Control Plans of the Yellow River* in July 2008, and the Hekoucun reservoir is being built currently. Under the influences both of climate change and human activities, flood control situation has changed significantly. Also, based on thoughts developed recently, such as *People-oriented*, *Maintenance of healthy life of the Yellow River* and so on, basic ideas to design flood control schemes has changed, and economic and social developments put forward new and higher requirements for flood management. Thereafter, as *Extreme Large Flood Control Schemes in the Yellow River* has not suited for current situation, based on present flood control situation and *Flood Control Plans of the Yellow River*, it is rather necessary to study on key technological issues of designing flood control schemes of the Yellow River and design *Flood Control Schemes of the Yellow River*, planning to solve main problems in a certain time and supporting for flood control decision-making.

2 Current main problems of flood control of the Yellow River

2.1 The middle and lower Yellow River

Under the influences of many factors, such as climate change, human activities and so on, characteristics of floods occurred since the 1990s have changed significantly, comparing with that in the 1950s or 1960s. The changes lie in the decreases of flood frequencies, flood magnitudes, flood durations, values of base flows before peak flow, ratio of water volumes upper Hekouzhen station to other parts, and etc. Meanwhile, effected by factors, such as changes of water and sediment conditions and restrictions of productive embankments, the flow carrying conditions of watercourses in the lower Yellow River have also changed remarkably. Firstly, the main water courses have been deposited and reduced, and which results in the decrease of the flow carrying capability. Before the flood period of 2002, the minimum bankfull discharge is only 1,800 m^3/s, and it increases to be about 4,000 m^3/s after operation of Xiaolangdi reservoir. Secondly, bankfull water level is higher than both sides of banks, which caused the harmful *Secondary Perched River* phenomenon and serious *Small flood-Big disasters* situation. At the same time, the floodplain areas of the lower Yellow River, where nearly 1,900,000 people live, are not only channels for flood flowing but also the homestead of these 1,900,000 people. In *Complex Plans of the Yellow River*, the floodplain areas of the lower Yellow River are planned as important flood detention and sediment deposition areas. Requirements from Economic and social developments and water resource managements for people's livelihood put the flood control of the floodplain areas be the focus of operations, and make common and small flood operation be an bottle-neck problem. In 2011, *Complex management plans for floodplain areas in the lower Yellow River* has been checked up, according to which the safely developing of the floodplain areas and composition policies for inundation have been prescribed and changes of the flood control situation of the floodplain areas may be caused.

Currently, the interval between Xiaolangdi reservoir and Huayuankou station, with 16,000 km^2 area, has been not controlled by any reservoirs. Since floods of this region commonly rise to peak flow quickly and the forecast periods are usually short, floods will put threats to the safety of the lower Yellow River once occurred. Though, there are no large floods in about recent 30 years since 1982, under the context of climate change, since the occurring probability of extreme large storms, it is evidently possible for large floods to occur in the lower Yellow River.

2.2 The upper Yellow River

The main flood control projects in the upper Yellow River include Longyangxia reservoir, Liujiaxia reservoir and embankments located along Lanzhou city, Ningxia and Inner Mongolia autonomous regions. Currently, there are 24 cascade reservoirs/hydropower stations downstream of Longyangxia reservoir. Joint operation of Longyangxia reservoir and Liujiaxia reservoir can undertake the flood control responsibility for Lanzhou city and cascade reservoirs/hydropower stations built located between Longyangxia reservoir and Qingtongxia reservoir mainly, and for Ningxia and Inner Mongolia autonomous regions and cascade reservoirs/hydropower stations being built secondarily. The main problems of flood control in the upper Yellow River are as follows: ①since the number of cascade reservoirs/hydropower stations downstream of Longyangxia reservoir is too large and flood control standards are different, operation of flood control projects are rather complicated; ②caused by changes of water and sediment situation, reservoir operation etc., water courses in Ningxia and Inner Mongolia autonomous regions have been deposited and reduced seriously and the flow carrying capability is remarkably small; ③embankments in Inner Mongolia autonomous region are weaken, whose ability against floods is very low.

2.3 Main tasks

Xiaolangdi reservoir is the most important reservoir in the lower Yellow River, with 4.05×10^9 m^3 flood control storage capacity, which undertake flood control responsibility for decreasing floods with peak flow higher than 10,000 m^3/s at Huayuankou station, and floods with peak flow lower than 8,000 m^3/s are not controlled. As stated above, after the operation of Xiaolangdi reservoir, the flood control situation in the lower Yellow River has changed significantly, comparing with in 1980s during the design period. During the former operation period, since the flood control storage is satisfactorily big, floods with peak flow lower than 8,000 m^3/s at Huayuankou station should be controlled, in order to reduce the inundation loss of the floodplain areas. It can be seen that the flood control modes of Xiaolangdi reservoir for common and small floods are different during the former operation period from during the design period. Nowadays, the Xiaolangdi reservoir has been into its later sediment-retaining period and the flood control storage capacity will reduce from 8.8×10^9 m^3, observed in April 2010, to designed 4.05×10^9 m^3, gradually along with being deposited in a period from now on. During this period, how to control common and small floods turn into a problem, which is thirsty to be solved. As common and small floods in the lower Yellow River are influenced by human activities, such as Longyangxia and Liujiaxia reservoirs in the lower Yellow River, water conservation projects in the middle Yellow River and so on, it is necessary to analyze the hydrographs and characteristics of floods at Huayuankou station, supporting for operation study for common and small flood control. Therefore, common and small flood control modes will make influences on heavy flood control, and joint operation study of flood control projects system during the later sediment-retaining period of Xiaolangdi reservoir is also very meaningful. In summary, for the lower Yellow River, there are 3 issues to be studied: characteristics analysis of common and small floods of Huayuankou Station under current projects condition, operation modes of Xiaolangdi reservoir and the joint operation modes of the flood control projects system during the later sediment-retaining period of Xiaolangdi reservoir.

The Longyangxia reservoir in the upper Yellow River, with the deigned flood control water level 2,594 m, was designed without functions for flood control either of cascade reservoirs/hydropower stations being built during flood periods, or of Ningxia and Inner Mongolia autonomous regions. In Summary, for upper Yellow River, there are 2 issues to be studied: ①Since the current flood control water level is 2,588 m, whether the flood control storage capacity between 2,588 m and 2,594 m is satisfactorily enough to ensure the safety of cascade reservoirs/hydropower stations being built during flood periods mainly, and consider flood control of Ningxia and Inner Mongolia autonomous regions secondarily? ②How about it if the designed flood control water level is reached?

3 Key technological issues of designing flood control schemes of the lower Yellow River

3.1 Characteristics analysis of common and small floods at Huayuankou station under influences of current projects

3.1.1 Influences of current projects on floods at Huayuankou station

Since 1950s, a large number of reservoirs and water conservation projects has been built, whose operation changes both the natural hydrographs in channels and underlying surfaces of basins, and has caused changes of conditions for runoff generation and confluence. Therefore, it is necessary to analyze changes of characteristics of floods and sediment under influences of current projects, and present characteristics of floods influenced by projects. Under current situation, hydrographs at Huayuankou station are mainly influenced by Longyangxia and Liujiaxia reservoirs in the lower Yellow River and water conservation projects in the middle Yellow River. Thus, hydrographs not influenced should be transformed under current conditions. In a special research for *Complex Plans of the Yellow River*, *Change research of common flood in the Yellow River*, the influences of Longyangxia and Liujiaxia reservoirs and water conservation projects on floods at Tongguan station, and hydrographs influenced by current projects were derived. In this study, firstly, based on results referred above, floods at Tongguan station transformed under current conditions were taken as inflows to Sanmenxia reservoir and then influences of water conservation projects between Tongguan and Huayuankou stations on floods were analyzed. Secondly, the storage and regulation of the Sanmenxia, Xiaolangdi, Luhun and Guxian reservoirs located in the Sanmenxia-Huayuankou interval were transformed originally, and then hydrographs of big reservoirs and water conservation projects in the upper and middle Yellow River transformed under current conditions and hydrographs of big reservoirs in the middle Yellow River transformed originally at Huayuankou station and Sanmenxia-Huayuankou interval can be obtained, used for analysis of both characteristics of common and small floods at Huayuankou station and flood control operation indexes of Xiaolangdi reservoir.

3.1.2 Magnitudes definition of common and small floods at Huayuankou station

Floods derived as stated above are real floods, which should be used in flood control projects operation. The magnitudes of peak flows at Huayuankou station with common probabilities, which are 8,000 m^3/s and 10,000 m^3/s of probabilities 33.3% and 20%. According to flow carrying capability of main water courses and embankments, in the light of magnitudes definition by the government, based on design floods researches, magnitudes of common and small floods was defined as between 4,000 m^3/s and 10,000 m^3/s, where 4,000 m^3/s is the flow carrying capability of main water courses during a long period, and 10,000 m^3/s is the minimum flood safe discharge after subtracting discharges from the Changqing and Pingyin basins.

3.1.3 Characteristics of common and small floods

(1) The occurring probability of common and small floods with peak flows between 4,000 $m^3/s$and 10,000 m^3/s is about twice a year. Among floods with peak flows between 4,000 m^3/s and 10,000 m^3/s, floods with peak flows between 4,000 m^3/s and 6,000 m^3/s is of 62%, floods with peak flows between 6,000 m^3/s and 8,000 m^3/s is of 28%, and floods with peak flows between 8,000 m^3/s and 10,000 m^3/s is of 10%.

(2) Common and small floods usually occur during periods from May to October. Among occurring numbers of the floods, the number of floods occurring during periods from May to June is of 5%, the number of floods occurring during periods from July to August is of 67%, the number of floods occurring during September is of 20% and during October is of 8%. Floods with peak flow magnitudes between 8,000 m^3/s and 10,000 m^3/s mainly occur in July or August. Peak flow magnitudes of floods occurring in September or October are usually lower than 8,000 m^3/s.

(3) About 80% of main peak flow water volumes of common and small floods are sourced from basins upstream of Tongguan station, and the lower magnitudes of floods, the higher percent of floods sourced from basins upstream of Tongguan station. For floods with peak flows in 4,000 ~ 6,000m^3/s, 84% of main peak flow water volumes are sourced from Tongguan station, 71% for floods with peak flows between 6,000 m^3/s and 8,000 m^3/s, and 74% for floods with peak flows between 8,000 m^3/s and 10,000 m^3/s.

(4) Among floods with peak flows between 6,000 m^3/s and 10,000 m^3/s at Huayuankou station, the peak flows within Xiaolangdi-Huayuankou interval of floods sourced mainly from Sanmenxia-Huayuankou interval, are usually higher than 4,000 m^3/s.

(5) Because of operation of big reservoirs in the upper Yellow River on floods during flood periods and industrial water supply increase, water sources flow into the middle Yellow River decrease evidently, which directly cause the flood magnitudes at Tongguan station decrease and the propor-

tion of with hyperconcentrated floods increase comparing with that during the period before Longyangxia reservoir was built. According to analysis of common and small floods occurred during 1987 ~ 2005, the ratio of hyperconcentrated to all floods with peak flows in 4,000 ~ 10,000 m^3/s is about 62%, where the ratio in 4,000 ~ 6,000 m^3/s is about 50% and most of floods with peak flows in 6,000 ~ 10,000 m^3/s are hyperconcentrated.

3.2 Inundation loss analysis of floodplain areas in the lower Yellow River

Inundation ranges and population of floodplain areas under floods of different magnitudes by Yellow River Institute of Hydraulic Research were calculated via flood confluence and disasters evaluation mode, based on the topography before flood period in 2009. Results showed that, under current situation, inundation loss of floodplain areas is lowest with peak flows of Huayuankou station smaller than 6,000 m^3/s, and increase quickly along with peak flows increase from 6,000 m^3/s to 8,000 m^3/s. Inundation ranges with peak flows 8,000 m^3/s is about 89% of with peak flows 22,000 m^3/s, and Inundation population 83%. In summary, inundation loss can reduce remarkably, if peak flows of Huayuankou station can be controlled smaller than 6,000 m^3/s.

3.3 Flood control operation modes for common and small floods during the later sediment-retaining period of Xiaolangdi reservoir

3.3.1 Phases division of flood control operation during the later sediment-retaining period of Xiaolangdi reservoir

Based on sedimentation reduction in the use of phasing and Xiaolangdi siltation of flood control storage capacity change, the flood control operation is divided into three phases The first phase is when the reservoir sedimentation is less than 4.2×10^9 m^3, flood control capacity is more than 2.0×10^9 m^3 below 254 m. The second phase is when the reservoir sedimentation volume is 4.2×10^9 m^3 to 6.0×10^9 m^3, reservoir flood control capacity is reduced more and often flood control operation water level is still not more than 254 m. The third phase is when deposition is greater than 6.0×10^9 m^3, often flood control operation may use flood control storage capacity above 254 m.

3.3.2 Index analysis of control operation for common and small floods.

(1) Requirement analysis of control storage capacities for common and small floods.

The flood control storage capacities of Xiaolangdi reservoir to control different magnitudes of floods were obtained, using designed and nearly 100 floods data. For floods at Huayuankou station with peak flow 10,000 m^3/s, the required flood control storage capacities are approximately 1.8×10^9 m^3, 0.9×10^9 m^3 and 0.6×10^9 m^3 to control the peak flows to be 4,000 m^3/s, 5,000 m^3/s and 6,000 m^3/s respectively at Huayuankou station. For floods at Huayuankou station with peak flow 8,000 m^3/s, the required flood control storage capacities are approximately 1.0×10^9 m^3, 0.6×10^9 m^3 and 0.3×10^9 m^3 to control the peak flows to be 4,000 m^3/s, 5,000 m^3/s and 6,000 m^3/s respectively at Huayuankou station. For floods at Huayuankou station with peak flow 7,000 m^3/s, the required flood control storage capacities are approximately 0.6×10^9 m^3, 0.3×10^9 m^3 and 0.2×10^9 m^3 to control the peak flows to be 4,000 m^3/s, 5,000 m^3/s and 6,000 m^3/s respectively at Huayuankou station. Results show that the lower the magnitudes of common and small floods at Huayuankou station and the smaller the peak flows to be controlled, the bigger the required flood control storage capacities should be.

(2) Control discharge for common and small floods.

The occurring frequency of floods with peak flows in 4,000 ~ 10,000 m^3/s at Huayuankou station is rather high. These floods are commonly met in reservoir operation and the long time influences of reservoirs on erosion and deposition of water courses and floodplain areas downstream should be considered in reservoir operation. According to operation computing results for common and small floods at Huayuankou station, peak flows at Huayuankou station can be controlled to be not higher than 6,000 m^3/s, operated by reservoirs in the middle Yellow River. Considering decrease inundation loss of floodplain areas and analysis results of flood magnitudes, the adequate control discharges at Huayuankou station are within 4,000 ~ 6,000 m^3/s. During the whole later sediment-retaining period of Xiaolangdi reservoir, the control ability of common and small floods will decrease gradually, as the sedimentation amount increases and the flood control storage capacity decreases. Therefore, flood control discharges at Huayuankou station should be adjusted as the sedimentation amount varies. When the sedimentation amount of Xiaolangdi reservoir is less than 4.2×10^9 m^3, within $4.2 \times 10^9 \sim 6.0 \times 10^9$ m^3 or more than 6.0×10^9 m^3, the flood control discharges should be 4,000 m^3/s, 5,000 m^3/s or 6,000 m^3/s respectively.

3.3.3 Control operation study for common and small floods

The sediment concentration of the Yellow River is rather high. Most of common and small floods at Tongguan station are hyperconcentrated, and hyperconcentrated common and small floods occur twice a year at Huayuankou station. In this study, the adjustment with a long series, sediment accumulated effect for many years is calculated, long-term impact of different flood control operation on the reservoir and channel and floodplain in downstream river is analyzed.

Operation mode of hyperconcentrated flood is the key issue of the Xiaolangdi reservoir flood control operation. Therefore, the development of the whole control scheme and hyperconcentrated vent open, non-hyperconcentrated control vent of two major programs operation. The whole control scheme analyzed four schemes to control Huayuankou discharge at 4,000 m^3/s, 5,000 m^3/s, 6,000 m^3/s and 4,000 ~ 6,000 m^3/s. After a comprehensive analysis, that control of 4,000 ~ 6,000 m^3/s is slightly better than the other three schemes, and recommend the program of this control scheme as a recommend fully-controlled manner. Full control recommended programs and hyperconcentrated flood vent open , non-high sandy controlled discharge program were analyzed , the results show that the sediment retention period of hyperconcentrated open vent scheme is about 17 years, significantly better than 4,000 ~ 6,000 m^3/s scheme (the length of the sediment retention period is about 14 years). Sediment trapping sedimentation reduction ratio of the two programs has little difference, the hyperconcentrated open vent is slightly better; the program the hyperconcentrated open vent is significantly higher than the control recommended program when discharge at Huayuankou is more than 6,000 m^3/s.

A comprehensive comparison of reservoir and river sedimentation reduction, the length of the reservoir storage period, the downstream floodplain disaster reduction and floodplain comprehensive management plan and other factors, from the role of long-term play Xiaolangdi reservoir and role played by the floodplain detention settling, the final recommendation is use the program open vent of high sediment concentration flood as often flood operation.

3.4 Joint operation of flood control projects system during the later sediment-retaining period of Xiaolangdi reservoir

3.4.1 Xiaolangdi reservoir

(1) Major floods in the upstream of the Sanmenxia reservoir.

Analysis and comparison of two scenarios whether considering control modes of common and small floods were taken in this study. Results show that the capacity of Xiaolangdi reservoir is bigger considering control modes than not. However, since control modes can save evacuation time for people in floodplain areas, considering reality and for safety of flood control operations, the common and small flood control modes are recommended during the later sediment-retaining period of Xiaolangdi reservoir.

(2) Major floods in the downstream of the Sanmenxia reservoir.

Major floods in the downstream of the Sanmenxia reservoir are mainly sourced from the Sanmenxia-Huayuankou interval, where the Xiaolangdi—Huayuankou interval and Xiaolangdi-Guxian-Huayuankou interval are the main sourced basins. As water flows are rather heavy downstream the reservoir, firstly, Xiaolangdi, Luhun and Guxian reservoirs are used to store floods upstream, and reduce water volumes into the downstream. Also, the forecast periods are short and the sediment concentration is low, in order to save evacuation time for people in floodplain areas and decrease inundation loss, operation should be as follows: during rising phrases of floods, Xiaolangdi reservoir should take control operation modes for common and small floods; once the reservoir storage reaches the flood control storage capacity or the discharge of the Xiaolangdi-Huayuankou interval is more than or equal to 9,000 m^3/s, Xiaolangdi reservoir should control discharges at Huayuankou station to be not higher than 10,000 m^3/s.

3.4.2 Sanmenxia reservoir

(1) Major flood happened in the upstream of the Sanmenxia reservoir.

During the normal period of the Xiaolangdi reservoir, the operation mode of the Sanmenxia reservoir is first done as free discharging and then as controlling discharging. During the former operation period of the Xiaolangdi reservoir, since the flood control storage is satisfactorily big, the operation mode of the Sanmenxia reservoir is done as free discharging to reduce the inundation loss of the floodplain areas. During the later sediment-retaining period of the Xiaolangdi reservoir, the feasibility of free discharging operation mode of the Sanmenxia reservoir should be studied first. The water storage capacity of the Xiaolangdi reservoir under the free discharging operation mode of Sanmenxia was analyzed, when the sedimentation is about 42×10^9 m^3, the common and small floods does not control and the Dongpinghu flood detention area is putted into operation. Results showed that, the Xiaolangdi reservoir is in dangerous for the water level is higher than 275 m when a 10,000-year return period flood occurred. It puts forward that the free discharging operation mode of Sanmenxia reservoir is infeasible during the later sediment-retaining period of Xiaolangdi reservoir.

Considering different sedimentation amounts of the Xiaolangdi reservoir, the opportunity of the operation mode for the Sanmenxia reservoir was analyzed, which is first done as free discharging and then as controlled discharging. Results showed that, the free discharging first and controlling discharging later operation mode of the Sanmenxia reservoir should be used when the recurrence of floods reaches or exceeds the standard of 50 years recurrence.

(2) Major floods occurring in the downstream of the Sanmenxia reservoir.

The Sanmenxia reservoir needs to be putted into operation when floods of a 100-year return period occurred during the normal period of the Xiaolangdi reservoir. At the former operation period of the Xiaolangdi reservoir, the operation of Sanmenxia reservoir is open vent. When the sediment amounts of the Xiaolangdi reservoir reaches to 4.2×10^9 m^3, if the operation of Sanmenxia reservoir is as free discharging, the water level of the Xiaolangdi reservoir will reaches to 277.93 m which is above the checked flood level. So the Sanmenxia reservoir should be putted into operation. Through analysis and comparison of three scenarios, those are Sanmenxia reservoir is putted into operation when the water level of the Xiaolangdi reservoir reaches to 260 m, 263 m and 265 m. The results showed that the earlier Sanmenxia reservoir is putted into operation, the lower the water level of Xiaolangdi reservoir is. Based on the study of the opportunity of Sanmenxia reservoir being putted into operation, the recommended operation is Sanmenxia reservoir being putted into operation when the water level of the Xiaolangdi reservoir reaches to 263 m or the probability is of a 200-year return period.

3.4.3 Dongpinghu flood detention area

(1) Major flood happened in the upper stream of the Sanmenxia reservoir.

Dongpinghu flood detention area need to be putted into operation when a 100-year return period flood occurred during the normal period of Xiaolangdi reservoir. At the later sediment-retaining period of Xiaolangdi reservoir, the opportunity of putting into operation is analyzed first when the sediment amounts of Xiaolangdi reservoir reaches to 4.2×10^9 m^3. Through analysis and comparison of three scenarios, those are Dongpinghu flood detention area is putted into operation when the water level of the Xiaolangdi reservoir reaches to 262 m, 263 m and 265 m. The results showed that the three scenarios are not show very different when floods of 10,000 years return period occurred. The Dongpinghu flood detention area is putted into operation is later, the storage of Xiaolangdi reservoir is bigger and the flood of the lower Yellow River is smaller. When the water level of the Xiaolangdi reservoir reaches to 272 m, 263 m and 265 m, the return period of Dongpinghu flood detention area being putted into operation is 100 years, 200 years and 300 years. Based on the study of the opportunity of Sanmenxia reservoir being putted into operation, the recommended operation is the Dongpinghu flood detention area is putted into operation when the water level of the Xiaolangdi reservoir reaches to 263 m.

(2) Major floods occurring in the downstream of the Sanmenxia reservoir.

Due to the flood happened in the downstream of the Xiaolangdi, Luhun and Guxian reservoirs cannot be controlled, Dongpinghu flood detention area need to be putted into operation when floods of a 30-year return period occurred.

3.4.4 Luhun, Guxian and Hekoucun reservoirs

During the later sediment-retaining period of Xiaolangdi reservoir, flood control operation modes of Luhun, Guxian and Hekoucun reservoirs will follow their designed modes, and will turn into flood control modes for the lower Yellow River when the forecasted peak flows reach 12,000 m^3/s. In this study, the subsiding sequence is taken as first for branches, then for main stream. And the subsiding sequence for branches is first for Luhun reservoir, then for Guxian reservoir, and last for Hekoucun reservoir.

4 Key technological issues of designing flood control schemes of the upper Yellow River

4.1 Study on operation modes of Longyangxia and Liujiaxia reservoirs under current projects condition

The flood control water level of the Longyangxia reservoir has been kept at 2,588 m in recent years. Since the Longyangxia reservoir is a multi-year regulating reservoir, its water level has not reached the flood control water level at the beginning of the flood season. Therefore, the flood control capacity under 2,594 m (design limited water level) is used to provide the safety of the project construction downstream, and also provide the flood-control safety of the Ningxia-Inner Mongolia reaches.

In recent years, the flood control standards of under-construction projects downstream of the Longyangxia reservoir are set from 10-year-flood to 200-year-flood. Due to different features of the various constructions, they are requiring different reservoir control vent flow of Longyangxia and Liujiaxia. Furthermore, because the storage which Longyangxia used to provide the safety of the project construction is limited, the flood control capability of Longyangxia and Liujiaxia is not big enough to control the flood in design standard of constructions. In most years, the two reservoirs can only give partial flood control capability and cannot guarantee the construction safe in the flood season.

Under current projects condition, if the project constructions downstream of the Liujiaxia reservoir require the Longyangxia and Liujiaxia reservoirs small outlet discharges, the Longyangxia and Liujiaxia reservoirs can not only ensure the safety of the under-construction project but also ensure the flood-control safety of the Ningxia-Inner Mongolia reaches.

4.2 Flood control regime of the upper Yellow River after Longyangxia reservoir and Liujiaxia reservoir operated on design modes

When the planned projects are all completed, Longyangxia Reservoir and Liujiaxia Reservoir will be operated on design modes. And the two reservoir's flood control task is to guarantees the projects and Lanzhou city safe. Currently the flood control design standard of the dike at downtown of Lanzhou is 100-year-flood, and the design flood peak is 6,500 m^3/s based on data of Lanzhou hydrological station. If Longyangxia reservoir and Liujiaxia reservoir are operated on design modes, they can keep the Lanzhou flood peak value under 6,500 m^3/s.

The flood standard of the yellow river dike in Ningxia Province and Inner Mongolia Province is from 20-year-flood to 50-year-flood. Different standards at different sections perform. After the joint flood control operation of Longyangxia Reservoir and Liujiaxia Reservoir, the 20-year-flood of Lanzhou station can be reduced from 6,440 m^3/s to 6,050 m^3/s, and the 50-year-flood can be reduced from 7,410 m^3/s to 6,310 m^3/s. The operation of the two reservoirs in normal mode can reduce the natural flood of the Ningxia-Inner Mongolia reaches, which is avoided the flood mainly by dike.

5 Conclusions and suggestions

5.1 Conclusions

(1) Under the current engineering conditions, the magnitude of normal and small flood at Huayuankou station is about 4,000 ~ 10,000 m^3/s, considering the influence of big reservoirs in the Sanmenxia-Huayuankou interval and restoring the flood to the original state. Statistically, these floods occur twice a year and about 69% happens in July and August. About 80% of the main peak amount is from areas above Tongguan station and 62% of these floods are carrying large amount of sediment.

(2) Under the current engineering conditions, the inundated beach area is small if the flood peak at the Huayuankou station is under 6,000 m^3/s. while if the flood peak increase from 6,000 m^3/s to 8,000 m^3/s, the inundated area will increase more intensely. The inundated area with flood peak of 8,000 m^3/s is about 89% of that of 22,000 m^3/s, and inundated people about 83%. So, the inundated area will be effectively decreased if the flood peak at Huayuankou station is controlled to be lower than 6,000 m^3/s.

(3) According to the sedimentation amount of Xiaolangdi reservoir, the flood management operation, during the later period of sediment detention, is divided into three stages, which are the stage before the sedimentation amount of 4.2×10^9 m^3, sedimentation amount between 4.2×10^9 ~ 6.0×10^9 m^3, and sedimentation amount more than 6.0×10^9 m^3. the controlling flood magnitude are respectively 4,000 m^3/s, 5,000 m^3/s, 6,000 m^3/s during these three stages. To carry out the normal and small flood prevention and raise the flood prevention standard, the smaller of the flood at Huayuankou is controlled, the larger of flood storage capacity of Xiaolangdi is needed. At the last sediment detention of Xiaolangdi, the controlling discharge during the normal and small flood is between 4,000 ~ 6,000 m^3/s, the flood storage capacity of Xiaolangdi is between 1.8×10^9 m^3 ~ 6.0×10^9 m^3. For the normal and small flood at Huanyuankou station, the operation mode is taken as two ways, which are free discharging for flood with large amount of sediment and controlling discharging for flood without large amount of sediment.

(4) Joint operation mode of downstream flood controlling system is as follows. For flood mainly coming from upper-streams, the operation of Sanmenxia is taken as free discharging firstly. When the flood water level reaches or exceeds the standard of 50 years recurrence, the operation mode of Sanmenxia is first done as free discharging and then as controlling discharging. While the water level of Xiaolangdi reservoir reached to 263 ~ 266.6 m, enlarge the discharge as free discharging or maintain the water level in the reservoir, and put the Dongpinghu detention area at downstream into operation to divide the flood. For flood mainly comes from downer-streams, Xiaolangdi reservoir is put into controlling operation firstly. When the water level of Xiaolangdi reservoir reaches 263 ~ 269.3 m, the discharging amount of Sanmenxia equals to the discharging amount of Xiaolangdi. If the prediction flood peak reaches to 12,000 m^3/s, the Luhun, Guxian and Hekoucun reservoirs are taken as the designing flood protection mode. When floods at Huayuankou station subsides under 10,000 m^3/s, the next subsiding sequence is taken as first for branches, then for main stream. And the subsiding sequence for branches is first for Luhun reservoir, then for Guxian reservoir, and last for Hekoucun reservoir.

(5) Under the current conditions of joint operation of Longyangxia and Liujiaxia reservoirs, the construction safety during flood season is only can be satisfied to some extent but not for all. The maximum extent that could be done is only to control the discharge under 2,500 m^3/s during the flood of 10-year recurrence, in order to take the flood protection for Ningxia-Inner Mongolia rea-

ches into consideration.

5.2 Suggestions

(1) The flood storage capacity of Xiaolangdi reservoir losses faster during post sediment detention stage. The flood control storage capacity may occupy the long-term effective storage when the normal and small flood occurs. Meanwhile, according to the flood protection mode of free discharging for flood with large amount of sediment and controlling discharging for flood without large amount of sediment, the inundating risk of the beaches is high. The suggestion is to strengthen the safety guard construction in floodplain areas, to carry out the compensation policy for floodplain areas, and so on to cope with the flood protection problems in floodplain areas.

(2) The main problem of flood protection for the lower Yellow River is sedimentation accumulation. The contradiction between flood protection requirement of floodplain areas and maintaining of the long-term effective capacity of Xiaolangdi reservoir is very obvious after the sedimentation capacity is full. The suggestion is to construct Guxian reservoir as early as possible, in conjunctive operation with Xiaolangdi, to maintain the effective capacity for long.

(3) The influence of human activity to flood and sediment of Yellow River is serious. The flood characters and the magnitude of design flood influenced by engineering works is the basis for watershed planning and engineering design work, while the study of it is still not profound and need more energy to be paid on the flood research.

(4) The design flood for Lanzhou hydrological station and Ningxia to Inner Mongolia reaches is examined in different period and there exists difference in the aspect of data basis and analyzing methods. Unclear duty allocation may be a problem for the obscure in flood magnitude when flood damage occurs. So, it is suggested that a study on the design flood for upper streams and other basic flood protection problems are carried out.

Acknowledgments

This study is financially supported by the National Department Public Benefit Research Found of Ministry of Water Resources (200901017).

Analysis and Countermeasures on Usual Problems of Diversion Sluices in the Lower Yellow River

Xu Jinjin[1], *Su Maorong*[2] and *Li Zhenquan*[1]

1. Water Supply Bureau, YRCC, Zhengzhou, 450003, China
2. Water Supply Bureau, YRHB ,YRCC,Zhengzhou, 450003, China

Abstract: Sluice gate of water supply is important for water supply project in the lower Yellow River, they were mostly built in the 1970s and the 1980s, there have been some different levels of defects in many existent sluices. The defects have not only influenced the benefits of the sluices, have but also affected flood control safety, economic and ecological development along the lower reaches of the Yellow River. Objectively to evaluate situation of sluices and analyze dangerous situations by detection and identification of sluices have been carried out, in order to take the corresponding measures. For the sake of healthy running, analyzed more than 40 sluices safety appraisal result and counted the dangerous type and number of them, it is finally summarized that at present the main damage types of sluices are leakage, over running age of machinery equipment, uneven settlement, concrete disease, the instability of lock chamber and so on. Analyses of reasons of the defects are the programming and construction, due to dam heightening or channel change, aging, and management and running. Based on the above reasons, and proposed the corresponding countermeasure, such as faster updating equipment, strengthening routine observation of settlement displacement, improvement of identification and standardized operation, more frequency of security identification to some serious defects in many existent sluices, strengthening danger control and reinforcement and rebuilding investment and so on.

Key words: the lower Yellow River, sluice, usual problems analysis, countermeasures

The lower reaches of the Yellow River go through Henan Province and Shandong Province, and is their most important customer of water resources. Water supply by the lower Yellow River can reach the areas including Hebei Province, Tianjin Province, Anhui Province and so on, across the three large basins of the Huaihe River, the Yellow River and the Haihe River. The Yellow River provides the important water resources to agricultural production, city life, especially Central Plains Economy Zone of the Huanghuaihai Plain. Because the Yellow River is the high suspended sediment depositing riverbed with the characteristic of its riverbed higher than the ground outside the levees, water of the Yellow River can extract from sluice gate by gravity. Therefore, sluice gate is important for water supply project in the lower Yellow River. Whether diversion headwork gate is safe, it is not only related to the development of career of water supply of the Yellow River, but also affects economic and ecological development along the lower reaches of the Yellow River. Therefore, sluices in the lower Yellow River were mostly built in the 1970s and the 1980s. Through long time operation, there are some the hidden danger in a part of sluice, and it directly affects the safety of flood control and water supply. It becomes an important problem how we have a very good solution to the common dangerous problems of sluice gate.

1 The basic situation of sluices of the lower Yellow River

There are 188 sluices managed by the Yellow River Water Conservancy Commission (YRCC) directly. Of them, 96 of diversion gates are withdrawn water from the Yellow River, with a design diversion discharge of 4,086 m^3/s; release sluices are 12, with a design flood diversion discharge of 28,800 m^3/s; others are 80 sluices on the Qinhe River, Beizhan dike and Beijin dike. Based on some correlated reference data and field investigations, sluices of the lower Yellow River are mostly built in the 1970s and the 1980s. After 30 ~ 40 years of operation, there are a variety of dangerous problems. The 40 sluices of the lower Yellow River had been carried out water sluice safety judgment from 2003 to 2011, of which the Class-4 consists of 7 sluices, Class-3 of 24 sluices, and Class 2 of 9 sluices.

2 The major problems of the lower Yellow River sluices

Analyses of the data surveyed and the result of water sluice safety judgment, there are 8 classes of major problems of the lower Yellow River sluices: ① because of the low flood control standard, 39.3% of sluices can not meet the current standard; ② the 20.9% of sluices are in the instability of seepage and the sluice foundation or pier wall soil failure generated by seepage; ③ there exists the aging of mechanical and electrical equipment and severe damage with 79% of sluices; ④ the 15.1% of sluices are in unaccording with the need of standard for uneven settlement and lock chamber unsteadiness and anti-slide stability safety coefficient; ⑤ the, 70.8% of sluices are the ageing and damaged severely of lock chamber concrete; ⑥ the 50.5% of sluices are damaged severely for energy dissipation and erosion protection; ⑦ the 9.8% of sluices can not meet the seismic demands; ⑦ the 19.9% of sluices are befallen severe sediment deposition problems.

3 Analysis of cause of the problem

3.1 On the planning problem

Design of sluices is sub-standard for some historical reasons and condition of economy and technology at that time. As water level exceeds design standard, there will be some problems such as the height deficiency of the pier, stable and structure and foundation strength of sluices, dam and seepage control.

3.2 Some problems in terms of design

Although designers consider the influence of sluice safety and the use of many factors as much as possible in sluice design, and adopt various measures on the structure, there are some difference in the use of the actual situation and the original design because of limited technical. In addition, quality defects of engineering are left in aspect of unreasonable of foundation scheme, structural system selection slip and the calculation method of difference etc. For example the Sanyizhai Sluice (appraised as a third kind dangerous sluice), it was founded in 1958, lied in the Lankao County. Due to cause excessive vibration, after sluices were put into operation there were 288 cracks in pier, the gate bottom, and hoist beam. The sluice in 1974 and 1990 has conducted two reconstructions, but the problems have not been fundamental solution.

3.3 The poor quality of construction

The quality of works is poor, and there is a serious security problem, due to historical reasons and the construction technology limited by the time. Project quality does not meet the national standards and requirements. Such as due to construction of soil compaction is not close, the Dayuzhang Sluice found water seepage in 1996. Seepage become serious because do not take timely measures. In 1997 the sluice was grouted, but the effect is not good.

3.4 Natural aging

Sluices are inevitable to occur with aging and local damage diseases, a dropping safety, and making project of normal play affected, and during operation due to air, load, freeze-thaw, water, pollution, wind, waves, rain and snow long-term effects, together some of the selected material and engineering environment incommensurate. As the majority generator of flood control sluice, hoist and other equipment have no buying parts available, unable to repair, can not ensure the normal opening and closing of the sluice for the production of 1970s, out products. In the Liaocheng City electromechanical equipment of the Weishan Sluice is ageing seriously. Number 3 hoist brake band and the brake wheel contact area does not meet the requirements, only 60% of the contact surface, affecting the brake performance, brake when the brakes are released, the brake band and the brake wheel can not be completely out, causing severe wear of brake band. In addition, the sluice embedded parts of serious corrosion, has formed a relatively concentrated pit and increased operation friction, and effected operation efficiency.

3.5 Application of management problems

Many of sluices are the lack of proper control way and complete operation procedures. Water project is lacking investment; repair and reinforcement of sluices are not timely and thorough, facilities management and measures are not perfect, dated and backward, management personnel quality is low. Standardized management system is not perfect, law enforcement supervision is weak. The sluices in the Yellow River Basin due to lack of maintenance funds for repair, have appeared phenomenon of sluice gate such as serious concrete carbonization, leakage reinforcement, crack, traffic bridge overloading deformation, wear, and water leakage, etc.. Such as the Sitouzhuan Sluice in the Changyuan County has seriously affected the play of benefit of sluice, because of seal damage, serious water leakage, iron gate slide wear and serious corrosion. The Hegangkou Sluice in Kaifeng City had seriously affected the flow capacity and safety use, due to the serious erosion at upstream paving and downstream energy dissipation pool .

3.6 No timely safety evaluation due to lack of funds

Sluices of the lower Yellow River mostly relatively remote location, manage difficult. Repair and maintenance funds can not meet the requirements, which cause disrepair of aging of hoist room, leaky roof, doors and windows, vandalism and other decadent damage problems. Due to the lack of identification funds, most sluices of the Yellow River basin have not carried out the security identification on time according to the requirements of the "Sluice Safety Appraisal Management Approach" issued by the MWR.

4 The harmfulness of the sick-dangerous sluices

4.1 Threat to flood control

These problems of seepage induced by osmotic deformation, the gate body of uneven subsidence, aging backward mechanical and electrical equipment not only hinder the normal function of flood control sluice, but also the safe production and the flood brought great hidden dangers. In recent years according to the analysis of settlement and displacement data observed, there are obvious uneven subsidence. Once the Yellow River flood, the sluices body under high water head pressure, the uneven settlement will be aggravated, and the gate body balance and stability have great threat. The sluices are built in the embankment and the shore. Every dangerous of sluices are vulnerable spots and hidden trouble. Dangerous sluices exist not only on cost of manpower and financial resources in the flood season, but also threat on the flood control and the safety of people's live.

4.2 Influence of the Yellow River sluices on the play of benefit

The reasons of sluices hidden sick-danger are long-term lack of repair sluice management funds, technology and other aspects. Some sluices have been sick for many years running, in addition to the industry and agriculture and the residents living water supply, also undertaken the important task of flood control. The perennial river deposition resulted in a riverbed of continuous elevating, caused diversion channel silted before the sluices. So a lot of sluices can not reach the design standard, must reduce the standard operation, serious influence the benefits of the sluices.

5 Countermeasures and suggestions

5.1 To strengthen the replacement of electromechanical equipment

In rebuilding electromechanical equipment keys are check power supply and capacity, to improve power supply reliability, to set a reserve power supply, and to power supply design, transformer capacity and wiring group selection, box-type transformer station selection and layout, electrical wiring, electrical protection. The mechanical and electrical equipment of mostly sluices are having never been built to update, screw of hoists corrosion, opening and closing device has been more than the required service life, mechanical and electrical equipment should be replaced quickly.

5.2 Strengthening the routine observation

Strict implementation of "Regulation for technical management of sluice", some aspects increase the observation frequency reasonably under special weather conditions such as vertical displacement observation, piezometric tube, water level, flow pattern et al. Reorganization timely observation results, master the state of engineering operation, problems can be found timely, to ensure the safe operation of the project.

5.3 Improving the norms of the management and operation

Maintenance of sluices is a technical management work. There must be a complete set of scientific management system. It is necessary to establish and perfect the regulations of sluice construction. To ensure safety running from the system and management with rule-based and conventions. So that enhance the work efficiency and management level, effectively promote the sluice management to be institutionalized, scientific, standardized construction, and ensure the normal operation of the engineering.

5.4 To strength the sluice safety identification

Security identification of sluices should be regular, it will become a systematic work. And according to the actual situation of safety identification of partial sluices, the identification frequency should be increased to ensure the safety of long-term operation. Strictly carry out safety appraisal according to the security identification procedures, and entrusted that the units with the corresponding quality work in the field of safety monitoring and consistent with the calculated, in order to ensure the safety of the appraisal work of quality, monitoring and reviewing reliability of results.

5.5 Accelerate the sick-dangerous sluice reinforcement and reconstruction as well as reconstruction work

It is great important that danger sluices require reinforcement; ensure project safety and normal operation. In the macro, it should be full of good price plan of the sick-dangerous sluice project, implement the responsibility system, strengthen the investment and management of reinforcement of funds, and improve the quality of personnel management. In microcosmic, according to the actual situation, different sluices with different methods ensure project safety and normal operation.

When carry out reinforcement research it should first ascertain the present situation, problem identification, and analysis of the reasons. According to the project and local characteristics, research the comprehensive reinforcement measures. Due to different conditions with the climate, natural environment, economic and technical development level, reinforcement methods are diverse, not copy. Reinforcement plan should reflect advanced, scientific and economy. Whether the survey, design, construction, management and other aspects, it should seriously take use new technology, new methods, new materials, and new technology, effort to improve the content of science and technology.

6 Conclusions

Early prevention is better than the best "treatment". According to the practice, anti-aging and dangerous prevention measures of many sluices should be carefully considered and implemented, in the planning stages of design and construction. At the same time in the using stage it should carefully check, maintenance and repair, so that slight injury do not further development. So the sluices safety period will greatly extend. Sluices management is crucial in safety operation of sluice. This requires staff to be explored ceaselessly and consider the in practice, and formation has certain practical significance, with the economic and social development mode of management.

References

Niu Yunguang. Elementary Analysis on the Problems and Countermeasures of the Sick-dangerous Sluices[J]. Hydropower Engineering , 2004(1):39 -43.

Ren Xuhua, Liu Li. Analysis and Prevention Measures of Sluices Diseases [J]. Soil Test and Safety Monitoring of Dam, 2003(6).

Zhan Liming. Study of Sluices Damage Forms and Reinforcement Measures [C]// Dangerous Reservoirs and Sluices Reinforcement of Professional and Technical Papers in China. Beijing: China Water Power Press, 2001.

Analysis of the Characteristics of "11 · 9" Autumn Flood in Weihe River

Lin Xiuzhi, *Hu Tian*, *Li Yong* and **Yang Chunxia**

Yellow River Institute of Hydraulic Research
Key Laboratory of Yellow River Sediment Research, MWR, Zhengzhou, 450003, China

Abstract: Weihe River is the largest tributary of the Yellow River and is critical for the safety of the whole Kuan-chung Plain. In September 2011, there had been three successive heavy rainfalls which made water of many tributaries in the Weihe River Basin raise above the warning line and the largest flood since 1981 appeared in the section below Xianyang. At that time the maximum peak discharge at Lintong Station was 5,410 m^3/s and the corresponding water level was 359.02 m, the highest level since the founding of the station in 1961. Meanwhile, Huaxian station witnessed a peak discharge of 5,260 m^3/s, the sixth largest peak flow since the establishment of the station in 1934 with a maximum flood level of 342.70 m. Based on field survey and relevant analysis, this article concludes that the flood has characteristics as heavy peak discharge, low sediment concentration, high water level, long duration, low peak clipping etc., which lay a good foundation for further treatment and flood control in the Weihe River.
Key words: flood, flood level, flood control, Weihe River

1 Overview

In September 2011, there had been three successive heavy rainfalls in Weihe River Basin ("11 · 9" autumn flood for short). From 8 am September 3rd to 8 am 20th, average cumulative rainfall of the whole Weihe River Basin in Shaanxi Province was 245.1 mm. The successive heavy rainfalls led to flooding in many rivers where water level raised above the alert line. Moreover, the heaviest flood since 1981 appeared in the reach below Xianyang. The flood peak at Lintong Station was 5,410 m^3/s and the corresponding water level was 359.02 m, the highest value since the founding of the station in 1961, which was 0.68 m higher than the "03 · 8" Weihe River flood peak and 0.44 m higher than the "05 · 10" one. Meanwhile, Huaxian station witnessed a peak discharge of 5,260 m^3/s, the sixth largest peak flow since the establishment of the station in 1934 with a maximum flood level of 342.70 m, which was only secondary to the "03 · 8" flood and 0.38 m higher than the "05 · 10" flood in terms of water level.

Based on field survey, this article makes a preliminary analysis on the rainfall, the flood source, the flood routing and the water level of the flood, and further compares those aspects with the "03 · 8" and "05 · 10" Weihe River flood so as to conclude the flood routing feature of the flood and lay a good foundation for treatment and flood management of the river.

2 Rainfall regimes and flood source

2.1 Precipitation and its distribution

Wei River flood in September 2011 was mainly consisted of water from the section after Linjiacun and the middle and lower reaches and the rainfall lasted from 8 am September 3rd to 8 am 20th with a duration of 17 d. Total rainfall in the basin is 15×10^9 m^3 and the corresponding average area-rainfall is 245 mm. The area which received rainfall more than 100 mm is 60,594 km^2, accounting for 91.7% of the total basin area and the area which received rainfall more than 200 mm is 42,395 km^2, accounting for 64.1% of the total basin area.

There are three distinct precipitation processes. The center of the first round of rainfall from

September 3rd to 8th was in the urban district of Baoji City and the rainfall in the North Slope of Qinling was more than North Weihe River. The center of the second round of rainfall from September 9th to 14th was in the southeast of Xi' an and the south of Weinan. In general, rainfall in the south of the Wei River Basin was more than the north and the east was more than the west. The heaviest rainfall was the third round from September 16th to 20th and the center was in the southwest of Xi' an and southeast of Baoji. There was generally more rainfall in the south than the north and in the center than the east and west.

2.2 Comparison of precipitation characteristics between "03 · 8" and "05 · 10" flood

2.2.1 Duration of precipitation

"03 · 8" heavy rain lasted from August 24th to October 13rd, 2003 with a duration of more than 50 d which were the longest consecutive rainy days. By comparison, "05 · 10" heavy rain lasted from September 17th to October 4th with a duration of 17 d and several short intervals, which was also a consecutive raining process; "11 · 9" heavy rain lasted from September 4th to 20th, 2011 with a duration of 17 d and two short intervals, which was a consecutive raining process and was akin to "05 · 10" (see Tab. 1).

Tab. 1 Comparison of precipitation characteristics between "03 · 8" and "05 · 10" flood

Precipitation Characteristics		"03 · 8" flood	"05 · 10" flood	"11 · 9" flood
Duration (day)		51	17	17
Rainy days with daily rainfall > 50 mm		8	5	7
Point rainfall (mm)	Maximum 6 h precipitation	72	103	77.6
	Maximum 12 h precipitation	73	126	156.4
	Maximum 24 h precipitation	116	138	166.4
	Maximum 72 h precipitation	144	213	229
Area rainfall (mm)	Maximum daily precipitation	44.2	33.3	61.3
	Maximum 3-day precipitation	76	76.3	93.6

2.2.2 Comparison of point rainfall

It can be seen from maximum 12 h, 24 h, 72 h precipitation that "11 · 9" has the largest point rainfall and the amount of the point rainfall in each period respectively reaches 156.4 mm、166.4 mm and 229 mm. Within three floods, "03 · 8" flood is the minimum one, with 73 mm, 116 mm and 144 mm of the point rainfall in each period. "05 · 10" flood is in the second place with 126 mm, 138 mm and 213 mm of the point rainfall in each period. From the perspective of maximum 6 h precipitation, "05 · 10" Flood has the largest maximum 6 h precipitation. Its maximum 6 h precipitation reaches 103 mm. The second one is 77.6 mm of "11 · 9" Flood and the last one is 72 mm of "03 · 8" flood.

2.2.3 Comparison of maximum area rainfall within one day and three days

The average area rainfall of "03 · 8", "05 · 10" and "11 · 9" floods within one day respectively reaches 44.2 mm, 33.3 mm and 61.3 mm. The average area rainfall of these three floods within three days reaches respectively 76 mm, 76.3 mm and 93.6 mm.

2.2.4 Comparison of rainstorm day

"03 · 8", "05 · 10" and "11 · 9" respectively has 8 d, 5 d, and 7 d of rainstorm day with daily rainfall above 50 mm.

Based on the above analyses, "11 · 9" Weihe River flood is larger than "03 · 8" and "05 ·

10" floods in maximum point rainfall above 12 hours and in average area rainfall within one day and three days. The number of rainstorm day within "11 · 9" Weihe River flood is less than that of "03 · 8" flood but is more than that of "05 · 10" flood.

3 Hydrological and sediment features of "11 · 9" Flood

3.1 Flood peak features

From September 6th to 26th, 2011, three floods took place in lower Weihe River (Fig. 1); peak features of each station are shown in Tab. 2. Due to less inflow of Jinghe River, the main source of the three floods is the main streams in middle reaches of Weihe River below Linjiacun and Nanshan tributaries.

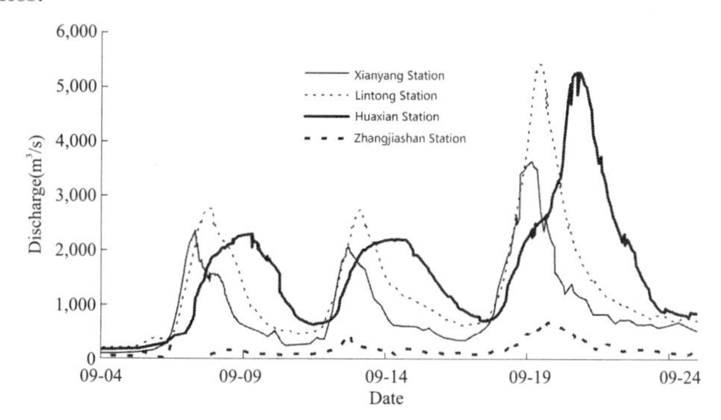

Fig. 1 "11 · 9" flood process of the lower Weihe River

Tab. 2 The peak characteristic value of "11 · 9" flood at each station in lower Weihe River

The name of station	September 6th ~ 11th		September 12th ~ 17th		September 18th ~ 26th	
	Peak discharge (m^3/s)	Occurrence time (the date and time)	Peak discharge (m^3/s)	Occurrence time (the date and time)	Peak discharge (m^3/s)	Occurrence time (the date and time)
Xianyang	2,350	09 – 07 6:42	2,050	09 – 12 16:36	3,630	09 – 19 3:42
Zhangjiashan (Jinghe River)	349	09 – 07 8:00	462	09 – 12 17:36	669	09 – 19 17:00
Lintong	2,750	09 – 07 17:00	2,730	09 – 13 2:00	5,410	09 – 19 10:00
Huaxian	2,290	09 – 09 6:00	2,190	09 – 14 5:00	5,260	09 – 20 20:00

During the process of the flood peak of the first and second floods spread from Xiayang to Lintong, the afflux of tributaries of south bank and Jinghe River of north bank increases the peak discharge. The peak discharge from Lintong to Huaxian has attenuated with an obviously expanded peak curve. Due to the inflow of Nanshan tributary in the reaches below Xianyang, the peak discharge of the third flood at Xianyang station is 3,630 m^3/s and reaches 5,410 m^3/s at Lintong station and 5,260 m^3/s at Huaxian station. In this flood process, the peak curve of Xianyang, Lintong and Huaxian is similar.

3.2 Hydrological and sediment features of floods

Tab. 3 shows that within "11 · 9" three floods, the first and second floods have a similar runoff and sediment and the third flood has the largest runoff and sediment. The total runoff and sediment of the three floods is respectively 2.935 $\times 10^9$ m^3 and 27.2 $\times 10^6$ t. The sediment concentration of the three floods is not too high with the largest one of 23.4 kg/m^3. The average sediment concentration of the three floods is only 9.27 kg/m^3. This flood process in lower Weihe River belongs to non-saturated sediment transport process, which tends to result in river erosion in lower Weihe River and favors the formation of the main channels in lower Weihe River.

Tab. 3 Runoff and sediment eigenvalue of "11 · 9" flood at Huaxian Station

Station name	Item	September 6th ~ 11th	September 12th ~ 17th	September 18th ~ 26th
Huaxian	Discharge ($\times 10^8$ m^3)	6.45	7.72	15.18
	sediment ($\times 10^8$ t)	0.075	0.068	0.129
	Maximum sediment concentration (kg/m^3)	23.4	12.9	12.8
	Average sediment concentration(kg/m^3)	11.62	8.75	8.48

4 Flood routing and flood levels

4.1 Peak clipping rate and flood level

The flood routing has been influenced by runoff, sediment, and river channel boundary conditions. In this process, great differences exist in its peak clipping rate and travel time. Tab. 4 shows that in "11 · 9" flood process, the travel time is 34 h from Lintong to Huaxian, which is longer than that of "03 · 8" flood and shorter than that of "05 · 10" flood. The peak clipping rate of "11 · 9" flood is similar to that of "05 · 10" flood and its flood peak has little attenuation.

Tab. 4 Comparisons of 2003, 2005, And 2011 Flood Routing from Lintong Station to Huaxian Station

Flood	Station name	Flood peak time	Peak discharge (m^3/s)	Peak clipping rate (%)	(h)
"03 · 8"	Lintong	2003 - 08 - 31 10:00	5,100	30.6	25
	Huaxian	2003 - 09 - 1 11:00	3,540		
"05 · 10"	Lintong	2005 - 10 - 2 13:48	5,270	7.4	43.7
	Huaxian	2005 - 10 - 4 9:30	4,880		
"11 · 9"	Lintong	2011 - 09 - 19 10:00	5,410	2.8	34
	Huaxian	2011 - 09 - 20 20:00	5,260		

From the perspective of maximum water level, the maximum water level of "11 · 9" flood at Xianyang, Lintong and Huaxian stations is higher than that of "05 · 10" flood. The maximum water level at Xianyang and Huaxian stations is lower than that of "03 · 8" flood, but the maximum water level at Lintong station is 0.68 m higher (Tab. 5).

Tab. 5 Comparison of the maximum water level along the reach of lower Weihe River

Station	Maximum water level (m)			Water level difference(m)	
	2003	2005	2011	"11 · 9" and "03 · 8"	"11 · 9" and "05 · 10"
Xianyang	387.86	385.78	386.56	-1.30	0.78
Lintong	358.34	358.58	359.02	0.68	0.44
Huaxian	342.76	342.32	342.7	-0.06	0.38

4.2 Changes in water level under the same discharge

The relations between water level and discharge at Xianyang, Lintong, and Huaxian stations have been plotted in Fig. 2 ~ Fig. 4. Based on Fig. 2, in "11 · 9" autumn flood in Weihe River, discharge of 3,000 m^3/s at Xianyang station corresponds to water level of 386.5 m, which is 0.49 m lower than that of 2003 but 0.78 m higher than that of 2005 under the same flow discharge. From figure 3, discharge of 3,000 m^3/s at Lintong station has a corresponding water level of about 357.48 m, which is 0.44 m lower than that of 2003 but 0.08 m higher than that of 2005 under the same flow discharge. Discharge of 5,000 m^3/s at Lintong station has a water level of 358.87 m, which is respectively 0.54 m and 0.30 m higher than that of 2003 and 2005 under the same flow discharge. From Fig. 4, discharge of 3,000 m^3/s at Huaxian station corresponds to water level of 342.04 m, which is 0.48 m lower than that of 2003 but 0.34 m higher than that of 2005. From the perspective of water level under the same flow discharge, for 2011 autumn floods, discharge of 3,000 m^3/s at three stations have lower water levels than those of "03 · 8" flood but higher than those of "05 · 10" flood. However, for discharge of 5,000 m^3/s, only Lintong station has a higher water level than that of "03 · 8" flood under the same flow discharge. Such unusual phenomenon needs further analyses.

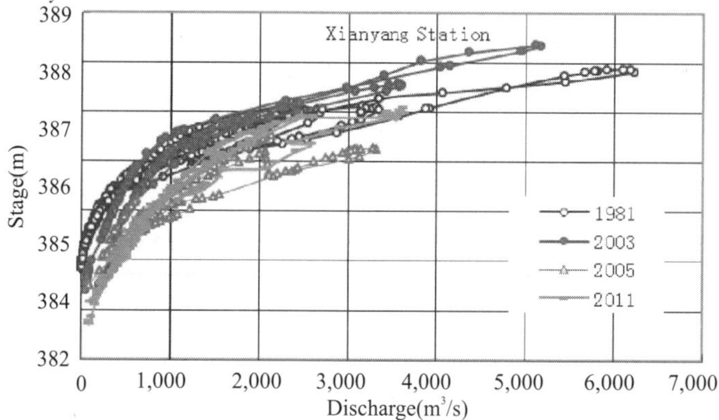

Fig. 2 Stage-discharge relation graph of Xianyang Station in lower Weihe River

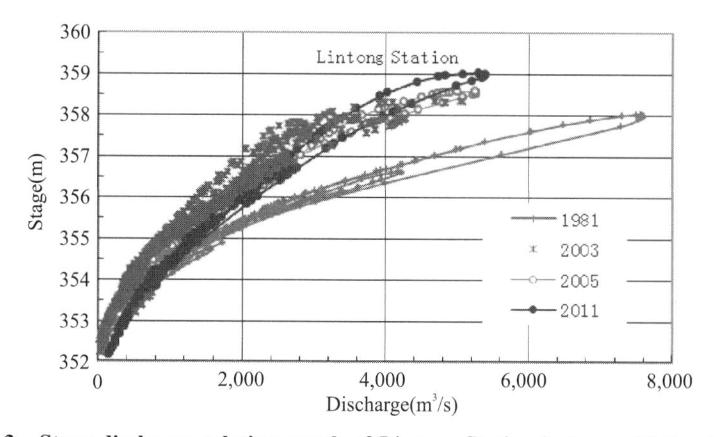

Fig. 3 Stage-discharge relation graph of Lintong Station in lower Weihe River

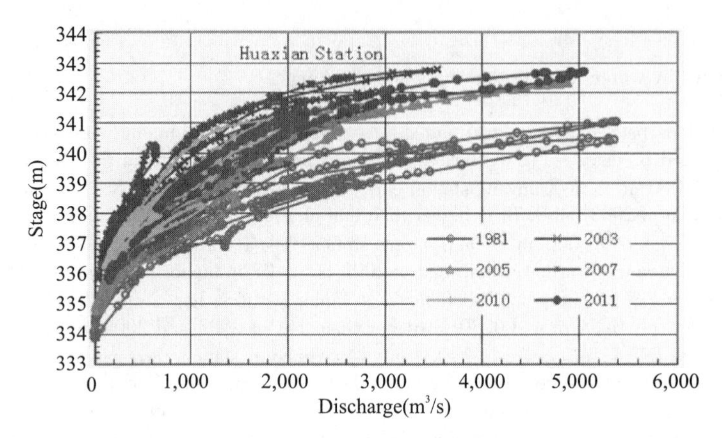

Fig. 4 Stage-discharge relation graph of Huaxian Station in lower Weihe River

5 Summary

(1) As the precipitation areas of "11 · 9" autumn floods are mainly located in Baoji in middle reaches of Weihe River and Qinling mountains in southern part of Weihe River, which are the clear water originating regions. Thus, 2011 floods in Weihe River have low sediment concentration. The lower reaches of Weihe River belong to non-saturated sediment transport areas, which will result in river channel erosion and favor the formation of main channels in lower Weihe River.

(2) As "11 · 9" floods transport along the lower Weihe River, after the peak clipping and attenuation of the first two floods, the third flood has few changes in peak pattern and has a low peak clipping rate. From the performance of the water level under the same flow rate, discharge of 3,000 m^3/s at Xianyang, Lintong and Huaxian stations within "11 · 9" autumn floods has a lower water level than that of "03 · 8" flood but its water level is higher than that of "05 · 10" flood. However, only at Lintong station, discharge of 5,000 m^3/s under the same flow has a higher water level than that of "03 · 8" flood. This unusual phenomenon may be resulted from the impact of channel boundary conditions. All these need further analyses.

References

Yellow River Institute of Hydraulic Research, YRCC. The Consultation Report of 2003 Yellow River[M]. Zhengzhou: Yellow River Conservancy Press, 2003.

Yellow River Institute of Hydraulic Research, YRCC. The Consultation Report of 2005 Yellow River[M]. Zhengzhou: Yellow River Conservancy Press, 2005.

Study on the Fuzzy Evaluation of Water Right Transfer Effect for Yellow River Irrigation Area in Ningxia and Inner Mongolian Autonomous Region

Feng Feng[1,2,3] ,***Zheng Jin***[3] and ***He Hongmou***[1]

1. Yellow River Institute of hydraulic Research, Zhengzhou, 450003, China;
2. Tsinghua University, Beijing, 100086, China
3. Yellow River Conservancy Technical Institute, Kaifeng, 475003, China

Abstract: Aiming at solving the result – evaluation problem existed in the finished water – rights transfer program for the Yellow River irrigation area, the author has constructed a multi – layer & semi – structured evaluation index system and established a multi – layer and multi – objective fuzzy optimization model based on fuzzy optimization theory and entropy concept. Making the indexes' relative membership degree vector of the first layer (input layer) as the basic data to constitute the judgment matrix, then the entropy weight can be obtained by the utility degree reflected by the index – variation degree. As to the weight vector of intermediate layer, making the output (the relative entropy weight optimal membership degree vector) of the first layer (input layer) to constitute the judgment matrix, the author, giving the consideration to the decision – maker's preference and the attribute of objective data, adopts different weight to calculate the integrated evaluation on the basis of the top – layer output.

Applying the model to the water – rights transfer project of Yellow River irrigation area of Ningxia Hui Autonomous Region and Inner Mongolian Autonomous Region, and then the example – verification and result – analyzing of the results give the technical support for further applying water – rights transfer project reasonably.

Key words: Yellow River Basin, irrigation area of Yellow River, water – rights transfer, fuzzy optimization, effect evaluation

1 Introductions

Yellow River Basin belongs to water resources shortage basin and the imbalance between water supply and demand is conspicuous. Ever since the unified dispatching management of the main stream of Yellow River in March, 1999, the water resource using capacity in Ningxia, the Inner Mongolia Autonomous Region nearly surpasses its water allocation plan each year, so the newly constructed industrial items was restricted by the water quota restrictions as well as the development of economy and society. Under the condition of the fixed available water quantity, we could only resolve the supply – demand contradiction by means of saving water and improving water use efficiency. Therefore, Water right transfer pilot work began in Ningxia Hui Autonomous Region and Inner Mongolian Autonomous Region in 2003 and the new idea "investment on saving water, meanwhile transferring the water rights" was raised, which means that saving – water engineering in irrigation area should be reconstructed by the investment of industrial enterprises, then the saved water quantity should be transferred to industrial enterprises. A new way that can be used to solve the water – using problem in economic society development in arid area has been explored. Up to the end of 2010, the water transfer project totally have been amounted to 35 items, mainly relating to thermal power, coal chemical industries and coal mining, etc, in which the installed thermal power project is 26,980 MW, the saved water quantity reaches $2.696,5 \times 10^8$ m^3, and the transferred water quantity reaches 239.17 m^3, and the investment of water right transfer project reaches 1.27×10^9 yuan.

The water – right transfer project strongly supports the economic society development of Ningxia and Inner Mongolian Autonomous Region. The water – right transfer project of Yellow River has been carried out for 7 years based on optimizing water – using structure and making full use of water resources, and it has fastened the economic society development and effectively alleviated the contradiction between water - demand and water - supply, and the water - wasting phenomenon, which improved the local ecological environment, promoting the construction of water - saving society and harmonious development between economic society and ecological environment, gaining obvious economic benefits, ecological benefit and social benefit as well. But up to now, there isn't a mature theory and method to evaluate the implementing effect on water - transfer items, so the reasonable evaluation index system and evaluation method for evaluating the effect of water right transfer is urgently needed.

The effect - evaluation of water transfer project is a kind of scientific assessment for economic, ecological and social effect generated by water right transfer project of Yellow River irrigation area, which can supply a scientific basis for developing and utilizing the potential of Yellow River Irrigation Area. Meanwhile, it is a complicated multi - layer & multi - objective evaluation process influenced by many factors and perplexing relationship which leads to the great variation in different conditions, such as: area, climate, engineering and background. So, the evaluation model is constructed based on the multi - layer fuzzy optimization theory, and according to the semi - structural index system composed by the quantitative index and qualitative index, and water right transfer characteristics of Yellow River irrigation area as well. As to the multi - layer and multi - objective evaluation, whether the determination of index weight is reasonable directly relates to the final results of the evaluation. According to the weight assignment method, weight can be divided into subjective weights and objective weights, and the former reflects the preference of decision - makers whereas the latter represents the contribution of data on decision - making. This article introduces entropy method to determine the objective weight of all - layers & all - objectives, and combining its objective weights at the interlayer; meanwhile, the article analyzes its feasibility, rationality and applicability based on fuzzy optimization model selection verified by the example of the Yellow River irrigation area in Ningxia and Inner Mongolian Autonomous Region.

2 Improved multi - level fuzzy evaluation model based on entropy weight

2.1 The construction of multi - level & semi - structure evaluation Model

The multi - level & semi - structure fuzzy optimization theory raised by Chen Shouyu is used for evaluating the Water resources utilization effect of Yellow River irrigation area.

The multi - level system can be divided into H layers, in which the highest level is H. If the lowest level, ie the first layer has number M target relative membership degree has to be inputted into several parallelized unit system (layer 2) where the multiple target relative membership degree have to be inputted into each unit system, and each target has different weight vector, then to calculate the output of each unit, i. e. schemes' relative membership degree vector.

$$u'_i(u'_{i1},u'_{i2},\cdots,u'_{i3})$$

Then $u'_i(u'_{i1},u'_{i2},\cdots,u'_{i3})$ constitute the i input of the third layer in a certain unit, as shown in Fig. 1.

$$u'_{ij} = (r_{ij})$$

Calculate $u'_{ij} = (r_{ij})$ as per this principal from the layer 1 to layer H till the top layer. As there is only one unit system in the top layer, so the top - layer's output can be obtained, i. e. schemes' relative membership degree vector.

$$u = (u_1,u_2,\cdots,u'_n)$$

Each scheme can be optimized and comprehensive evaluated according to the results.

2.2 Evaluation model and calculation process

The absolute limitation doesn't exist in flood utilization evaluation, which owns the intermediary transition characteristic and belongs to fuzzy concept. According to the multilevel fuzzy optimization theory, the index weight vector can be obtained by entropy method and the risk - benefit comprehensive evaluation model for flood utilization is constructed as well. The calculation process is as follows:

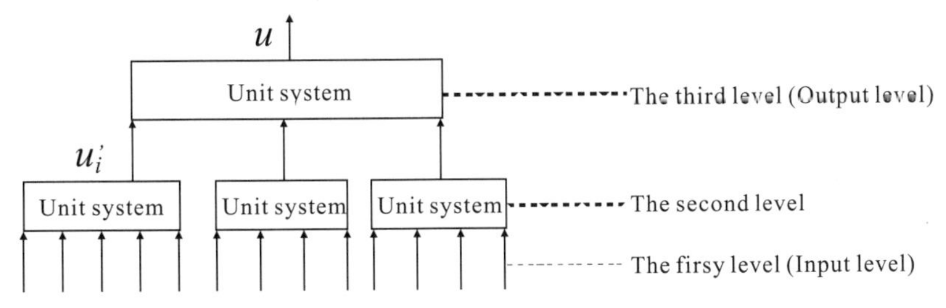

Fig. 1 3 – lay system of fuzzy optimization

(1) Construct the characteristic value matrix with n schemes and m evaluation indexes $X = (x_{ij})_{m\times n}(i = 1,2,\cdots,m;j = 1,2,\cdots,n)$.

(2) Determine target relative membership degree matrix R, and calculate its normalized judgment matrix B. According to the index type, the – bigger – the – better index shall be calculated as per formula (1), and the – smaller – the better index shall be calculated as per formula (2).

$$b_{ij} = \frac{x_{ij} - \min_j x_{ij}}{\max_j x_{ij} - \min_j x_{ij}};\ \forall j \tag{1}$$

$$b_{ij} = \frac{\max_j x_{ij} - x_{ij}}{\max_j x_{ij} - \min_j x_{ij}};\ \forall j \tag{2}$$

In formula, b_{ij} is the i row and the j column in normalized judgment matrix; x_{ij} is the index value of the i row and the j column; i is the matrix's row number, i. e. index number; j is the number of columns of a matrix, i. e. evaluation scheme number.

(3) Introduce entropy method to calculate index weight vector. In information theory, entropy reflects the information disordering degree, which means the smaller the value, the smaller the disorder degree. So the information entropy can be used to evaluate the order degree and its effectiveness of the obtained system information, that is: the judgment matrix constituted by the evaluation index values can be used to determine the index weight, which can eliminate human disturbance by calculating the weight of each index, and information entropy is used to determine the weights carried by index data. According to the concept of "entropy", the entropy of evaluation index can be determined as:

$$H_i = -\frac{1}{\ln n}\left[\sum_{j=1}^{n} f_{ij}\ln f_{ij}\right] \quad i = 1,2,\cdots,m;j = 1,2,\cdots,n$$

$$f_{ij} = \frac{1 + b_{ij}}{\sum_{j=1}^{n}(1 + b_{ij})} \tag{3}$$

In formula: H_i is entropy of index i; f_{ij} is parameters of entropy calculation.

Calculate the entropy weight vector of evaluation index.

$$\omega_i = \frac{1 - H_i}{m - \sum_{i=1}^{n} H_i},\ \sum_{i=1}^{m}\omega_i = 1 \tag{4}$$

In formula: ω_i is entropy weight vector of No. i index.

(4) Calculate the optimal relative membership degree vector of each level in different

schemes.

$$u_{hj} = \begin{cases} 0; h < a; or\ h > b_j \\ (d_{hj}^2 \cdot z_j)^{-1}; a_j \leqslant h \leqslant b_j, d_{hj} \neq 0 \\ 1; d_{hj} = 0 \end{cases} \tag{5}$$

$$h = a_j, a_{j+1}, \cdots, b_j; j = 1, 2, \cdots, n$$

$$d_{hj}^2 = \sum_{i=1}^{m} [\omega_i (b_{ij} - s_h)]^2; z_j = \sum_{k=a_j}^{b_j} (d_{kj}^2)^{-1} = \sum_{k=a_j}^{b_j} (d_{hj}^2)^{-1}$$

In formula: u_{hj} is relative membership degree vector of level h corresponding to the j scheme ; i is matrix line i, totally m lines, same as the index number; j is matrix column j, totally n columns, same as the evaluation – scheme number; h is the number of evaluation level; d_{hj}、z_j is calculation parameters; s_h is standard value vector corresponding to level h.

(5) Calculating level characteristics value.

$$H = h \cdot u_{hj} \tag{6}$$

In formula: H is level characteristics value of scheme j.

3 Evaluation examples

3.1 The selection of evaluation indexes

Through several years' practice on industry feeding – back agriculture, agriculture supporting industry, the Yellow River water right transfer has realized agricultural and industrial win – win and made an initial success in Ningxia and Inner Mongolian Autonomous Region, and it will definitely give a positive guidance and play an exemplary role to the industrial and agricultural development of western water – shortage region. Although Yellow River water right transfer is just at the beginning period, its enormous economic benefits, social benefits and ecological benefits can not be ignored. Select the evaluation indexes from economic effect, ecological effect and social effect to construct the evaluation index system which can be used to evaluate the implementing effect ever since Ningxia, Inner Mongolia water – right transfer. The index system for evaluating implementing effect of water – right transfer in irrigation area should be constructed firstly, in which the top layer is the objective level, i. e. the comprehensive effect; the second layer is the standard layer, including economic effect, ecological effect and social effect; the third layer is the index layer being constituted by a certain number of interrelated indexes. The economic effect indexes are mainly concerning the following aspects of channel water resource: utilization degree, water – transferring volume and generating benefit, which consists of channel utilization coefficient, canal utilization coefficient, the increased net benefit of water resources and the newly increased industrial output. The factors such as: the variation of groundwater level, wetland area, and vegetation growth should be taken into account for ecological effect indexes. And the social effects are mainly considered from effective – utilizing of water resources, improvement of irrigation engineering, guarantee the farmers' water rights and the social – consciousness on saving water, etc. , and it mainly includes the indexes as: engineering reconstruction ration of irrigation district, investment proportion of water – saving project, excess water – utilizing ratio of transition region, etc.

Combining the water – right transfer characteristics of irrigation area and considering the indexes' availability, independence and the limitation of the data, the multi – layer and multi – objective semi – structural index system for assessing water – right transfer effect of Ningxia, the Inner Mongolia autonomous irrigation region is constructed, which includes 11 quantitative indexes and 6 qualitative index, shown in Tab. 1.

Tab. 1 Indices of benefit evaluation of water resources in a certain irrigation area in Shaanxi Province

Sub system		Index name	Symbol	Different irrigation index data			
				No water right transfer	Hexi irrigation area	Hedong irrigation area	Southern Bank of Yellow River irrigation
				scheme 1	scheme 2	scheme 3	scheme 4
1	Economic effect index	channel - water utilization coefficient	I_1	0.71	0.77	0.77	0.77
		canal - water utilization coefficient	I_2	0.43	0.60	0.60	0.60
		the increased net - benefit of water resources ($\times 10^6$ Yuan)	I_3	0	17,9167.7	93,434.2	498,114.4
		The newly increased industrial output value($\times 10^8$ Yuan)	I_4	0	29.6	15.5	218.9
		water - charge reducing per mu (Yuan/mu)	I_5	0	2.84	1.20	11.00
2	Economic effect index	the actual water - transfer volume($\times 10^4$ m^3)	I_6	0	3,490.0	1,820.0	12,037.6
		the declining of underground - water level /m	I_7	0	0.5	0.3	1.0
		variation of lake & wetland	I_a	Qualitative index			
		vegetation - growth influence	I_b	Qualitative index			
		Saline & alkali soil distribution area	I_c	Qualitative index			
3	Social effect index	reconstruction ratio of irrigation area(%)	I_8	0.0	6.0	22.9	70.0
		reconstruction ratio of irrigation area (%)	I_9	0.0	10.0	10.0	4.5
		Water saving project financing ($\times 10^4$ Yuan)	I_{10}	0.0	14,809.7	6,359.5	117,845.0
		the exceeding water utilization ratio of transition area (%)	I_{11}	27.5	8.5	8.5	12.8
		improving industrial wateruti-lization structure	I_d	Qualitative index			
		guaranteeing the farmers' water utilization right	I_e	Qualitative index			
		strengthen water - saving consciousness	I_f	Qualitative index			

3.2 Relative membership degree matrix determination

The subsystem 2 has the following qualitative indexes: lake & wetland area variation I_a, vegetation growth effect I_b, saline & alkali soil distribution area I_c, and subsystem 3 has the qualitative indexes as follows: improving industrial utilization structure I_d, guaranteeing farmers' water – utilization right I_e, and strengthen water – saving consciousness I_f. According to the modal operator and relative membership degree table, the relative membership degree matrix of each qualitative index can be found out.

$$ {}_2R_2 = \begin{bmatrix} 1.000 & 0.905 & 0.818 & 0.667 \\ 1.000 & 1.000 & 1.000 & 0.905 \\ 0.667 & 0.905 & 0.905 & 1.000 \end{bmatrix} $$

$$ {}_2R_3 = \begin{bmatrix} 0.481 & 0.818 & 0.818 & 0.905 \\ 0.739 & 0.905 & 0.905 & 1.000 \\ 0.429 & 0.905 & 0.905 & 1.000 \end{bmatrix} $$

3.3 The Calculation about Entropy Weight Vector Indexes of the First Layer (Input Layer)

The following is the calculation of weight vectors of the three subsystems' index. Among all the qualitative indexes, all the indexes are the ones which belong to the bigger the better type except the two indexes: the declining of underground – water level I_7 and the exceeding water utilization ratio of transition area I_{11}, which belong to the smaller the better one. The relative membership degree of the qualitative indexes is determined by the data value which constitutes judgment matrix which combines quantitative index to constitute normalized judgment matrix B of each subsystem according to (1) & (2).

$$ B_1 = \begin{bmatrix} 0.0000 & 1.0000 & 1.0000 & 1.0000 \\ 0.0000 & 1.0000 & 1.0000 & 1.0000 \\ 0.0000 & 0.3597 & 0.1876 & 1.0000 \\ 0.0000 & 0.1352 & 0.0708 & 1.0000 \\ 0.0000 & 0.2582 & 0.1091 & 1.0000 \\ 0.0000 & 0.2899 & 0.1512 & 1.0000 \end{bmatrix} $$

$$ B_2 = \begin{bmatrix} 1.0000 & 0.5000 & 0.7000 & 0.0000 \\ 1.0000 & 0.7147 & 0.4535 & 0.0000 \\ 1.0000 & 1.0000 & 1.0000 & 0.0000 \\ 0.0000 & 0.7147 & 0.7147 & 1.0000 \end{bmatrix} $$

$$ B_3 = \begin{bmatrix} 0.0000 & 0.0857 & 0.3271 & 1.0000 \\ 0.0000 & 1.0000 & 1.0000 & 0.4500 \\ 0.0000 & 0.1256 & 0.0540 & 1.0000 \\ 0.0000 & 1.0000 & 1.0000 & 0.7737 \\ 0.0000 & 0.7948 & 0.7948 & 1.0000 \\ 0.0000 & 0.6360 & 0.6360 & 1.0000 \\ 0.0000 & 0.8179 & 0.7179 & 1.0000 \end{bmatrix} $$

According to (3), (4), the entropy weight vector of the indexes of the first layer shall be:
ω_1(0.153, 0.153, 0.156, 0.197, 0.175, 0.165);
ω_2(0.238, 0.244, 0.283, 0.235);
ω_3(0.164, 0.147, 0.189, 0.135, 0.123, 0.116, 0.125).

3.4 Calculating the relative membership degree vector of first layer subsystem

Applying (5), (6) to calculate the output of the first layer, the relative membership degree vector of economic effect, ecological effect and social effect shall be as follows respectively:

${}_1u_1$ = (0.000, 0.766, 0.731, 1.000);

${}_1u_2$ = (0.703, 0.725, 0.740, 0.416);

${}_1u_3$ = (0.000, 0.610, 0.623, 0.825).

3.5 Calculating the weight vector and rank feature value vector

According to the multilevels – multitargets – multigrades fuzzy optimization theory, the output of the first layer subsystem is the input of second layer subsystem. Applying (1)(2) to make the input of the second layer i. e. relative membership degree matrix R of each subsystem as the judgment matrix to constitute the normalized matrix B.

$$B_3 = \begin{bmatrix} 0.000 & 0.766 & 0.731 & 1.000 \\ 0.703 & 0.725 & 0.740 & 0.416 \\ 0.000 & 0.610 & 0.623 & 0.825 \end{bmatrix}$$

$$B_3 = \begin{bmatrix} 0.0000 & 0.7660 & 0.7310 & 1.0000 \\ 0.8858 & 0.9537 & 1.0000 & 0.0000 \\ 0.0000 & 0.7394 & 0.7552 & 1.0000 \end{bmatrix}$$

Apply (3), (4) to calculate entropy weight vector, equivalent in weight vector and subjective weight vector respectively, and apply (5), (6) to calculate rank feature value vector; Tab. 2 shows the results. According to the output of the top layer, i. e. the principal of the maximum relative membership degree and the minimum rank characteristic value, the scheme 3 for implementing results of water – rights transfer in Yellow River irrigation area is the optimal, i. e. the comprehensive results of Ningxia Hedong irrigation area is optimal.

Tab. 2 Weight vectors of each subsystem and rank feature value of output layer

Calculation method	Weight vector of each subsystem/w			Rank feature value vector/H				Optimal scheme
	subsystem1	subsystem2	subsystem3	Scheme 1	Scheme 2	Scheme 3	Scheme 4	
Entropy weight vector method	0.321	0.358	0.320	3.372	1.931	1.884	2.823	3
Equivalent weight vector method	0.334	0.333	0.333	3.473	1.959	1.924	2.724	3
The subjective weight vector method	0.400	0.400	0.200	3.260	1.879	1.846	2.934	3

4 Comparative analyses of evaluation results

4.1 Comparison of evaluation results

According to the 5 rank values among relative membership degree vector of standard value, S = (1 0.8 0.6 0.3 0) correspond to "excellent, good, satisfactory, barely adequate, and fail", Tab. 3 shows the results obtained by 5 methods.

Tab. 3 Comprehensive evaluation value and ranking of each scheme

Calculation method	Comprehensive evaluation value				ranking of each scheme				Optimal scheme
	subsystem 1	subsystem 2	subsystem 3	Scheme 4	Scheme 1	Scheme 2	Scheme 3	Scheme 4	
Entropy weight vector method	0.488	0.814	0.823	0.635	satisfactory	good	good	satisfactory	3
Equivalent weight vector method	0.505	0.808	0.815	0.655	satisfactory	good	good	satisfactory	3
The subjective weight vector method	0.548	0.824	0.831	0.613	satisfactory	good	good	satisfactory	3

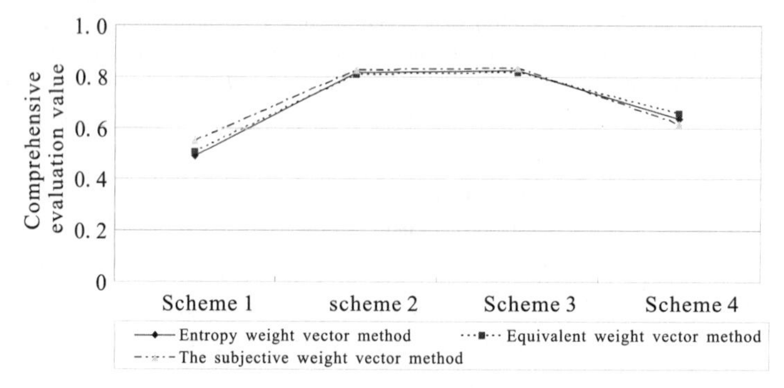

Fig. 2 The evaluation – results comparison of different weight vector

4.2 The analysis on evaluation result

As shown in Fig. 1, the evaluation results are completely identical regardless of the entropy weight vector method, the equivalent weight vector method and the subjective weight vector method. The evaluation results of scheme 1 is "satisfactory" deviating to "barely adequate"; that of scheme 2 & 3 are good and that of 4 is satisfactory. As to the implementing results of Ningxia and Inner Mongolia water – right transfer project, the comprehensive results on economic, ecological and social evaluation of Hedong irrigation area (scheme 3) are the best. In the No. 1 subsystem – evaluation, the economic result of the Southern Bank of the Yellow River irrigation area of Inner Mongolia (scheme 4) is the best; in the No. 2 subsystem – evaluation, the ecological result of Ningxia Hedong irrigation area (scheme 3) is the best; and in the No. 3 subsystem – evaluation, the social effect of the southern bank of the Yellow River irrigation area of Inner Mongolia (scheme 4) is the best.

As to the multi – layer and multi – objective comprehensive evaluation, the basic viewpoint of entropy method is that the bigger the differences between the indexes, the more important they are, and then the entropy weight is comparatively bigger. The data – reflecting contribution degree is reasonable. Eliminating the human factors in index – weight determining, the weight of each index of the 3 subsystem in the first layer is determined by the entropy weight vector of the information entropy which is objective and reasonable. When we compute weight vector in the intermediate layer, the subjective weight vector always reflects decision – makers' preference. According to the evaluation model, the equivalent weight vectors of the three subsystems are equally important. The three subsystems' weight vectors of the second layer can be obtained by three different methods and the last four schemes' results are almost the same which shows that weight variation had little impact on the results.

5 Results

Aiming at the water – transfer issue of Ningxia and the Inner Mongolia Autonomous Region, the semi – structure index system for evaluating economic, ecological and social effect is constructed based on fuzzy optimization theory and entropy weight concept, which adopts multi – method for determining the weight vector, and considering the decision – makers' preference and objective data attribute as well, at last, the top output will be used for program evaluation. Taking water – right transfer program of Ningxia, the Inner Mongolia Autonomous Region for example, the comprehensive evaluation on economic, ecological and social evaluation of Hedong irrigation area (scheme 3) is the best. The objective and fair evaluation for water – right transfer program of the Yellow River irrigation area can be obtained by this model, which also lays a theoretical basis for further water – right transfer program, sustainable utilization and economic, ecological and social effect and it is meaningful for the reasonable water – right transfer, supporting facilities improvement, water – saving reconstruction, sustainable planning and scientific management.

References

Huang Xiuqiao, Gao Feng, Wang Xianjie. Water – saving Irrigation and Sustainable Utilization of Water Resources in 21 Century[J]. Irrigation and Drainage,2001(3):1 –5.

Shao Dongguo,Liu Wuyi,Zhang Xianglong. Review of the Researches on Theory and Technology of Regulation for High Efficient Utilization of Water Resources in the Irrigation and Drainage System[J]. Transactions of the CSAE,2007,23(5):251 –257.

Feng Feng, Xu Shigou. Benefit Evaluation System of Water Resources Based on Fuzzy Optimization Theory [J]. Advances in Science and Technology of Water Resources, 2008(2):35 –38.

Chen Shouyu. Theories and Application of Fuzzy Optimization Recognition in Complex Water Resources System [M]. Jilin: Jilin University Press, 2002.

Chen Shouyu. Theories and Methods of Variable Fuzzy Sets in Water Resources and Flood Control System [M]. Liaoning: Dalian University of Technology Press, 2005.

Zhang Xianqi, Liang Chuan. Application of Fuzzy Matter – element Model Based on Coefficients of Entropy in Comprehensive Evaluation of Water Quality [J]. Journal of Hydraulic Engineering, 2005, 36(9): 1057 –1061.

Richard Iovanna, Charles Griffiths. Clean Water, Ecological Benefits, and Benefits Transfer: A Work in Progress at the U. S. EPA [J]. Ecological Economics, 2006(9):473 –482.

Jiang Ming, Lu Xianguo, Xu Linshu. Flood Mitigation Benefit of Wetland Soil – A case Study in Momoge National Nature Reserve in China [J]. Ecological Economics, 2006(11):2 –7.

Existing Problems and Countermeasures of Water Level Elevation in the Taihu Basin

Mao Xinwei[1], *Xu Weidong*[1] and *Cheng Yinda*[2]

1. Taihu Basin Hydrology and Water Resource Monitoring Bureau, Wuxi, 214024, China
2. Wuxi Branch Bureau of Jiangsu Hydrology Bureau, Wuxi, 214000, China

Abstract: Taihu Basin is a typical plain river network region, and the high precision in water level record is required. But precision of water level can not meet the basin's water management demand, owing to lack of equal land subsidence, unified height datum and enough reliable leveling calibration and measurement starting survey points. Based on the analysis of problems of water level elevation system in Taihu Basin, This article proposed that related water management departments should pay attention and increase investment to build a basin - wide, high - grade leveling measure net covering all hydrological station with reliable high - grade leveling starting survey points. The department also should develop a unified basin elevation management software system to manage water level elevation, and improve precision of basin water level record to provide reliable basis for water management.

Key words: Taihu Basin, water level, elevation system, leveling

1 Introduction

Taihu Basin lies in the middle of Chinese coast area, south of the Yangtze River delta, with area of 36,985 km^2, belongs to Jiangsu, Zhejiang, Shanghai, Anhui, etc. in administrative division. The bottom elevation of main lakes in Taihu Basin is around 1.0 m (as Zhenjiang - Wusong datum), ground elevation of the plain is between 2.5 m and 10.0 m, mostly between 3.0 m and 5.0 m. Taihu Basin is a typical plain river network, with sensitive water level because of reciprocating flow in river network. 0.01 m error in water level would present unfavorable effects to flood prevention, drought resistance and water resources scheduling, cause water resource contradictions in upstream - downstream and provincial boundary (Taihu Basin Authority, 2010).

2 Main problems of water level elevation system in Taihu Basin

2.1 Water level elevation can not been accurately reflected due to various height datum

Water level data is the foundation of hydrologic work. Water level is elevation of water surface, as relative elevation of free water surface to sea - level or other datum at certain time and location. Height datum is the starting level of ground elevation, all the water level records of hydrology stations in Taihu Basin is the data above their height datums (Ge Teng and Ma Ruifen, 1999). There are multifarious height datums in various systems without being unified in Taihu Basin at present.

Hydrology stations in Jiangsu, Zhejiang, Shanghai each apply their own water level height datum: Zhenjiang—Wusong height datum in southern Jiangsu, 1985 National Height Datum (first stage) in Zhejiang, Sheshan—Wusong height datum in Shanghai, frozen Wusong height datum in the lower Yangtze River Bureau of and the Yangtze Estuary Bureau of The Yangtze River Water Resources Commission (YRWRC). Water level height datums of some hydrology stations have not been connection surveyed with 1985 National Height Datum, brought difficulty to water resource scheduling and flood prevention, owing to lacking of conversion values and contradiction for various conversion relation among different height datums. For example, there is significant height datum deviation in the five hydrology stations around Taihu Lake in average water level calculation, because they belong to different provinces and different local hydrology departments and with different height datum. So the result can not represent exact water level of Taihu Lake.

At present, conversion relations among different height datums are as follows (Xu Juhua,

Jiang Benhai and Yao Chuguang, 2007):

1985 National Height Datum (first stage) + 1.658,7 m = Sheshan – Wusong height datum;

1985 the Yellow Sea elevation Datum (first stage) + 1.926 m = Zhenjiang – Wusong height datum (slightly different in different area);

Zhenjiang – Wusong height datum –0.264 m = Sheshan – Wusong height datum

2.2 Serious contradiction in water levels due to continuing land subsidence

Due to geological conditions and groundwater overexploitation, there is serious land subsidence in Taihu Basin. According to statistics data, the groundwater depression funnel in Suzhou – Wuxi – Changzhou area is as deep as 80 m and covering 5,483 km^2, with cumulative depression over 1 m in the center (Changzhou 949 mm, Suzhou 1,682 mm, Wuxi 1,100 mm); there are 8 depression funnels in Hangzhou—Jiaxin—Huzhou area, among which, cumulative depression over 0.81 m in Wuzheng, 0.62 m in Wangjiangjin since 1976; average cumulative depression speed is about 10.3 mm/a since 90's. Although the speed of land subsidence has eased up due to local government's measures as closing deep groundwater well and ban on groundwater exploitation, there is still serious land subsidence in local area. The speed of land subsidence is various in different areas. The government has not organized high – grade leveling calibration and measurement regularly in recent years, especially in serious land subsidence areas, which affects consistency and reliability of basin water level data, and leads to unreasonable phenomenon as water flows upwards in flood presentation and water resource scheduling because of contraction between upstream and downstream water levels of some river channel.

2.3 Low precision of leveling calibration and measurement due to insufficient reliable leveling starting points

Taihu Basin has high requirement of precision of leveling calibration and measurement because of subdued topography and low surface slope of river net, thus high – grade bench mark as starting point of water level elevation calibration and measurement is necessary (Dou Chunhong, 2011). The first class leveling starting points, promulgated by nation and been used all along, are only I Wu – Hang 1 main benchmark, I Wu – Hang hidden 16 benchmark, Sheshan benchmark, Washan benchmark, Qinshan base, and I Qin – Hang benchmark 108. These starting points locate mainly on eastern Taihu Basin. They are insufficient in the north and south of Taihu Basin, and especially rare in the west and southwest, as shown in Fig. 1. This situation has caused low control capability of leveling survey net and low precision, which can not meet the need of water level elevation's calibration and measurement of the basin for long distance in geodetic height determination.

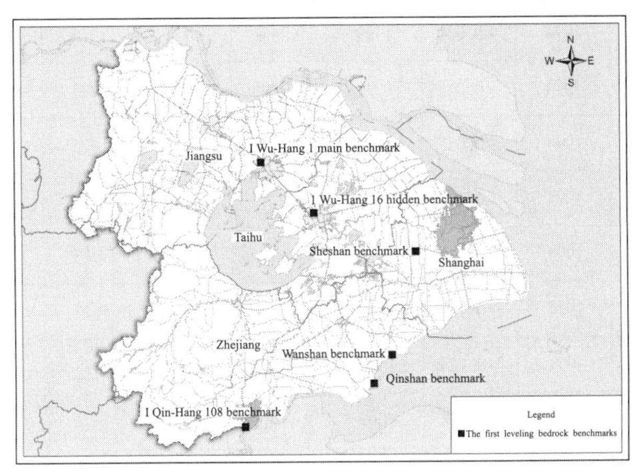

Fig. 1 Distribution map of the first class leveling bedrock benchmarks in Taihu Basin

3 Present situation of calibration and measurement of water level elevation in Taihu Basin

The basin authority of the Ministry of Water Resources (MWR) and local water conservancy departments have pay high attention to the problems in water level elevation in Taihu Basin. Since catastrophic flood in 1999, Taihu Basin Authority (TBA) of MWR has organized a second class leveling for important hydrologic stations in the basin, proposed correction coefficients to the hydrologic stations in serious land subsidence areas, as around Taihu Lake, around Wangyu River and around Taipu River; in 2009, TBA combined with Zhejiang Bureau of hydrology, has organized a second class leveling for important hydrologic stations in the areas as around Tai Lake, around Wangyu River, around Taipu River and northern Zhejiang, made rectification to part of hydrologic stations in serious land subsidence areas. In the meantime, each rock foundation markstone has been established on Suzhou Zhangqiao and Wuxi Dapu as the start point for water level elevation survey. Correction coefficients proposed in the last two leveling as shown in Tab. 1.

Tab. 1 Correction coefficients for part of important hydrologic stations in Taihu Basin

Station	Correction coefficients (1999)	Relative year	Correction coefficients (2009)	Relative year	Station	Correction coefficients (1999)	Relative year	Correction coefficients (2009)	Relative year
Wangting (Taihu)	-0.405	1994	-0.184	1999	Chensu	-0.33	1994	-0.220	1999
Wangting (da)	-0.418	1994	-0.148	1999	Dapukou	0	1994	-0.182	1999
Linqiao	-0.251	1994	-0.037	1999	Baishao hill	-0.04	1994	-0.061	1999
Changshu	-0.030	1994	-0.042	1999	Duhill pass	0	1994	-0.049	1999
Xiangchen	-0.053	1994	-0.019	1999	Taipufate (down)	-0.04	1994	-0.073	1999
Fengqiao	-0.082	1994	-0.046	1999	Jiaxing (hang)	-0.33	1975	-0.145	2005
Kunshan	-0.039	1994	-0.033	1999	Wangjian gjing	-0.62	1975	-0.356	2005
Xukou	-0.015	1994	-0.080	1999	Jiashan	-0.12	1975	-0.020	2005
Xishan	-0.002	1994	-0.080	1999	Xitang	-0.20	1975	-0.191	2005
Guajingkou	-0.036	1994	-0.075	1999	Chongde	-0.61	1975	-0.118	2005
Chenmu	-0.015	1994	-0.057	1999	Wuzhen	-0.81	1975	-0.138	2005
Pinwang	-0.132	1994	-0.173	1999	Hangchang bridge	-0.15	1975	-0.100	1999
Wangyugate (UP)	+0.033	1994	-0.081	1999	Xiaomeikou	-0.04	1975	-0.070	1999
Qinyang	-0.52	1994	-0.120	1999	Nanxun	-0.15	1975	-0.180	2005
Ganlu	-0.25	1994	-0.114	1999	Shuanglin	-0.09	1975	-0.130	2005
Southgate of Wuxi	-0.07	1994	-0.058	2007	Deqing	-0.02	1975	-0.090	1999
Luoshe	-0.61	1994	-0.117	2007	Jiapu	-0.07	1975	-0.070	1999

Local hydrologic departments have organized hydrologic stations to conduct level calibration within their own jurisdication area regularly. But owing to different starting points and inconsistency time, especially without unified requests and standards in correction for land subsidence and various correction time, there is no guarantee in consistency and reliability for the chaos water level data. The problem has not been solved fundamentality.

4 Countermeasures and thoughts

4.1 On organizing regularly basin - wide high - class calibration and measurement of water level

The last two leveling works organized by TBA since 1999 did not covered the whole basin but only a few areas, parts of stations have not been corrected, which brought contradiction between upstream and downstream of the basin. The basin authority should establish a basin - wide leveling measure net with local water conservancy departments to cover all hydrological stations; and make combined high - class leveling (above second class leveling); unite adjustment; grasp the impact of land subsidence on water level elevation, survey all the water level of hydrology stations in the basin, and propose correction coefficients in time.

4.2 On building reliable high - grade leveling starting points

River net area has high requirement of precision of water level, an amount of high - class leveling start points are needed for leveling work. For present situation, with insufficient and unreasonable distribute high - class leveling start points, a strong leveling start points net should be built by increasing high - class leveling start points and hydrology station ' s water level check points (Zhang Luocheng; Roger C K Chan & Wu Chucai, 2003). More bedrock benchmarks in serious land subsidence area are also needed in sensitive hydrology area, important hydraulic engineering and areas of bedrock benchmarks in sub - standard distribution (Ref: "The first and second class leveling measurement standard") to improve precision. In the meantime, embed basic leveling benchmarks near hydrology stations, determine geodetic height of basic leveling benchmarks by bedrock benchmarks, and verify zero points of water level elevation of hydrology stations before and after flood in time.

4.3 On development of a unified basin elevation management software system

Although different provinces use various water level height datum, there is difficulty in unitized 1985 National Height Datum. So a unified basin elevation management software system should be developed to solve this program. On the basis of results from combined basin - wide high - class leveling, conversion relationship among different height datums should be verified and validated after calculation and analysis, to query information among different height datums of different stations in Taihu Basin by database software, server software (C/S) and browser software (B/S). It would support water management in a simply and visually way.

5 Conclusions

Precision of water level data is the fundament of water management, which directly affect flood and drought control, water resource management, water resource protection and important water conservancy project construction in Taihu Basin. The problem of water level elevation in Taihu Basin has always been exiting, but can not been solved without implement funds. Related water management departments should pay attention and increase investment to take measures to unify management of water level elevation, provide reliable basis for water management decision.

References

Taihu Basin Authority. Water Transfer Experiment from the Yangtze River to Taihu[M]. Beijing: China Water Power Press, 2010.

Ge Teng, Ma Ruifen. Height Datum of the Yellow River Basin Needs to be Unified[J]. Yellow River, 1999, 21(11): 15 - 16.

Xu Juhua, Jiang Benhai, Yao Chuguang. Generality of Wusong Height Datum in the Middle and Lower Yangtze Main Stream[J]. Yangtze River, 2007, 38(10): 85 – 88.

Zhang Luocheng, Roger C K Chan, Wu Chucai. Land Subsidence Problem and Its Control in Taihu Basin of South Jiangsu Province due to Overexploitation of Underground Water[J]. Journal of Lake Science, 2003 15(3): 257 – 262.

Dou Chunhong. Application of Digital Level DiNi12 in Standard Calibration in Hydrological Stations of Taihu Basin[J]. Journal of Zhejiang Water Conservancy and Hydropower College, 2011 (2): 68 – 71.

Construction Plan for Flood Detention Areas in Lower Yellow River

Liu Juan*, *Cui Changyong*, *Cui Meng* and *Cai Chunxiang

Yellow River Engineering Consulting Co. , Ltd. , Zhengzhou, 450003, China

Abstract: Dongping Lake flood detention area and Beijindi flood detention area are the two most important flood detention areas in lower Yellow River. Due to long – term insufficient investment, the engineering, non – engineering and safety construction in these areas are far lagging behind, which causes serious problems. In order to improve the operation safety of the flood detention areas as well as the safety of people's lives and property in these areas, construction plan is put forward based on the analysis of construction situation and existing problems. The construction plan includes the construction contents and scale of engineering measures, and that of non – structural measures. In addition, three resettlement schemes, namely resettlement of moving outside, nearby resettlement and temporary evacuation, are introduced to reduce the flood risk of residents in the flood detention areas. The corresponding population scale of each settlement scheme is analyzed, along with the construction contents and scale of the supporting safety facilities.

Key words: lower Yellow River, the flood detention area, Dongping Lake flood detention area, Beijindi flood detention area, construction plan

1 Overview of lower Yellow River flood detention areas

In the lower Yellow River, the Dongping Lake and Beijingdi flood detention areas are set up for flood diversion and flood loss mitigation (see Fig. 1). Dongping Lake detention area is located across Liangshan, Dongping and Wenshang county of Shandong Province, where the wide Yellow River channel turns into a narrow one. Due to the important location, it is considered as the key project to guarantee the flood security of the following Aishan narrow river. It has an area of 627 km^2 and a flood detention capacity of 3.05×10^9 m^3, which is divided into two parts of the New Lake (418 km^2) and Old Lake (209 km^2) by the secondary dam. The Old Lake is connected with the Daqing River and Dawen River so it contains water throughout the year. The projects of this detention area include the surrounding embankment, secondary embankment, separation embankment at the mountain pass, the flood diversion gates and outlet gates. The surrounding embankment stretches from Xuzhuang gate to Wujiama with a length of 88.300 km which includes 10.471 km embankment serving as embankment for both the river and lake, from Xuzhuang gate to Guonali in Liangshan. The embankment elevation is 48 m and the face towards lake is covered by dry masonry revetment with a crest elevation of 46.0 m. The secondary embankment stretches from Linxin gate to Xie River estuary with a length of 26.731 km and an elevation of 48.0 m. The side towards the Old Lake is covered with stone revetment. There are three flood diversion gates named Shiwa gate, Linxin gate and Shilipu gate, and two outlet gates named Chenshankou gate and Qinghemen gate. To ensure the safety of flood detention area, Sihai gate was built in 1998 for discharging flood into Nansi Lake. After that, two gates were built in 2003. One is Baliwan gate built to connect the Old Lake and New Lake, the other is the Pangkou backwater gate which is located at the gateway of the Northern discharging system.

Beijindi flood detention area refers to the area between the Yellow River bank and Beijingdi bank, which locates across Puyang City of Henan Province and Liaocheng city of Shandong Province. It is established in 1951 by the Government Administration Council for dealing with the super – standard flood in Lower Yellow River. It has an area of 2,316 km^2 and a volume of 2.07×10^9 m^3. The projects of Beijingdi flood detention basin include Beijindi embankment, Qucun flood diversion gate, Zhangzhuang sluice and flood control works. Beijindi embankment is the original remote embankment of the Yellow River. After the establishment of Beijingdi flood detention area, it became the grade 1 embankment of the flood detention area. It has experienced three large – scale reconstructions, and the current length is 123.3 km.

2 Problems in flood detention area construction

2.1 Engineering problems

There are serious problems resulting from long – term inadequate investment. One prominent

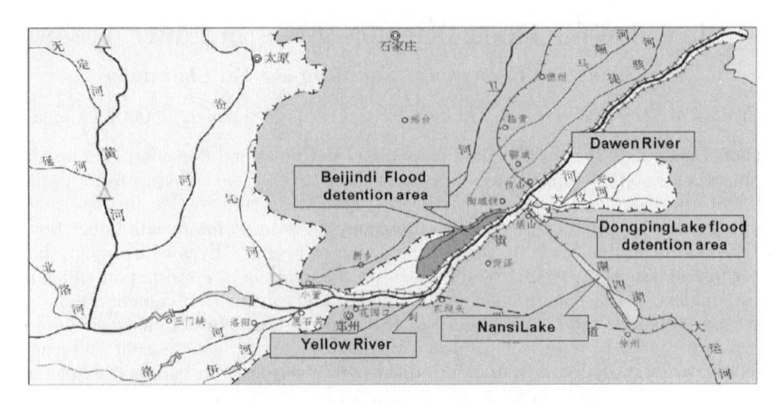

Fig. 1 Distribution of flood detention areas in lower Yellow River

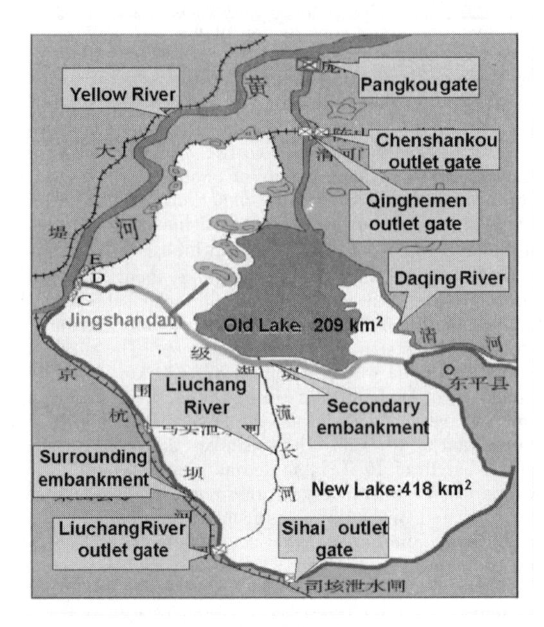

Fig. 2 Layout of Dongping Lake flood detention area

problem is that the construction cannot reach the designated standard. For example, the embankment of Dongping Lake flood detention area is insufficient in height, and its slope lacks effective protection. Once in operation, it is almost impossible to be used as planned. Another problem is that the flood diversion and outlet sluice are inconvenient in operation because of aging equipments. Qucun Gate, for instance, is the flood diversion sluice of Beijindi flood retention area. However it has never been used since it was built so far. Due to lack of maintenance, the gate is facing serious problems such as leaking hoist room, broken steel wire rope, the aging hoist motor, large quantity of sedimentation in stilling pool and gate slot.

2.2 Non – engineering problems

Some communication and warning facilities of the flood detention area have been discarded, and the control system of flood diversion gate are aging. Furthermore, the flood detention areas lack necessary equipments for acquisition and transmission of hydrological and project condition information. In addition, the operation of gates relies on traditional method with low efficiency. These existing problems have significant effect on flood control decision – making which will result in missing the best opportunity of flood diversion. For instance, the network bandwidth of Dongping Lake is so narrow that it cannot meet the dynamic images transmission and multimedia application requirements needed in the flood control conference online. Qucun diversion gate of Beijindi flood detention area has similar problems. The control system of this gate consists of 56 surface control cabinets and other control equipments. All these equipments are controlled manually, which leads to inaccurate elevation control and the delay of information feedback.

2.3 Safety construction problems

The construction of safety facilities are way lagged behind which leads to inadequate security facilities and evacuation roads. Therefore, the life security and main property of most residents cannot be guaranteed during the operation of flood detention area. The low standard of most safety facilities already constructed cannot meet the demand to reduce property loss as well. Take Dongping Lake as an example, according to the planning, only 1% of all the people can to be resettled safely in case of flood diversion, which means that 99% of local residence will be transferred to other places.

3 Construction content and scale of flood detention

3.1 Dongping Lake flood detention area

3.1.1 Engineering measures

Surrounding embankment of Dongping Lake is ranked as the first grade embankment. According to the standards of first grade embankment, 55,53 km stone slope and 2.45 km slope foot of surrounding embankment needs to be reinforced. In order to improve the reliability of transportation, the surface of 77.83 km levee top and 6.5 km road connecting the villages and levee is to be hardened. 77 flood control houses are to be built with an area of 9,240 m^2. 26.73 km secondary embankment is planned to be strengthened to comply with the fourth grade standards.

Other embankment improvements include reinforcement of the separation levee at mountain pass with the length of 2.39 km, the construction of 22.25 km dual – use levee that serves for river and lake, levee top surface hardening with the length of 13.28 km.

As for flood diversion and discharge projects, water flow measurement facility is installed for accurate control of the diversion flow of the flood diversion works. The drainage improvement project planning is based on the principle that discharging flood into Yellow River is the main option while discharging flood into Nansi Lake via Liangji canal is happened when needed. The projects of discharging into Yellow River involve the expansion of Pangkou Sluice, excavation of silt beach between the sluice and the Yellow River main channel, as well as the improvement in the management house, transformer room and other facilities. In addition, the excavation of 7km upstream approach channel of Chenshankou sluice and Qinghemen sluice. In order to discharge flood from Old Lake into Southern area, the 17.6 km reach of Liuchang River is to be enlarged and excavated, and a 4.3 km channel is to be excavated to meet the requirements of flood discharge. 200 m waterway needs to be dredged and some protection projects needs to be built for bettering the southern engineering supporting system.

3.1.2 Non engineering measures

Non engineering measures involve reconstruction of information network for flood control between Dongping Lake Management Bureau and its subordinate departments as well as construction of mobile emergency team for flood control and corresponding equipments.

Communication network with the transmission bandwidth for 20 ~ 30 m is constructed between Dongping Lake Management Bureau and Yellow River Shandong Bureau, and that between Liangshan, Dongping and Tianan is 10 m. As for the computer system, the original computer system of Dongping Lake is to be improved while the computer system of Tai'an base is to be newly built. The corresponding software and hardware are to be equipped for better services. In addition, the communication network and the office network are to be built between Tai'an and Yellow River Shandong Bureau, between Tai'an and County Bureaus, as well as between Tai'an and the government of "two cities and three counties". Other information network construction lies in the communication of Dongping county base. Mobile emergency team for flood control and corresponding equipments needs to be constructed in Yellow River Shandong Bureau.

3.1.3 Safety facilities

Safety facilities construction in flood detention area is mainly to improve life and property safety for residents in this area. Three resettlement schemes, namely resettlement of moving outside, nearby resettlement on platform and temporary evacuation, are introduced to reduce the flood risk of residents. The flood risk analysis result, the resettling capacity outside, the diameter of the arable land and the geological location are the four main factors taken into consideration when choosing the proper resettlement schemes. .

In Dongping Lake flood detention area, there are 306.1 thousand residents needed to be resettled. Three villages located within 1.5 km from embankment and not on the hillside, shall be resettled outside with the population of 5.1 thousand. As for the 27.1 thousand people who live to the west of Jinshan dam and along the secondary embankment in Old Lake, although they are in high risk region, they shall be resettled nearby on the platform due to insufficient resettlement capacity. The total platform area is 2.79×10^6 m^2.

There are 274 thousand residents left which includes 24.7 thousand residents of Old Lake and 249.3 thousand residents of New Lake. They have to be evacuated provisionally considering all the factors. According to the village distribution and existing road system, the planning road system for provisional evacuation consists of 200.38 km road and 86 bridges. In the Old Lake, three evacuation roads are to be built with a total length of 13 km. While in the New Lake 23 roads need to be built with a total length of 187 km, including 80.6 km reconstruction road and 106.78 km new road.

3.2 Beijindi flood detention area

3.2.1 Engineering measures

Engineering construction of Beijingdi flood detention area focuses on the construction and maintenance of Yellow River backwater area. The embankment slope reinforcement stretches from embankment 103 + 400 to Zhangzhuang gate with a length of 19.94 km. As for the flood diversion and discharge facilities, engineering improvement involve renovation in gate hoist room of Qucun sluice, reinforcement in control embankment and transformation of standby power. At the same time, underwater dredging at the Zhangzhuang gate, and building the150 m embankment near the Yellow River. In order to mitigate the potential risks of embankment in this area, 13 culverts and water gates are to be removed and 4 will be improved while another 13 culverts and water gates will be reconstructed.

3.2.2 Non engineering measures

Based on the existing problems, flood diversion sluice control center will be established at Qucun sluice including 7 subsections concentrated controller and 56 hoister control stations. As for Zhangzhuang gate, automatic monitoring and control system will be installed, as well as the construction of special information and computer network system for flood control.

3.2.3 Safety facilities

Due to the lower flood risk, the scheme of temporary withdrawal is used in the area with a total resettlement population of 1 million which are mainly distributed at the lower reach of backwater area. The supporting safety facilities are comprised of 120 km road and 38 bridges for retreat.

4 Conclusions

As an integrated part of flood – control system, flood detention areas alleviate flood control pressure of flood protection zones by reserving excessive flood. On the other hand, the residents in flood detention area expect to have booming economy and higher living standard. The flood detention areas are always in such a dilemma of balancing flood diversion and economical development demand. Hence, it is necessary to launch flood control construction on the base of current problem. All these construction contents and scale discussed above provide scientific guidance for the flood detention area construction in the next stage. They are also beneficial to improve the safe operation of flood detention areas, as well as the safety of people's lives and property in flood detention areas and flood protection areas.

References

Cui Changyong, Liu Juan, Liu Shenyun. The Division of Flood Prone Areas in Dongping Lake Flood Detention Area and the Safety Construction Schemes[J]. Yellow River, 2012(2): 21 - 22.

Fu Xiang, Liu Ning, Ji Changming. Research on Flood Insurance Mode in Flood Storage and Retention Area of China[J]. Yangtze River, 2005(8): 71 - 74.

The Study on Building the Plain Reservoir Relying on Flood Control Project of Jindi River Downstream

Song Ning[1], *Chen Chen*[1], *Chen Pihu*[2] and *Zhang Shuhong*[2]

1. Water Bureau of Yanggu County, Yanggu, 252300, China
2. The Yellow River Yanggu Bureau, Yanggu, 252300, China

Abstract: Jindi River is an important branch of the Yellow River downstream, and the watershed is located in flood detention of the North Jindi River of the Yellow River, which is an important part of flood control system of the Yellow River downstream. In 1999, the first improvement engineer project of the main stream of Jindi River began to run, and several heavy floods happened. The flood confluenced quickly, but it was quite hard to be drainaged into the Yellow. And the countercurrent was long. After the Zhangzhuang sluice was rebuilt, floor elevation was 40.0 m (the Yellow Sea level, the same below), while the level of beach face of the entrance to the Yellow was 43.0 m. It was becoming harder and harder for flood of Jindi River to be drained into the Yellow River, meanwhile there were many troubles for the Zhangzhuang sluice, which made the flood prolonged stay within the 20 km riverbed of the downstream, forming impound about 5.0 m to 6.0 m deep and 1.00×10^8 m^3, for 9 months long. The field in the river road could not be cultivated and the water can't be used, which make great loss to the population. Firstly, the water can't be used, and drown the field. Secondly, it can't make full use of the water resource. Then we can use the present project, with less investment. If building plain reservoir based on flood control project of Jindi River, it can save investment, be efficient, and can solve water shortage. Another advantage is to transfer water to the Yellow River at dry season from other reservoir, to clean riverway in order to keep the Yellow River healthy and play its role in economic development.
Key words: the Yellow River, Jindi River, Plain Reservoir, water resource, study

Jindi River that originated in the Huaxian, Henan Province, is an important branch of the Yellow River downstream, flows northeast from west to south, and drains into the Yellow River from the Zhangzhuang sluice of Taiqian County in Henan Province, which is an inter-provincial river crossing 12 counties about Xinxiang, Hebi, Anyang, Puyang in Henan Province and Liaocheng in Shandong Province and so on. Jindi River basin, to the north with the Weihe River, Majiahe River, Tuhaihe River of adjacent, to the south with the Yellow River embankment and natural rock drainage basin adjacent, to the west of the People's Victory Canal Irrigation District east of Seven Mile Camp, to the east with Zhangzhuang country at the junction of the Taiqian County in Henan Province with Yanggu County in Shandong Province that drains into the Yellow River. And the basin is wider at the top that the maximum width of more than 60 km, is long and narrow triangle about 200 km from east to west, lying south-west and northeast, and the basin area of 5,047 km^2, the arable land of 5.281×10^6 mu and population of 2.88×10^6, among the basin area of 115 km^2 and arable land of 114,000 mu in Shandong.

Jindi River is from Gengzhuang of Huaxian in Henan to Zhangzhuang of Taiqian County entering the Yellow River that the stage is Main River Channel about 158.6 km long, height difference of nearly 30 m, the average gradient of 1/5,000, the gentle slope of 1/10,000 in the downstream. In 1999, the first improvement engineer project of the main stream of Jindi, its governance includes that main stream of the river on the original excavation 131.3 km (including river Wrecker) from Huaxian Wuyemiao Village in Henan to Taiqian Zhangzhuang sluice, South small embankment enforced 49.2 km and the North dike enforced 22.6 km, the new and converted crossing bridge 8, the new South embankment culvert 2, to overhaul Zhangzhuang pumping station and so on. After completion, the flood control standards once in 20 years and drainage standards reached in 3 years in the Jindi River.

1 The necessity of building the Plain Reservoir in Jindi River downstream

It is necessary to build the Plain Reservoir relying on flood control project of Jindi River downstream for three reasons. Firstly, there will be plenty of water storage. Over the years the most runoff of more than 2.00×10^8 m^3 were stored in Jindi River downstream resulting in a lot of land flooded and serious economic losses. Secondly, the vast areas of water are short in the Yellow River water or groundwater, a large number of references and mining have formed a large funnel area, sewage infiltration, to result in a groundwater qualitative difference. Furthermore, it can save a lot of investment for the country by use of a large number of flood control projects in the Jindi River downstream.

1.1 The runoff situation of Jindi River

The actual measurement Jindi River runoff (Mean annual runoff) is as follows. Wuyemiao sluice is 1.108×10^8 m^3, Puyang station is 1.64×10^8 m^3, Fanxian station is 2.22×10^8 m^3, and Kongcun Station 0.405×10^8 m^3. Measured runoff annual variation for example Puyang annual maximum runoff is the best path flow 53.8 times. And runoff distribution is uneven for Puyang station that the flood season (July to October) Proportion of the year accounted for 68.3%, Fanxian Station for 75%. Runoff from each month that the rest of the month runoff in Puyang station was small except for 8,9 months between 1956 and 1987, especially in a particular year there were 3 months appeared full month of zero flow such as 1957, 1959, 1965, 1979, 1981. And it has also appeared in the same situation at Fanxian in 1979.

1.2 Difficulties of Jindi River runoff into the Yellow River

Since Zhangzhuang sluice has been renovated in 1999 because of uplift of the riverbed of the Yellow River, converted from 37.00 m to 40.00 m, improve the 3 m, while the sluice before Jindi bed elevation less than 37.00 m. But as the Yellow River riverbed is silting and high, it was becoming harder and harder for Zhangzhuang sluice outflow by the backwater of the Yellow River. Under Zhangzhuang sluice, surface elevation reached 43.00 m than Zhangqiu in Jindi riverbed 5 m more. See the flood situation since recent years, the flood retention time of the Jindi River end was significantly prolonged. In 2000, although Zhangzhuang sluice drain off water, the floods were prolonged stay for 6 months. By the 3 - year drainage standard, the Zhangzhuang sluice Pro Jindi water level is 43.22 m corresponding while the Yellow River flow 2,000 m^3/s in 2004. In other words, when the Yellow River flow is greater than 2,000 m^3/s, the flood under 43.22 m can't be excreted into the Yellow River, the cuddle embankment depth of about 4m in 104 +400. When the Yellow River flow reached 7,000 m^3/s and water level along the Yellow River reached 45.64 m in Zhangzhuang sluice, which is higher 0.04 m than the flood control standards once in 20 years, so Jindi flood will not be able to discharge into the Yellow River.

But many problems exist by using Zhangzhuang sluice discharged into the yellow for small mention rows, higher operating costs and so on. The Zhangzhuang sluice was founded in 1980. Its role is to remove the stagnant water in the lower Jindi Triangle area about 293 km^2, and in conjunction with Zhangzhuang sluice of row Jindi to the flood. The station is designed drainage standards reached in 3 years, discharge capacity of 64 m^3/s, the total installed capacity of 6,400 kW (8×800 kW). Now Zhangzhuang station belongs to the stage of Henan Province, the county government management, drainage should be reported to the approval of the flood control headquarters of Puyang. The pump station is energy - hungry and flood - long stay at the end of Jindi River impacts Yanggu of Shandong seriously and impacts Henan less. In the case of power shortage in flood season and taking into account local interests, station use is very restricted and can't fully play its role.

1.3 Inadequate water resources in Jindi River downstream and surrounding areas

The Jindi River downstream is related to four counties in two provinces for Liaocheng, Yanggu, Shenxing in Shandong, and Fanxian, Taiqian in Henan, its main resources come from the Yellow River, but pass-by water resources are scarce. According to statistics, the region for many years average annual rainfall reached 606.4 mm is about more than three times of annual evaporation of 2,100 mm. Precipitation characteristics is that time and space are unevenly distributed resulting in drought and floods alternately. , To take Liaocheng for example, the average annual rainfall is 566.7 mm equivalent to 4.648×10^9 m^3 of water, which can be formed of surface runoff 3.69×10^8 m^3 and infiltration recharge of 9.27×10^8 m^3, and the uneven distribution of precipitation mostly concentrated in 7 ~ 9 months that it is accounting for 75.10% of the total annual precipitation, a annual evaporation is 1,316 mm, the drought index 2.3. The city's water resources is 1.32×10^9 m^3, 243.6 m^3 per capita, the only fairly 10% of national per capita. Currently water scale and quota are calculated by various departments, the total water demand at Liaocheng have been 3.368×10^9 m^3 by 2000, water consumption increased to 4.064×10^9 m^3 by 2020. According to the present situation of water resources, water shortages amounted to 1.198×10^9 m^3 by 2000, water shortage amounted to 1.894×10^9 m^3 by 2020, and with the development of the regional economy, the gap will increase year by year. In recent years especially the drought year of 1997, because of the lack of water resources, to regional rainfall in, the shortage of water resources, a serious shortage of surface water resources, resulted in acute water table brush down, the city's funnel area will be expanded to 3,899 km^2, large amounts of sewage intrusion, part of the groundwater was seriously polluted. It's resulted 290,000 hm^2 of crop production or no, more than 4.3×10^9 yuan in direct economic losses only 1997, 0.49×10^6 people drinking water difficulties, to the people's production and life brought serious difficulties.

2 The feasibility of building the plain reservoirs in Jindi River downstream

Firstly, the water can't be eliminated as quickly as possible in Jindi River downstream, and drown the field, which makes great loss. Secondly, the water shortages of Jindi River downstream brought serious losses to industrial and agricultural production. Now it's good condition that the coming of Jindi River on the upper and middle can meet water requirements. After the north Jindi River and South small embankment are reinforced as work as a reservoir, cited revitalization of Jindi River water renovated as used engineering, Zhangzhuang sluice can be used as excretion engineering. The use of existing works can save many investments, and ensure 1.00×10^8 m^3 of water to supply the water demand areas of Jindi River downstream in water demand season, to alleviate tension water for Jindi River downstream.

2.1 The inflow of the upper and middle Jindi River can meet water storage requirements

The actual measurement Jindi River runoff (Mean annual runoff) that Wuyemiao sluice is 1.108×10^8 m^3, Puyang station is 1.64×10^8 m^3, Fanxian station is 2.22×10^8 m^3. Based on the above analysis, Fanxian station is 2.22×10^8 m^3 that can meet assurance 1.00×10^8 m^3 water of the Jindi downstream.

2.2 The channel storage capacity in the lower Jindi River

Jindi River downstream that watercourse under 104 + 000 is 20 km long, the average width of 550 m, stagnant water area of 11 km^2, average water depth of 5 m, the storage of river channel of 1.1×10^6 m^3. In other words, only using channel storage in the lower Jindi River can meet the storage capacity without requirement of arable land. If to build a plain reservoir with the depth of 2.5 m, land resources can be saved with over 240 km^2.

2.3 The relationship between the water storage of the lower Jindi River and the flood control of the Yellow River

Jindi River flows directly into the North Jindi flood detention district. It has a direct impact on the amount of flood diversion, so it is necessary for flood diversion analyzes in the North Jindi detention basin and the Yellow River flood happened.

Basin shape is wider at the top, flat terrain, wide river, many low – lying, flood detention storage, to flood hydrograph Jindi obesity. A flood lasted generally more than 8 d, two consecutive rainfall bimodal flood lasted up to 13 d. Such as the august 1963 flood, the peak flow at Zhangzhuang sluice was 735 m^3/s, flood volume was 6.5×10^9 m^3, which lasted 22 d.

Based on the measured rainfall and measured flood data of the Yellow Rivers Station and Jindi basin, the heavy rain with Jindi River basin above the mouth of the Yellow River Garden does not occur at the same time.

After analysis, when a thousand – year flood happens at the Huayuankou station, the corresponding 12 – day flood in Jindi River is only 1.04×10^8 m^3, slightly larger than three – day rainstorm flood volume of 1.01×10^8 m^3 in three – year flood at Zhangzhuang Station. When a hundred – year flood happens at the Huayuankou station, the corresponding 12 – day flood in Jindi River is only 0.85×10^8 m^3, less than three – day heavy – rain flood volume in three – year flood at Zhangzhuang station. The Yellow River flood and Jindi River flood aren't generally suffered, small and medium – sized flood suffered Jindi River little water. Then when devastating flood is forecasted in the upper and middle Yellow River, on one hand, as the river water regime with Zhangzhuang sluice to drain into the Yellow River in a timely manner, on the other hand to open Zhangqiu sluiceway and discharge water to the Grand Canal and Tuhai. In addition, it doesn't matter without discharge because flood routing through Jindi inflow of 7.00×10^8 m^3 in North Jindi River detention basin, now only 1.00×10^8 m^3 water, is far less than the amount of water to flood routing. Therefore, building the plain reservoir in the Jindi downstream does not affect the Yellow River flood control.

2.4 The relationship between water storage and drainage in the Jindi River downstream

The low – lying physical feature and the main channel management of the Jindi River downstream speed up convergence speed of the Jindi River downstream storage, extend the duration of the high water level of the end of the detention basin. By drainage 3 – year standard, the Zhangzhuang sluice level is 43.22 m, prolongs stay for 4m depth. Under the 104 + 400; by the flood control standards once in 20 years, the water level is 45.6 m, 6.5 m in depth. It is a natural water depression and will not affect the drainage of the Jindi flood water.

2.5 Available project for building the plain reservoir in Jindi River downstream

2.5.1 The North Jindi

The North Jindi on Huoxiangtou of Henan Puyang from Gaodikou of Shenxian County to the Taochengpu of Yanggu County in Shandong is 123.34 km in length, and is a major barrier of North Jindi River detention basin. Among Shandong length 83.4 km, corresponding stake is form 39 + 936 to 120 + 335, 0 + 000 (120 + 335) to 3 + 000. To protect the part village of Henan Taiqian and Shandong Yanggu, there are two sections of North dike with the North Gimjae phase in the detention area. One is from Yanggu Ziludi to Jindouying, stake for 76 + 000 ~ 79 + 260, length 3,260 m; another is from Yanggu Hualianchi to Liugai, stake for 83 + 000 ~ 104 + 400, length 21,400 m. Shandong section of North Jindi include: the long dike 83.4 km; a vulnerable spot 13 containing 83 of the dam Coast; culvert 8.

2.5.2 South small embankment

South small embankment from Zhangzhuang sluice to Gaodikou, is 80 km in length. In order to discharge upstream floods, protect the people's production and life in the Delta region, in 1965, the 50km south embankment was built from Gucheng to zhangzhuang sluice by the flood control standards once in 20 years during the Jindi governance. South embankment from Gucheng to Gaodikou is formed on the original Plain Reservoir Embankment basised. Because the South dike which is located in Henan and Shandong provinces border, is no unified management institutions, resulting in serious damage to dikes. A long time, the damage has been scoured by wind and rain and human activity, embankments incomplete. Masses destroyed the levee phenomenon have occurred, for example built kiln along the embankment, spate of breaches soil, water diversion and drainage of Pa embankment, flat embankment planting and destroyed dike building. Levee is thin with more gap, and even a few kilometers of some embankment sections were razed to the ground and lose the flood control capacity.

In 1999, South embankment between Zhangzhuang sluice and Menglouhe sluice in Gucheng (49 km length of embankment sections) was reinforced by the flood control standards once in 20 years. The section from Gucheng to Gaodikou has been met the flood control standards once in 20 years without reinforced. And it is the necessary renovations for the certain gap segment and embankment sections residue combined with river excavation.

2.5.3 Zhangzhuang sluice

Zhangzhuang sluice is located at Wuba town, Taiqian County, Henan Province (left bank of the Pro Huang embankment Stake for 193 +981), is a multi-sluice for withdrawal of water, retaining, intrusion. It is the original from Jindi River to entrance the Yellow River, so when Jindi River floods can discharge into the Yellow River, when the Yellow River floods flow backward from the entrance. In 1949, Mingnian built along the Yellow River embankment to intercept Jindi River entrance yellow, by the Zhangqiu sluice, drainage into the Jinghang Canal; when happened the large floods, Temporary drainage into the Yellow River to clawed along the Yellow River embankment. Since building North Jindi River detention basin in 1951, there is no withdrawal of water works. Until 1962, when the North Jindi River detention basin temporarily abandoned, the State Development and Planning Commission have agreed to construction zhangzhuang sluice. When the water level is higher than that of the Yellow River, Jindi River releases on; When the water level is lower than that of the Yellow River, Jindi River sluice closes retaining, to prevent intrusion.

The sluice was designed in 1963, its discharge ability to 1,000 m^3/s in the use of the North Jindi detention basin. When draining the Yellow River flood into the detention area by the Zhangzhuang sluice, the flow capacity is 1,000 m^3/s. In 1999, Zhangzhuang sluice was converted, retained the Yellow River intrusion and flood diversion withdrawal of water two functions, elevated sluice bottom from 37 m to 40 m.

2.5.4 Water drainage culvert

North Jindi River downstream have eight water (drainage) culverts, most built in the sixties and seventies. In 1984 and 2000, Zhangqiu sluice and Mingdi sluice have been renovated. Gaodikou sluice that belonged Jindi Authority jurisdiction was built in 1996, after revocation of the Jindi Authority, was placed under the management Liaocheng Bureau. Sluice in the Tab. 1.

Tab. 1 The North Jindi culvert Statistics at Liaocheng

culvert and sluice name	Pile number	Engineering scale						Underground water(m)	Build date	notes
		Hole Qty	Hole width (m)	Hole height (m)	Hole height (m)	Design flow (m^3/s)	Irrigation area ($\times 10^6$ mu)			
Total						129.53	191.6			
Gaodikou	40 +110	3	2.4	2.4	46.5	30	63	50.8	1996.10	
Dongchi	40 +552	1	2.8	2.6	55	10	20	50.56	1979.11	Old Dongchi sluices have been abandoned
Daokou	55 +280	2	2.8	2.6	55	20	42.3	48.6	1979.11	Old Daokou sluices have been abandoned
Zhongzimiao	73 +146	2	2	2	42	7	22	46.4	1972.9	Under the flood control standards once in 20 years
Mingdi	83 +650	2	2.5	1.8	114	10	14.8	48.89	2000.8	
Zhaoshengbai	92 +012	4	2	2.9	25	22.53	5	45.44	1960.8	under the flood control standards once in 20 years
Balimiao	101 +940	3	1.2	2	25	15	10	45.24	1959.12	under the flood control standards once in 20 years
Zhangqiu	113 +750	1	2.4	2	107.8	15	14.5	48.61	1984.11	

3 The problems of building the plain reservoir in Jindi River downstream

3.1 Uneven Inter – annual Inflow in the upper and middle Jindi River

Measured annual runoff changed obviously. For example, annual maximum runoff at Puyang is as 53.8 times as the minimum one. Runoff distribution is uneven for Puyang station that the flood season (July to October) proportion of the year accounted for 68.3%, Fanxian Station for 75%. Jindi River runoff away from both the inter – annual or runoff during the year, to purpose of abundance and drought by the water of building plains relying on Flood Control Project of Jindi River downstream.

3.2 Serious water pollution in the upper and middle Jindi River

According to recent test results by Liaocheng and Yanggu county environmental protection department, the water quality of Jindi is in five categories, and seriously impacts on the surrounding environment, also affects the Jindi flood water to the Yellow River.

For it is an inter – provincial river, the Jindi River basin management is more difficult. Many small paper and small chemical result in the Jindi pollution, impact on water storage. If the investment turned into plain reservoir storage can't be applied, that would be a national cause great waste. So it is recommended that the Yellow River Conservancy Commission will co – ordinate two provinces of Henan and Shandong, increase Jindi River pollution control efforts, close the excess of the standard discharges from factories and businesses etc, to radically improve Jindi River pollution problems from the waste water to water resources, to create a good environment for the use of water resources.

3.3 Anti – seepage problem in Plain Reservoir construction

Building the plain reservoirs is faced with a common problem that is seepage control. In 1960, because Great Lakes dam was taken no seepage control measures and the poor quality of construction of the dam, when Dongping Lake reservoir test build, the water didn't come to design water level due to seepage deformation serious, had to stop. Also it causes serious losses as the water table elevation and a large area of arable land salt less. So building the plain reservoirs are faced with a problem that is seepage control. The downstream river catchment areas in Jindi is the north Jindi and south small embankment. The south small embankment takes Vertical Plastic processing to prevent water penetration deformation and temporary embankment Land salt reduction, resulting in losses..

3.4 Recommendations of inter – provincial water disputes

Jindi River is located at the junction of Shandong and Henan as an inter – provincial river. There are more inter – provincial water disputes so that Yellow River Conservancy Commission set up a bureau – level Jindi Authority before to coordinate the provinces and deal with water disputes. Due to the influence of various factors coordinating role, later Yellow River Conservancy Commission withdrew the Authority and converted part of management powers to the local department, while the North Jindi flood – control project is still managed by Liaocheng and Puyang Yellow River Bureau. To resolve inter – provincial water conservancy disputes and prevent pollution of Jindi River and the problem of flood control, to unify planning and management, it is recommended that the Yellow River Conservancy Commission recover the management of Jindi not only easy coordination and unified management, but also conducive to the mediation of inter-provincial water disputes.

4 Investment estimate and economic benefit analysis

4.1 The estimate of total static investment (see Tab. 2)

Tab. 2 The estimate of total static investment (Unit: $\times 10^4$ yuan)

Project name	Dikeroad	Gap recovery	Leveefill residues	Grass protection	Vegetation Net Slope protection	Concrete mesh slope protection	Stone pitching	Pressure grouting
Invest	5,004	5	480	245	195	282	342	4
Project name	Splitting Grouting	Mixing Pile Cutoff wall	Wave forest	Vulnerable spot	Culvert sluice	South dike heightening and consolidation	Total	
Invest	765	536	317	689	1,214	1,078	11,156	

Conservation funding of North Jindi River flood control projects each year is 8.2×10^6 yuan.

4.2 Economic benefit comparison

Normally it will cost about 7.00×10^8 yuan to build a plain reservoir, but it need just more than 1.00×10^8 yuan to build the plain reservoir relying on flood control project of Jindi downstream, saving investment nearly 7 times, and reducing the loss of land over 240 km^2. Also to reinforce flood control projects will improve the project's flood - control ability and settle the material foundation for the Yellow River flood control.

5 Conclusions

In summary, it is necessary and feasible to construct plain reservoir relying on Jindi downstream flood control project. Firstly, the inflow of Jindi River can meet water requirements. secondly, after the Zhangzhuang sluice was rebuilt, it was becoming harder and harder for flood of Jindi River to be drained into the Yellow River, meanwhile there were many troubles for the Zhangzhuang Sluice. So the field in the river road couldn't be cultivated, the water can't be used, which makes great loss to the population. Thirdly, for the inadequate water resources of Jindi River downstream and the surrounding areas and the great demand water of Liaocheng and surrounding areas, it is difficult to meet the needs of their water resources only relying on the Yellow River. Fourthly, use of a large number of flood control projects in the Jindi river downstream can save a lot of investment for the country, improve the flood strength, and laid the material foundation for the Yellow River flood control. Fifthly, it will not affect the Yellow River flood control and Jindi drainage function. Lastly, reinforced fencing work of the reservoir and the use engineering can meet the needs of building plain reservoirs. So building plain reservoir is an economic and effective engineering for the country, and to solve the Jindi pollution problems and inter - provincial water disputes are the key points.

Flood Routing Characteristics and Flood Control Situation in the Lower Yellow River since Xiaolangdi Reservoir Operation

Wan Qiang*, *Jiang Enhui* and *Zhang Linzhong

Yellow River Institute of Hydraulic Research, YRCC Key Laboratory of Yellow River Sediment Research, MWR, Zhengzhou, 450003, China

Abstract: At present, with the operation of Xiaolangdi Reservoir, the river cross – section form in Lower Yellow River is adjusted and some new flood routing characteristics appears. Based on the analysis of the erosion and deposition variation and cross-sections form, cooperating with research of recent physical model flood experiment, the new flood characteristics adjustment is discussed in this paper. It is considered that the channel flow capacity has increased for erosion since the first Water and Sediment Regulation while the capacity of Gaocun – Sunkou – Aishan channel increases less than other reaches. So the flood crest cutting decreases in the meandering channel while it increases relatively in Gaocun – Sunkou – Aishan channel, the flood routing duration is shortened, while the proportion of routing duration in Gaocun – Sunkou – Aishan channel occupying whole duration is more than other channel. The water level at the same discharge falls wholly while it falls at Gaocun and Sunkou less than other stations. Therefore, when the flood routes to the lower channel fast, at the same time, the flood crest cutting is less than ever, the water volume routing to the down flow increases obviously. When flood arrive the channel below Gaocun, the disaster situation would be aggravated.

Key words: Xiaolangdi Reservior, water and sediment regulation, Lower Yellow River, flood routing

1 Introduction

The Lower Yellow River (LYR) channel is the worldwide famous suspended river crossing the Huang – huai – hai plain. The flood disaster in the LYR is the serious adversity of Chinese people, the levee crevasse and reach modification occurs frequently, the loss is enormous, the research on the flood characteristic is of vital important in the Yellow River management. Since the 1970's, with the change of water and sediment condition and deposition in the channel, the secondary suspend river is more serious and the threat of flood is more and more severe.

Since Xiaolangdi Dam was put into use in October 1999, water and sediment regulation has been carried out 14 times by 2010. With the adjustment of river channel form, flood routing characteristics changes and new challenge is taken out. Based on the analysis of cross-sections form and flood factors and research of recent physical model flood experiment, the new flood characteristics adjustment is discussed in this paper, and could supply some technical base for flood control.

2 Erosion and deposition and adjustment of river boundary in LYR

2.1 Characteristics of channel erosion and deposition in LYR

According to statistics, about 1.06×10^9 m^3 sediment has been eroded in the LYR since 2000, the whole channel is scoured, but the erosion intension of each reach is different, it appears that scour at ends is relatively severe and is little at the middle reach. Erosion mostly occurred in the reach upper Gaocun and the erosion amount in the reach is 76% of the total amount. The erosion amount in Sunkou – Aishan and Luokou – Lijin is only 3.3% and 9% respectively. Otherwise, erosion in upper part (upper Sunkou) and deposition appears in lower part (lower Sunkou) in non-flood season in LYR.

2.2 Adjustment of cross section and improvement of channel drainage ability

The river bed in LYR is downcutting since the 1st Sediment and Water Regulation and the regime coefficients decrease in various degree, from 6 ~ 18.7 before Sediment and Water Regulation to 4.1 ~ 12.1 at present. The depth increases 1.5 m and 400 m in width of the cross section in Xiaolangdi – Huayuankou. The situation is similar in Huayuankou – Jiahetan. The river regime is more stable in Jiahetan – Gaocun and the river width increases less than 50m in average. River bed downcutting is principal. Due to stronger earth erosion resistance and bank for production, the width does not change but the depth increases in lower Sunkou.

Corresponding to the adjustment of cross section form, the channel drainage ability increases obviously. There was only 1,800 m^3/s of bankfull discharge in some part of LYR in 2001, while it recovered to over 4,000 m^3/s after Sediment and Water Regulation in 2011. The bankfull discharge increases by 1,100 ~ 3100 m^3/s in LYR and most increase occur in upper Gaocun (more than 3,000 m^3/s at Huayuankou, Jiahetan and Gaocun station). Increase amount is 1,100 ~ 1,600 m^3/s in Gaocun – Aishan and less than in upper Gaocun. At present, the bankfull discharge is more than 6,000 m^3/s in upper Jiahetan, more than 5,000 m^3/s in Jiahetan – Gaocun and more than 4,000 m^3/s in Sunkou – Aishan.

2.3 Vertical section form

The vertical section form is downcutting in the erosion process, and the gradient ratio change is not obvious, it appears decrease current in the upper reach and weaker change in the lower reach. The gradient ratio decreases from 0.26‰ ~ 0.27‰ in 1990s to 0.25‰ in 2008 before flood in Tiexie – Huayuan kou. It maintains 0.2‰ in Huayuankou – Jiahetan. It decreases from 0.16‰ to 0.14‰ in Jiahetan – Gaocun. The gradient ratio is 0.1‰ in Aishan – Lijin without change.

3 Characteristics of flood routing in Water and Sediment Regulation

With the increase of channel drainage ability, the flood discharge rises year after year. The sediment concentration is low and is fine sand ejected by density current.

3.1 Change of flood discharge along the river

Generally speaking, the crest discharge is always curtailed when flood routes to the downstream along channel, but in different terms and discharge levels, the degree of curtailment is different. In the 1990's, deposition and shrinking in the LYR made the larger probability of flood overbank and obvious flood curtailment. After operation of Xiaolangdi Reservoir, the LYR channel has been eroded by clear water, the channel drainage ability has been strengthened, and the probability of flood overbank decrease and action of flood crest reduction of channel also decrease. Due to water and sediment regulation under the personal intervention and control, the flood routed basically in the channel, therefore, the crest reduction of the beach land was very weak.

Comparing the flood reduction ratio between Water and Sediment Regulation of 2008 and flood in the year of 1994 at the same discharge rate (4,200 m^3/s at Jiahetan), it finds that the ratio of 2008 is lower than that of 1994. It is because that the small and middle flood in 1994 was over the bank of atrophic channel, which made the crest reduction effect larger; while the flood of 2008 is not overbank. The crest reduction of 1994 was 33.8% while that of 2008 was 8.8% in Huayuankou – Gaocun. In Gaocun – Sunkou, crest reduction of 1994 was 2.56% while that of 2008 was 1.2%. That is to say, from 1994 to 2008, the crest reduction decreases 2.84 times in Huayuankou – Gaocun, while it decreases 1.13 times in Gaocun – Sunkou. So the crest reduction effect has decreased in the whole LYR while the effect has increased relatively in Gaocun – Sunkou.

In addition, it is worthy to be mentioned that the increase phenomenon of flood peak appears more than once in LYR since the 1st Water and Sediment Regulation in 2002. It is because that the roughness ratio has decreased greatly for the reservoir sediment flushing and the subsequent flow velocity increases the subsequent folds on the frontal flow. The frequent flood peak increase makes the flood forecast very difficult and it is one of uncertain factors of flood control.

3.2 Flood routing time

Flood overbank is the main reason of flood routing time extending. Small and middle scale flood occurs in LYR since the 1st Water and Sediment Regulation. Flood route basically in the channel except flood in 2002 and 2003, so the routing time has reduced. At the same time, hydraulic radius of channel is increased for the erosion of riverbed and the average velocity of flood rises. So the flood routing time had been shortened from the 1990's to 2003 and it is basically stable since 2004. the routing time of flood of 2,500 ~ 4,000 m^3/s level is shortened by 254 h from 2002 to now in the whole LYR and 110 h shortened in Xiaolangdi – Gaocun.

Tab. 1 contrast of flood routing duration in LYR (Unit: h)

Occur period	"96 · 8" flood	2002	2003	2004.7 flood	2004.8	2005	2006	2007	2008
Gaocun—Sunkou	121	146.5	44	24	15.5	12	26	15	13
Xiaolangdi—Lijin	406.7	354.1	173.9	116.7	92	108	122.8	117	100.6

The proportion of the flood routing time to the whole time in LYR is relatively large, it is 30% in 96.8 flood, 41% in 2002 and 25% in 2003. The phenomenon gives counterevidence that the proportion of overbank is larger in that reach.

3.3 Water level change

Corresponding to the increased channel drainage ability of LYR, the water level in same discharge (3,000 m^3/s) of 2010 is lower than that of initial stage of water and sediment regulation (1999). The reduction degree in Huayuankou – Gaocun is larger than that in Gaocun – Sunkou and the water level at Gaocun and Sunkou stations is still higher.

Tab. 2 3,000 m^3/s discharge water level change (Unit: m)

Year	Huayuankou	Jiahetan	Gaocun	Sunkou	Aishan	Luokou	Lijin
1999	93.74	75.35	63.55	49.10	41.60	31.10	13.60
2010	92.10	73.34	61.64	47.73	40.70	29.86	12.75
Water level difference	-1.64	-2.01	-1.91	-1.37	-0.90	-1.24	-0.85

4 Recent physical model experiment result

The flood forecast experiment in Xiaolangdi – Taochengpu was carried out in July 2008 at Yellow River Institute of Hydraulic Research, YRCC. The "59 · 8" flood process and terrain before flood of 2008 were adopted for the initial conditions of the experiment. The flood peak at Huayuankou is 10,000 m^3/s. The experiment reflects the big flood routing and situation of overbank and provides the evidence for the similar flood routing in the present channel.

Based on the experiment result, the ratio of crest reduction in Gaocun – Sunkou is 16.1% in the experiment, it is more than the ratio (14.8%) of 96.8 flood in that reach, so it is indicated that the flood overbank situation is very serious. The time of flood routing in the experiment is 40.1 h in Gaocun – Sunkou, it occupies 45.7% of the whole flood routing time in Xiaolangdi – Sunkou, that is similar to the ratio of 45.8% in 96.8 flood. Contrast with the prototype of 96.8 flood, because the riverbed is higher than the experiment channel, the water level is higher too. The experiment water level of same discharge falls for the channel downcutting after 8 times of Water and Sediment Regulation. The water level descends 8 cm at Gaocun station and 14 cm at Sunkou station.

Corresponding to the flood routing characteristic, the flood overbank situation in the experiment is different from that before Water and Sediment Regulation. Such as, the water volume overbank in upstream reach is smaller than that of downstream. The beachland of Gaocun – Sunkou is almost submerged. The water volume stay in the beach is about 3.01×10^8 m^3 and the sediment is 1.537×10^8 t.

5 Conclusions

The river channel erosion intensity shows big in both ends and small in the middle. The regime coefficient decreases in different degree. The channel drainage ability increases obviously but it increases less in Gaocun – Aishan. The vertical gradient ratio shows decreasing trend in upstream of LYR and changes less in downstream. Therefore, comparing present flood routing characteristics with that before the operation of Xiaolangdi Reservoir, obvious differences are as follows:

(1) Due to the increase of channel drainage ability, the flood overbank probability and the crest downcutting effect decrease in the meandering channel in LYR. The crest downcutting effect relatively enhances in GaoCun – SunKou because the channel drainage ability is smaller (bankfull discharge is less than 4,000 m^3/s).

(2) The flood routing time has been shortened since the Water and Sediment Regulation. The flood routing time in 2008 is reduced by 254 h than that of 2002. The ratio of flood routing time in Gaocun – Sunkou occupies heavier than that of other reaches. It reflects that the channel drainage abililty has increased in recent years and the overbank probability in GaocunSunkou is still larger relatively.

(3) The water level in same discharge has fallen since the 1st Water and Sediment Regulation for the channel erosion. But the wash load near Sunkou reach is less than other reaches and occupies 3.3% of the whole wash load in LYR. The water level at Gaocun and Sunkou stations is still higher.

The recent flood forecasting experiment proves the change of the flood routing characteristics, too. Therefore, under the present channel terrain condition, the flood crest cutting is less than ever as soon as the flood routes to the lower channel fast, the water volume routing to the downflow increases obviously. When the flood arrive the channel below Gaocun, the disaster situation would be aggravated. The concerned department should attach importance to it.

References

Shang Hongxia, Sun Zanying, Li Xiaoping. Analysis on Erosion and Deposition Effect in Lower Yellow River After Operation of Xiaolangdi Reservoir [R]. Zhengzhou: Yellow River Institute of Hydraulic Research, YRCC, 2007.

Shang Hongxia, Su Yuanqi, Zhang Min. The Contrast Analysis on Channel Regulation of LYR in Sediment Storeage Period of Different Reservior [J]. Water Resources and Water Engineering Journal, 2008(8).

Jiang Enhui, Zhao Lianjun, Wei Zhilin. The Mechanism and Verification of the Flood Peak Increase in LYR [J]. Journal of Hydrolic Engineering, 2006(12).

Li Yong, Zhang Xiaohua, Hou Suzhen. The Overbank Flood Routing Pattern in LYR and 96 · 8 Flood Routing Characteristic [J]. Yellow River, 1998(5).

Zhang Linzhong, Cao Yongtao, Liu Yan, et al.. The Flood Forecast Physical Model Experiment in Xiaolangdi—Taocengpu[R]. Zhengzhou: Yellow River Institute of Hydraulic Research, YRCC, 2008.

Discussing on Water Price Renewing in China

Jin Yinjuan[1], *Yang Yisong*[1], *Bian Yanli*[1] and *Zhang Yousheng*[2]

1. Yellow River Institute of Hydraulic Research, YRCC, Zhengzhou, 450003, China
2. Forestry Bureau of Wuqi County, Wuqi, 717600, China

Abstract: With the socio-economic development and the population increasing, water shortages will become a very important factor restricting China's social and economic development. Now, the pressure from water resources shortages becomes greater and greater, in the meanwhile, waste of water resources are widespread in domestic water, industrial water, water used in agriculture and ecological use. This paper suggests that establishing a scientific and rational water price system and making full use of economic instruments can increase awareness and use of water resources for water-saving efficiency. This not only helps people enhance their awareness of water-saving, relieve the pressure of water resources shortage in China, but will also help to decrease the farmers' burden and to carry out sustainable development in China.

Key words: water resources, water price, water fee

With the development of society and economy and the population increasing, the shortage of water resources will be an important factor for restricting our country social economy development. It was estimated that water resources in China is lower than 8,000 × 10^8 m^3 (Qian Zhengying, Zhang GGuangduo, 2001, Chen Minjian, Huo Weicheng, 1998). And it might be lower than 7,000 × 10^8 m^3 according to the result by the Chinese Academy of Sciences sustainable development strategy research group. In 2030, the amount of water resource used for domestic, industrial and agricultural purpose could be about 6,535 × 10^8 m^3, and it could be up to 6,600 × 10^8 m^3 if including the ecological water (the Chinese Academy of Sciences sustainable development strategy research group, 2007). The shortage of water resources will be more and more serious on one hand, and the waste of water resources are common in every water consume section on the other hand (Yang Yisong, Bian Yanli, 2010). The causes of the wasting water resources is not only because people have lack of the awareness of saving water, but also because the water resources fee is too low. This article wants to provide reference for decision-making to alleviate the pressure of water resources shortage by making a reasonable water resources fee system, which is making full use of economic levers and taking into account the farmer burden and can improve water-saving awareness of people and utilization efficiency of water resources.

1 The status of water resources consume in China

The yearly water consumption was increasing year by year from 2005 to 2009, and the total water consumption increased from 5,633 × 10^8 m^3 in 2005 to 5,965.2 × 10^8 m^3 in 2009, with an average annual growth rate of 1.45% (National Bureau of statistics of China, 2010). In all water consumption section, including domestic water, industrial water, agricultural water and ecological water use, the domestic water was different from the others, which was keeping increasing by more than 2.6% annually. If we look at the water used share, the amount of water used for agriculture tops, more than 60%. It was 3,723.1 × 10^8 m^3 in 2009, the 62.4% among the total water used. And the industrial water followed, nearly to 25%. It was 1,390.9 × 10^8 m^3 in 2009, the 23.3% of total water use. The domestic water used in that year was 748.2 × 10^8 m^3, 12.5% of total, and ecological water use was only 103 × 10^8 m^3, 1.7% of the total (the details shows in Tab. 1).

Tab. 1 The status of water resource consumption from 2005 to 2009 in China

Sector & amount		Year				
		2005	2006	2007	2008	2009
Domestic	Amount($\times 10^8$ m^3)	675.1	693.8	710.4	729.3	748.2
	Percent of the total(%)	12.0	12.0	12.2	12.3	12.5
Industrial	Amount($\times 10^8$ m^3)	1,285.2	1,343.8	1,403.0	1,397.1	1,390.9
	Percent of the total(%)	22.8	23.2	24.1	23.6	23.3
Agricultural	Amount($\times 10^8$ m^3)	3,580.0	3,664.4	3,599.5	3,663.5	3,723.1
	Percent of the total(%)	63.6	63.2	61.9	62.0	62.4
Ecological	Amount($\times 10^8$ m^3)	92.7	93.0	105.7	120.2	103.0
	Percent of the total(%)	1.60	1.60	1.8	2.0	1.7
Total	Amount($\times 10^8$ m^3)	5,633	5,795	5,819	5,910	5,965.2
	Percent of the total(%)	100.0	100.0	100.0	100.0	100.0

2 The current water price and the status of water resources fee in China

2.1 The current water price

In China, collecting water fee has begun in 1980. and now all of the country including the all the provinces, cities and autonomous regions has made the policy of Water resources fee collection and management and collects water fee year by year. It was very important for improving the saving and use of water resources and maintaining the water environment by collecting water resource fees. The water price is different in different provinces, cities and autonomous regions because of the status of water resource and the development of local economy, which is from 0.001 yuan/m^3 to a few yuan/m^3. And it is lower than 1 yuan/m^3 in China except the Beijing, Tianjin and Shandong province. At the same time, the standards of water resource fee collection are various greatly, and it is not obviously related with the water resources conditions and the local economic development.

2.2 The status of water resources fee

From 2005 to 2009, the water resource fee collected in industrial water collection shared about 45% ~ 50% of the total amount levied, wherein the share for hydroelectric use accounts for about 10% ~ 12% of the total collection, and for power water, about 8% of total levy. The share of living water is 25% ~ 35%, agricultural irrigation water is less than 2%. The annual charge to industrial, life uses grow by more than 15% annually. But it was less than 2% for agricultural water used, which consumed shares more than 60% of the total water resource.

3 Discussion on the reform of current water price

3.1 The general principle of water price reform

The main problems of the current water price system are mainly focusing on the different standards ,low price and low levy rate. It is not only difficult to achieve the optimal allocation of water resources, to promote the conservation and protection of water resources, but also causes the waste of water resources. For example, agricultural water resource fee is too low, and it is deferred or exempted in some provinces. This can slow down the farmer's burden, promote farmer to plant grain crop and ensure national food safety, but in the same time it, to some extend, encourages the serious waste of water resource by flood irrigation in some places. It is a must to adjust water price for saving water resources and allocation of water resource. In the No. 1 policy document of the Central Committee in 2011, it puts forward the overall principle: It must carry out progressive charge system of water resource on the excess water use in the industry and the service industry gradually and adopt different water price between the high water consumption industries and the lows. It is a must to reasonably adjust the water price for the domestic use and promote ladder water price system gradually. It may promote the comprehensive reform of agricultural water price of agricultural irrigation and drainage project according as the principle of promoting water resource saving, reducing farmer water fee and guarantee the benign operation of Irrigation and drainage engineering. And to discuss the progressive charge system of water resources on excess water use and ladder water price system in agricultural water while the fee of the management and operation of agricultural irrigation and drainage project is supported by the financial subsidies.

3.2 The method of Water quota and water price determination

How to determine a reasonable water quota is the foundation of achieving the system of progressive water price on the excess water use and the prerequisite of managing water resources effectively, achieving optimal allocation of water resources, and promoting the conservation and protection of water resources by the economic lever. The reasonable water quota should not only reflect the supply and demand contradiction of water resource, but also promote the saving of water resources and the development of economical society and reduce the burden of water saving users and farmers. Therefore, it is necessary to determine the different water quota and perform different price based on the different condition in different region.

3.2.1 Determination of domestic water quota

The domestic water use was rising from 2005 to 2009 in China, with an average annual growth rate of more than 2.6%. The water price was still low though it had risen certainly. The reasonable water quota may be the average water amount used in the last year. It applies the lower price when the water consumption is under the water quota, and execute high price when the water consumption is above the water quota.

3.2.2 Determination of agricultural water quota

It is very difficult to determine the agricultural water quota because there are different water demands in different region when planting the same crop. So it can be determined by the central water conservancy research institutions and the central agricultural research institutions and be verified by further study among different provinces ,cities and autonomous regions. It is free for agricultural water use when the consumption is under the quota and execute high price when the water consumption is above the water quota. Thus, it is beneficial to reduce the economic burden of the farmer and ensure national food security on one hand, and on the other hand it is also conducive to save water resources and achieve efficient use of water resources.

3.2.3 Determination of industrial water quota

The industrial water quota can be determined by the relevant state departments on the basis of China's industrial energy-saving emission reduction and the state industrial development planning. It executes reasonable low water price when the consumption is under the quota and the executes punitive progressive water price on the excess water amount if the consumption is above the quota. For promoting saving water resource, the priority should be given to the industries which want to a further expansion of the scale if their water consumption is under the quota and their water use process reach or exceed the international advanced level.

3.3 Strict the collection of water resources fees

It is very difficult to determine a reasonable water price in the water price reform. However, all efforts might be in vain when not seriously and strictly comply with the standard of water collection even if it had a reasonable water price system. It is very difficult in the water fee collecting work because it connects with the national water conservancy departments system and the national government systems. So it is a must to establish the water fee collection management systems by the national water conservancy departments at all levels and uses the local chief executive responsibility system for ensuring that water collection.

References

Chen Minjian, Huo Weicheng. Discussion on China Water Resources Utilization[J]. Journal of Natural Resource,1998:22 – 25.

Chinese Academy of Sciences Sustainable Development Strategy Research Group. China Sustainable development Strategy Report 2007—Water: Governance and innovation[M]. Beijing:Science Press ,2007.

National Bureau of Statistics of China. The 2010 China Statistical Yearbook[M]. Beijing: China Statistics Press,2011.

Qian Zhengying, Zhang Guangduo. China Sustainable Development Strategy of Water Resources Comprehensive Research Report and Various Special Reports[M]. Beijing:China Water Conservancy and Hydropower Press,2001.

Yang Yisong, Bian Yanli. Discussing on the Problems and Methods of the Water Right Transform in Yellow River Basin [J]. China Water Resources,2010(21):16 – 17.

The Research on the Taihu Quantity Control of Pollution Discharge in Typical Moderate Precipitation Year

Zhang Hongju[1,2], ***Shang Zhaoyi***[3], ***Gan Shengwei***[4] and ***Yuan Hongzhou***[5]

1. State Key Laboratory of Lake Science and Environment, Chinese Academy of Sciences, Nanjing, 210008, China
2. Taihu Basin Authority of Ministry of Water Resources, Shanghai, 200434, China
3. East China Normal University, Shanghai, 200062, China
4. Hydrology & Water Resources Supervision & Measurement Bureau, Wuxi, 214024, China
5. Shanghai Investigation, Design & Research Institute, Shanghai, 200434, China

Abstract: The article determines the sectional water quality protection goal of Taihu Lake of 2012 and 2020 based on the recent goal of "Master Plan for Comprehensive Management of Water Environment in the Taihu Lake Basin" and analyses of the current water quality of Taihu Lake and the river around the Lake. The article establishes a 2D numerical water quality model integrated the rainfall condition of the year 1992 and uses the model to calculate the river controlling concentration of pollutant which can meet the lake quality protection goal. The calculating results meet the lake pollutant request in the discharge condition of 2005.

Key words: Taihu Lake, shallow lake, limited total pollutant discharge

1 Introduction

In 2008, The State Council definitely proposed the water quality conservation goal of Taihu Lake in the approved *Master Plan for Comprehensive Management of Water Environment in the Taihu Lake Basin* (MPCMWETLB). The water quality goal is to improve water quality of Taihu Lake to Grade Ⅴ in 2012 and inhibit eutrophication. Simultaneously, *Water Function Zoning of Taihu Basin* requires that to improve water quality of overall water to Grade Ⅳ and some water area to Grade Ⅲ in 2020; ulterior, improve water quality to fulfill water quality goal in whole lake.

Pollutant of Taihu Lake is mainly from rivers around the lake. In order to achieve the stage goals, pollution load that entered Taihu Lake shall be controlled. This article researches the pollution load to put forward the comment on limitations of total pollutants discharge. According to historical monitoring data, pollutants in rivers around the lake are more intensive and water environment risk is higher in moderate precipitation year type, thus the study picked a moderate precipitation year 1992 for imitation.

2 Study area

2.1 General situation

Taihu Lake is located on the southern margin of Yangtze River Delta, and it extends across Jiangsu and Zhejiang Provinces (Fig. 1). Owning an area of 2,338 km^2, Taihu Lake is the third biggest fresh water lake of China. Taihu Lake is a big shallow lake with the mean depth of 1.89 m and the maximum depth of 2.6 m. The distance between the north and south is 69 km, that between the east and west is 34 km on average, and the lakeshore line is 405 km.

2.2 Water quality of rivers around Taihu Lake

Overall water quality of Taihu Lake in 2010 was evaluated as Inferior to Grade Ⅴ. 0.3% of the water area was in Grade Ⅳ, 18.8% in Grade Ⅴ and the rest 80.9% was inferior to Grade Ⅴ. The main indexes which fail to meet Grade Ⅲ were TN, TP and BOD_5. From north to south, west

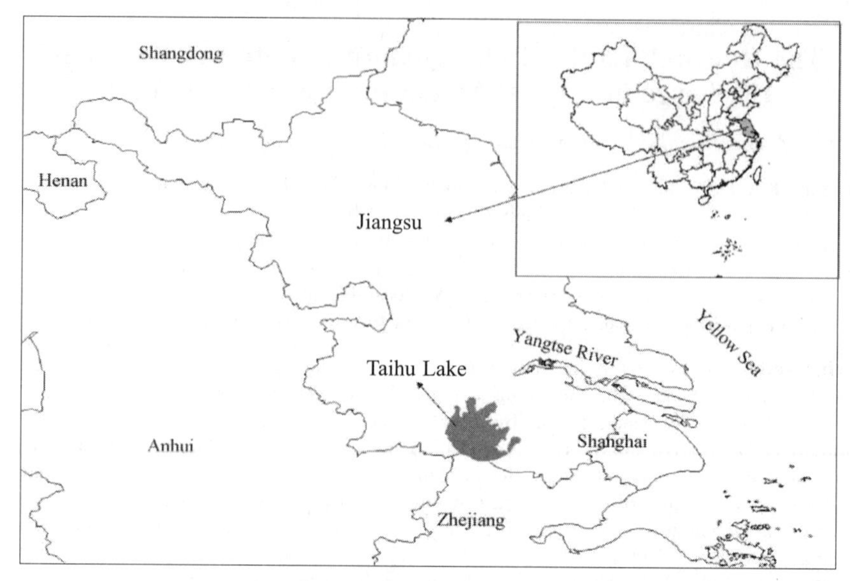

Fig. 1 Location of Taihu Lake in Yangtze River Delta

to east, water quality trends of the lake districts were improving, among which Wuli Lake, eastern littoral area and Dongtaihu Lake were the best, while Zhushan Lake and western littoral area, which were inferior to Grade Ⅴ, were the worst.

Rivers around Taihu Lake are mainly located in three cities (Wuxi, Changzhou and Suzhou) of Jiangsu Province and Huzhou City of Zhejiang Province. There are 22 rivers in all, 15 in Jiangsu Province and 7 in Zhejiang Province. In 2010, water quality levels of 7 rivers were evaluated as Inferior to Grade Ⅴ with 6 in Jiangsu and 1 in Zhejiang. The main indexes that fail to meet Grade Ⅲ were NH_3, BOD_5 and COD.

3 Research on limitations of total pollutants discharge of Taihu Lake

3.1 Definition

The definition of *Limitations of Total Pollutants Discharge of Taihu Lake* is the maximum amount of pollutant that are carried by rivers to enter Taihu Lake in different water quantity condition under the premise that completing water quality conservation goal.

3.2 Technical route

Firstly, the article determined the water quality conservation goal of each lake district based on water quality conservation goals in year 2012 and 2020 of MPCMWETLB, and integrating with current water quality of Taihu Lake. Secondly, the article chose a typical precipitation year type, and calculated water quality consistencies of rivers when achieving water quality conservation goals using a 2D numerical water quality model. Thirdly, in accordance with water quality conservation goals of *River Function Zoning Plan*, the article determined the limitation of total pollutant discharge of Taihu Lake in a moderate precipitation year. The technical route is shown as Fig. 2.

3.3 Water quality conservation goals

As a large, wide and shallow lake, spatial difference of water quality is obvious, water quality

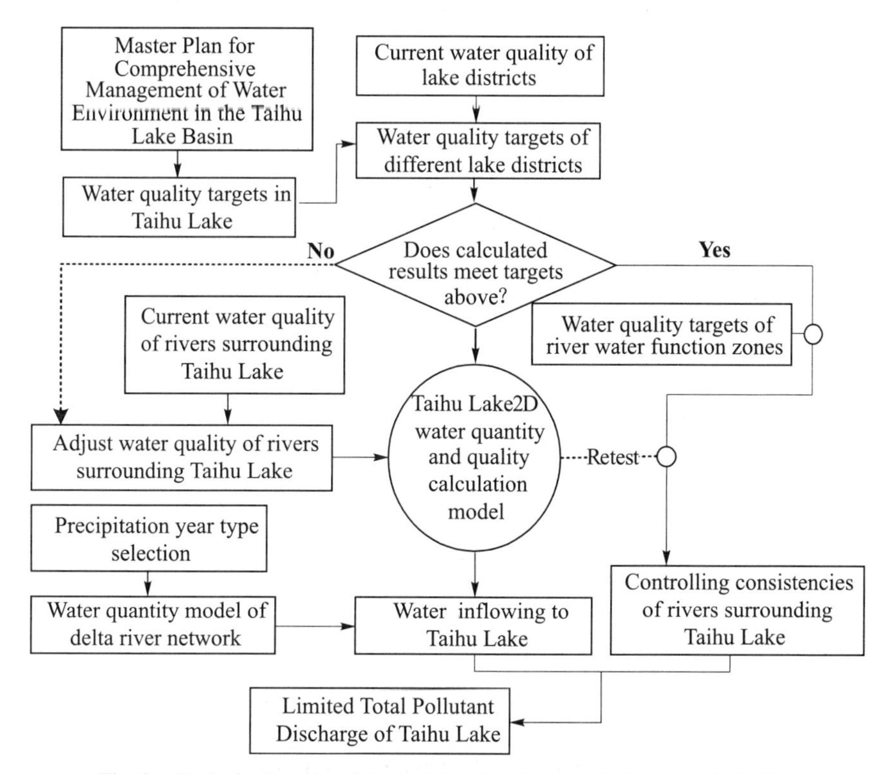

Fig. 2 Technical route of limited total pollutant discharge calculation

of northern – west district is inferior and water quality of east district is comparatively good. Therefore, requests toward water quality of rivers flowing into different districts of Taihu Lake vary a lot.

Water quality conservation goals of year 2012 and 2020 determined in MPCMWETLB are listed in Tab. 1. Through area weighted calculation method, the goals can be distributed to each lake district (Fig. 3).

Tab. 1 Water quality conservation goals of Taihu Lake

Index	COD_{Mn}		NH_3-N		TP		TN	
Target Year	2012	2020	2012	2020	2012	2020	2012	2020
Consistency	4.50	4.00	0.46	0.45	0.070	0.050	2.00	1.20
Grade	Ⅲ	Ⅱ	Ⅱ	Ⅱ	Ⅳ	Ⅲ	Ⅴ	Ⅳ

30 consistency controlling points are settled in Taihu Lake (Fig. 4). Water quality consistencies of different lake districts are calculated by mean arithmetical value of controlling points, and the average consistency of whole lake is computed by area weighted calculation method of different lake districts.

3.4 Method

3.4.1 Calculation of water quality

Establish a 2D water quantity and quality mathematical model. The fundamental equations are:

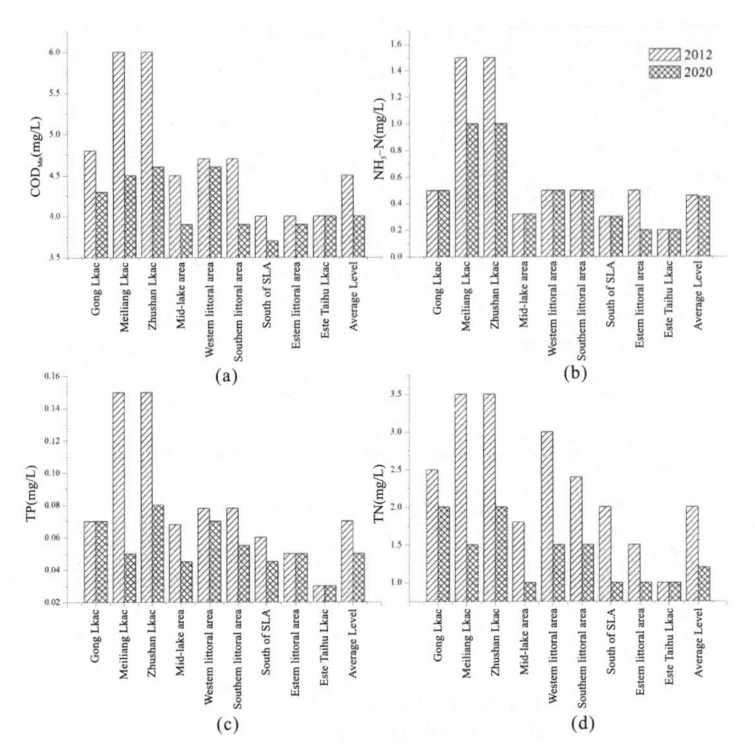

Fig. 3 Water quality targets distribution of different lake districts

$$\frac{\partial z}{\partial t} + \frac{\partial (hu)}{\partial x} + \frac{\partial (hu)}{\partial y} = q$$

$$\frac{\partial u}{\partial t} + u\frac{\partial u}{\partial x} + u\frac{\partial u}{\partial y} + g\frac{\partial z}{\partial x} = S_{fx} + fu$$

$$\frac{\partial u}{\partial t} + u\frac{\partial u}{\partial x} + u\frac{\partial u}{\partial y} + g\frac{\partial z}{\partial y} = S_{fy} - fu$$

In the equation, z stands for water level; t is time; h means water depth and $h = z - z_B$; z_B is elevation of lakebed; u is velocity in x – direction; u is velocity in y – direction; q is source and sink of lake surface precipitation, evaporation and lakebed leakage; f stands for coriolis acceleration and f equals for $2\omega \sin \varphi$, in which ω is spin velocity of the earth and φ is latitude of Taihu Lake (N31°10′); S_{fx}, S_{fy} are tangential stresses in x and y directions. Unit of z, h, z_B, is meter; unit of t is second; unit of u, and q is m/s; unit of ω is rad/s; and unit of S_{fx} and S_{fy} is N/m^2.

The calculation equation is:

$$\frac{\partial hC}{\partial t} + \frac{\partial hUC}{\partial x} + \frac{\partial hVC}{\partial y} = \frac{\partial}{\partial x}(hE_x\frac{\partial C}{\partial x}) + \frac{\partial}{\partial y}(hE_y\frac{\partial C}{\partial y}) + \frac{hS}{86,400}) + S_w$$

And C is consistency of some water quality index; E_x and E_y are coefficient of diffusion in x and y directions; S is the bio – chemical reaction of water quality; S_w is external source and sink of some water quality index.

The article generalized the study area Taihu Lake into 2,545 calculation cells by square mesh in the unit of 1,000 m multiplying by 1,000 m. Based on observation station data of rivers surrounding Taihu Lake, the article generalized rivers surrounding Taihu Lake to 22 rivers.

The model used water volume in and out Taihu Lake, lake surface evaporation and precipitation in 2002 as calibration information. Result showed that calibration errors are 20% for COD_{Mn},

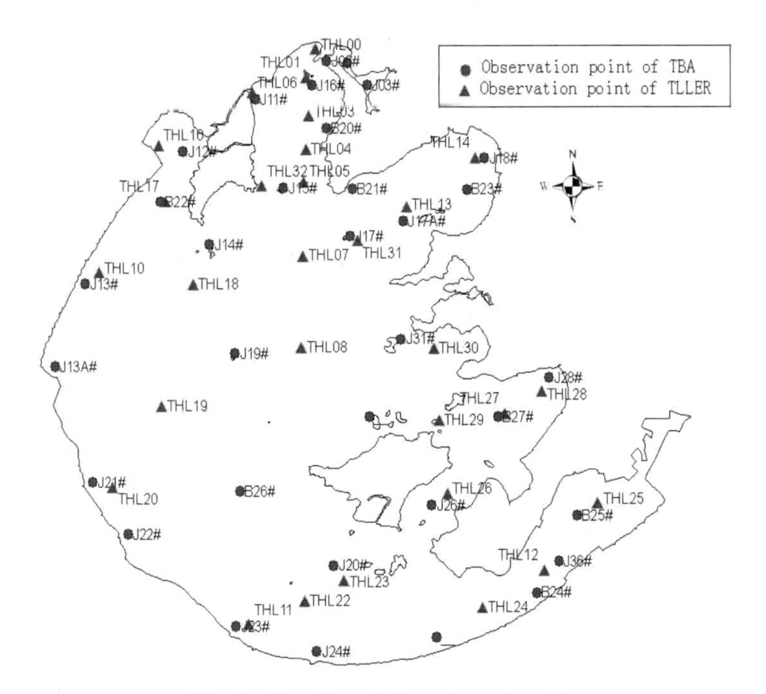

Fig. 4 Location of the measuring sites

50% for NH_3-N, 58% for TP, and 40% for TN, which satisfied with the fundamental calculation demand.

3.4.2 Calculation of limited total pollutant discharge

Using the water inflow and outflow process of rivers surrounding Taihu Lake calculated by river network water volume model, the article computed pollutant consistencies of inflow by Taihu 2D water quantity and quality model, to get controlling consistencies of pollutants discharge into Taihu Lake and the limited total pollutant discharge respectively.

The calculation equation of limited total pollutant discharge is:

$$M = Q * C_{control}$$

In the equation, M is limited total pollutant discharge, Q is water volume flowing into Taihu Lake, and $C_{control}$ is controlling consistencies of pollutants.

3.5 Calculation condition

3.5.1 Designed hydrologic condition

Considering the current water quality of rivers surrounding Taihu Lake and operability of management, this study used precipitation process in 1992 year as designed hydrologic condition. In 1992, total precipitation of Taihu Delta was 1,018.3 mm, and precipitation frequency was 78%. It was a moderate precipitation year type. Under such designed condition, water volume flowing into Taihu Lake in 1992 was 9.085×10^9 m^3, and water volume flowing out of Taihu Lake was 6.972×10^9 m^3.

3.5.2 Water quality parameters

The article settled pollutant comprehensive degradation parameters based on calibration results. In this study, the parameters of COD_{Mn}, NH_3-N, TP and TN were 0.005/d, 0.025/d, 0.04/d, and 0.006/d.

3.5.3 Initial condition

Based on recent water quality of Taihu Lake, settling the initial condition of water quality, the article settled initial water level as 2.60 m, and calculation time step as 900 s.

3.6 Result and analysis

3.6.1 Result

In order to fulfill water quality protecting target in different precipitation year, water quality targets of rivers surrounding Taihu Lake must be in step with water quality targets of lake districts in consideration of current water quality of rivers surrounding Taihu Lake, and the accessibility of targets shall be considered. In this study, under the foundation of calculated inflow pollutant consistencies, current water quality of rivers surrounding Taihu Lake and water quality targets of water function zones were computed to calculate inflow consistencies. After calculation, under the 1992 precipitation year type, pollutant controlling consistencies of rivers discharge into Taihu Lake are listed below: controlling consistencies for COD_{Mn} are from 4.5 mg/L to 6.0 mg/L in 2012, and from 4.0 mg/L to 5.5 mg/L in 2020; controlling consistencies for NH_3-N are from 1.0 mg/L to 2.2 mg/L in 2012, and 1.0 mg/L in 2020; controlling consistencies for TP are from 0.12 mg/L to 0.25 mg/L in 2012, and from 0.12 mg/L to 0.18 mg/L in 2020; controlling consistencies for TP are from 2.8 mg/L to 4.5 mg/L in 2012, and from 2.2 mg/L to 3.0 mg/L in 2020 (Tab.2).

Tab.2 Calculation results of pollutant controlling consistencies of rivers surrounding

Corresponding river districts	Rivers surrounding Taihu Lake	2012				2020			
		COD_{Mn}	NH_3-N	TP	TN	COD_{Mn}	NH_3-N	TP	TN
Gong lake	Wangyu River	5	1	0.15	2.8	4	1	0.12	2.2
	Other rivers except Wangyu River	5.5	1.5	0.18	3	5.5	1	0.15	2.8
Meiliang lake	Liangxi River, Wujin River, Zhihu River	6	2	0.2	4	5	1	0.18	3
Zhushan lake	Caoqiao River, Yincun River, Taige Canal	6	2.2	0.25	4.5	5	1	0.15	3
Western littoral area	Chendong River, Dapu River, Shaoxiang River	5.5	1.5	0.15	4	5.5	1	0.13	2.5
Southern littoral area(SLA)	Changxing River, Hexi new River etc.	5	1.2	0.12	4	5	1	0.12	3
South of SLA	Xitiao River, Xiaomei River, Changdou River, Daqiao River etc.	4.5	1	0.12	2.8	4.5	1	0.12	2.8

Based on pollutant controlling consistencies of inflows surrounding Taihu Lake, and considering water volume discharged into Taihu Lake, the model calculated limited total pollutant discharge in 1992 precipitation year type. The total amount for COD_{Mn} in 2012 is 47,513 t per year, in 2020 is 43,389 t per year. For NH_3-N it is 12,532 t per year in 2012, and 9,074 t per year in 2020. For TP it is 1,462 t per year in 2012, and 1,172 t per year in 2020. For TN it is 31,919 t per year in 2012, and 23,629 t per year in 2020. Distribution plan in detail is listed in Tab. 3.

Tab. 3 Limited total pollutant discharge of Taihu Lake in different periods

(Unit: t)

District		2012				2020			
		COD_{Mn}	NH_3-N	TP	TN	COD_{Mn}	NH_3-N	TP	TN
Jiangsu Province	Wuxi	15,212	4,917	444	13,024	13,892	3,559	357	9,642
	Changzhou	12,239	2,252	316	5,335	11,177	1,630	252	3,948
	Yixing	7,709	2,089	325	5,305	7,039	1,514	262	3,927
	Sum	35,160	9,258	1,086	23,664	32,108	6,703	871	17,517
Zhejiang Province	Changxing	1,450	586	43	955	1,324	424	34	709
	Huzhou	10,904	2,689	333	7,300	9,957	1,947	267	5,403
	Sum	12,353	3,274	376	8,256	11,281	2,371	302	6,112
Total		47,513	12,532	1,462	31,919	43,389	9,074	1,172	23,629

3.6.2 Rationality analysis

Inputting the calculated results into Taihu model to recheck rationality, calculation results indicate that all indexes fulfill corresponding water quality targets, which means pollutant controlling consistencies and limited total pollutant discharge results are rational (Tab. 4).

Tab. 4 Average water quality of Taihu Lake calculated by model

(Unit: mg/L)

Index	2012		2020	
	Target	Calculated consistency	Target	Calculated consistency
COD_{Mn}	4.50	4.01	4.00	3.89
NH_3-N	0.46	0.38	0.45	0.34
TP	0.070	0.048	0.050	0.045
TN	2.00	1.21	1.20	1.01

4 Application analysis

The article used water inflow and outflow process of rivers surrounding Taihu Lake in recent years to test calculated results. In 2005, precipitation frequency was 75% and water volume inflowing to Taihu Lake was 7.91×10^9 m^3. This situation was roughly equivalent to situation of 1992 and had representativeness.

4.1 Pollutant controlling consistencies test of rivers surrounding Taihu Lake

The article inputted water volume inflowing to Taihu Lake of 2005 into model and settled pollutant controlling consistencies of 2012 as boundary. Calculation results are listed in Tab. 5. The results show that all indexes in 2005 can fulfill targets in 2012. This indicates that under designed condition, the controlling consistencies can achieve water quality control requirement of Taihu Lake.

Tab. 5 Water quality calculation result under pollutant controlling consistencies condition

(Unit: mg/L)

Lake district	Targets in 2012	2005		
		Average	Highest	Lowest
COD_{Mn}	4.50	3.77	4.127	3.496
NH_3-N	0.46	0.45	0.638	0.276
TP	0.070	0.050	0.103	0.041
TN	2.00	1.26	1.652	0.985

4.2 Test of limited total pollutant discharge

After examining the rationality of limited total pollutant discharge by water quality evaluation of Taihu Lake, and comparing actual pollutant load discharge and pollutant consistencies in Taihu Lake of 2005 to limited total pollutant discharge and water quality target in lake body of 2012 under 1992 precipitation year type, pollutant load discharges of different indexes in 2005 are higher than those of 2012 under 1992 precipitation year type, and water quality misses the targets (Tab. 6). This shows that in condition of similar water volume inflowing to Taihu Lake, limited total pollutant discharge evaluation is in accord with consistency control evaluation.

Tab. 6 Comparison of pollutant load discharge and water quality consistency of Taihu in 2005

(Unit: t)

Index	Evaluation of total pollutant			Evaluation of water quality in Taihu Lake		
	Pollutant load discharge in 2005	Limited total pollutant discharge in 2012	Reach the mark or not	Water quality target of Taihu in 2012	Current water quality of Taihu in 2005	Reach the mark or not
COD_{Mn}	55,562	47,513	No	4.50	5.05	No
NH_3-N	22,031	12,532	No	0.46	0.47	No
TP	1,721	1,462	No	0.070	0.078	No
TN	41,892	31,919	No	2.00	2.48	No

5 Conclusions

Water quality of different districts in Taihu Lake is influenced by water quantity and quality of rivers surrounding Taihu Lake. Under current situation that non – point source pollution can hardly be controlled, limiting water quality consistency of inflows and total pollutant control is a main method. This research uses 2D Taihu water quality model as a tool, and calculates the pollutant consistencies of inflows and limited total pollutant discharge to fulfill lake water quality targets under a moderate precipitation year. This research is clear, operable, and provides examples for other districts.

Acknowledgements

The research was supported by MWR's special scientific fund for non – profit public industry No. 201101025.

References

Zhai Shuhua, Zhang Hongju. Water Quantity and Waste Load Variation of Rivers around Lake Taihu from 2000 to 2002[J]. Journal Lake Science, 2006, 18(3):225 -230.

TBA. Taihu Healthy Condition Report 2010[R].

State Council Approved. Master Plan for Comprehensive Management of Water Environment in the Taihu Lake Basin. 2008:38.

The Analysis on Current Situation of Zhengzhou Yellow River Diversion Water Supply Industry and the Discussion on Its Development Prospect

Xin Hong, *Jiao Haibo*, *Xue Xiping*, *Qin Ting* , *Cheng Feng*, *Zhang Chuang*, *Zhang Ruifang* and *Wang Xiying*

Zhengzhou Water Supply Sub – office of Water Supply Authority of Henan Yellow River Bureau

Abstract: This paper, according to the development state of Zhengzhou Yellow River diversion water supply and relevant data on water supply operation in recent years, analysized the current situation of Yellow River diversion water supply industry, discussed the development prospect of water supply, which has played a positive role in developing water supply work sound and fast in future. When analysing the arguments on Zhengzhou's economic development, ecological environment protection and sustainable utilization of water resources and other issues, the paper fully combined the current operation and management mode of Yellow River water diversion and provided necessary reference for more scientific decision – making, allocation, utilization of the Yellow River water resources. By water supply in urban area, irrigation water recharge, ecological water supplement and other water supply measures, it better achieves the optimal allocation and the efficient use of water resources and plays a good role in the sustainable development of social economy along the Yellow River and the improvement of ecological environment.

Key words: Yellow River diversion water supply, current situation, prospects, discussion

1 The current situation of Yellow River diversion water supply in the jurisdiction of Zhengzhou

Now, there are 15 water – taking projects from the Yellow River in the jurisdiction of Zhengzhou, 10 water – taking projects from surface water, and 5 water – taking projects from groundwater in beach area. In addition that the irrigation pumping station of the Xingyang Licun, with no water taking point, is under construction, other water – taking projects can work normally.

The characteristic of Zhengzhou Yellow River diversion water supply is that the proportion of city industrial water is large, while the proportion of agricultural water is small; the water supply for ecology rises considerably; the water diversion target is obviously insufficient. Before 2007, the quantity of Yellow River diversion water was about 200,000,000 m^3 per year; since 2008, it has been increased year by year.

Tab. 1 Water diversion quantity in recent six years

Years	2006	2007	2008	2009	2010	2011
Water diversion quantity	1.83	1.84	4.51	5.42	6.23	8.46

2 The control status of water supply

2.1 Perfecting the water diversion metering system

With the economic and social development and the faster pace of urbanization of Zhengzhou, the urban population of Zhengzhou city grows quickly and the water demand for urban environment, resident living and enterprises and orgnizations enlarged increasingly. How to fully and effectively use the limited water resource of the Yellow River has an important role in keeping sustainable development of economy and society. In comparison with other advanced regions, the Zhengzhou Yellow River diversion water supply industry is in the primary development stage, so its task is very formidable in scientific water use and control. The only road for the scientific development of Yellow River diversion water supply is to perfect the water diversion metering system and water pricing system in accordance with the development rules of market economy and in full combination with the governmental guidance.

Due to historical reasons, some water users of the Yellow River water diversion project in Zhengzhou has not yet installed or normally used metering equipment at present. They even continue to use traditional agreement to take water. With repeatedly great rectify and reform, the meterage parttern has been changed a lot. However, affected by local traditional factors, there are still a lot of week links in the metering, supervision, management procedure, while they are becoming perfect gradually.

In 2006, the Peach Blossom Valley's flood control project and East Dam's decurrent flood prevention path project were built. Zhengzhou Yellow River Bureau has built two sluices on Yellow River Diversion canal head of Peach Blossom Valley and East Dam, meanwhile has installed precise standard water – taking metrical instrument and they has been formally put into operation since August, 2007, which satisfies the supply side and the demand side on recognition of water diversion data, meanwhile realizes to distribute the Yellow River diversion water scientifically. All other sluices for the Yellow Water water diversion have been installed with this kind of water – taking metrical instrument one after another, which lay a good foundation to establish standard management, deepening, fine and scientific distribution system for the Yellow Water Diversion project in future.

2.2 Adjusting price to promote benign development of water supply project

In recent years, the water price of Yellow River Diversion canal head project in the downstream of the Yellow River changed three times. The National Development and Reform Committee informed to change the industrial water and city life water prices of Yellow River Diversion canal head project in the downstream of the Yellow River: before June 30, 2005, the industry and city life water supply price was from 0.039 to 0.046 yuan/m^3; from July 1, 2005 to June 30, 2006, the price was from 0.062 to 0.069 yuan /m^3; after the date of July 1, 2006, the price was promoted within the range from 0.085 to 0.092 yuan /m^3. The improvement of water price can push forward the optimal allocation of water resources greatly, alleviate tense situation on operation and maintenance costs of the water supply project and promote the water supply project's safe operation and healthy development.

At present, the water price of Yellow River water diversion channel – head project is still serious low, which has a great contrast with the price of water supply in the market. In addition, there are many intermedia links in the management. As a result, the raw – water unit of Yellow River water diversion has not enough money to improve water diversion condition and increase water supply gurantee rate futher.

Due to people's low water saving awareness, the waste on water is still serious. It is necessary to enlarge the reguatling action through price lever futher and improve the saving water awareness on Yellow River diversion water in whole society.

2.3 Establishing communication system and rationalizing the water supply relationship

In recent years, the Yellow River water supply department takes the initiative to go out and strengthen lateral linkage in order to use the limit Yellow River water rescource better. The department constantly enhances the communication with local government, water conservancy and irrigation department; establishs scientific management on ergent need for diversion water; meanwhile, holds conferences and forums regularly; uses media to promote the awareness that "water is commodity and people need to pay for the water"; strives for the support from local government on colleting water fee; negociate with the units concered to gain high value on it, which play an important role in revocering and enlarging the agricultural irrigation scale of the Yellow River water diversion.

2.4 Developing Zhengzhou Yellow River digital water supply

Zhengzhou Yellow River Bureau, adhering to the scientific and technological innovation to promote the development of water supply work, has put forward the goal for Zhengzhou Yellow River digital water supply and has developed "the metering and charging system on the Yellow River water diversion", "automatic flow measurement system for natural river canal" and "step – by – step type metering and charging centralized management system" cooperatively with associate corporate. These systems can overcome the problems specific to Yellow River, including deposit on river bottom and mainstream wandering, solves the problems of high sediment concentration and curved riverbed water diversion metering. It can reduce the influence of artificial factors, provide real – time data transmission, achieve remote automatic transmission function on water diversion data and realize data sharing on information platform of Zhengzhou Yellow River Bureau.

2.5 Strengthening management on sluice project and pulling off inter – regional water supply

The overall designed water diversion quantity for agricultural irrigation is 334.96 m^3/s in the jurisdiction of Zhengzhou and the designed irrigation area is 6,734,300 mu. As a result of the continuous development of urban and rural construction and the continuous decrease of the Yellow River water, serious aging of canal and the less of canal lining, the irrigation area reduces year by year.

According to the development of agricultural Yellow River diversion water, Zhongmu and Zhaokou irrigation areas are the main development objectives. In the spring of 2008, Zhengzhou Yellow River Bureau takes strengthening the management of water supply project as guidance, regards spring irrigation as an opportunity, treats restoring and expanding Zhongmu and Zhaokou Yellow River water diversion irrigation area's supplying water as goal and excavates approach channel before Zhaokou sluice successfully, which make the Zhao Kou channel – head sluice's water discharge reach 20 m^3/s and completely solve the poor flow problem of Zhongmu Zhaokou sluice and Zhaokou irrigation area from Yellow River diversion water for years. In 2012, Zhaokou sluice, with more than 30 m^3/s of water discharge, can supply Yellow River water to Zhongmu, Kaifeng and other regions with greater flow, which provide a running start to break through the bottleneck of water diversion problem on Yellow River diversion water for agriculture in Zhengzhou and make the Yellow River water diversion for agricultural irrigation in Zhongmu and Kaifeng full of vigor and vitality. Since 2010, the water diversion condition of Zhaokou sluice, close to the river mainstream, has been improved continuously, completely solving the approach channel problem that puzzled the sluice for many years. This makes Yellow River diversion water for agricultural irrigation of Zhongmu and Kaifeng full of vitality, which enhances the possibility to supply the Yellow River water to some counties, downtown of Zhoukou, Xuchang and Shangqiu through Zhaokou irrigation channel in future.

2.6 Strengthening communication with the irrigation management departments to facilitate

The largest irrigation area of Yangqiao reached 274,000 mu in history, taking up 75% of designed irrigation area. But in recent years, the water can not be guaranteed and irrigation area drops; in 2004, the actual irrigation area was only 50,000 mu. Farmers commonly use well for irrigation, resulting that much land is planted with drought crop. The Yellow River Bureau of Zhenghzhou takes initative to conmuicate with irrigaton management department and strengthens their relationship. As the investment on reform of irrigation project and on construction of water conservancy works is enlarged, the water for agriculture rise sharply; it promotes the water supply gurantee rate to divert water and mangage water in scientific way; therefore, the irrigatin area enlarges substantially.

2.7 Benefit analysis

2.7.1 Economic efficiency

According to calculation, the well irrigation cost double higher than the Yellow River water irrigation. The Yellow River irrigation saves time, energy, has no mechanical cost and the water from the Yellow River contains organic matter for soil and it can prevent the ground from hardening. It can recover 10,000 mu irrigation area, we can save more than 400,000 yuan irrigation cost directly, meanwhile we can restore underground water recharge area of about 7,000 mu. It has great effect in soil improvement by using the Yellow River water for irrigation; it can also recharge the groundwater, improve underground funnel effectively, recover and expand irrigatin area.

2.7.2 Social benefits

Through a series of engineering measures, such as, transforming water supply project of irrigation area, providing supporting building, adopting administrative, economic, legal measures, optimize the allocation of water resources of irrigation area, provide abundant water sources to industry, agriculture, urban and rural life and environmental improvement in irrigation area and provide basic condition to improve the people's life in irrigation area and ensure social stability. The irrigation area can make better use of the Yellow River water, which can not only improve the soil hydrological condition, but also improve air humidity and temperature of irrigation area, thus improve agricultural microclimate. It has a positive sense to reduce soil erosion, improve ecological environment and promote harmonious development of ecology.

What's more, the irrigation area can also save a lot of money and expand its irrigation area. It plays an irreplaceable role to improve the local industrial and agricultural production and natural appearance. It can benefit the improvement of water resource conditions, promote the harmonious development of agroecological environment, help virtuous circle of ecological environment, improve water supply condition of irrigation works in irrigation area, change the water resources shortage in irrigation area, enhance the ability to resist natural disaster in agriculture and provide reliable conditions for full - scale development of farming, forestry, stock raising and fishery; it has an important sense to improve the appearance of irrigation area, promote the development of industrial and agricultural production and improve the people's life in irrigation area. In addition, the improvement of production and life condition will provide guarantee for the stable and unified society, which has obvious social benefits.

3 Analysis on status of water diversion growth in recent years

3.1 Domestic water

The actual intaking quantity of domestic water in 2011 was 201,400,000 m^3. With the increasingly expanded unban scale of Zhengzhou and the growing population, the domestic water demand for urban area would be increased but not with dramatic rises.

The First Sluice for Dongdaba Canal of Diverting Water from Yellow River supplies water for Sino - French Raw Water Co., Ltd., providing drinking water and industrial water for 1,400,000 people of Zhengzhou City. The output of water supply would reach 85,000,000 m^3 by 2015.

Shifo Water Plant plans to restore the water intaking wells burst by water flow of Yellow River, which would help the water intaking quantity reach 28,000,000 m^3 by 2015.

Dongzhou Water Plant mainly provides water for east district of Zhengzhou, and its water intaking quantity would be increased to 32,850,000 m^3 by 2015.

3.2 Industrial water

There are two industrial water intakes in the area under administration, one of which is the Tanxiaoguan water intake wells group supplying water for Zhongfu Industrial Co., Ltd. with the planning water consumption increased to 10,000,000 m^3 by 2015. The other one is Gubaizui Water Pumping Project, which provides water for Aluminum Corporation of China Henan Branch. As the annual production of Chinalco is continually growing along with the yearly increased water intaking quantity, the original permitted water intaking quantity has been unable to satisfy the production and life demands, the planning water consumption would be increased to 25,000,000 m^3 by 2015.

3.3 Agricultural water

Along with the intensified reconstruction project in irrigation districts, the agricultural water consumption would be gradually increased and so for the agricultural irrigation area depending on Yellow River which would be expanded from 200,000 mu in 2012 to 300,000 mu by 2015. The total annual agricultural water consumption for the two irrigation districts of Yangqiao and Sanliuzhai Sluices would be increased to 160,000,000 m^3. Owning to the intensified reconstruction project in Zhaokou Irrigation District, the agricultural irrigation area would be increased year by year.

3.4 Ecological water

The ecological water would be distinctly increased after the water distribution project in ecological water system of Zhengzhou City is completed with the estimated water consumption increased to 1,000,000,000 m^3 by 2015, which indicates that the contradiction of water supply from Yellow River would be increasingly prominent.

4 Predictions for the water demand in recent years and the development prospect

Thanks to the national investment in irrigation and water conservancy projects and the urban development in recent years and years ahead, the agricultural and ecological water consumption will be considerably increased; and water intakes with larger increased water consumption mainly include Huayuankou Sluice, Yangqiao Sluice, Sanliuzhai Sluice, Zhaokou Sluice and the newly added water intake projects.

4.1 Water diversion forecast of Huayuankou Sluice

Huayuankou Sluice is mainly responsible for the water supply for Zhengzhou Three Rivers and One Canal Project and Longhu Regulating and Storage Project. The water supply status in next few years is as follows: providing water for the three rivers and one canal, with 5 times of changing water every year, changing water flow of 5 m^3/s for Dongfeng Canal, Jialu River and Jialu River Branch, 2 m^3/s for Suoxu River, and the annual total water demand of 37,100,000 m^3; frequent water supply of 1 m^3/s, with daily operation of 12 hours and annual water demand of 11,230,000 m^3; annual total evaporative capacity of 4, 780, 000 m^3; annual total osmotic quantity of 1,640,000 m^3. The annual water demand of the four rivers and canals is 54,750,000 m^3.

Huayuankou Sluice also supplies water for Zhengzhou Longhu Regulating and Storage Project which has been finished and starts to conserve water currently (however the downstream engineering is still under construction). There are two major targets for Longhu Regulating and Storage Project: one is to recover the irrigation area in Huayuankou Irrigation District, improve the irrigation water conditions for areas of Yangqiao and Sanliuzhai irrigation districts which can not be fully irrigated, promote the agricultural development in irrigation districts and guarantee the grain production security; the other goal is to improve the eco – environment of surroundings as the urban ecological water of Zhengzhou while providing irrigation water for irrigation districts. With a total storage capacity of 26,800,000 m^3, the Longhu Regulating and Storage Project has normal water level of 85.5 m and the water surface area of 5.6 km^2.

If Huayuankou Sluice assumes the responsibility of supplying water for Zhengzhou Three Rivers and One Canal along with Longhu Project in the coming years, the annual water diversion of this sluice would be increased to 200,000,000 m^3.

4.2 Water diversion forecast of Yangqiao Sluice and Sanliuzhai Sluice

To meet the demands of Zhengzhou ecological water and Zhongmu Watertown construction, by 2015, the water consumption for ecological environment of Yangqiao Sluice would reach 180,000,000 m^3 and that of Sanliuzhai Sluice would reach 120, 000,000 m^3. Only the water diversion of Yangqiao Sluice and Sanliuzhai Sluice is respectively increased to 260,000,000 m^3 and 200,000,000 m^3 in the next couple years could the growing water demands for agriculture and ecological environment of Zhongmu Irrigation District depending on Yellow River be satisfied.

4.3 Water diversion forecast of Zhaokou Sluice

The actual irrigation area of Zhaokou Sluice would be gradually expanded from the original 12 counties (districts) of 4 cities (Zhengzhou, Kaifeng, Xuchang and Zhoukou) to 15 counties (districts) of 5 cities (the mentioned 4 cities plus Shangqiu city). At the same time, the agricultural water consumption of Zhaokou Irrigation District would be increased year by year, and it is expected to reach 1,300,000,000 m^3 in the coming years.

4.4 Newly added water intake projects

4.4.1 The First Sluice for Gongyi Yulian Water Supply Canal of Diverting Water from Yellow River with annul planning water intaking quantity of 2,700 m^3 would be mainly responsible for providing domestic water for the residents in New East Zone of Gongyi City, environmental and farmland irrigation water for Eastern Industrial Park of Gongyi.

4.4.2 Xingyang Guodian Water Intake Project with annul planning water intaking quantity of 410 m^3 hasn't been built yet.

4.4.3 Niukouyu Ecological Water Intake Project has been well planned with annul planning water intaking quantity of 400,000,000 m^3, which is expected to start construction in Oct. 2012. Once this water intake is established, not only the water supplying demands of Zhengzhou Yiju Health City, Yiju Vocational City and other cities in western district could be satisfied, but also the support capacity of domestic and ecological water consumption for part of central urban area of Zhengzhou could be guaranteed, which would play a vital role in the construction of Zhengzhou ecological water system.

4.4.4 As the spare water intake station of Taohuayu Sluice, Taohuayu Intake Pumping Station aims at ensuring the security and reliability of the first - level water source area and guaranteeing the normal water supply of western district of Zhengzhou unaffected. The pumping station was put into construction in Apr. 5 this year, and its water intaking quantity would reach 100,000,000 m^3 after the project being completed. Based on the investigation and analysis for water diversion status of each water intake in recent years, the knowledge about water users' demands in the coming years and the water consumption predictions, by 2015, water diversion from Yellow River would be increased to 2,000,000,000 m^3 in Zhengzhou to meet the water demand of urban development.

5 Conclusion

The permiteed annual water intaking quantity from Yellow River of Zhengzhou in 2012 approved by the superior has bot been able to satisfy the demands of regional development for the water resources of Yellow River because of the special geographical position Zhengzhou is located; thus, water diversion from Yellow River of Zhengzhou enjoys broad space for development. The development of water supply from Yellow River plays a significant role in effectively alleviating intense conflict of water shortage in some regions of Henan province, raising the guaranteed rate of urban and industrial water application, expanding the agricultural irrigation area, expanding the water supply space with continuous innovation, developing the water application for urban ecological environment, and improving ecological environment. Water supply from the Yellow River of Zhengzhou is characterized by a broad development prospect due to the extraordinary natural environment and regional advantages that Zhengzhou owns, which is not only greatly beneficial to the social and eco – environment effects, but also vital for ensuring the economic and social sustainable development of districts along the Yellow River and moreover crucial for supporting their comprehensive construction for a well – off society.

Study on the Promoting Effect on Protection of Hengshui Lake after the Water Diversion from Yellow River

Zhang Hao[1], *Guo Yong*[1] and *Miao Pingping*[2]

1. Water Resources Protection Bureau of Haihe River Basin, Tianjin, 300170, China
2. Hebei Reserach Institute of Investigation & Design of Water Conservancy & Hydropower, Tianjin, 300250, China

Abstract: Hengshui Lake Wetland is an important water source of the North China Plain, which playing an important role in regulating the climate of the areas surrounding Beijing and Tianjin, improving the ecological environment and flood control. To protect the Hengshui Lake Wetland, China began the Yellow River Diversion Project in 1994. The implementation of Yellow River Diversion Project has great significance in the maintenance of normal production, improving the surrounding environment and maintaining the ecological balance. To study the trend, this paper compared the data of the Hengshui Lake wetland water quality and quantity before and after the Yellow River Diversion project, using Mann – Kendall test and spearman correlation coefficient method. Against the practical problems in the Hengshui Lake Wetland, corresponding protection measures were put forward.

Key words: water diversion projection from Yellow River, Mann – Kendall test, spearman method

1 Introduction

Wetlands is the world's largest and most widely distributed and the most productive natural environment, with important ecological functions such as keeping water, water purification, flood and drought controlling, climate regulation, soil erosion controlling, reduce environmental pollution, maintenance of biological diversity and so on, and also to provide mankind with a wide range of properties and leisure travel venues. Wetlands have significant economic value and social value. Healthy wetland ecosystem is an important component of national ecological security and important foundation of sustainable economic and social development. Hengshui Lake Wetland is located in the economic circle around Beijing and Tianjin, the Bohai Sea economic circle and the Yellow River Economic Cooperation. Hengshui Lake is an important water source and also plays an important role in regulating the climate of the Beijing—Tianjin region, improving the environment, controlling flood and drought. In recent years, due to the construction of a large number of water conservancy projects in Hengshui Lake Basin and global climate change, the amount of water into Hengshui Lake decreased and dried up frequently.

Han from the reality of Hengshui Lake investigated and analyzed the animal husbandry and fishery sources of the present situation of the corresponding control measures. (2007). Li and Hao analyzed and evaluated the Hengshui Lake water environment situation (2009). Zhang analyzed the principle and means of recovering ecosystem of Hengshui Lake wetlands by using outer drainage area to supply water, concrete questions about the recovery of marsh were discussed (2010). Jiang et al. used uncertainty models to evaluate the water quality of Hengshui Lake and the water quality data including the water level, temperature, dissolved oxygen concentration. But the present study is less in the impact of Yellow River Diversion water diversion on the Hengshui Lake.

2 Study area

Hengshui Lake (Fig. 1) commonly known as "thousand hectares of depressions", is located in Hengshui City, Hebei Province between Taocheng and Jizhou. The largest east – west width is 22.28 km and the largest north – south length is 18.81 km. The control area of Hengshui Lake is

1,654 km^2, there are five rivers flowing into the lake, in addition, six flood channels are distributed in the surrounding. Hengshui Lake is a wetland nature reserve taking reservoir constructed wetlands as a base, biodiversity conservation as a key, resourcing characteristics of rare birds as a task. Hengshui Lake is also an important base for the protection of birds in China and an important place to carry out biodiversity conservation, environmental pollution monitoring.

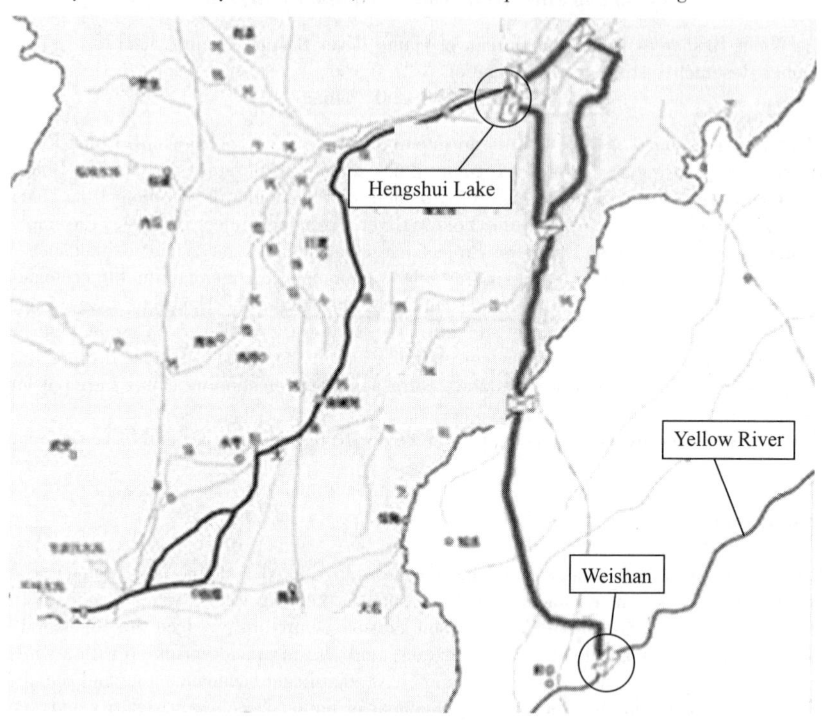

Fig. 1 Yellow River Diversion route

The Hebei Yellow River water transfer trunk built in 1994 is the main project of Yellow River to Hebei and Yellow River to Tianjin. Main water diversion channel starts by the Weishan County of Shandong and enters in the Linxi County of Hebei Province. Through the Xinkai River, Erzhi River, Donggan River and Qingliang River to reach Lian county gate, total length is 151.9 km, involving eight counties of Xingtai and Hengshui City. Since the implementation of the project in 1994, Hengshui Lake had transferred total 626×10^6 m^3. It was playing an important role in maintain production and living, improving the environment, maintain ecological balance.

3 Methods of trend analysis

Sequence changes affect the hydrological trend analysis of hydrological characteristics, modeling and prediction, thereby affecting the efficient use of water resources management and issues such as decision – making. Therefore, the study of trends in hydrological series has a very important feature. This paper used Mann – Kendall test and spearman method to study the trend.

Mann – Kendall trend test is through taking the U – values compared to the assumptions of the confidence level. When $U > 0$, there is an upward trend in the sequence, when $U < 0$, there is a downward trend in the sequence, when $U = 0$, there is no trend in the sequence. When $|U| \geqslant U_{0.05} = 1.96$, there is significant changes in the trend series, otherwise, when $|U| < U_{0.05} = 1.96$, there is no significant changes.

Suppose the time series of hydrological is:

$$X_t(t = 1,2,\cdots,i)$$

$$P - \sum_{i=1}^{n} r_i \tag{1}$$

where, p is duality number.

$$r_i = \begin{cases} 1, & X_i < X_j \\ 0, & X_i \geqslant X_j \end{cases} \qquad j = 1,2,\cdots,i \tag{2}$$

τ determined according to p, then come to var(τ) and U:

$$\tau = \frac{4p}{n(n-1)} - 1 \tag{3}$$

$$var(\tau) = \frac{2(2n+5)}{9n(n-1)} \tag{4}$$

$$U = \frac{\tau}{var(\tau)^{1/2}} \tag{5}$$

Spearman method calculated the rank correlation coefficient r_s by Eq. (6). When $r_s < 0$, there is a downward trend, otherwise, there is an upward trend. If $r_s \geqq w_p$, there is significant changes in the trend series.

$$r_s = 1 - [6(\sum_{i=1}^{n} d_i^2)]/[N^3 - N] \tag{6}$$

$$d_i = x_i - y_i \tag{7}$$

where, N is the number of cycles, x_i is the serial number of variables arranged from small to large, y_i is variable serial number arranged in chronological order.

4 Water quantity impact by the Yellow River Diversion

From November 11, 1994 to April 9, 2008, a total of 12 years experience 13 water transfer, the total number of Yellow River water diversion into Hengshui Lake was 626×10^6 m^3 (Fig. 2).

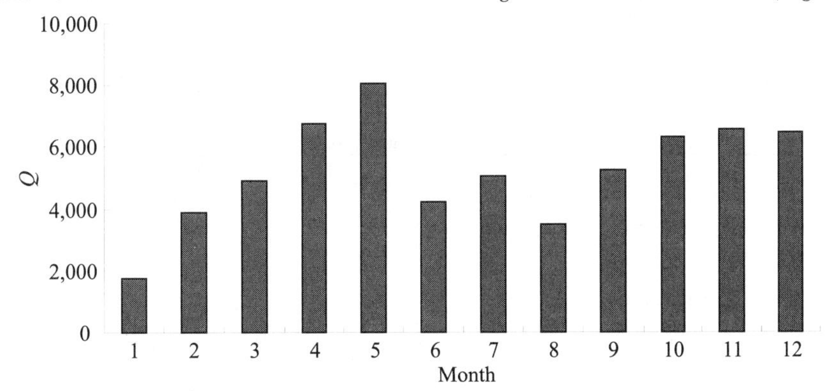

Fig. 2 Quantity of water diversion from Yellow River over the years ($\times 10^4$ m^3)

After water diverted from Yellow River, the average water level of Hengshui Lake rose from 18.39 m to 19.74 m, rose 7.34%. Assurance rate of above 20 m water level rose from 36.84% to 86.67%, equivalent to 135.23%. The Hengshui Lake average maximum water level was 9.55 m before diverted Yellow River water, this data was 20.57 m after the water transfer. The implementation of Yellow River diversion project made the Hengshui Lake's average water level increases up to 1.02 m and did not dry up after the case (Tab. 1).

Tab. 1 Comparison of hydrological factors before and after the water diversion from Yellow River

Statistical indicators	Before	After	Change rate
Annual average water level	18.39 m	19.74 m	7.34%
Mean square deviation	7.23	2.27	-68.60%
Frequency of dry	68.42%	0	-100%
Assurance rate of the highest level greater than 20 m	36.84%	86.67%	135.23%
Maximum amplitude of annual water level	2.69 m	2 m	-25.65%
Amplitude of average water level	1.31 m	1.47 m	12.21%

Analyzing the trend of the highest level, Mann - Kendall method results were: $U = -0.105$ in 1975 ~ 1993, It was shown that there was a downward trend in the highest water level of Hengshui Lake; $|U| < U_{0.05} = 1.96$, the downward trend was not significant. In 1994, $U = 0.054,7$, it was shown that there was an upward trend in the highest water level of Hengshui Lake, $|U| < U_{0.05} = 1.96$, the upward trend was not significant. Spearman method results were: in 1975 ~ 1993, $r_s = -0.236$, it was shown that there was an downward trend in the highest water level of Hengshui Lake; $|r_s| < w_{0.05} = 0.9$, the downward trend was not significant. In 1994, $r_s = 0.773$, it was shown that there was an upward trend in the highest water level of Hengshui Lake; $|r_s| < w_{0.05} = 0.9$, the upward trend was not significant.

5 Water quality impact by the Yellow River Diversion

Used integrated pollution index to analysis water quality before and after the previous water diversion(Tab. 2, Fig. 3)

$$P_n = \frac{1}{n} \sum_{i=1}^{n} \frac{C_i}{C_{si}} \tag{8}$$

where, P_n is integrated pollution index, n is the number of items of pollutants, C_i is measured concentrations of the i - th pollutant, C_{si} is evaluation criteria of the si - th pollutants (refer to GB3838—2002).

Tab. 2 Comprehensive pollution index of classification

P_n	Degree of pollution
<0.2	Very clean
0.2 ~ 0.4	Clean
0.4 ~ 0.7	Light pollution
0.7 ~ 1.0	Medium pollution
1.0 ~ 2.0	Heavy pollution
>2.0	Serious pollution

Analyzing the trend of integrated pollution index after the water transfer from Yellow River, Mann - Kendall method results were: $U = -0.183$, it was shown that there was an downward trend in the integrated pollution index of Hengshui Lake; $|U| < U_{0.05} = 1.96$, the downward trend was not significant. Spearman method results were: $r_s = -0.865$, it was shown that there was an downward trend in the integrated pollution index of Hengshui Lake; $|r_s| < w_{0.05} = 0.9$, the downward trend was not significant.

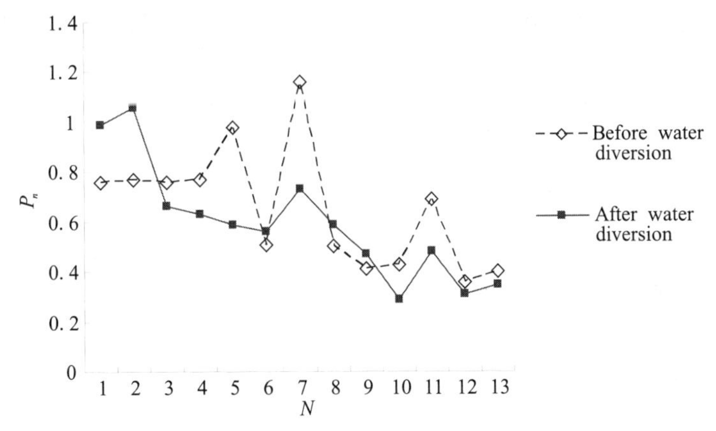

Fig. 3 Comparison of comprehensive pollution index before and after the water diversion

6 Conclusions

In water quantity, the Yellow River water diversion into the lake timely ends the frequent dry situation. However, Hengshui Lake eco – environmental water demand needs further study to determine and the Hengshui Lake water supply protection measures should be explored. In water quality, integrated pollution index of Hengshui Lake presented a downward trend. However, after the diversion of Yellow River, rushes and other pollutants flowing into Hengshui Lake will cause a series of impacts on Hengshui Lake, which should be studied further.

In order to safeguard the health of Hengshui Lake, the following aspects should be focused on:

(1) Divert water resources to protect the Hengshui Lake water quantity. In dry years, water transfer time should be extended to protect the ecological Hengshui Lake water.

(2) Increase the environmental management efforts and investment in environmental protection. The main sources of pollution are aquatic plant overgrowth and import channel runoff, therefore, on the one hand, plants should be harvested within the lake and sediment should be removed timely, on the other hand water quality into the lake should be improved, especially measures should be taken to avoid secondary pollution of the Yellow River water on the way.

(3) Improve the management of Hengshui Lake. To monitor the health of Hengshui Lake regularly and fixedly, monitoring project should be involved in water quality, the nutritional status, biodiversity and so on. Through timely, accurate and detailed data collection, future work can build Hengshui Lake health warning system.

References

Han jiugao. Hengshui Lake Pollution of Agricultural Production and Control Measures[J]. Journal of Anhui Agricultural Science ,2007,35(9):270 – 270.

Li Haitao, Hao Yueying. Analysis of Hengshui Lake Water Resources Situation[J]. Water Sciences and Engineering Technology,2009(5):10 – 12.

Zhang Xuezhi. Principles and Methods of Wetland Ecosystem Restoration of Hengshui Lake[J]. South to North Transfers and Water Science and Techonlogy,2010,8(1):122 – 125.

Jiang Chunbo, Zhang Mingwu, Yang Xiaolei. Water quality evaluation for the Hengshui Lake wetland in northern China[J]. Tsinghua University(Science & Technology),2010,50(6):848 – 851.

Donald H Burn, Mohamed A, Hag Elnur. Detection of Hydrologic Trends and Variability[J]. Journal of Hydrology,2005,(255):107 – 122.

The Classification Theory of Compensation Mechanism for Agricultural Water Transfer

Dai Xiaoping[1], ***Jiang Qiong***[2], ***Zhang Zhanning***[3], ***Xu Hui***[1], ***Zeng Wenxue***[4] and ***Zhang Panhui***[6]

1. College of Water Conservancy, North China University of Water Conservancy and Electric Power, Zhengzhou, 450011, China
2. Immigrant Settlement Planning Center of Sichuan Province, Chengdu, 610031, China
3. Nianyushan Reservoir Administration Bureau, Xinyang, 465350, China
4. Henan Provincial 1st Water Conservancy Engineering Bureau, Zhengzhou, 450004, China
5. Henan Water Conservancy Investment Co., Ltd., Zhengzhou, 450008, China
6. Luoyang Water Resources Surveying & Designing Co., Ltd., Luoyang, 471000, China

Abstract: The compensation mechanism for agricultural water transfer (CMAWT) is an important method to reduce and eliminate the negative influence of agricultural water transfer. It is important to summarize the global experiences of CMAWT for setting up the reasonable CMAWT of China. This paper established the classification theory of CMAWT and put it into the evaluation of CMAWT in China and Japan. The result shows that the CMAWT of China is converting from bureaucratic type to market type. The CMAWT of Japan is mainly autonomy type. The CMAWT of Japan is more autonomy than the CMAWT of China. The case study shows that this theory can classify the CMAWT of different countries.

Key words: the compensation mechanism for agricultural water transfer (CMAWT), classification theory, China and Japan

1 Background

With the shortage of water resources, the agricultural water transfer is increasing quickly in China. The compensation mechanism for agricultural water transfer (CMAWT) is necessary to reduce and eliminate the negative influence of agricultural water transfer. It is important to compare the CMAWT of China to other countries to find out the direction of setting up the reasonable CMAWT of China. The classification system for water management and water use has been set up before . And the compensation mechanism for agricultural water right transfer has been set up too . But there is no classification theory of CMAWT. So, this paper tried to establish the classification theory of CMAWT and put it into the evaluation of CMAWT in China and Japan.

2 The classification theory of CMAWT

2.1 Related concepts

2.1.1 The agricultural water transfer

Agricultural water transfer is the behavior of transferring the use right of agricultural water from agricultural water user to other users in agricultural industry or other industries.

2.1.2 The compensation mechanism for agricultural water transfer (CMAWT)

The compensation mechanism for agricultural water transfer (CMAWT) is the mechanism of releasing the negative impacts of agricultural water transfer by compensating the cost of agricultural water saving and the lost of the losers.

2.1.3 The compensation bodies and compensation objects

The compensation bodies are the beneficiary in agricultural water transfer. The compensation objects are the losers in agricultural water transfer.

2.1.4 The compensation approaches and compensation methods

The compensation approaches are the combination of compensation bodies and compensation objects. The compensation approaches include the government compensation and the beneficiary compensation. The government compensation is the compensation of government to the compensation objects. The beneficiary compensation is the compensation of other beneficiaries except government to the compensation objects. The government compensation happens when other beneficiaries don't have the ability to compensate the compensation objects by themselves.

2.2 The stakeholder analysis of the compensation of agricultural water transfer

2.2.1 The government

The government plays an important or unimportant role in different kind of agricultural water transfer. The government may be the pusher, supervisor or participant in different agricultural water transfer. Generally, the government is the beneficiary in the agricultural water transfer. Otherwise the government will forbid the agricultural water transfer.

2.2.2 The agricultural water supplier

The agricultural water supplier is mainly the administration bureau of irrigation district. The administration bureau of irrigation district is responsible for water intake and water distribution. The agricultural water suppliers may be the beneficiary or losers in the agricultural water transfer.

2.2.3 The agricultural water user

The agricultural water users are the subjects who own the agricultural water right and transfer water out. The agricultural water users include the water user association and the farmers who use water. The water user association is the basic water distributor. The farmer is mainly the direct agricultural water user, and they can directly get the water too. The agricultural water users may be the beneficiary or losers in the agricultural water transfer.

2.2.4 The agricultural water accepters

The agricultural water accepters can be divided into the accepters in the agricultural sector, the industrial sector, the municipal department and the ecological environment by the properties of water users. And the agricultural water accepters can be divided into the individual accepter, the enterprise accepter and the government. The agricultural water accepters are definitely the beneficiary in the agricultural water transfer, otherwise the agricultural water transfer will not happen.

2.2.5 The third parties

The third parties are the subjects who are impacted by the agricultural water transfer though they don't participate in the agricultural water transfer. The third parties include individual, enterprise, local government and ecological environment. For example, the agricultural water user, enterprise, local government and ecological environment in the downstream may be impacted when the agricultural water is transferred out of agricultural sector for the decrease of irrigation return water. Generally, the third parties are the losers in the agricultural water transfer.

2.3 The theoretical framework in the classification of CMAWT

The model of CMAWT can be analyzed from the view of vertical form and horizontal form.

2.3.1 The vertical form

The vertical form mainly reflects the relationship of government and other stakeholders. The influencing factors of the vertical form include the forms of property rights, the participation level of stakeholders and the compensation approaches. The vertical form of CMAWT can be divided into the bureaucratic type, the autonomy type and the market type.

(1) The bureaucratic type

The agricultural water users only own the use right of agricultural water and they can't transfer the use right in the bureaucratic type. And there is no participation of agricultural water users and the third parties in this type. The water amount transferred and the term of transfer is mainly decided by the government. The compensation amount and compensation method is also mainly decided by the government.

(2) The autonomy type

The agricultural water users own the use right of agricultural water and they can transfer the use right under some conditions. Under the guide of government, the agricultural water users and the third parties can participate in the agricultural water transfer, and they can put forward their opinions on the decision of transfer amount, the term of transfer, and the compensation methods. Generally, the compensation fee is paid by the government and other bonefishes in the agricultural water transfer.

(3) The market type

The agricultural water users own the ownership or relatively complete use right and assignment right of agricultural water. The agricultural water transfer has changed to agricultural water right transfer. The compensation fee has changed to the transfer fee. The transfer price is decided by the negotiation between the both sides of transfer. The third parties participate in the agricultural water right transfer in different extent, and give opinions to the transfer. The transfer fee is only paid by the buyer. The compensation fee to the third parties is paid by the both sides of water right transfer. The government only supervises the water right transfer.

2.3.2 The horizontal form

The horizontal form is the compensation methods which reflect the operation form of compensation in agricultural water transfer. It includes the fund compensation and material compensation.

Fund compensation is the compensation with cash to decrease or eliminate the loss of losers.

Material compensation is the compensation with material or project to improve the development ability of losers, e. g. the investment to the agricultural water saving project, eco-restoration project and so on.

2.3.3 The mode of CMAWT

The mode of CMAWT is composed by the vertical form and horizontal form of agricultural water transfer. All of the modes of CMAWT are showed in Tab. 1.

Tab. 1 The mode of CMAWT

form	bureaucratic type	autonomy type	market type
material compensation	bureaucratic—material	autonomy—material	market—material
fund compensation	bureaucratic—fund	autonomy—fund	market—fund

We can see from Tab. 1 that the mode of CMAWT can be divided into six categories. We can preliminary classify the mode of CMAWT in different countries with this method. But the accurate classification of the mode of CMAWT is difficult. So the quantitative analysis method of the classification theory is developed.

2.4 The quantitative analysis method of the vertical form of CMAWT based on the method of AHP

The evaluation index system of the vertical form of CMAWT (see the Fig. 1) can be set up based on the influencing factors.

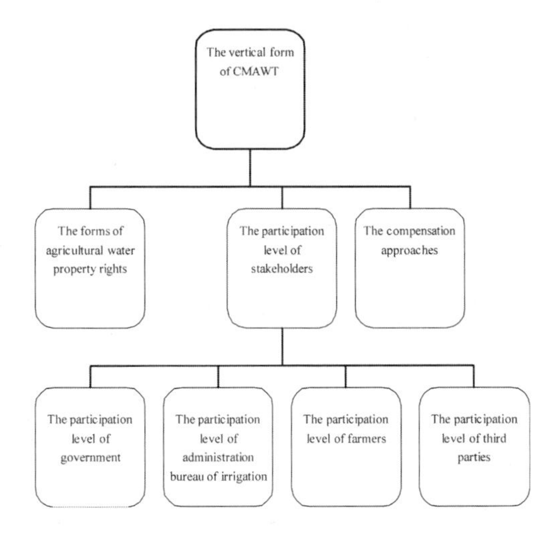

Fig. 1 the evaluation index system of the vertical form of CMAWT

2.4.1 The interpretation and weighting of indexes

(1) The forms of agricultural water property rights

Tab. 2 The valuation of indexes in classifying the vertical form of CMAWT

No.	Indexes		Scoring Criteria
1	The form of agricultural water property rights		only own use right = 0, own use right and assignment right = 0.5, own the property right, use right and assignment right = 1
2	participation level of stakeholders	government	leading = 0, partially participating = 0 ~ 1, no participating (supervision) = 1
		administration bureau of irrigation district	no participating = 0, partially participating = 0 ~ 1, completely participating = 1
		farmers	no participating = 0, partially participating = 0 ~ 1, completely participating = 1
		third parties	no participating = 0, partially participating = 0 ~ 1, completely participating = 1
3	compensation approaches		completely compensated by the government = 0, completely compensated by the beneficiary = 1, compensated by the government and the beneficiary = the ratio of compensation fee compensated by the beneficiary

Agricultural water right is the property right, use right and assignment right of irrigation water. Generally, the water right which the agricultural water users own is becoming more and more complete as the type of CMAWT changes from bureaucratic type to market type.

(2) The participation level of stakeholders

The agricultural water accepters definitely participate in the water transfer whatever the type of CMAWT. So the participation level of the agricultural water accepters is not considered. The participation level of government becomes smaller as the type of CMAWT changes from bureaucratic type to market type. And the participation level of other stakeholders becomes bigger as the type of CMAWT changes from bureaucratic type to market type.

(3) The compensation approaches

The compensation approaches change from the government compensation to the beneficiary compensation as the type of CMAWT changes from bureaucratic type to market type.

In a word, the vertical form of CMAWT can be classified by these indexes. The valuation of indexes is showed in Tab. 2.

2.4.2 The weight of indexes

(1) The importance of the first grade indexes

The importance of the first grade indexes is decided by the method of analytic hierarchy process (AHP) (see Tab. 3).

Tab. 3 the judgment matrix of the first grade indexes

the first grade indexes	A	B	C
B	3	1	5
C	1/3	1/5	1

Note: A: the forms of agricultural water property rights; B: the participation level of stakeholders; C: the compensation approaches

The consistency check of the judgment matrix shows that the matrix has good consistency ($CR = 0.033 < 0.1$).

The weight of indexes which are calculated with the method of normalization row mean is: (A, B, C) = (0.291, 0.605, 0.103).

(2) The importance of the second grade indexes

The importance of the second grade indexes is also decided by the method of AHP (see Tab. 4).

Tab. 4 The judgment matrix of the second grade indexes

The second grade indexes	a	b	c	d
a	1	1/2	1/5	1/3
b	2	1	1/3	1/2
c	5	3	1	2
d	3	2	1/2	1

Note: a: the participation level of government; b: the participation level of the administration bureau of irrigation district; c: the participation level of farmers; d: the participation level of the third parties

The consistency check of the judgment matrix shows that the matrix has very good consistency ($CR = 0.005 < 0.1$).

The weight of indexes which are calculated with the method of normalization row mean is: (a, b, c, d) = (0.087, 0.164, 0.471, 0.278).

(3) The final importance of all of the indexes

The final weight of the indexes is calculated by multiplication of the weight of the first grade and second grade indexes (see Tab. 5).

Tab. 5 the weight of the indexes

indexes		weight
the forms of agricultural water property rights		0.291
the participation level of stakeholders	the participation level of government	0.053
	the participation level of the administration bureau of irrigation district	0.099
	the participation level of farmers	0.285
	the participation level of the third parties	0.168
the compensation approaches		0.103

(4) The judgment of the evaluation result

The evaluation result is a number between 0 ~ 1. The mode of the CMAWT is close to the bureaucratic type when the evaluation result is close to 0. Conversely, the mode of the CMAWT is close to the market type when the evaluation result is close to 1.

3 The case study—the analysis of the CMAWT of China and Japan

3.1 The change of the CMAWT of China and it's characteristics

The CMAWT of China is occupancy compensation before the introduction of the theory of water right and water market into the water resource management. The basis of occupancy compensation is the "compensation method of occupying the irrigation water and irrigation and drainage projects" which is published by the Ministry of Water Resources together with the Ministry of Finance and the State Planning Commission. According to the regulation, the loss of the occupancy of the irrigation water and irrigation and drainage projects should be compensated and the compensation should match the loss. The farmers only own the use right of agricultural water and they are not allowed to transfer it in this period. The participators include the agricultural water accepters, the government and the administration bureau of irrigation district. The government plays the leading function in this kind of transfer. And there are no participation of farmers and the third parties. The compensation amount is decided according to the regulation. The compensation fee is mainly paid by the government. The compensation methods include the fund compensation and the material compensation.

Many agricultural water right transfer cases have appeared after the introduction of water rights theory into the water resource management since 2000. The agricultural water right transfer cases can be divided into the following three types:

The background of the appearance of the first type is the decrease of irrigation area and the increasing need of water in cities. So, much agricultural water is transferred to the city. The city received water begins to sign agreement of water right transfer with the administration bureau of irrigation district after the introduction of water right theory. The transfer amount, transfer period and transfer price is determined in the agreement. The typical case o f this kind of water transfer is the water right transfer between Dongyang city and Yiwu city. Under the guidance of the government, the transfer is carried out with the negotiation of the administration bureau of irrigation district and the agricultural water accepter. Both of the farmers and the third parties don't participate in the agricultural water transfer in these cases. The compensation fee is mainly paid by the water accepter. The compensation method is the fund compensation.

In the second type, the saved agricultural water is transferred out with the investment of water accepters on the agricultural water saving. The typical case is the water right transfer in the upstream of the Yellow River. The regulation of "Some opinions about the water right transfer" is published by the Ministry of Water Resources in 2005. According to the regulation, the water use right can be transferred in some limiting conditions. This kind of water transfer is promoted by the government. The agricultural water accepters and the administration bureau of irrigation district are the main participators in the transfer. But the farmers and the third parties still don't participate in the transfer. The compensation amount is decided by the negotiation of the beneficiary and the administration bureau of irrigation district. And the compensation fee is paid by the government and the agricultural water accepters. The compensation methods include the material compensation and the fund compensation.

The third type happens in the area in which the agricultural water right has been clearly distributed. The farmers transfer their agricultural water right freely. The typical case of this kind of transfer is the water right transfer with water-ticket in Zhangye. The participators in this kind of transfer are the farmers. The government only supervises the transfer instead of participates in the transfer directly. The compensate methods include the material compensation and the fund compensation.

3.2 The characteristics of CMAWT in Japan

The agricultural water right in Japan can be divided into the customary water right and approved water right. The agricultural water right is only assigned to the water user association. The agricultural water right which farmers own are the use right. This agricultural water use right can't be transferred. The saved agricultural water is transferred by the investment of water user in the industry and the water supply enterprise into the agricultural water saving. The typical case of this kind of water transfer is the tone-center water transfer.

The government not only promotes the agricultural water transfer but also invests greatly into the improvement of irrigation system to save water. The compensation fee is paid by the government and the beneficiary. The farmers can effectively participate in the water transfer because many irrigation districts are managed by the water user associations. Though there is no participation of the third parties, the government has published some regulations to avoid the loss of the third parties.

3.3 The classification of the CMAWT of China and Japan

The CMAWT of China and Japan is evaluated according to the classification theory of CMAWT. The result is showed in Tab. 6.

Tab. 6 the classification of the CMAWT of China and Japan

country	period	Typical water transfer case	Indexes						core	mode
			The forms of agricultural water property rights	The participation level of stakeholders				Compensation approaches		
				Government	Administration bureau of irrigation district	Farmer	Third parties			
China	Before 2000	Occupancy compensation	0	0	0.2	0	0	0	0.020	bureaucratic + material /fund
	After 2000	The water right transfer between Dongyang city and Yiwu city	0.5	0.5	1	0	0	0	0.271	autonomy + fund
		The water right transfer of Nanan irrigation district	0.5	0.5	0.5	0	0	1	0.325	autonomy + material
		The "water-ticket" transfer in Zhangye city	0.5	1	0	1	0	1	0.587	market + material / fund
Japan	1992 ~ 2003	The tone-center water transfer	0	0.5	1	0.8	0.2	0.53	0.442	autonomy + material

The score of the CMAWT in China have changed from 0.02 to 0.586 with time. This shows that the mode of CMAWT in China is changing from bureaucratic type to market type. Most of the compensation mode of water transfer cases are the type of the "autonomy + material/fund". The score of the CMAWT in Japan is 0.442, which shows that the mode of CMAWT in Japan is autonomy + material. The score of CMAWT in all of the water transfer cases analyzed in China except the "water-ticket" transfer in Zhangye city are lower than the score of CMAWT in the case of Japan. This shows that the degree of autonomy of the CMAWT in Japan is higher than that in China. The reason is that there is lack of the participation of farmers in the water transfer in China. The score of CMAWT in the "water-ticket" transfer in Zhangye city is only 0.587, which shows that the market degree is low in the water transfer. The reason is that the compensation method in this case is material compensation. And there is no participation of the third parties and the administration bureau of irrigation district. The scale of water right transfer is also small.

4 Conclusions and prospects

The classification theory and quantitative evaluation method of classification theory of CMAWT is developed in this paper. The mode of CMAWT contains the aspects of the vertical form and the horizontal form. The case study shows that this method can classify the type of CMAWT. The CMAWT of China is gradually improved. And the CMAWT of China has a certain gap compared with the CMAWT of Japan.

The research still has some limitations for the case study is limited in China and Japan. The classification of the CMAWT of other countries should be further studied to improve the classification method.

References

Chen Jing. The Basic Concepts and Classification of Water Management System [J]. China Water Resources, 2001(3): 25-26.

Dai Xiaoping. Research on the Compensation Mechanism and Models for Agricultural Water Rights Transfer [D]. Hohai: Hohai University, 2010.

Compensation Method of Occupying the Irrigation Water and Irrigation and Drainage Projects (2012. 5. 13) http: // info. nanning. gov. cn/n1066/n40217/n64945/n64986/n65235/3969563. html.

Some Opinions about the Water Right Transfer [DB/OL]. China Water Resource News (network edition) (2005. 1. 11) http: // 2004. chinawater. com. cn/newspaper/2b/20050314/200503140059. asp.

Yutaka Matsuno, Nobumasa Htcho, Soji Shindo. Water Transfer from Agriculture to Urban Domestic Users: a Case Study of the Tone River Basin, Japan[J]. Paddy Water Environment, 2007 (5): 239-246.

Reflection on Restructuring China's Water Right System —Based on an Analysis of the Current Situation of Water Resource Utilization in the Yellow River Basin

Min Xiangpeng[1], *Xu Yuchang*[1,2], *Li Xiehui*[1] and *Wang Zhaojin*[1,2]

1. Center for Yellow River Civilization and Sustainable Development, Henan University, Kaifeng, 475001, China
2. Guizhou Minzu University, Guiyang, 550025, China

Abstract: Currently, the Yellow River Basin is facing a most serious water resource crisis. The contradiction between supply and demand of water resources in the Basin is being aggravated by water pollution and water waste. The reasons for this are many, among which China's faulty water right system is a crucial one. This paper, with the Yellow River Basin as a point of penetration for study, commences with an exposition of the defects of China's water right system, i. e. the ambiguous concept of water right, the void of water right subject, and wrong positioning of the government role. Next, through an analysis of the concept of water right, water right subject system and government role positioning, the basic intension of water right is defined and the role of the government in handling water right - related matters is correctly positioned. Finally, using foreign water right management systems and their advanced experience for reference and applying Coase Theorem, and based on China's actualities, the water right allocation, China's water right management system and operating mechanism are elaborated with a view to rebuilding a water right system that accords with China's real conditions.
Key words: water right, water right system, system restructuring

As China's second longest river, the Yellow River is an embodiment of 5,000 - year - old Chinese civilization, hence dubbed affectionately by Chinese people as "Mother River". With the growth of the country's economy and modern industries, water diversion and water consumption in the Yellow River Basin are on the constant increase, making the contradiction between supply and demand of water resources in this region more intense. The Yellow River is confronted with a severe water shortage crisis, and water pollution and water waste make things even worse. As a consequence, the Yellow River water quality has been seriously deteriorated, and the function of water body is weakening or even losing. Here, taking water resource utilization in the Yellow River Basin as a point of penetration for discussion, this paper attempts to disclose the problems existing in China's water right system at great length in hopes of contributing to rebuilding a water right system which accords with China's actualities.

1 The characteristics of water resource utilization in the Yellow River Basin

1.1 Disorderliness

One of the serious problems on the utilization of the Yellow River water resources is water waste. A major reason for this is that the water resources utilization is in an disorderly state, which manifests itself as follows: unstatutoriness of the guideline, changeability of the governance system, randomness of the policy making, and multiplicity of the water resources users.

Over a long time, we have been lacking in a clear guideline and sound governance system for the Yellow River water resources utilization, and the policies on water governance are of a random nature. Coupled with the relative shortage of the Yellow River water resources due to its great demand, multiplicity and dispersion of its users, lack of unified planning and standardized operation, all these have led to a low rate of utilization of the Yellow River water resources and a most serious water waste.

1.2 Irrationality

Another problem on the utilization of the Yellow River water resources is its irrationality. In recent years, along with the rapid social and economic development, especially with the acceleration of urbanization and the growth of modern industries, the sewage and pollutants discharged into the Yellow River have increased significantly. Besides, may a water resources project has been constructed without subjection to scientific evaluation, and sewage disposal facilities lag far behind the times. All these have resulted in the quality of the Yellow River water decreasing drastically and the water body functioning much worse than before. Another manifestation of the irrationality in the Yellow River water resources utilization is lack of scientific operation and of overall planning of industrial layout along the banks of the Yellow River based on an evaluation of water quality. At the same time, inheriting a development pattern of low input, high consumption and heavy pollution, a legion of enterprises in the Yellow River Basin are accused of high water consumption and sewage, blaming for a serious damage to the body of the Yellow River and worsening water quality. According to statistics, in 2010, the Yellow River Basin sewage discharge reached 4.361×10^9 t, of which city residents living sewage discharge 1.159×10^9 t, second industrial sewage discharge 2.863×10^9 t, the tertiary industrial sewage discharge 0.338×10^9 t, accounting for 26.57%, 65.65% and 7.78% of the total respectively.

1.3 Inherent shortage of water resources

Apart from being the second longest in China, the Yellow River is also the only one running through the west to the east in North China. Nevertheless, owing to the relatively arid climate in its geographical location, the runoff water of its tributaries only accounts for 2.2% of the totality of all the rivers throughout the country, ranking the fifth among the country's seven great rivers. The average annual natural runoff of the Yellow River is 5.8×10^{10} m^3. It is forecast that, despite water saving in irrigation in the Yellow River Basin and related areas, the Yellow River basin will face a water shortage of 4×10^9 m^3 in 2010, and by 2030 the shortage will reach 1.1×10^{10} m^3, with a even worse water shortage in arid years. The situation is grim for the Yellow River in respect of water shortage, as the water demand has surpassed the load capacity of the River water resources, and it is still on the increase.

In short, with regard to the utilization of the Yellow River water resources, the disorderliness accounts for an overdue consumption, the irrationality for water pollution and the inherent shortage for a serious contradiction between water supply and demand. To tackle these problems, while we should make some technical efforts, China's water right system needs restructuring.

2 The defects of China's current water right system

2.1 The ambiguity of the concept of water right

The concept of water right is at the core of the water right system as well as the starting point at which the water right system is logically studied. Despite its fixedness and vulnerability, water resource is a necessity to sustain life, development and environment. Although "water right" is a long-standing term widely used in China's legislation and judicial practice, it has an ambiguous connotation in law till now. At present, the legal practitioners hold divergent views even on the concept of water right, which include the "usufruct only" doctrine, the "ownership plus usufruct" doctrine and the "multiple rights" doctrine. The ambiguity of the concept of water right inevitably leads to the void of water right subject and wrong positioning of government role in water resource management.

2.2 The void of water resource owners

Both China's Constitution and Water Law clearly provide that water resources are owned by the State and that the water of ponds and reservoirs belonging to rural economic collectives are owned by such collectives. From the perspective of legislation alone, the ownership of water resources seems to be clear and definite in China. However, both "state" and "rural economic collective" are set concepts with fixed but abstract meanings only. They have no specific legal personality, so they cannot act as legal bodies in legal operation or market bodies in economic activities. Only through specific executing bodies can they exercise their ownership in legal and economic activities; otherwise, their identities as legal and economic bodies are a mere formality.

As mentioned above, a legal person in the abstract sense can exercise its ownership in legal and economic activities only through a certain specific executive organ. So is the case with the state which enjoys the ownership of water resources. Therefore, China's Water Law clearly specifies that, "For water resources the State applies the system under which management of river basins is combined with management of administrative regions." "The administrative department for water resources under the State Council is responsible for unified management of and supervision over water resources throughout the country." "The administrative departments for water resources under the local people's governments at or above the county level shall, within the limits of their specified powers, be responsible for unified management of and supervision over the water resources." The Water Law is so stipulated as to make the administrative department for water resources under the State Council and the administrative departments for water resources under the local people's governments at or above the county level enjoy the ownership of water resources, but they impede the exercise of the ownership instead.

Accordingly, the empowerment of the administrative departments for water resources by the Water Law is without any nature of civil affairs, and it follows that the country's stipulation on the ownership of water resources remains only in name and the civil subjects exercising such ownership are still non-existent.

2.3 Malposition of the government role

Except for "water resources being owned by the State" and "unified management and supervision over water resources by the administrative department for water resources under the State Council and those under the local people's governments at or above the county level" as provided in China's Constitution and Water Law, there is not any stipulation on the exercise of water resource ownership. Such stipulations confuse administrative power and ownership and neglect the double identity of the State both as manager and as owner of water resources, causing disposition of government role as well as the void of water resource owners. This, in turn, has created a phenomenon of high cost and low efficiency in China's market operation in addition to a serious disorder in water resource management.

To sum up, as a result of the ambiguous concept of water right in China's water right system, water resources owners only in name and government role in water resources being dispositioned, China's water right operational mechanism lacks rational guidance and the administration and supervision of water resources are in great disorder, which, in turn, has led to a serious loss and pollution of China's water resources. Therefore, it appears to be of vital importance to establish an efficient water right operational mechanism and management system which meet market conditions.

3 The restructuring of China's water right system

3.1 The setting of water rights

In formulating any system, the setting of the subject's rights and obligations is essential, or rather, designing a system is nothing but setting the subject's rights and obligations. So, the author here proceeds from the setting of water rights with a view to rebuilding a scientific water right system in China.

3.1.1 The principles for setting water rights

As an embodiment of legal principles applied to setting water rights, the principles for setting water rights must be observed in the whole process of setting water rights. These principles are of great significance to the scientific, effective, just and rational setting of water rights. They are intrinsic to water right system itself. In my point of view, the principles for setting water rights are quite a few, but the ones of high value come as follows:

(1) Stability.

The stability of water rights is crucial to water right system. "A stable water right system can promote the investment on exploitation and protection of water resources. Stable water rights are a kind of assets, a warrant for creditability and appendages as well as an important factor for asset evaluation. In addition, stable and certain water rights and their effects help to understand current economic situation so as to prevent social unrest." According to some scholars, the stability of water rights, as a major principle of water right system, an be traced back to the law of ancient Rome.

(2) Effective and favorable use.

This is another important principle established in the Dublin Principles. As one bearing on people's livelihood, water resources are valuable in that they are rare but available. Accordingly, water laws of most countries in the world provide that water resource owners and users shall utilize water resources effectively in order to ensure the holding and exercise of water rights. For example, the law in the western states of the United States stipulates that water rights, when not in use, or not used in accordance with the permit or license, are at a risk of loss. According to The Dublin Principles, the principle of effective and favorable use is for the following purposes: ①to avoid acquisition of water resources out of speculation or waste (in actual uses); ②its end use must be accepted by the social public; ③prohibition of water abuse (i. e. to make appropriate and effective use of water) and ④as compared with other uses, this use must be rational.

(3) Effective restriction.

As a kind of resource for public utility, water resources are consisted in conservation of ecological environment exclusive of competition and monopoly for the interest of the social public, which requires the government to interfere and control. The non – effective and non – profit exercise of water rights , therefore, should be strictly restricted. The restrictions range a wide scope including effective use, prevention of harmful effect, compensation payment, preventive evaluation, designated management, remedies and payment of public management, in diverse forms such as restriction by license, by ban, by heavy penalty and pre – restriction and post – restriction. A typical example of post – restriction can be found in Water Law of Germany, which specifies that water rights will be canceled in case the license holder does not use, change use or not need it any longer. Whatever form it may take, the restriction is usually intended to protect the interests of the third party and the environment as well as effective utilization of water resources.

(4) Priority.

Water has manifold, desirable uses. It can be regarded as a kind of property in law and at the same time it is rare in terms of economy. In this sense, water can be a commodity, a commodity of unusual kind, because it bears on the environment, economy and society in an unique way. Owing to its multiple unique role, people has a variety of needs of it. In order to strike a balance between these needs, we should establish a priority system concerning water rights. One characteristic all countries have in common in their water law is that priority is established among different water uses to appropriately allocate water resources when in shortage, to grant water rights if applicants are in fierce competition, or to ensure fair access to water rights. For instance, priority is given to uses or occupations of water resources for the purpose of daily drinking.

3.1.2 Applying Coase Theorem to the water right setting

The Coase Thorem is a theory on the relationship among design of ownership system, transaction costs and efficiency of resource allocation in modern economics on ownership. It is also as an essential part of modern economics on ownership. The bulk of the theory is contained in Ronald Coase's paper entitled "The Problem of Social Cost" which was published in 1960. The term "Coase Theorem," however, was not advanced by Ronald Coase himself. It was drawn by Williamson, Steven Cheung, de M Gates and other scholars on the basis of summing up the main idea of the aforesaid paper, and was first adopted by Stiegler in his masterpice *Pricing Theory*. According to a widely accepted version of Coase thorem, provided that the ownership is clear and transaction costs are zero or very few, whoever is entitled to this ownership at the outset, an effective final result with a Pareto optimal resource allocation can be achieved through market equilibrium.

Most Chinese scholars will refer to Coase Theorem when studying water right system, but for the purpose of analyzing water right transactions rather than its setting. Admittedly, a few scholars have applied Coase Therom to the setting of water rights, but they havent done so thoroughly. Although Coase Therom is about the relationship among design of ownership system, transaction costs and efficiency in resource allocation, the key problem it addresses should be the arrangement of property right system and economic efficiency rather than transaction costs. Ronald Coase once said, "Once the market transaction cost is taken into account, the initial delimitation of legal rights will surely take effect on the operation of the economic system. A certain adjustment on rights may generate more value than other arrangements. " As far as Coarse Therom is concerned, I think it is conducive to the setting of water rights as follows:

Firstly, water rights should be clear, specific and certain. Water rights being clear means the ownership or usufruct of water resources must be clearly specified, they may go to enterprises, individuals or other organizations. Water rights being specific means they must be allotted to specific persons, that is to say, the water right subject should have a specific legal personality and actual interest, they are not vague or just in name. In addition, the above two aspects of water rights should be effectively carried out, with considerable certainty and stability. These three aspects are not only a basic requirement of water rights as a legal one, but a prerequisite for water resources to be traded on the market.

Secondly, under the condition of non – zero transaction cost, the initial setting of water rights is of critical importance, as it bears on the efficiency in water resource allocation and even overall social welfare. Therefore, in initially setting water rights, we must consider how to reduce transaction costs, the most effective way of reducing transaction costs being allocating water rights to those who can make the best use of them. By doing so, not only are the transaction procedures simplified, but optimal utilization of water resources is achieved. Accordingly, although water belongs to the public domain in most countries, the constitutions and state – or provincial – level legislations often grant the ownership or usufruct of water resources to certain natural persons or enterprises and this receives protection of property law.

Thirdly, If the initial setting of water rights fails to make water resource allocation achieve Pareto's optimization, market transactions can help make it, but they can only erase part of social loss caused by the inappropriate setting of water rights or due to changed conditions after the initial setting of water rights. From this, we can conclude that in the system of Coase Theorem, water resource trading system is an inherent part of water right system and that, as a useful, effective alternative to improving water right allocation, water right transaction should be attached importance to.

In the final analysis, Coase Theorem is about the relationship among the arrangement of property right system, transaction costs and optimization of resource allocation. It shows that an arrangement of appropriate property right system for water resource allocation is crucial to improving economic benefits and social welfare. This property right system for water resources should be definite and clear, and its initial setting is of vital importance. Transaction costs should be minimized, and if the initial setting of water rights fails to Pareto's optimization of water resource allocation, improvements can be made by virtue of water right trading system.

3.2 The management system of water rights

Water resources are rare and susceptible to reduce, yet indispensable to sustaining life, development and environment. Considering the unique characters of water resources, we must conduct effective management on water rights in line with the principle of restricting water rights. It is, therefore, indispensable to establish a scientific and sound water right management system. To establish such a system, an important thing is arrangement of its top – tier scheme, i. e. the arrangement of its management body and rules. However, at present, water resources owners only in name and mispositioning of government role pose a big barrier on the way to establish such a system. To overcome this obstacle, I think we must first distinguish between the double identities of the state as water right owner and as water right manager.

3.2.1 The differentiation of double identifies

In all countries where water resources belong to the state, the state has double identities both as water right owner and as water right manager. These two identities are different in nature: ①They are different in the property of right. The ownership the state enjoys as an economic body is a private right representing property rights in essence, whereas the right to manage it possesses as a manager is an administrative power, which belongs to the domain of public right. ②They differ in value orientations. As a public manager, the state should have political, economic and social targets such as political stability, national security, macroeconomic equilibrium, equal market competition, etc. As an owner, it should pursue optimal interests of water resource utilization. ③They adopt different means of intervention. As a public manager, the state intervenes in the market in a manner it behaves like an outsider as giants market transactors and competitors. By contrast, as an owner, the state participates in market activities through its assets, it is not only a market intervener, but a market transactor and competitor. Furthermore, they are distinct from each other in source, means of exercise, rule of operation and so on.

3.2.2 The separation of double identities

In countries practicing market economy, the two identities as a manager and as an owner are usually separated, that is, setting up a special organ to exercise ownership and making out particular rules for water right operation. In the process of our country's economic restructuring, the separation of these two identities can be made through detachment of organs and setting different rules.

By "detachment of organs", it means to detach the administrative organ from the organ in charge of ownership in order to change the state's dual identities. To this end, the function of ownership enjoyed by water resource authorities should be transferred to special agencies. This helps to achieve a complete separation of the state's dual identities. But in practice there exist at least two knotty problems: ①The important status and enormous amount of our country's water resources make it necessary to establish a powerful agency or even more to exercise the ownership. Whether the newly established agency can successfully assume its power or not depends on how its status is defined. If it is listed as a environmental department, the problem will remain unsolved that the function of ownership is dominated by that of management. If it is set on or above the same level of the government, not only is this at variance with common practices of other countries, but it is extremely costly to establish a colossal function system of ownership. ②In modern countries, the functions of the state at once branch out and merger, and new functions are arising at the same time. If we try to separate functions of water rights only through "detachment of organs", expansion of agencies will occur, which is unfavorable to the merger of functions.

By "setting different rules", it means that one and the same organ is allowed to possess dual functions as a public manager and as an owner, but different rules of operation are set respectively for these two functions. Each function is performed by the state in accordance with the specific rule set for itself. This is conducive to the merger of functions and the streamlining of agencies, as it avoids the detachment of organs. Its effect, however, is dependent upon a legal environment in which act in accordance with law is esteemed. In default of act in accordance with law, the two functions will not able to performed pursuant to their respective rule and according to their respective rule.

Based on the above analysis, we should conduct organizational construction and system design through "detachment of organs" and "setting different rules," in the light of the state's dual identities as a public manager and as an owner. The basic idea is, at the beginning of the establishment of the management system, we can implement "setting different rules" only and when conditions are ripe, we can adopt "detachment of organs".

3.3 The operating mechanism of water rights

A consensus has already been reached worldwide that water is a commodity. But "commercialization of water resources will be an arduous task in countries where Islam, Hindu or Chinese philosophy are preached". The trading of water as a commodity eventually will lead to the existence of water market, which is an important mechanism and means ensuring sound operation of water rights and optimal allocation of water resources. How, then, can we set up a market mechanism for the operation of water rights? In my opinion, the following should be addressed carefully.

3.3.1 A prerequisite for the water market: the system for water rights and water transactions

As a consumer goods requiring a high cost of maintenance, water is usually not traded by itself, but through its ownership certificate, namely, water rights. There are two obstacles in respect of system in China's water marketization for the present: One is that water right is not scientifically defined, the other is water right trading is not provided in China's Water Law. Therefore, we must provide a prerequisite in terms of system for water marketization by scientifically defining water rights, initially configuring water rights in a clear – cut manner and standardize water right trading by virtue of law.

3.3.2 The formation of water prices on the market

To charge or price on water is no easy job, as it is very hard to ascertain the concrete price representing the value of price. According to water resource experts, water has a relatively small economic value in respect of marginal profit. The value per unit of water for the initial use of a city may be very high, but that for the consequent uses may be quite low. In addition, water market is a "bleak" one in itself, with a rather small trading volume. It can never be counted as a standardized market in that water trading features clear and anonymous contractedness, prompt exchange and completeness at one time. Some people, accordingly, argue that water market is not a competitive one in fullness and does not necessarily reflect all trading costs.

At present, Charging or pricing on water mainly take the following forms worldwide: One is to pay water cost or protect the environment through charging fees. That is to say, charging or pricing depends on the cost to be paid or that used to protect the environment. So charging is made on water resource management cost, water resource related services and the cost of water resource protection, so as to ensure water resource regeneration in case of water resources being polluted owing to the deterioration of the environment. Another is to charge according to the opportunity cost of water, which is intended to link charge with the actual conditions of different areas, taking social and political conditions of varying users. This is a main source of money for the administration and investment of water administrative organs of the country. A third one is to charge on the basis of the value of the assets exploited by users. The actual amount of charging is evaluated on the basis of economic returns generated by the assets.

3.3.3 The oversight of water market

While we are cultivating water market, we can never neglect the oversight of water market lest the environment and society be harmed. However, in our country, there is no water market, nor oversight of water market. According to the consensus upon the Dublin Principles and the U.S. legislation and its legal practice, water market oversight mainly include: ①the principle of dependent rights, which aims at preventing land speculation through prohibition of land transfer without regarding water rights as dependent ones of land; ②any transfer shall be subject to the approval of legal court, legislative body or administrative organ (what particular organ it is depends on the law of the relevant state); ③to publicize the intention of transfer, which may provide interested parties or water right holders a collection of documents for expressing disapproval; ④to issue permits to reallocation or use, which subjects reallocation and use to the existing and emerging rules, including confirmation of completion of projects and profitable use; ⑤without prior approval, water rights may lose (fine is imposed on slight offences in some states); ⑥to place some restriction on the transfer of past consumptive uses; ⑦any transfer shall not harm the interests of all other uses, including minors, who are entitled to maintaining the condition of waters. Harms may arise out of one or more of the following: water amount change, time change, storage amount change, ways of diversion, water quality change, backflow loss, diversion place, etc. ;⑧to regulate water uses by reducing or preventing harm; ⑨compensation and payment of expenses; and ⑩to place restriction on transfer of past uses.

Besides above regulations, there is also "public interest review", which is applied to reviewing "the external effect of public interests" in examining applications of water right transfers. It includes: ①the impact of water uses on economic activities; ② the effect on fish, wild beasts and wild fowls as well as on public entertainment; ③the effect on public health; ④the opportunity cost of water use; ⑤the adverse effect on other people; ⑥the intention and capacity of water use; ⑦ the impact on access to public waters and routes; ⑧for hydraulic purposes; and ⑨ special local factors. Correspondingly, a redistribution is not permissible if it affects basic health, environment in violation of safety codes. Nevertheless, extra conditions such as requirement of reducing public worries may be placed on a redistribution, which can make it permissible. Though the general public have no doubt on the authenticity and legality of public welfare, they are suspicious about the manners of public interest review and its organizations. While showing no discrepancy on its administrative and judicial nature, some authoritarians argue that these manners and organizations should include water plans and public participation. Others for the sake of public interest review are assessment on the effect of transfer and evaluation on the environment of the transferee's place, on the levy basis, and on the impact upon local economy of the transferee's place.

References

The Yellow River Water Conservancy Committee of Ministry of Water Resources. The Bulletin of The Yellow River Water Resources for 2010[R]. 2011.

Miao Changhong. Sixty Years' Exploitation and Harnessing of the Yellow River[M]. Beijing: The Science Press, 2009.

Beck, Robert E Ed. Water and water rights, Vol. 1 - 3, Charlotte Sitwell:1991.

U. S. National Water Commission. Water Resources Plan[M]. Springfield: the United States Ministry Commerce, 1972.

Lin Guanzheng. On Slackened Control of Water Resources and Water Right System[M]. Beijing: China Economy Press, 2007.

Technical Advisory Committee for GWP:Miguel Solanes Fernando Gonzalez - Villarreal, 1996.

Study on the Han River Mode and its Implementation Way Based on the South – to – North Water Transfer Project

Peng Zhimin and *Tang Pengfei*

Hubei Academy of Social Sciences, Wuhan, 430077, China

Abstract: The South – to – North Water Transfer Project bring a series of negative effects in economic – social development and ecological environmental systems on the middle and lower reaches of Han River such as significant changes in water quantity and quality, spatial and temporal distribution, flow velocity and water level have undergone, reduction in capacity of its environment, and their development right, with huge ecological benefits, economic and social benefits to China. After careful study, careful calculation and multi – party discussion, the paper proposes the general framework of the Han River ecological compensation mode and implementation ways.
Key words: inter – basin water transfer, Han River Mode, ecological compensation, regional coordinated development

1 Introduction

The distribution of water resources in China is very unbalanced with its vast territory. Its southeast part lies in subtropical monsoon climate zone, abundant in water resources, while the majority of northwest and northern regions belong to the arid and semi – arid regions, very poor in water resources. Out of match among water and land resources, the industrial layout and population distribution leads to great difference on possession of water resources and its utilization in basins. Lots of facts, water resources utilization of Haihe River Basin as 90% , more and more rivers including Yellow River become seasonal, declining groundwater levels, drought as normal, show that water shortage has become a bottleneck in north China's economic and social development and ecological environment improvement. Inter – basin water transfer is inevitable based on this despite of problems. Large – scale, long – distance inter – basin water diversion projects are also expected to be constructed, with the urbanization of the population, economic development and water demand of the intensification.

Inter – basin water transfer is huge system engineering, not only related to various aspects of the environmental protection, economic development and social progress, but also affected current and future development of the water diversion area, receiving area and the affected zone. In one sense, whether a inter – basin water transfer project can be built or not, depends on the results of multi – stakeholder's game. In accordance with the requirements of the scientific concept of development, inter – basin water transfer project must adhere to the principle of people – oriented, co – ordinate harmonious development between urban and rural, regions, economic and social, man and nature.

2 Theory and practice of the river basin ecological compensation

2.1 Study of the river basin ecological compensation

Internationally, the concept of ecological compensation is not used as frequently as the general concept is PES: payments for ecological/environmental services, whose meaning is close to domestic ecological compensation. Many ecologists and economists have made important contributions to the ecological compensation research and theories of ecological economics, environmental economics, resources economics have laid foundation in the mechanisms of ecological compensation, especially the value assessment of ecosystem service function, externality theory and public goods theory.

The rapid economic development enhances the development and utilization of river, accordingly, the researches on river basin ecological compensation have caught the widespread attention, which mainly focus on the aspects as follows.

Firstly, researches on the overall framework of basin ecological compensation, which are devoted to its connotation, theoretical basis, as well as the principle, type, standard, mechanism and its policy design and other aspects from the theoretical level and the overall framework.

Secondly, ecological compensation researches in terms of interregional and inter - basin water transfer. Most of the researches in this field probe into the typical case of the South - to - North Water Transfer Project, while a few deal with case studies, such as the practice of and research on transferring Yellow River water to Tianjin and Liaocheng City, Shandong Province.

Thirdly, researches on the impacts of South - to - North Water Transfer Project on the middle and lower reaches of Han River and ecological compensation research. Whereas some are the impact studies of water transfer on ecological, economic, social perspectives, based on the investigation of current situation, hydrology and water resources, economic theory in the middle and lower reaches of Han River, others contribute to the theoretical analysis of water compensation mechanism, the basic principle and necessity as well as compensation measures in South - to - North Water Transfer Project.

As the theory of river ecological compensation has just emerged in China, problems arise such as diverse concepts, multiple standards, and restrictions to foreign theory introduction, inadequate case analysis and poor maneuverability, not to mention the inadequate studies on inter - basin water ecological compensation and relevant laws, regulations as well as policy.

2.2 Practice and exploration of domestic and international ecological compensation

In the world, developed countries such as the United States, Australia and Canada, as well as those vast countries, inclusive of Russia, India, Pakistan, South Africa have accumulated many successful experiences in river basin ecological compensation on account of early development and utilization of rivers. Among these countries, there is one important common point of attaching great importance to the issue of environmental protection, in the planning and management of the inter - basin water transfer project. For example, the United States has constructed an advanced market system of river basin ecological compensation and adopts various measures to protect water resources including bills of exchange mechanism in order to standardize the management of river basins in the form of market. Australia attempts to promote the provincial integrated river basin management through the federal government subsidies. South Africa incorporates the river - basin preservation and recovery actions with the initiative of helping the poor, spending about $ 170 million in hiring the vulnerable groups to protect river - basins, improve water quality, and increase the water supply. However, most states combine river - basin protection compensation with forest environmental services and in the implementation of the corresponding compensations, government only plays an intermediary role, giving way to the market mechanism.

The practice of river basin ecological compensation began in 1980s in China, when the concept of ecological compensation theory has just been introduced while the mechanism and system were not perfect, so exploration to the problem stranded in this period. Shanghai became the pioneer in exploring COD emission trading. During 1987 ~ 2002, Minhang District has implemented a total of 40 water pollution discharge transaction. Since twenty - first Century, as the environment problems get more and more serious and ecological compensation theory gradually matures, both the central and the local governments successively have made a series of laws concerning ecological environment protection and ecological compensation. Some areas made a variety of attempts and beneficial explorations on the ecological compensation according to these regulations, from which derive some typical models of the domestic ecological compensation, such as Ziya River Basin Ecological Compensation withholding mode, Dongyang - Yiwu water right transaction mode, Sanjiangyuan and Danjiangkou reservoir area ecological compensation mode and so on. In the light of the subjects of implementation and the different operational mechanisms, these models can be divided into two types: the government compensation and the market compensation. At present, government compensation, as one of the most important ways of compensation, has been applied most extensively and achieved obvious effects while the market compensation, sporadic and scattering in the local area is still in the exploration stage with respect to river basin ecological compensation. As a useful complement to the governmental compensation, market mode is expected to play a more and more important role in the basin ecological compensation.

2.3 The trend of China's river basin ecological compensation

Firstly, inter - basin water transfer is one of the inevitable choice to solve water imbalance based on unbalance between domestic water distribution and layout of the national economy. As Chen Lei, Minister of Water Resources of China, pointed out in 2010 at the National Water Resources Planning Program Work Conference , we should tackle the problem of water supply in areas of water resources shortage in the way of the rational allocation of interregional and inter - basin water resources. Secondly, absorbing and drawing lessons from experience worldwide, river basin compensation will become an important guarantee for the national ecological environment and sustainable development in view of the uncertainties of long - term impacts on ecological environment. Thirdly, from the point of view of the planning and constructing inter basin water transfer projects, formidable industrial upgrading and water - saving society construction, the scarcity of water resources and the rising in the water values, inter - basin water ecological compensation standard will also rise gradually, taking it into account that the rate of progress of national economy will be relatively high for a long time.

3 Influence to the middle and lower reaches of the Han River and its losses

3.1 General situation of south - to - north water transfer project and the lower reaches of the Han River

The middle route of the south - to - north water diversion project, which supply water to the Huang - Huai - Hai Plain from the Danjiangkou reservoir, across the Han River, Huaihe River, Yellow River, Haihe River Basin, an average annual water diversion scale of 95×10^8 m^3, mainly to solve the North China including Beijing, Tianjin, Hebei and Henan provinces (municipalities) with severe water shortages. According to the price level of the third quarter of 2004, static investment of the first - stage construction is 201.3×10^9 yuan, of which the main project was 136.5×10^9 yuan, pollution control engineering (soil conservation) was 70×10^9 yuan. The main project of this project is composed of three parts of water engineering, the main canal project and the Han River treatment. Among them, the middle and lower reaches of Han River treatment engineering is to alleviate the ecological and environmental problems brought about by the project to the middle and lower reaches. However, this engineering can only solve some of the problem in some river sector and part time of Han River In order to mitigate its fundamental negative impact, it is necessary to establish a long - term mechanism.

Han River Basin covers the northwest and central part of Hubei Province, including 28 counties, municipalities, districts involved in the Administrative Region of the total area of 6.79×10^4 km^2, accounting for 36.5% of the total land area. Han River region is an important economic corridor in Hubei Province, automotive, electrical, mechanical, chemical, building materials, electronics, textile, food - led industry to flourish, is the province's automotive industry corridor, equipment manufacturing, textile and garment production base, the province's major commodity agricultural base, occupies an important position in modern agriculture development. Han River Basin gathering a wealth of elements of resources, the amount of water than the rich, resource - rich land, mineral range is one of the important mineral area in Hubei Province. Its deep historical and cultural heritage is an important birthplace of the Chu Culture, is the province's important ecological and cultural tourist destination.

3.2 Influence on the middle and lower reaches of Han River

As the south – to – north water division goes, 24.5% of the storage of runoff water from the Danjiangkou Reservoir will be diverted, which will have a significant impact on socio – economic and ecological environments of the middle and lower reaches, giving rise to new social problems and environmental and ecological issues.

3.2.1 Environmental impact

The south – to – north water transfer project first brings problem to the ecological environment in the middle and lower reaches of the Han River, which is mainly presented the following aspects:

Firstly, from the hydrological perspective, the south – to – north water transfer project will reduce both the long – term average river flow and flow velocity as well as the water level in the important hydrological stations in the middle and lower reaches of Han River, especially in the middle reaches of the Han River area (from Danjiangkou to ZhongXiang section), which with no compensation of water from the Yangtze River, become severely affected by south – to – north water transfer project.

Secondly, from the perspective of water quality, the potassium permanganate, ammonia nitrogen and phosphorus concentration in Han River and tributaries will increase, leading directly to the water environmental capacity deterioration. The four treatment projects are just for the compensation of the water, which will not ease the negative effects the south – to – north water transfer project will bring to the water quality in the middle and lower reaches of Han River.

Thirdly, from the perspective of ground water resources, the south – to – north water transfer project may produce adverse effect on Xiangfan regional level 1 terraces and second terrace front which have water and material exchange relationship with Han River, manifested in the falling of the underground water level and the increase of the land supply to the river.

3.2.2 Economical and social impacts

In terms of economic and social considerations, the Middle Route Project of Han River to bring some favorable factors, such as increased flood protection standards of the Han River region, to improve the irrigation of some areas of the Han River. However, after the water transfer makes the Han River water supply and demand balance contradiction is more prominent, the loss in terms of economic, industrial, financial, safety of drinking water and development permissions is obvious.

Middle and lower reaches of agricultural production has been greatly affected the impact of agricultural irrigation facilities, agricultural planting structure, fishery production, sustainable agricultural development, The benefits of water consumption leading industry affected small businesses the water industry will be eliminated, the industrial restructuring is difficult, which will promote new industrial layout. The constraints of the development of these industries will directly affect the speed of economic development, resulting in reduced revenue. At the same time, seriously affecting the effectiveness of the shipping industry and shipping companies, shipping guaranteed rate is reduced by 30% ~ 60%. Urban and rural drinking water safety, the majority of waterworks, industrial and mining pump station on the water intake modification or extension of a direct impact on the Hanjiang River in 28 counties (cities, districts), urban and rural water supply. Rural and urban development has been built on the local impact in major projects and projects to build the proposed project effectiveness and development is restricted; impact on town planning layout and urban structure, regional development is limited, the right to development, development expectations and the scale was significantly lower than before the state is not diverting water.

3.3 The composition and calculation of the ecological compensation in the middle and lower reaches of Han River

Ecological compensation in the middle and lower reaches of Han River is mainly composed of loss of water environmental capacity, loss of economic social value and environmental functions value.

It is estimated that ammonia nitrogen water environmental capacity loss reaches 1.044×10^4 t/a in the middleand lower reaches of Han River compared to 2007, the year before the water transfer. Adopting the shadow engineering method and considering the COD management cost and transaction price, we can predict the economic values of COD and the ammonia nitrogen environmental capacity loss every year. The core of the social and economic value loss is the direct social and economic value loss caused by reduction of water resources, measured by the System Dynamics method, combined with the willingness to pay for water resources to study the value impact of water resources on the social economic system in the middle and lower reaches of Han River. As to the transformation of waste plants and agricultural pump station, the closing of enterprises, shipping economic loss, we should measure the cost according to the direct losses or project demand. In addition, the quantification and calculation of the development right of an area influenced by large construction projects is an important theoretical problem, which leaves much room to study, especially as far as inter - basin water transfer project is concerned, as the book Han River mode: inter - basin ecological compensation mechanism only considered the influence of lack of water in the GDP in spite of its exploration into the quantification of development right.

4 The connotation and framework of the Han River mode

4.1 Connotation of the Han River mode

The main connotation of the Han River mode of ecological compensation is that; ①the main body composed of the water diversion area, receiving area and the affected zone; ②extending several provinces geographically and involving different departments practically; ③the method of more comprehensive and integrated, including central fiscal transfer payment and horizontal fiscal transfer payment, ④a large amount, a wide range and a long term.

4.2 Characteristics of the Han River mode

Firstly, Stakeholders turn into three parts from two. The traditional mode thought that stakeholders of water diversion were only composed of the water transfer area and receiving area. However, the Han River mode involves three stakeholders which are the water diversion area, the receiving area and the affected zone, which is put forward first time while it often became "the blind spot" of ecological compensation and vulnerable groups. The water diversion area includes Danjiangkou reservoir area and the upstream region, the receiving area is the region where people use the water, and the affected zone is the middle and downstream of Han River.

Secondly, the allocation of water rights has been redefined. Based on the needs of water quality and aquatic ecosystem protection, the river basin ecological compensation for water source redefines the emission rights of waste in water source and different sections of the downstream, compensates the main protector of water quality and aquatic ecosystem of the source area, and charges the beneficiaries and the vandals of water quality and aquatic ecosystem of the downstream. Different from this mode, the Han River Mode needs to re - define not only the emission rights, but also the water - drawing right and the water - using rights in other rivers.

Thirdly, the dominant sectors of the eco - compensation are complex. The ecological compensation of upstream and downstream is often dominated by the environmental protection department, through taking the form of monitoring water quality of sections to ensure the implementation of ecological compensation. The opportunities produced by the Han River mode lie in that the South - to - North Water Transfer Project is the cross - basin water transfer project with the largest scale and affection so far within our country. It involves a lot of areas and departments, which forms a complex interest relationship. Thus, more widely consultation among central government departments, the water diversion area and receiving area is required to obtain further recognized effects.

Fourthly, the function of ecological compensation is more comprehensive and the accounting standards have been improved. The goal of the Han River mode is not only to protect the water quality of the water source, but also the environmental capacity of the middle and lower Han River, the water demand of socio – economic, shipping. In addition, the Han River mode has also focused on the consideration of the hidden cost of the affected zone such as the development permission, making the inspection of compensation function more comprehensive.

Fifthly, the river basin ecological compensation has a wider spatial and temporal scale. In the ecological compensation of the Han River model, not only the inter – provincial (Hubei Province and the receiving area) river basin ecological compensation but also the river basin ecological compensation among municipalities and counties in the province is involved . What the inter – provincial river basin ecological compensation emphasizes is the communication and coordination between provinces, which needs the regulation of the central government building a platform for the inter – provincial river basin ecological compensation and clearing the responsibility and obligation between Hubei Province and other provinces in the receiving area.

Sixthly, the compensation mode is more diversified. The influence of South – to – north water transfer Project on the middle and lower Han River can't be reflected by water quality monitoring of a section of the river basin fully, and the economical and social impact caused by water diversion must also taken into account. Therefore, the compensation for the middle and lower reaches of Han River cannot just rely on a certain index. The evaluation and allocation of the multi – value items only relying on the government is apparently inadequate, so we can reach a set of compensation standard or agreements through the market means or negotiations as a supplement.

5 Path of Han River mode

5.1 Starting the program of formulating relevant laws and regulations

Han River mode will be of greater significant with the construction and completion of a number of large inter – basin water transfer project including the south – to – north water transfer project. This calls for the legislation on inter – basin water transfer and related issues at the national level, and promoting the establishment of the ecological compensation standard, and enhance the improvement of the fiscal and taxation law and corresponding revision. Local governments also needs to promote the relevant local laws and regulations in order to ensure water safety and river health, set up and perfect the laws concerning inter – basin water transfer and related issues, integrate the ecological compensation and incorporate the implementation of environment compensation and water right compensation. Meanwhile, passive and unusual compensations should be transformed into active and usual compensations. In addition, the system to assure the proper allocation and use of the ecological compensation funds as well as the relevant local laws on finance and tax should be promoted.

5.2 Promotion of integration configuration in water intake right and sewage right

The middle route of south – to – north water transfer project not only sets higher demands for the water quality of water source but may also have some negative effects on the water quality in the middle and lower reaches of Han River. The integration configuration of water intake right and sewage right is the prerequisite and basis of Han River mode.

5.3 Speeding up the formulation of water allocation scheme and the construction of the regional water right

Water allocation scheme is a part of water right system and core to implement water right system. Therefore, speeding up the formulation and implementation of water allocation scheme is the basis of scientific configuration of initial water right in all areas in the basin. Based on this, I suggest the Ministry of Water Resources should make relevant regulations as soon as possible and clarify the guidance principle, scope, procedure and types of the initial water right allocation in the basin and establish a complete set of water right allocation system in the areas. Meanwhile, the Ministry should also make the regulations concerning water right transfer between areas and funds allocation and funds use after making profits in order to ensure the smooth implementation of ecological compensation in interbasin water transfer.

5.4 Establishing the assessment system of ecological compensation

According to the basic theory in economics, an important factor to judge the value of water resources in an area is the opportunity cost of the local water resources. With the development of social economy, water resources gradually become the local rare resources and the opportunity cost gets higher and higher. Therefore it is necessary to set up the corresponding assessment system so as to assess the amount of compensation on ecological environment on a regular basis and adjust the amount of compensation. In addition, environment and climate changes on the globe will also influence the ecological compensation in Han River basin. The middle route of south - to - north water transfer project aims at the situations of water resources in which the south has plentiful in water but the north is inadequate. But when both areas influenced by climate changes, suffer from the same water problems, for example both are dry or flooding, the water transferring mode will vary accordingly in the south - to - north water transfer. As a result, it is necessary to adjust the ecological compensation regulations and amount of compensation.

5.5 Establishing the accounting technical standard for ecological compensation in interbasin water transfer

At present, with more and more cases on ecological compensation in interbasin water transfer, the related researches on ecological compensation accounting and values of water resources are maturing. Therefore, departments concerned may consider to make the technical guidance and principle on ecological compensation accounting first, then publicize the accounting technical standards when it is the right time, which standardize the ecological compensation accounting on a comprehensive basis from the principle, type, index and procedure and offer a objective reference to the following negotiations regarding ecological compensation in interbasin water transfer.

5.6 Perfecting the engineering management system and system in interbasin water transfer

The water - transferrable principle with regard to the water transfer area, the water - transferring principle concerning the water receiving area and the principle of not influencing the third party should be established and we should manage the water transfer in the interbasin according to the features of the water transfer project and the scale. We are supposed to set up a united management organization for interbasin water transfer, clarify the duty of the departments and implement supervision and management on various stages of water transfer such as planning, water intaking, using water and compensation. We should clarify the compensation mechanism to offset the effects of ecological environment, and losses of people, groups and enterprises that develop and use water resources in the water source and lower reaches in the basin. In addition we should make explicit the compensation mechanism for all kinds of losses along the route of the water transfer project.

5.7 Overcoming the barrier of horizontal transfer payment

At present, the governmental transfer payment is the onefold longitudinal transfer payment system with no official, standard and horizontal transfer payment system, although the central government has been encouraging the provincial governments to help one another. We can attempt to set up the "provincial financial fund of balance" between the provinces and a special organization which is independent from the local governments and responsible for its operation. For example, we can launch a special fund and it is the fund that is in charge of the overall arrangement of the ecological compensation transfer payment in interbasin water transfer. The fund is promoted by the local governments and operated in the form of market.

5.8 Establishing the long - term mechanism of ecological compensation

At the level of the central government, the financial transfer payment system should be established to make ecological compensations in the middle and lower reaches of Hanhe River. The standard of ecological compensation accounting should be publicized as the basis of ecological compensation. At the level of local governments, it is necessary to come up with relevant systems concerning conservation and use of water resources, make related management initiative and take effective measures. Meanwhile the local governments should formulate the regulations on the management and use of ecological compensation funds so that it can play the active role in the conserving the water ecological environment and bringing happiness to the people on both sides of Hanhe River.

References

Peng Zhimin, Zhang Bin. Han River Model: the New Interbasin Ecological Compensation Mechanism[M]. The Everbright Daily Press, 2011.

China 21st Agenda Management Center. Ecological Compensation: International Experience and Chinese Practice[M]. Beijing: Social Science Literatures Press, 2007.

Wang Jinnan, Zhuang Guotai. Ecological Compensation Mechanism and Policy Plan[M]. Beijing: Chinese Envrionmental Sciences Press, 2006.

Li Shantong, et al. South - to - north Water Transfer Project and Chinese Development[M]. Beijing: Economic Science Press, 2004.

Ruan Benqing, Xu Ffengran, Zhang Chunling. The Reserch Progress and Practice of Ecological Compensation in the Basin[J]. University Journal of Water Resources, 2008.

Chen Wei, Liu Yulong, Yang Li. Features of Ecological Compensation in the Basin and Strategies [J]. Water and Electricity Energy Science, 2010.

Li Hao, Liu Tao, Huang Wei. The Across - regional Conflict Motivation in Water Resources and Coordination[J]. Natural Resources University Journal, 2010.

Zhao Xia. Setting up the Ecological Compensation Mechanism in the Middle and Lower Reaches of Han River and Strategies[J]. Hydrauli Economy, 2010.

Peng Zhimin. Integrated Management and Regional Coordinated Development of Han River Based on the South - to - North Water Transfer, Proceedings of the Fourth Yangtze Forum[M]. Wuhan: Changjiang Press, 2011.

The Variation Characteristics of Ice Flood and its Influencing Factors in the Inner Mongolia Reach of the Yellow River

Wang Ping, *Li Ting*, *TianYong* and *Hu Tian*

Yellow River Institute of Hydraulic Research, Zhengzhou, 450003, China

Abstract: This paper analyzed the variation characteristics of ice flood and its influencing factors in the Inner Mongolia reach of the Yellow River on the basis of the field data. The research results indicate that ice flood in the Inner Mongolia reach of the Yellow River has the characteristics of increasing along the river, rapidly rising and steeply falling, high peak and small flood volume, short duration and high water level. The reach below Sanhuhekou is the main section where ice flood peak gets magnified. Since Liujiaxia Reservoir and Longyangxia Reservoir are put into operation, the flood peak and volume of ice flood obviously increase at Toudaoguai section of Yellow River with its highest water level in an upward trend. The increase of the channel-storage increment in the reach between Sanhuhekou and Toudaoguai is the main reason that leads to the increase of ice flood peak and volume. The reaches below Sanhuhekou break up late with relatively intensive ice breakup dates, intensive release of channel storage increment and great ice flood peak increments. During the ice breakup process, the unstable rise of temperature has great impact on ice flood. The rapidly increase in temperature during the ice breakup process will lead to the breakup of these reaches in a short period, intensive release of the channel - storage increment and higher ice flood peak. The slowly increase in temperature will result in dispersed ice breakup of these reaches, separate release of the channel-storage increment and lower ice flood peak. Since the 1990s, high winter temperatures lead to the reaches above Sanhuhekou breaking up in advance and prolonging the ice breakup process. Thus, although the channel - storage increment increases in the reach between Bayangaole and Sanhuhekou, the ice flood peak increment at the section does not increased. River channel sedimentation and decreased carrying capacity are the main reason that leads to the rise of the water level of ice floods.

Key words: ice flood, flood control, channel-storage increment, the Inner Mongolia Reach of the Yellow River

1 Overview

The Inner Mongolia reach of the Yellow River is located in the northernmost Yellow River basin, running from Mahuanggou of Shizui Mountain, Ningxia Province to Yushuwan, with a total length of 840 km (Fig. 1). From Shizuishan to Bayangaole, the reach flows south to north. The river reach runs northeast from Bayangaole to southeast Sanhuhekou, then turns to Toudaoguai by the direction of west to east (slightly by south) and finally below Toudaoguai in general flows to the south. The region that the reach flows through has an obviously continental climate. In winter, the weather in this region is dry and cold. Each year the frozen period of this reach lasts about four months. Generally, in the middle and last 10-days of November, floating ice begins to occur in the river. In the first 10-days of December, the river begins to freeze up and in the middle and last 10-days of March; the river begins to break up of the ice (Fig. 2). As this river reach spans a wide latitude zone, the difference in thermodynamic factors that are formed by the difference of latitude results in drift ice and freeze-up of the Inner Mongolia reach firstly occurring in the reach between Sanhuhekou and Toudaoguai and then extending from this section to upstream and downstream. Usually, ice breakup firstly occurs in the reach above Sanhuhekou, and then extends from upstream to downstream. Freeze-up from downstream to upstream and ice breakup from upstream to downstream will lead to special ice flood conditions such as ice jam in freeze-up and ice breakup periods. In that case, the water level of river channel increases rapidly and then ice flood disaster happens. Especially in the ice breakup period, the channel-storage increment releases intensively and the flow along the river increases, which jointly forms the ice flood peak. The river current carries a large amount of shattered ice pieces to downstream. In the region, when the shattered ice pieces pass, the river water level rises and the ice cracks. The increasing density of the ice, flow resistance and rising water level threat the security of dikes. With the same discharge, the water level of ice flood peak is usually higher than that of summer floods. There is a greater risk in dike breakdown in the ice flood period. In history, more ice flood disasters occurred in the Inner Mongolia reach than that summer floods. It is known that "summer flood easily prevented and ice flood hardly defended". Based on the field data, the variation characteristics of ice flood and its influencing factors in the Inner Mongolia reach of the Yellow River are analyzed in the paper.

Fig. 1 The plan sketch of the Inner Mongolia reach of the Yellow River

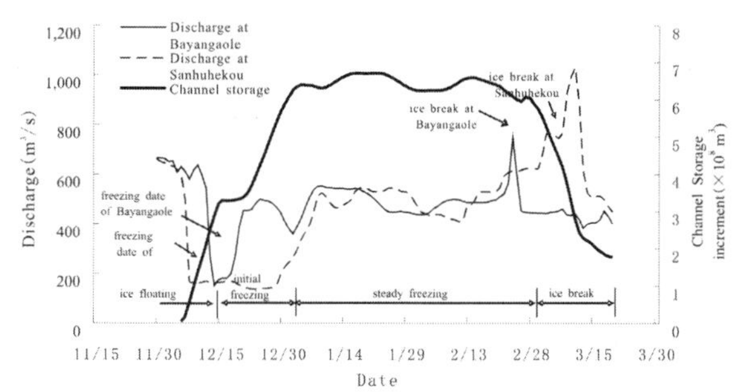

Fig. 2 The channel-storage increment and discharge process of ice season in the reach between Bayangaole and Sanhuhekou (2002)

2 The characteristics of ice flood

When the river ice breaks up, channel-storage increment will release along the way. The increase in the discharge along the way forms the ice flood process. As the ice floods are different from summer floods which are formed by upstream rainfall, ice floods have different characteristics. One is that ice floods increase along the way. But in flood season the floods in the Inner Mongolia reach have few convergent flows and distributaries along the way with no obvious increase or decrease in flood discharge along the way (shown in Fig. 3 and Fig. 4). The discharge of ice flood increases obviously in the reach below Bayangaole. According to the statistics on values of ice flood peak of all stations in the Inner Mongolia reach from 1955 to 2005, the average annual values of ice flood peak of Shizuishan station, Bayangaole station, Sanhuhekou station and Toudaoguai station are 815 m^3/s, 800 m^3/s, 1,352 m^3/s and 2,282 m^3/s, respectively (Tab. 1). The average value of ice flood peak generally does not increase in the reach from Shizuishan to Bayangaole. The average ice flood peak volume increases by 552 m^3/s in the reach from Bayangaole to Sanhuhekou. It increases by 930 m^3/s in the reach from Sanhuhekou to Toudaoguai. The maximum 10-days flood volumes at Bayangaole station, Sanhuhekou station, Toudaoguai station are 4.5×10^8 m^3, 6.9×10^8 m^3 and 9.4×10^8 m^3, respectively. Another characteristic is that the ice floods rise rapidly and fall steeply with high peak, small flood volume and short duration. The channel-storage increment releases intensively within a few days, which will lead to the ice floods rising and falling rapidly. The ice flood in the reach of Toudaoguai will generally last 10 days or less. The amount of its ice flood peak is generally above 2000 m^3/s. Sometimes, it can reach above 3,000 m^3/s. The maximum 10-day flood volume in the reach of Toudaoguai is between 5.00×10^8 m^3 and 1.5×10^9 m^3 (from 1955 to 2005), with average annual flood volume of 930×10^8 m^3. The third characteristic is that the ice flood has a high water level. In ice breakup period, although the ice sheet melts and cracks, a large amount of shattered ice bodies usually existing in river channel and shoal, which leads to a relatively large flow resistance. Under the same flood discharge, the water level of ice flood peak is apparently higher than that of the flowing freely period (Fig. 5), which brings a great threat to ice flood control.

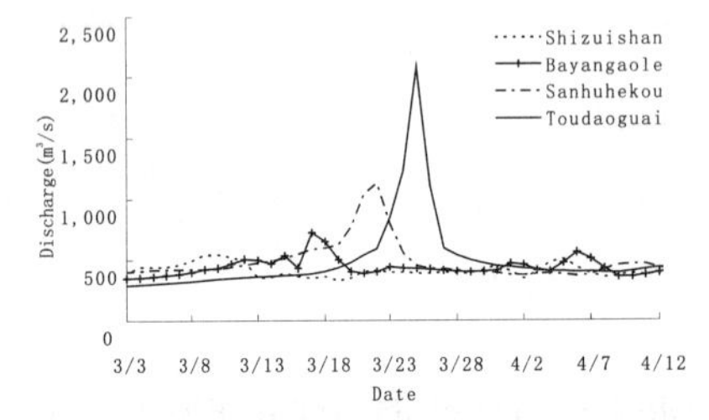

Fig. 3 The ice flood process in March, 1978

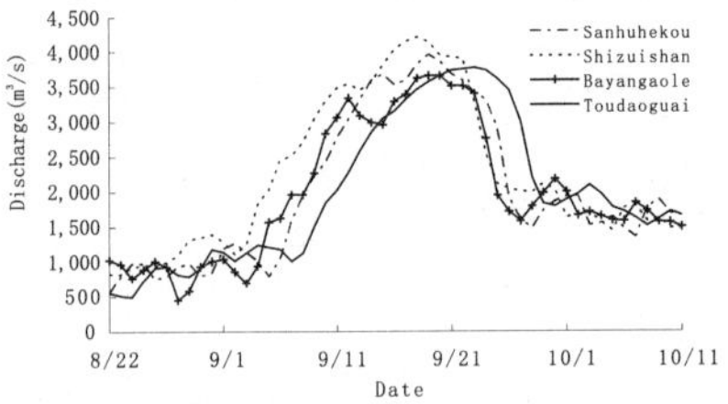

Fig. 4 The flood process in September, 1978

Tab. 1 The eigenvalue of flood peak and volume of all stations in ice flood seasons from 1955 to 2005

Program	Shizuishan	Bayangaole	Sanhuhekou	Toudaoguai
Peak Maximum	1,700	1,890	2,220	3,500
discharge Minimum	422	452	708	1,000
(m^3/s) Average annual	815	800	1,352	2,282
Average Maximum 10-dayS flood volume ($\times10^8$ m^3)		4.5	6.9	9.4

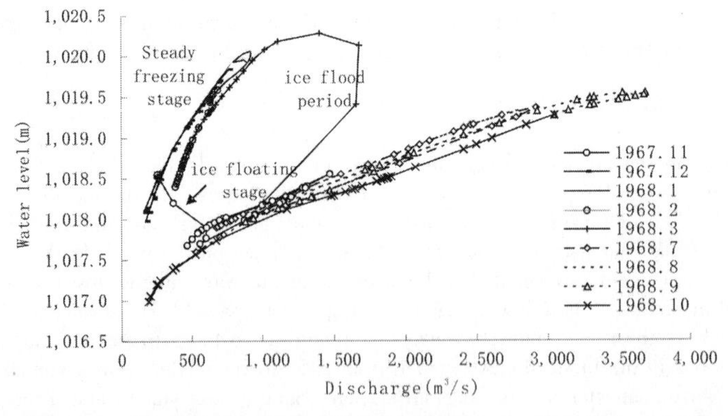

Fig. 5 The stage-discharge relation at Sanhuhekou section of Inner Mongolia

3 The variation of ice floods in different periods

Ice flood of different periods presents its variation characteristics. As reservoirs have been built successively in the upper Yellow River, especially Liujiaxia Reservoir with strong regulation (operated in 1968) and Longyangxia Reservoir (operated in 1986), the operation of these reservoirs changes natural runoff and sediment of the Inner Mongolia reach. The statistical analysis of ice flood is conducted based on the completion time of reservoirs. Ice flood has the following variation characteristics.

(1) The discharge of ice flood peak at Toudaoguai station increases while the discharge of ice flood peak at Shizuishan station, Bayangaole station and Sanhuhekou station decreases (Fig. 6, Tab. 2). As only Liujiaxia Reservoir puts into operation (from 1969 to 1987), annual average ice flood peak at Toudaoguai station increases by 350 m^3/s (from 1955 to 1968) while the discharge of ice flood peak at Shizuishan station, Bayangaole station and Sanhuhekou station decreases by 190 m^3/s, 201 m^3/s and 241 m^3/s respectively. As Longyangxia Reservoir puts into operation (from 1987 to 2005), the average ice flood peak of these stations slightly decreases with small difference, compared with that when only Liujiaxia Reservoir operates.

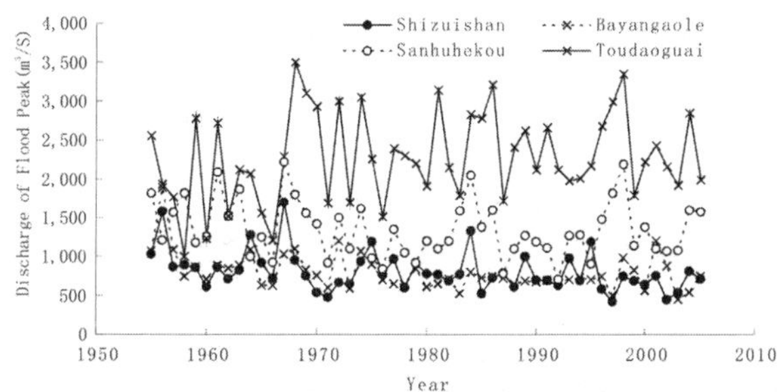

Fig. 6 Discharge of the ice flood peak at each station of the Inner Mongolia reaches (1950 ~ 2010)

Tab. 2 Average ice flood peak and interval increment at each station in different periods

Periods (year)	The average flood peak (m^3/s)				Interval ice flood peak increment (m^3/s)			maximum 10-days flood volume of Toudaoguai ($\times 10^8 m^3$)
	Shizuishan	Bayangaole	Sanhuhekou	Toudaoguai	From Shizuishn to Bayangaole	From Bayangaole to Sanhuhekou	From Sanhuhekou to Toudaoguai	
1955 ~ 1968	972	955	1,540	2,091	-17	585	551	7.51
1969 ~ 1986	782	754	1,299	2,441	-27	545	1,142	9.08
1987 ~ 2005	719	722	1,266	2,326	4	544	1,060	10.93

(2) Ice flood peak increment in the reach from Sanhuhekou to Toudaoguai increases obviously. From 1955 to 1968, the annual average ice flood peak increment is 551 m^3/s. The ice flood peak increment in the period of 1969 ~ 1987 and 1987 ~ 2005 is 1,142 m^3/s and 1,060 m^3/s respectively, increasing by almost one-fold. Slight changes occur in the annual average ice flood peak increment in the reach between Shizuishan and Bayangaole, and between Bayangaole and Sanhuhekou.

(3) In the reach of Toudaoguai the ice flood peak with small discharge decreases while the ice flood peak with medium and high discharge increases. In the period of 1969 ~ 1987, the ice flood peak with the discharge of less than 1,500 m^3/s does not occur while the ice flood peak with a large discharge of more than 3,000 m^3/s increases obviously (Tab. 3). In the period of 1987 ~ 2005, the ice flood peak with a large discharge of above 3,000 m^3/s decreases while no ice flood peak

with the discharge of below 1,500 m^3/s occurs. Meanwhile, the ice flood peak with the discharge of 2,000 ~ 3,000 m^3/s increases apparently.

(4) The volume of ice floods increases in successive periods. The maximum 10-days flood volume to represent the volume of the ice flood is chosen for analysis. In the period of 1955 ~ 1968, before the construction of reservoirs, the average maximum 10-days flood volume in the reach of Toudaoguai is 7.51×10^8 m^3. In the period (1969 ~ 1987) when only Liujiaxia Reservoir operates, the average maximum 10-days flood volume in the reach of Toudaoguai increases to 9.08×10^8 m^3. After Longyangxia Reservoir put into operation (1987 ~ 2005), the average maximum 10-day flood volume in the reach of Toudaoguai increases to 1.093×10^9 m^3.

(5) The water level of the ice flood peak presents an upward trend. Since the mid-1980s, the highest water level of the ice flood peak emerges in an upward trend (Fig. 7). In 2008, the water level of the ice flood peak reaches 1,021.11 m. Compared with the water level of 1986, it increases by 1.47 m.

Tab.3 Frequencies of ice flood peak of different discharge stage of Toudaoguai in different periods (Unit: times)

Periods (year)	Discharge stage (m^3/s)				
	<1,500	1,500 ~ 2,000	2,000 ~ 2,500	2,500 ~ 3,000	>3,000
1955 ~ 1968	3	4	3	3	1
1969 ~ 1986	0	5	5	3	5
1987 ~ 2005	0	5	8	5	1

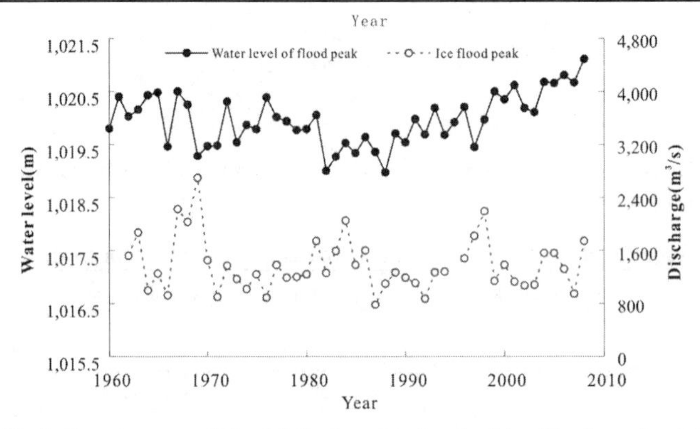

Fig. 7 Variation process of the highest water level of ice flood peak at Sanhuhekou

4 The main factors influencing the ice flood

Channel-storage increment and temperatures are two prerequisites for forming ice floods. The ice flood is formed by the release and convergence of the channel-storage increment along the way. The channel-storage increment is the water source of ice floods while temperature is an objective condition of forming ice floods. In additions, the thawing situation is also an important factor influencing the ice flood peak. For example, the discharge of flood peak formed by the intensive release of water volume under the violent breakup is larger than that of flood peak under the tranquil ice breakup. However, the thawing situation is mainly influenced by temperatures. Besides, the variation of channel erosion and deposition direct impacts on the water level of ice floods. Thus, the paper mainly analyses the influencing factors of ice flood peak from three aspects of channel-storage increment, temperature, and channel conditions.

4.1 Channel-Storage Increment

As a great number of factors influence ice floods, based on the variation of the channel-storage increments in the Inner Mongolia reach and the changing discharges of ice flood peak in Toudaoguai reach in the period of 1950 ~ 2010 (Fig. 8), in ice flood season the inter-annual variation of the channel-storage increment may not synchronize with that of the discharge of the ice flood peak in Toudaoguai reach. The impact of channel-storage increment on ice flood peak is mainly reflected by the trend variation in relatively long periods. The discharge in non-flood season increases under the regulation of Liujiaxia Reservoir and Longyangxia Reservoir, which leads to the increase in channel-storage increment in ice flood season. The increase in channel-storage increment results in the increase in the average discharge and volume of ice flood peak, compared with the discharge and volume before the two reservoirs operate (shown in Fig. 9). Under the condition that the channel-storage increment increases in the period of 1987 ~ 2005, the average ice flood peaks still decrease, only the ice flood peak of 3,000 m^3/s decreases while the ice flood peak of 2,000 ~ 2,500 m^3/s and 2,500 ~ 3,000 m^3/s still increases (Tab. 3). In the period when only Liujiaxia Reservoir being put into operation, the annual average maximum 10-days flood volume in Toudaoguai increases by 1.57×10^8 m^3, compared with the volume before the reservoir operates. As Liujiaxia Reservior and Longyangxia Reservior operate jointly, the annual average maximum 10-days flood volume in Toudaoguai reach increases by 1.85×10^8 m^3, compared with the volume when Liujiaxia Reservoir operates separately.

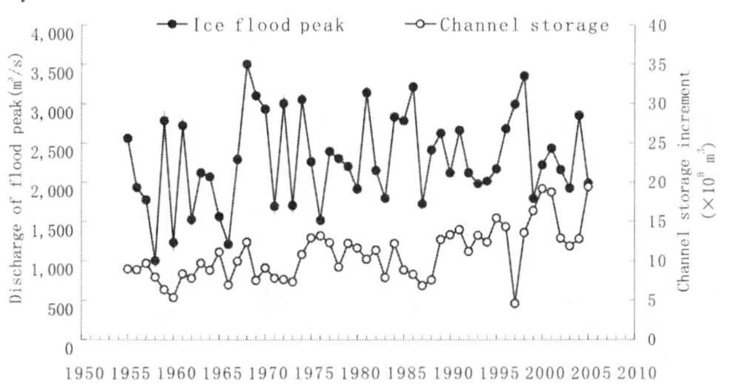

Fig. 8 Changes of maximum channel-storage increment in ice flood season and the discharge of ice flood peak of the Toudaoguai reach in the Inner Mongolia reach from 1950 to 2010

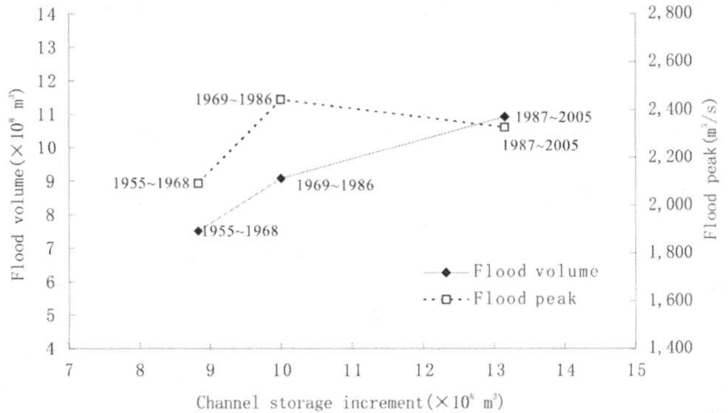

Fig. 9 Changes of the channel-storage increment and the average ice flood peak of the Toudaoguai reach in the Inner Mongolia reach in different periods

The distribution of the channel-storage increment along the way has a dramatic impact on ice flood peak. Fig. 10 illustrates that generally no flood peak increment appears in the reach between Shizuishan and Bayangaole. The increase in flood peak mainly emerges in the reach below Bayangaole, which indicates that the channel-storage increment in the reach below Bayangaole plays a key role in the formation of flood peaks. Before the construction of reservoirs, the flood peak increment in the reach between Bayangaole and Sanhuhekou is almost equal to that in the reach between Sanhuhekou and Toudaoguai. But the channel-storage increment in the reach between Sanhuhekou and Toudaoguai is less 1.29×10^8 m^3 in volume than that in the reach between Bayangaole and Sanhuhekou. In the period of 1969 ~ 1986, the channel-storage increments in the Inner Mongolia reach increase. Compared with that of the previous period, the channel-storage increment in the reaches between Shizuishan and Bayangaole and between Sanhuhekou and Toudaoguai respectively increases by 1.35×10^8 m^3 and 2.88×10^8 m^3 while the channel-storage increment in the reach between Bayangaole and Sanhuhekou decreases by 0.57×10^8 m^3. However, the flood peak increment in the reach between Shizuishan and Bayangaole does not increase while the flood peak increment in the reach between Bayangaole and Sanhuhekou decreases slightly. The flood peak increment in the reach between Sanhuhekou and Toudaoguai almost doubles. In the period of 1987 ~ 2005, the increase in the channel-storage increment mainly occurs in the reach between Bayangaole and Sanhuhekou. But the flood peak increment in this reach does not increase. The channel-storage increment in the reach between Sanhuhekou and Toudaoguai is generally equal to that of the previous period. The flood peak increment in this reach also changes little. Thus, the channel-storage increment in the reach between Sanhuhekou and Toudaoguai has a dominant influence on the formation of ice flood peak.

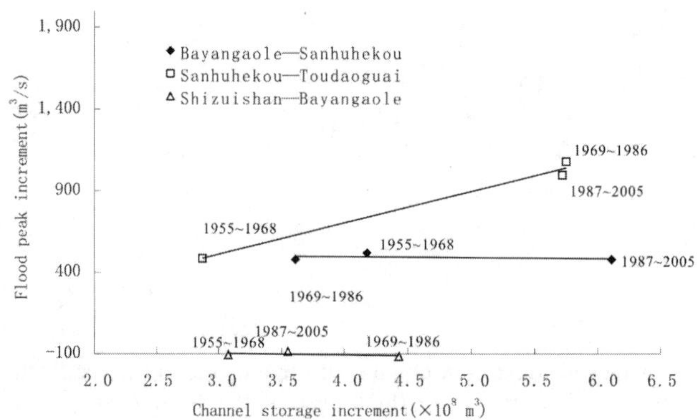

Fig. 10 Changes of the channel-storage increment and the average interval ice flood peak increment in the Inner Mongolia reach in different periods

4.2 Temperature

The impact of temperature on ice floods is mainly reflected by the uneven spatial and temporal changes in temperature, which will influence the ice breakup process and then affect the formation and development of ice floods. Due to the geographical location and special strike of the river channel of the Inner Mongolia reach, the rise of local temperature of different reach in ice breakup period does not synchronize. The temperature in the reach above Sanhuhekou rises early and then the river breaks up early while the temperature in the reach below Sanhuhekou rises late and then the river breaks up late (shown in Tab. 4). For the three periods, the average date difference between Shizuishan and Bayangaole is 9 d, 13 d and 16 d, respectively. The date difference between Bayangaole and Sanhuhekou is 3 d, 5 d and 9 d, respectively. The date difference between Sanhuhekou and Toudaoguai is 2 d, 0 d and -1 d (the ice breakup date of Sanhuhekou is later than that of Toudaoguai), respectively. The uneven temperatures and different ice breakup dates caused by uneven temperatures lead to the early and slow release of the channel-storage increment in the reach between Shizuishan and Bayangaole. Thus, its interval ice flood peak does not increase dramatically. But the reach between Sanhuhekou and Toudaoguai has a relatively intensive channel-storage increment and a higher ice flood peak increment.

Tab. 4 Date of the temperature above 0 ℃ and the ice breakup date at each station in ice flood period of the Inner Mongolia reach

Name of stations	The date of the temperature above 0 ℃ (month-date)			The ice breakup date (month-date)		
	1958 ~ 1968	1969 ~ 1986	1987 ~ 2005	1958 ~ 1968	1969 ~ 1986	1987 ~ 2005
Shizuishan	03 ~ 04	03 ~ 03	03 ~ 01	03 ~ 07	03 ~ 06	02 ~ 21
Bayangaole	03 ~ 08	03 ~ 11	03 ~ 07	03 ~ 16	03 ~ 19	03 ~ 09
Sanhuhekou	03 ~ 09	03 ~ 14	03 ~ 10	03 ~ 19	03 ~ 24	03 ~ 18
Toudaoguai	03 ~ 09	03 ~ 15	03 ~ 09	03 ~ 21	03 ~ 24	03 ~ 17

Different spatial changes in temperature determine the general characteristics of ice flood in the Inner Mongolia reach. The uneven rise of temperature in ice breakup period has a determining influence on ice floods. If in ice breakup period the temperature rises rapidly, the channel-storage increment releases intensively, the discharge along the way increases quickly with prominent flow dynamics, the violent ice breakup from upstream to downstream can be easily formed. If the temperature rises slowly, the channel-storage increment releases in a long period of time and each reach breaks up separately. In this case, the tranquil ice breakup occurs with small discharge of ice flood peak. The ice floods in 1981 and in 1980 are respectively the representation of the violent ice breakup and the tranquil ice breakup. In 1981, the temperature rises dramatically in the ice breakup period of ice flood (shown in Fig. 11). The temperature in mid-March is 3.3 ~ 4.4 ℃ higher than the annual mean temperature. The channel – storage increment intensively releases. Thus, the river reach from upstream to downstream breaks up one by one (Tab. 5). The rapid and progressive increase in discharges results in relatively a great ice flood (Fig. 12). The discharge of ice flood peak in Toudaoguai reach reaches 3,140 m^3/s. The ice flood peak is steep and the main part of the channel-storage increment has released for six days. The discharge of ice flood peak is 2.72 times of the average flood discharge . In 1980, in the ice breakup period of ice flood the temperature rises slowly with frequent rising and dropping (Fig. 3). The release of channel-storage increment lasts a period of 13 d (Fig. 14). Bayangaole station breaks up far earlier than other stations below it. Zhaojunfen station and Toudaoguai station break up earlier than Sanhuhekou station. Such ice breakup situation releases the channel-storage increment dispersedly. The flood peak discharges of the stations below Sanhuhekou emerge on the same day, which reduces the amplification effect of the ice flood peak discharge along the way. The discharge of ice flood peak in Toudaoguai reach is only 1,910 m^3/s, which is 1.61 times of average flood discharge.

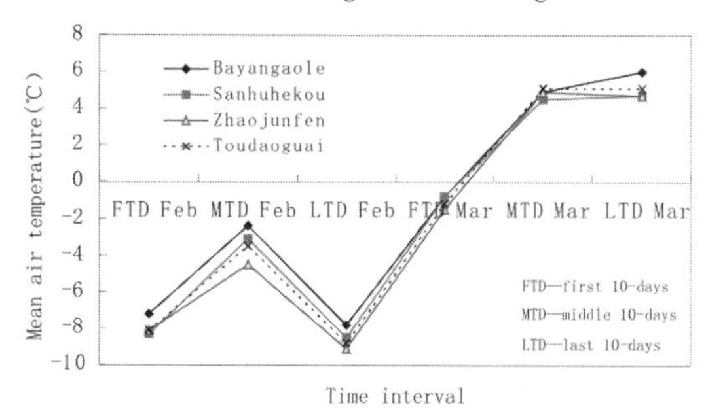

Fig. 11 Changes of mean temperature in 10-days duration in February and March, 1981

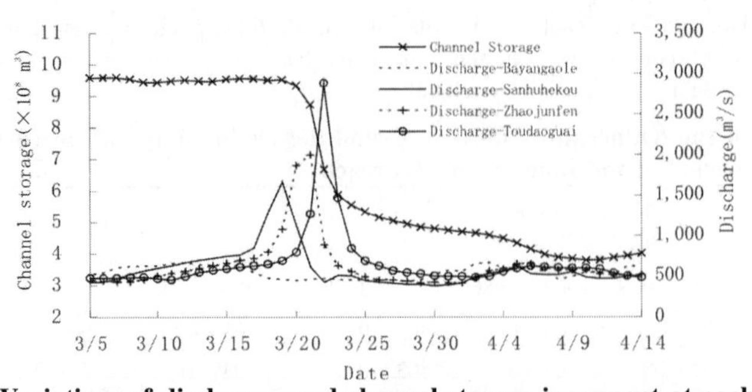

Fig. 12 Variations of discharges and channel-storage increment at each station in ice breakup period of 1981 in the Inner Mongolia reach

Tab. 5 Ice breakup date and characteristics of ice flood peak in Toudaoguai reach in typical years

Year	Ice breakup date (month-date)				Channel-storage increment ($\times 10^8$ m^3)	Ice flood peak of the Toudaoguai reach Q_{max} (m^3/s)	Average ice flood discharge of Toudaogua $\overline{Q}$(m^3/s)	$Q_{max}/\overline{Q}$
	Bayangaole	Sanhuhekou	Zhaojunfen	Toudaoguai				
1981	03 ~ 16	03 ~ 18	03 ~ 19	03 ~ 21	10.18	3,140	1,155	2.72
1980	03 ~ 19	03 - 28	03 - 27	03 - 27	11.6	1,910	1,187	1.61

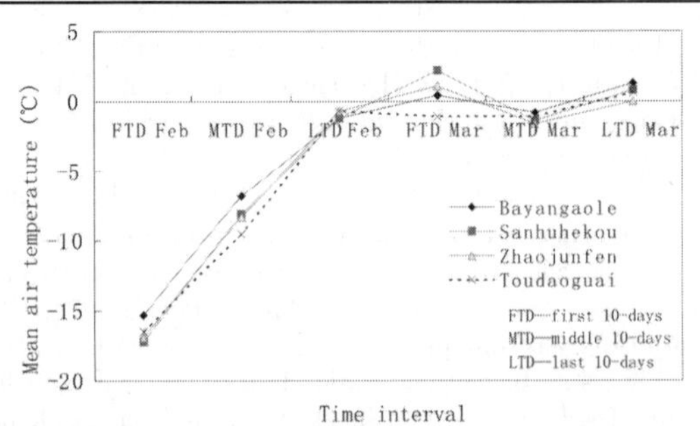

Fig. 13 Changes of mean temperature in 10-days duration in February and March, 1980

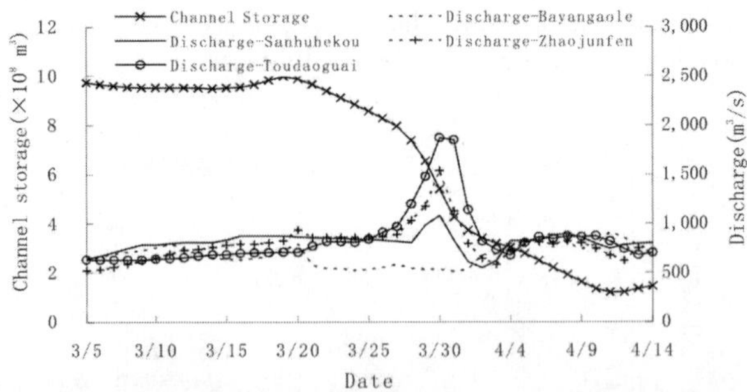

Fig. 14 Variation of discharges and channel-storage increment at each station in ice breakup period of 1980 in the Inner Mongolia reach

Since the 1990s, the winter temperature is a bit higher (Fig. 15). the average temperature in March at Bayangaole, Sanhuhekou, Baotou, and Toudaoguai stations is 1.55 ℃, 1.51 ℃, 2.18 ℃ and 0.76 ℃ respectively, higher than that of the 1980s. The variation leads to the early rise of temperature in ice breakup period and the advance of ice breakup date. The ice breakup date difference between Shizuishan reach and Bayangaole reach increases from 13 d to 16 d while the ice breakup date difference between Bayangaole reach and Sanhuhekou reach increases from 5 d to 9 d. The Sanhuhekou reach and Toudaoguai reach usually break up on the same day. In this case, Toudaoguai reach breaks up one day earlier than Sanhuhekou reach. The ice breakup process in the reach above Sanhuhekou has been further prolonged, which results in the prolonged duration of the release of interval channel-storage increment. Thus, during the ice breakup period, the channel-storage increment in the reach between Bayangaole and Sanhuhekou increases greatly but interval ice flood peak increment does not increase, compared with that of the previous period. That Toudaoguai reach breaks up earlier than Sanhuhekou reach leads to a low probability of a large discharge in Toudaoguai.

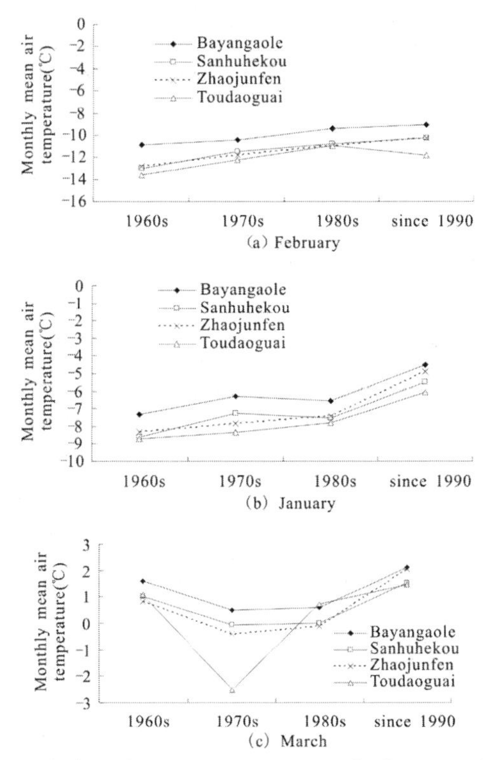

Fig. 15 Variation of average temperature in January, February and March in different decades

4.3 River Channel Sedimentation

Since the 1980s, due to reservoir regulation and decrease in natural runoff, severe channel sedimentation and shrinkage emerges in major alluvial reaches of the Yellow River, which reduces the capability of flood drainage and sediment transport. The research indicates that since 1987 annual average sedimentation in the reach between Bayangaole and Toudaogou reaches 0.514×10^8 t. The deposition mostly appears in main channel. Sediment deposition results in constant shrinkage in the cross section of river channel and continual reduction in carrying capacity of river channel (Fig. 16). By 2005, the bankfull discharge of Sanhuhekou station decreases into less 1,000 m^3/s. The reduction in carrying capacity of river channel becomes the main factor that leads to the water level of ice floods in an upward trend.

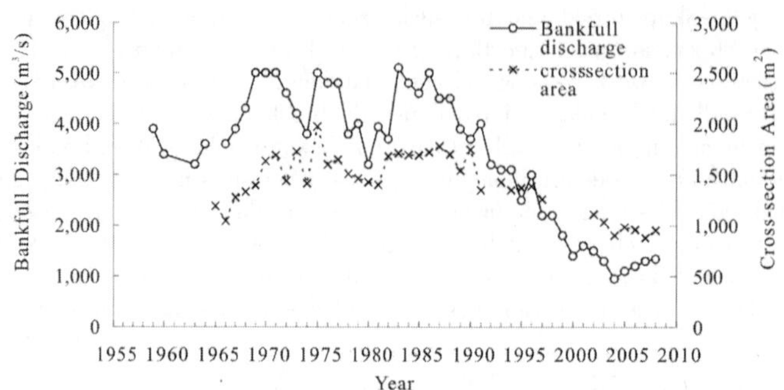

Fig. 16 Cross-section area of river channel and bankfull discharge at Sanhuhekou of the Inner Mongolia reach (1955 ~ 2010)

5 Conclusions

(1) The ice flood peaks in the Inner Mongolia reach of the Yellow River have the characteristics of increasing along the way, rapidly rising and steeply falling, high peak and small flood volume, short duration and high water level. The reach below Sanhuhekou is the main section where ice flood peak gets magnified. Since Liujiaxia reservoir and Longyangxia reservoir are put into operation, the flood peak and volume of ice flood obviously increase at the Toudaoguai section of the Yellow River with its highest water level in an upward trend.

(2) The increase of the channel-storage increment is the main reason that leads to the increase of ice flood peak and flood volume. Especially the variation of the channel-storage increment in the reach between Sanhuhekou and Toudaoguai exerts a strongest influence on the ice flood peak.

(3) In ice breakup period, the temperature rise early in the reaches between Shizuishan and Sanhuhekou and above Sanhuhekou. Thus, the ice breakup process lasts in a long period. As channel-storage increment release relatively early, the ice flood peak increment in these reaches is small. The reaches below Sanhuhekou break up late with relatively intensive ice breakup dates, intensive release of the channel storage volumes and great ice flood peak increments. Since the 1990s, high winter temperatures lead to reaches above Sanhuhekou breaking up in advance and prolonging the ice breakup process. Thus, although the channel-storage increment increases in the reach between Bayangaole and Sanhuhekou, the ice flood peak increment at the section does not increased.

(4) During the ice breakup process, the unstable rise of temperature has a great impact on ice flood. The rapid increase in temperature during the ice breakup process will lead to the ice breakup of these reaches in a short period, intensive release of the channel-storage increment and higher ice flood peak. The slow increase in temperature will result in dispersed ice breakup of these reaches, separate release of the channel-storage increment and lower ice flood peak.

(5) River channel sedimentation and decreased carrying capacity are the main reason that leads to the rise of the water level of ice flood peak.

Acknowledgements

This study is funded by the National Twelfth Five-Year Scientific and Technological Support Plan Subject (No. 2012BAB02B03). The Basic Research Fund for Central Public-interest Scientific Institution (HKY-JBYW-2009-9).

References

Liu Xiaoyan, Hou Suzhen, Chang Wenhua. Cause of Main Channel Shrinkage Occurred to the Inner-Mongolia Reaches of Yellow River [J]. Journal of Hydraulic Engineering, 2007, 40(9): 18 - 21.

Wang Ping, Hou Suzhen, Chu Weibin. The Incremental Variation of Channel Storage and its Influencing Factors of Inner-Mongolia Reach of the Yellow River in Ice Flood Season [J]. Yellow River, 2011, 33(9): 19 - 21.

Hou Suzhen, Chang Wenhua, Wang Ping. The Characteristics and Causes of Channel Shrinkage in the Inner Mongolia Reach of the Yellow River [J]. Yellow River, 2007, 29(1): 25 - 27.

Analysis of Water Supply Situation on Water Resources of Intake Area in Han – to – Wei River Water Transfer Project of Shaanxi Province

Zhang Yongyong[1], *Kong Fenfen*[2], *Shao Jina*[2] and *Huang Qiang*[3]

1. Yellow River Engineering Consulting Co., Ltd., Zhengzhou, 450003, China
2. Resettlement Bureau Yellow River Conservancy Commission, Zhengzhou, 450003, China
3. Xi'an University of Technology, Xi'an, 710048, China

Abstract: With the aggravation of contradiction between supply and demand of water resources, the fast development of social and economic in Guanzhong areas of Shaanxi province are encounter with serious restriction. The imbalance between supply and demand of water resource system would be largely alleviated, with the implementation of the Han – to – Wei River water transfer project. Combined with the Han – to – Wei River water transfer project, two water supply mode with transferable water quantity of 10×10^9 m^3 and 15.5×10^9 m^3 is discussed by the local water supply system and Han – to – Wei River Transfer supply system of the year 2020 and 2030, using the deficit rate and deficit index as the evaluating target, then the water situation on water resources of the benefited areas is discussed. The result indicates that when supplied singly by the local water supply system, the water imbalance between supply and demand of the intake area is serious. When supplied jointly by the local water supply system and Han – to – Wei River Transfer supply system, the annul deficit index would increase from 3.17 of the year 2020 to 8.18 of the year 2030 in scheme 1, and the annul deficit index would increase from 0.82 of the year 2020 to 3.31of the year 2030 in scheme 2 which is of significantly improvement than scheme 1. It can be seen that when the water transfer quantity is 1.55×10^9 m^3, it could meet the water demand of intake area in 2020, but up to 2030, the water shortage is obviously increasing, the contradiction between water supply and demand is still severe, therefore the new water source must be arranged to ease the water shortage situation, so as to promote the economic and social of Guanzhong areas continually and rapidly development.

Key words: Han – to – Wei River Water Transfer Project, deficit rate, deficit index, water supply situation

1 Introduction

Shaanxi Province is located in the east of Northwest China, it belongs to typical mainland monsoon climate. Bound to Qinling Mountain, its south and north respectively belongs to Yangtze River and Yellow River, the water resources therefore is plentiful in south but scarcity in north. To be specified, the water amount of Yangtze River is 31.4×10^9 m^3 while the water resources of Yellow River is 12.5×10^9 m^3, and the water shortage in Guanzhong areas of Yellow River is severe. Guanzhong areas is the strategic focus of economic and social development in Shaanxi Province, which is lack of large – scale mainstay regulation engineering, and the contradiction between water supply and demand is intension, water shortage is the main factor restricting economic and social continually development. In order to fully utilize the plentiful water resources of Yangtze River, Han – to – Wei River water transfer project is planed to implement, so as to ease the increasingly severe contradiction between water supply and demand. Combined with the Han – to – Wei River water transfer project, based on the evaluation index of deficit Rate and shortage index, the water supply situation of intake area is analyzed, which would have an important Practical significance and practical Value to efficiently utilize water resources and guarantee water resources safety.

2 General situation of water supply system of intake area

The local water resources for supplying consists of surface water, ground water and rain water reuse, which works mainly in surface water storage projects, such as Jinpen reservoir, Shibianyu Valley reservoir, Lijiahe reservoir, Shitouhe reservoir and Jianyu reservoir. Water supply system of the Hanhe River to the Weihe River engineering is called the diversion from the Hanhe River to the Weihe River engineering. Huangjinxia Reservoir is planned to built in the mainstream of Hanhe River, and pumping stations are constructed on the left bank of reservoir. The water withdrawn from Hanjiang River enters through the tunnel built in the Hanhe River on the left bank tributary of Sanhekou reservoir. After Reservoir Operation and through a long tunnel from the reservoir area northward through the Qinling Mountains, the water diverted enters Jinpen reservoir in Weihe River tributary. It shouldn't participate in Jinpen reservoir Operation, and then directs to the Guanzhong areas.

3 Evaluation index

3.1 Deficit rate

Deficit Rate could be defined as the percentage of total water shortage quantity in a certain period against the water demand within the same period. By analysis of the deficit rate of the water resources, the situation between the water supply and demand could be known in present stage.

$$DR = \frac{water\ shortage\ quantity}{water\ demand} \times 100\% \tag{1}$$

3.2 Shortage index

Shortage Index was defined by water conservancy project center of American army corps of engineers, which is square of deficit rate. The lower the shortage index, which is not representative that deficit rate or water shortage quantity is lower in each period, means that deficit rate is well - distributed in each period. So, shortage index could show the long - term water shortage situation of water resources.

$$SI_{year} = \frac{100}{N}\sum_{i=1}^{N}\left(\frac{yearly\ water\ shortage\ quantity}{water\ demand}\right)^2 = \frac{100}{N}\sum_{i=1}^{N}(yearly\ deficity\ rate)^2 \tag{2}$$

According to the practically demand, the monthly shortage index and the ten - days shortage index can be proposed.

$$SI_{month} = \frac{100}{N}\sum_{i=1}^{N}\left(\frac{monthly\ water\ shortage\ quantity}{monthly\ water\ demand}\right)^2 = \frac{100}{N}\sum_{i=1}^{N}(monthly\ deficity\ rate)^2 \tag{3}$$

$$SI_{ten-days} = \frac{100}{N}\sum_{i=1}^{N}\left(\frac{ten-daysly\ water\ shortage\ quantity}{ten-days\ water\ demand}\right)^2 = \frac{100}{N}\sum_{i=1}^{N}(ten-days\ deficit\ rate)^2 \tag{4}$$

where, SI_{year}, SI_{month}, $SI_{ten-days}$ are zero dimension unit, N is the total years of data.

4 Analysis of water supply situation on local water supply system

4.1 Simulation principle

According to the present situation of water resources utilization, the reclaimed wastewater and harnessed rain water is the first to be used, then water diverted and stored in the reservoir and the last to be used is the local groundwater. Simulation principles of water supply are as follows:

(1) Reclaimed wastewater: firstly meet the riverside ecological water requirement, then if there is more water left, to supply the industrial sector.

(2) Reservoir water: firstly meet the demand of ecological base flow in non - flood season. If there is more left, to supply irrigation district for agriculture, and finally to supply for the urban use.

(3) Among the urban water uses, domestic use comes first, then urban industrial use, finally urban ecological use.

(4) For improvement of ground water situation, the cap of used water is the permitted exploitable amount. Only in Special dry year can the amount go up but still within the maximum exploitable amount.

4.2 Water supply situation of local water supply system

The local water supply system in intake area consists chiefly of Jinpen reservoir, Shibianyu reservoir, Lijiaxia reservoir, Shitouhe reservoir and Jianyu reservoir. According to the simulation principle of water supply, this paper used ten - day runoff data in 1954 ~ 2005 to simulate regulation and calculated the ten - day water shortage rate and index in 2020 and 2030 level years, as shown in Fig 1. When only by local water supply system , water shortage rate is between 46% ~ 82% in 2020 and average annual water shortage rate is 64%. For the year 2030, water shortage rate is between 50% ~ 85%, average annual water shortage rate is 68%. It can be seen from the Fig. 1 that average water shortage rates of 2020 and 2030 showed little difference. The water shortage index is 44.21in 2020, the annual water shortage index in 2020 is 44.06, and the water shortage index is 49.56 in 2030, the annual water shortage index in that year is 47.54. The ten - days water shortage index and annual water shortage index are considerable. It means that water resources in benefited areas are in short supply, the situation of water shortage is very serious. To sum up, the local water supply system has been difficult to meet the water demand in 2020 and 2030 level years, and we must arrange for some water works to alleviate the gap.

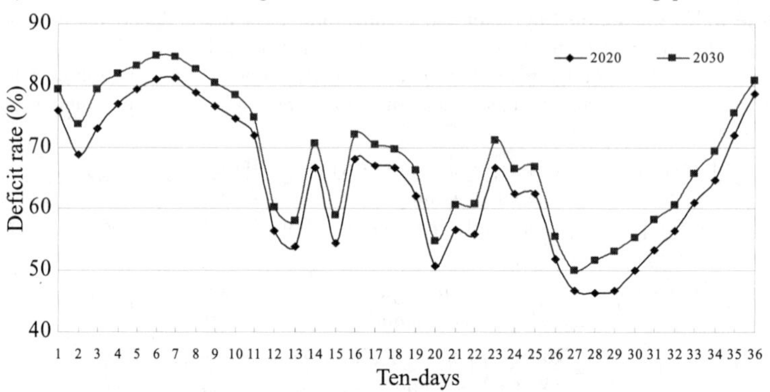

Fig. 1 The deficit rate of intake area only supplied by local supply system

5 Water supply situation of Han - to - Wei River Water Transfer Project

The water supply system of Han to the Wei River mainly gets water from the main stream of the Haijiang River and its tributary stream, Ziwuhe river. Through the united operation of HuangJinXia reservoir and Sanhekou reservoir the water is supplied to the intake area. The study sets transferable water amount of Han - to - Wei River Water Transfer Project as the boundary conditions, two diversion schemes are in simulation. One is transferable water is 1×10^9 m^3, another is transferable water 1.55×10^9 m^3,. Through the simulation, evaluation index is identified to analyze water supply situation on the water supply system of Han - to - Wei River Water Transfer Project.

5.1 Simulation dispatching model of reservoir

5.1.1 Simulation dispatching model

In the reservoir dispatching simulation, release of reservoir and water storage process are decided by the following two types:

$$R_t = \begin{cases} S_t + Q_t - E_t & S_t + Q_t - E_t \leqslant D_t \\ D_t & D < S_t + Q_t - E_t \leqslant D_t + V_{st} \\ D_t + L_t & S_t + Q_t - E_t > D_t + V_{st} \end{cases} \tag{5}$$

$$S_{t+1} = \begin{cases} 0 & S_t + Q_t - E_t \leqslant D_t \\ S_t + Q_t - E - D_t & D < S_t + Q_t - E_t \leqslant D_t + V_{st} \\ V_{st} & S_t + Q_t - E_t > D_t + V_{st} \end{cases} \tag{6}$$

where, R_t is release of reservoir in period t; D_t is water shortage in period t (here is life, industry and ecological comprehensive shortage of intake areas) ; Q_t is inflow in period t, E_t is water loss of evaporation and leakage in period t; L_t is surplus water in period t; S_t, S_{t+1}, are respectively reservoir storage at the beginning of period t and period $t+1$; V_{st} is active storage in period t.

5.1.2 Joint operation rules of Huangjinxia reservoir and Sanhekou reservoir

Huangjinxia reservoir and Sanhekou reservoir are the main regulation reservoir of Han – to – Wei River Water Transfer Project. The joint operation rules are as follows:

(1) the total water amount that can be supplied by the joint operation the two reservoirs can not be more than the water that can be diverted from the Hanhe River to the Weihe River.

(2) When Huangjinxia reservoir can meet the demand of water supply, but Sanhekou reservoir not, they joint operation supply water, Sanhekou reservoir transferring water from Huangjinxia reservoir supply intake area. If Huangjinxia reservoir may have surplus water, according to the supply water gap and dispatched storage capacity of intake area of Sanhekou reservoir, and the dispatched storage capacity of Huangjinxia reservoir, determine the transferable water, surplus water and water supply volume of Sanhekou reservoir. If Huangjinxia reservoir has no surplus water, according to the supply water gap of intake area of Sanhekou reservoir, determine the transferable water of Huangjinxia reservoir and the water supply volume of Sanhekou reservoir.

(3) When the two reservoirs supplying water are not enough, they respectively dispatch independent.

(4) When the two reservoirs are both enough, if Sanhekou reservoir may have surplus water, they respectively dispatch independent. If they both have no surplus water, they respectively dispatch independent. If Sanhekou reservoir has no surplus water, but Huangjinxia reservoir may have surplus water, according to the adjustable storage capacity of Sanhekou reservoir, determine the transferable water and surplus water of Huangjinxia reservoir.

(5) When Huangjinxia reservoir supplying water is not enough, but Sanhekou reservoir supplying water is enough, they respectively dispatch independent.

5.2 Water supply situation when Huangjinxia and Sanhekou reservoirs joint operation

Joint reservoir simulation dispatching is carried out by adopting 51 – year – runoff material of Huang jinxia and Sanhekou reservoir between 1954 and 2005. Water shortage rate and water shortage index of these two kinds of schemes are calculated respectively in the level year of 2020 and 2030 and the water supply situation of different scheme is analyzed.

5.2.1 Scheme 1: transferable water is 1.00×10^9 m^3

Water shortage rate and water shortage index in the level year of 2020 and 2030 are shown in Tab. 1 and ratio of water deficiency is demonstrated in Fig. 2. From the statistics in Tab. 1, we learn that: with the increase of yearly water demand, the shortage target of every ten days increased from 406 in level – year 2020 to 743 in level – year 2030. And multiple annul average rate of water shortage was up from 7.41% in level - year 2020 to 15.97% in level - year 2030. In terms of ten - days water supply, the percentage that water requirement is met dropped from 78% to 60%. The annual average water shortage raised from 2.044,4 $\times 10^8$ m^3 to 5.322,5 $\times 10^8$ m^3, which implies the tension between demand and supply appeared in quite a few years. The yearly water shortage index raised from 3.17 in level - year 2020 to 8.18 in level - year 2030, and the ten - days water shortage index increased from 4.94 to 11.22. It is known from the above that the contradiction between supply and demand is still outstanding and it is hard to meet the demand for water in every level year.

Tab. 1 The shortage target of every ten days of different level years under scheme 1

Level year	Yearly water demand ($\times 10^8$ m^3)	Total periods of ten days of Water shortage	Percentage of meeting water supply	Multiple water shortage quantity ($\times 10^8$ m^3)	Multiple average deficit rate	SI_{year}	$SI_{ten-days}$
2020	2.757,58	406	78%	2.044,4	7.41%	3.17	4.94
2030	3.332,52	743	60%	5.322,5	15.97%	8.18	11.22

Fig. 2 shows that: the ten - days water shortage rate is between 1.5% and 23% in 2020. Within the year, the average rate of water shortage is 7% and 14 ten - days rates of water shortage are above 7%. With the increase of water demand, in level - year 2030, the rate of water shortage is between 3.5% and 33%, within the year, the average rate of water shortage is 16% and 16 ten - days rates of water shortage are above 16%. As a result, along with the increase of water demand, the water shortage in intake area is also in synchronous increase, especially for some extra dry year, lack of water situation will be extremely serious.

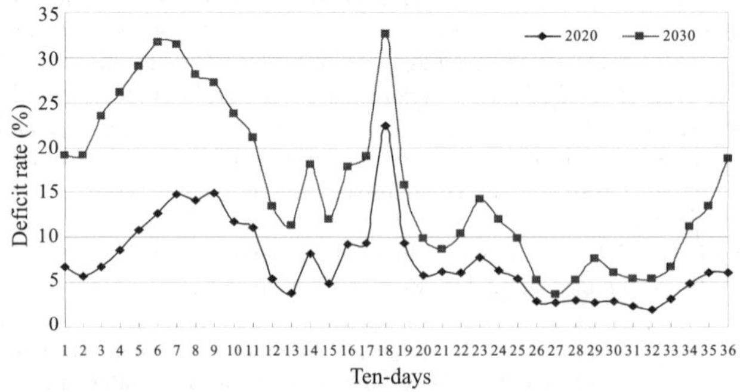

Fig. 2 The deficit rate of intake area supplied jointly by Huang Jinxia and San Hekou reservoirs under scheme1

5.2.2 Scheme 2: transferable water is 1.55 $\times 10^9$ m^3

Water shortage rate and water shortage index in the level year of 2020 and 2030 are shown in Tab. 2 and ratio of water deficiency is demonstrated in Fig. 3. Water shortage rate and water shortage index in the level year of 2020 and 2030 are shown in Tab. 2. From the statistics in Tab. 1, the water shortage index of scheme 2 is obvious improvement than scheme 1, because of the water transferable quantity increased, the total periods of ten days of meeting the water supply is obviously increased than scheme 1, and the total periods of ten - days of water is significantly reduced, shortage target of every ten days is 172 in level - year 2020, and reach level - year 2030 the shortage target of every ten days is 510. The multiple average water shortage deficit is 2.4%, the percentage of total periods of ten days of meeting the water supply is up to 91%, it basically meets the water demand, and up to level - year 2030, the multiple average water shortage deficit is only 8.4%, but percentage of total periods of ten days of meeting the water supply is drop to 72%, and compared with the scheme 1, the multiple average water shortage deficit and total periods of ten days of meeting the water supply are obvious improvement. The yearly water shortage index rise from 0.82 in level - year 2020 to 3.31 in level - year 2030, the ten - days water shortage index increased from 1.3 to 4.96, the water shortage degree is significantly increased, but it is obviously decreased. The multiple average water shortage quantity is only 0.665×10^8 m^3, the contradiction between water supply and demand is remission, but up to level - year 2030, the multiple average water shortage quantity is up to 2.83×10^8 m^3, the water shortage between water supply and demand is significantly increased than level - year 2020, however, it is obvious improvement than scheme 1, and the contradiction between water supply and demand of in take area is still severe.

Tab. 2 The shortage target of every ten days of different level years under scheme 2

Level year	Yearly water demand ($\times 10^8$ m^3)	Total periods of ten days of water shortage	Percentage of meeting water supply	Multiple water shortage quantity ($\times 10^8$ m^3)	Multiple average deficit rate	SI_{year}	$SI_{ten-days}$
2020	2.757,58	172	91%	0.665	2.4%	0.82	1.3
2030	3.332,52	510	72%	2.83	8.4%	3.31	4.96

Compared with the Fig. 2 and Fig. 3, in level - year 2020, the average rate of water shortage is only 2.4%, it descends 4.6% than scheme 1, the average rate of water shortage is significantly decreased, and the maximum water shortage deficit is obviously decreased. With the water demand increased, up to level - year 2030, the average rate of water shortage is 8.4%, but compared with average rate of water shortage namely 16% of scheme 1is obviously decreased, and the maximum water shortage deficit is obviously decreased than scheme 1, it is shown that scheme 2 can ease the tension situation of water supply as far as possible.

Through the above analysis, the water transfer quantity is 1×10^9 m^3 in level - year 2020, it could meet the water demand of intake area, and the contradiction between water supply and demand is remission, but up to level - year 2030, the situation between water supply and demand is still severe, the new water source engineering should be arranged to meet the continuously increasing water demand.

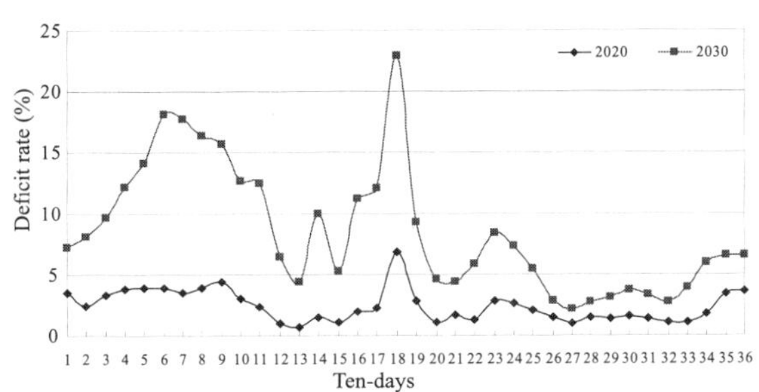

Fig. 3 The deficit rate of intake area supplied jointly by Huang Jinxia and San Hekou reservoirs under scheme 2

6 Conclusions

Through research on the water situation of local water supply system and the water supply system of Han - to - Wei River water transfer project. The main research results are as follows:

(1) When only depending on the local water supply system, the situation between water supply and demand is intension, and the water shortage situation is acute, it must quickly construct the Han - to - Wei River water transfer project to ease the water supply tension of local water supply system.

(2) When depending on Huangjinxia reservoir and Sanhekou reservoir jointly operation of the water supply system of Han - to - Wei River water transfer project, considering water transferable quantity with 1×10^9 m^3 and 1.55×10^9 m^3 two schemes, the water situation is analyzed, in scheme 1, the yearly water shortage index rise from 3.17 in level - year 2020 to 8.18 in level - year 2030,

the ten – days water shortage index increased from4. 94 to 11. 22, the contradiction between water supply and demand is highlight; in scheme 2, the yearly water shortage index rise from 0. 82 in level – year 2020 to 3. 31 in level – year 2030, the ten – days water shortage index increased from 1. 3 to 4. 96, it is significantly improvement than scheme 1.

(3) The water transfer quantity is 1.55×10^9 m^3 in level – year 2020, it could meet the water demand of intake area, but with the increasing water demand, up to level – year 2030, the water shortage is obviously increasing, the contradiction between water supply and demand is still severe, the new water source must be arranged to ease the water shortage situation, in order to promote the economic and social of Guanzhong area continually and rapidly development.

References

Juridical Person for the Research Center of Agricultural Engineering. Programming for the Water Operating Mechanism for the Jointly Application of Feicu and Shimen Reservoir[R]. Taiwan: National Taiwan Ocean University, 2007.

Huang Wenzheng, Zhang Dongxing. Research for the Allocation and Application of Water Resource in Xinzhu Area [D]. Taiwan: National Taiwan Ocean University, 1999.

Feng Guozhang. Research for the Water Simulating Operation Model of the Regional Jointly Operation and Water Supply System [J]. Water Resources and Hydropower Engineering, 2000, 30 (9): 15 – 17.

He Xinchun. Method and Model Research for the Profiting and Regulation Calculation of Multi – Reservoir Jointly Operation [J]. Guangdong Water Resources and Hydropower, 2000, 30(9): 15 – 17.

Veerakcuddy Rajasekaram, K. D. W. Nandalal. Decision Support System for Reservoir Water Management Conflict Resolution [J]. Water Resources Planning and Management, 2005, 131 (6): 410 – 419.